BIOFOAMS

Science and Applications of Bio-Based Cellular and Porous Materials

BIOFOAMS

Science and Applications
of Bio-Based Cellular and
Porous Materials

Edited by

SALVATORE IANNACE
CHUL B. PARK

CRC Press
Taylor & Francis Group
Boca Raton London New York

CRC Press is an imprint of the
Taylor & Francis Group, an **informa** business

Cover Figures by Enza Migliore

CRC Press
Taylor & Francis Group
6000 Broken Sound Parkway NW, Suite 300
Boca Raton, FL 33487-2742

First issued in paperback 2020

© 2016 by Taylor & Francis Group, LLC
CRC Press is an imprint of Taylor & Francis Group, an Informa business

No claim to original U.S. Government works

ISBN-13: 978-1-4665-6179-3 (hbk)
ISBN-13: 978-0-367-78336-5 (pbk)

Visit the Taylor & Francis Web site at
http://www.taylorandfrancis.com

and the CRC Press Web site at
http://www.crcpress.com

Contents

Preface

In the recent past, there has been a confluence of fundamental research and development work being conducted on biofoams and porous systems by a variety of very different research communities. A large number of innovative routes have been explored, generating new materials with wide-ranging technological applicability. As the need for cellular and porous materials with more complex structures and functions has increased, so has the ability to synthesize such systems with tuneable mechanical properties, well-defined pore sizes, pore wall functionalities, and controlled pore size distribution and interconnectivity.

Written for students, professors, and professionals, this book deals with bio-based materials for the development of biodegradable and sustainable polymeric foams, foams in food, foams in bio-medical applications, and bio-inspired foams. It represents a coordinated effort by international experts and provides an integrated, forward-looking perspective. By joining together the most active groups representing diverse fields of study, the editors offer a unified approach on biofoams. Scientists and technologists working in the fields of material science and technology, chemical and biomedical engineering, food technology, and polymers and composites may take advantage of such a unified approach, where the most relevant theoretical and experimental aspects governing the formation of cellular and porous materials are presented in the context of specific applications.

This book describes how gas bubbles can form and grow in a viscous medium leading to the development of products for the plastic (e.g., seats, packaging, insulating material), food (e.g., beers, mousse, ice cream, snacks, cakes), and biomedical (e.g., implants, artificial organs, products for surgery) industries. It also shows how to use renewable resources to develop more sustainable light-weight materials and discusses how scientists and engineers are inspired by nature when designing novel technologies for the production of complex and multifunctional structures.

OVERVIEW OF THE BOOK

Chapter 1 introduces the building blocks used in nature to create hierarchical structures in cellular solids. Bio-based polymers, including biodegradable polymers, used to develop sustainable foams and cellular structures for biomedical applications are also presented. Some of them, such as starch and cellulose, are well known and actively used in products today, while many others are known only in very specific sectors and remain underutilized despite the great potential.

Chapter 2 focuses on the methods employed to develop organic–inorganic hybrid materials derived by solgel reactions of alkoxysilane compounds, as well on the aspects related to the preparation of porous hybrid materials with particular attention to systems obtained from biopolymers.

In Chapter 3, the challenging issues related to the production of porous polymers are reviewed in an effort to pinpoint the requirements that should be met by prospect thermodynamic models to be used for describing and, ultimately, designing such processes. The role of external conditions of temperature, pressure, and composition of solvent or antisolvent is discussed, and the polymer foaming processes are categorized accordingly. The role of the glassy state or the (semi)crystalline state for stabilizing the final porous polymer structure is also discussed.

Chapter 4 shows that proper qualitative and quantitative description of polymer foaming process is a multiphysics problem involving mass and heat transport and momentum transfer, as well as sorption thermodynamics in a high-pressure polymer-penetrant system. There are several material and thermodynamic properties playing a major role, ranging from volumetric, rheological, and mass transport behavior to interfacial properties of the binary mixtures of polymer and foaming agent. The general problem of gas foaming where gas solubility and diffusivity, as well as the effect of gas concentration on interfacial properties and viscosity of the gas–polymer mixture, rule the whole process of bubble nucleation and growth is considered here. In particular, the focus is on the properties of gas–molten polymer mixture.

Chapter 5 discusses the cell nucleation mechanisms in PLA foaming by taking into account (1) the fundamental studies on heterogeneous cell nucleation, (2) the mechanisms of cell nucleation in PLA foams through micro-/nano-sized additives and crystals, and (3) the influence of crystallization on cell nucleation behavior of PLA foams through various foaming processing technologies such as bead foaming, extrusion foaming, and foam injection molding.

Chapter 6 presents a comprehensive study on microcellular foaming of PLA with CO_2 gas by the solid-state process. All aspects of the process including analysis and testing of the PLA foamed specimens are discussed: gas sorption/desorption, diffusion, foaming behavior, microstructure, crystallization, and mechanical properties. Specific guidance on the optimal processing conditions for food-packaging application is presented.

Chapter 7 describes the general approaches in the preparation of thermoplastic starch by using plasticizers and polymeric processing agents. Technologies developed for the production of starch foams for various nonfood applications are described with particular attention to the multiple roles of water molecules: as a solvent, a processing aids/plasticizer, and a blowing or co-blowing agent.

Chapter 8 describes the technologies used to prepare bio-based aerogel by supercritical carbon dioxide. Due to their importance in the selection of proper process condition, both thermodynamics and kinetics aspects related to aerogel preparation are discussed.

Chapter 9 summarizes recent progresses in the development of biocomposite foams based on clays of diverse nature, emphasizing aspects related to the role of the clay filler on the foam formation, as well as the active and passive properties of the resulting foams. The enhancement of mechanical properties or fire retardancy, adsorption behavior, and other specific functionalities derived from the clay and biopolymer synergy, as well as possible applications of these materials, is presented.

Chapter 10 focuses on the use of bio-based components for the synthesis of polyurethane foams. The chemical aspects and the mechanical and morphological properties of bio-based flexible and rigid polyurethane foams prepared by using vegetable derived oils, fats, and plant fillers are discussed.

Chapter 11 describes several technologies available for the production of foams based on thermoplastic polymers. Foaming technologies of thermoplastics are very broad, and it is not the purpose of this chapter to cover all of them. Some basic aspects, especially related to the use of bio-based thermoplastic polymers, are given here with the purpose of providing the background for those that are not familiar with foaming technologies.

Chapter 12 introduces the commonly used synthetic polymer scaffold fabrication methods, including solvent casting/particle leaching, thermally induced phase separation, electrospinning, gas foaming, and rapid prototyping. Recent progress regarding scaffold fabrication—such as combining different scaffold fabrication methods, combining various materials, and improving current scaffold fabrication methods—is discussed as well.

Chapter 13 focuses on the preparation and properties of polymer-based composite and hybrid scaffolds, which represent a very convenient solution for tissue repair and regeneration, providing a wider set of options and possibilities in implant design with tailored physiochemical and biological features. The crucial roles of chemical/physical properties of constituent materials, that is, degradation, mechanical properties, and bioactivity, on the biological function of porous scaffolds at micro, sub-micro, or nanometric scale are discussed.

Chapter 14 reviews current research trends on composite and nanocomposite materials for tissue engineering, including strategies for fabrication of the scaffolds with highly porous and interconnected pores. Cell–scaffold interaction using the colonization of stem cells and degradation of the scaffolds *in vitro* are discussed.

Chapter 15 examines the fundamental aspects related to the formation of gas bubbles in food and drink products. Bubbles are an integral part of many foods and beverages and the control of the foamed structure is crucial for the modification of food properties and for enhancing product quality of aerated food systems.

Chapter 16 is aimed to identify why food foams are attractive to the consumer and the manufacturer. It examines the prominent mechanisms behind food foam stability, their function, and finally highlights possible new trends. Examples are given of several different mechanisms and molecules that are commonly used to form food foams. The intention was to highlight the variety of structures and structural molecules that are so familiar to us but are still capable of producing the vast array of structures that we enjoy eating; for instance, a beer foam is inherently different from a bread but both use protein as their stabilizer.

Salvatore Iannace
Universita di Napoli Federico II

Chul B. Park
University of Toronto

Editors

Salvatore Iannace graduated in chemical engineering (1988) and earned his PhD (1994) from the University of Naples "Federico II," Napoli, Italy. He is the founder and director of the Thermoplastic Composites Manufacturing Laboratory at the Institute of Polymers, Composites and Biomaterials of the National Research Council of Italy. His principal research interests lie in the fields of biodegradable and sustainable multicomponent and multiscale foams with designed structural and functional properties. Dr. Iannace has published more than 300 articles, including 140 in international refereed journals, 18 book chapters, and 4 patents and delivered more than 150 conference papers. Iannace is also a member of the editorial board of the *Journal of Cellular Plastics*. He chaired the first and third International Conferences on Biofoams in Italy (2007 and 2011) and co-chaired the second and fourth editions (2009 and 2013) in Canada. He has been invited as keynote speaker at several major international conferences for his pioneering works on the gas-foaming behavior of biopolymers and their nanocomposites. He has been honored with several awards, including the Nanochallenge and Polymerchallenge Grand Prizes in 2007.

Chul B. Park earned his PhD from MIT in 1993. He holds the Canada Research Chair (Tier 1) in Microcellular Plastics at University of Toronto, Toronto, Canada. He is the founder and director of Microcellular Plastics Manufacturing Laboratory and the Centre for Industrial Application of Microcellular Plastics. He has worked on many foundational research topics that have greatly improved the understanding of the basic physical properties and mechanisms governing the plastics foaming process. He has published more than 900 papers, including 240 refereed journal articles, 2 books, and more than 30 patents and delivered more than 560 conference papers. He is a fellow of three academies: Royal Society of Canada, Canadian Academy of Engineering, and Korean Academy of Science and Technology. He is also a fellow of the American Association for the Advancement of Science, the American Society of Mechanical Engineers, the Canadian Society for Mechanical Engineering, the Engineering Institute of Canada, and the Society of Plastics Engineers. He is the editor-in-chief of the *Journal of Cellular Plastics*. He has international recognition in polymer foaming and has been honored with over 30 major awards.

Contributors

Luigi Ambrosio
Department of Chemical Science and Materials
   Technology
National Research Council
Rome, Italy

Shuichi Arakawa
Advanced Polymeric Nanostructured
   Materials Engineering, Graduate School
   of Engineering
Toyota Technological Institute
Nagoya, Japan

Pilar Aranda
Instituto de Ciencia de Materiales de Madrid
Madrid, Spain

Philip Cox
School of Chemical Engineering
Edgbaston, Birmingham, United Kingdom

Ernesto Di Maio
Department of Chemical, Materials
   and Production Engineering
University of Naples Federico II
Naples, Italy

Francisco M. Fernandes
Sorbonne Universités
Paris, France

Alistair Green
School of Chemical Engineering
Edgbaston, Birmingham, United Kingdom

Vincenzo Guarino
Department of Chemical Science and Materials
   Technology
National Research Council
Rome, Italy

Ryoji Ishiki
Medical Affairs
Toyota Technological Institute
Nagoya, Japan

Xin Jing
Polymer Engineering Center
University of Wisconsin-Madison
Madison, Wisconsin

Vipin Kumar
Mechanical Engineering
University of Washington
Seattle, Washington

Domenico Larobina
Institute of Polymers, Composites
   and Biomaterials
National Research Council
Naples, Italy

Marino Lavorgna
Institute of Polymers, Composites
   and Biomaterials
National Research Council
Naples, Italy

S.T. Lee
Research & Development
Sealed Air Corporation
Saddle Brook, New Jersey

Leno Mascia
Materials Department
Loughborough University
Loughborough, United Kingdom

Giuseppe Mensitieri
Department of Chemical, Materials
   and Production Engineering
University of Naples Federico II
Naples, Italy

Hao-Yang Mi
Polymer Engineering Center
University of Wisconsin-Madison
Madison, Wisconsin

Krishna V. Nadella
MicroGREEN Polymers Inc.,
Arlington, Washington

Mohammadreza Nofar
Mechanical and Industrial Engineering
University of Toronto
Toronto, Ontario, Canada

Masami Okamoto
Advanced Polymeric Nanostructured
  Materials Engineering, Graduate School
  of Engineering
Toyota Technological Institute
Nagoya, Japan

Costas Panayiotou
Department of Chemical Engineering
Aristotle University of Thessaloniki
Thessaloniki, Greece

Roberto Pantani
Department of Industrial Engineering
University of Salerno
Fisciano, Italy

Maria Giovanna Pastore Carbone
Department of Chemical, Materials
  and Production Engineering
University of Naples Federico II
Naples, Italy

Aleksander Prociak
Cracow University of Technology
Kraków, Poland

Maria Grazia Raucci
Institute of Polymers, Composites
  and Biomaterials
National Research Council
Naples, Italy

Eduardo Ruiz-Hitzky
Instituto de Ciencia de Materiales de Madrid
Madrid, Spain

Reika Sakai
Advanced Polymeric Nanostructured
  Materials Engineering, Graduate School
  of Engineering
Toyota Technological Institute
Nagoya, Japan

Sarah Santos-Murphy
School of Chemical Engineering
Edgbaston, Birmingham, United Kingdom

Martin G. Scanlon
Department of Food Science
University of Manitoba
Winnipeg, Manitoba, Canada

Giuseppe Scherillo
Department of Chemical, Materials
  and Production Engineering
University of Naples Federico II
Naples, Italy

Ciro Siviello
Department of Chemical, Materials
  and Production Engineering
University of Naples Federico II
Naples, Italy

Jim Song
College of Engineering & Physical Science
Brunel University
Uxbridge, London, England

Andrea Sorrentino
Institute of Polymers, Composites
  and Biomaterials
National Research Council
Naples, Italy

Luigi Sorrentino
Institute of Polymers, Composites
  and Biomaterials
National Research Council
Naples, Italy

Ioannis Tsivintzelis
Department of Chemical Engineering
Aristotle University of Thessaloniki
Thessaloniki, Greece

Lih-Sheng Turng
Polymer Engineering Center
University of Wisconsin-Madison
Madison, Wisconsin

Letizia Verdolotti
Institute of Polymers, Composites
    and Biomaterials
National Research Council
Naples, Italy

Bernd Wicklein
Department of Materials and Environmental
    Chemistry
Stockholm University
Stockholm, Sweden

1 Bio-Based and Bio-Inspired Cellular Materials

Salvatore Iannace and Andrea Sorrentino

CONTENTS

1.1 INTRODUCTION

Porous and cellular materials have long fascinated scientists and engineers. A great variety of cellular materials is present in nature. Their architectures as well as their unique properties have inspired researchers for the development of lightweight structures, energy-absorbing padding, or thermal insulating panels. Examples of cellular architectures in nature are wood, plant leaves and stems, trabecular bone, coral, cork, and sponges.

Nature has optimized the microstructure of these materials to perform a number of functions. Wood and cork have honeycomb-like cellular structures characterized by closed cells, while trabecular bones are foam-like cellular materials with open-celled structures. One of the most important functions of these structures is related to their specific mechanical properties. In many weight-sensitive applications, where high mechanical properties and low mass are requested, properties per unit mass are more important than absolute properties. Thanks to its high specific properties, wood has been used for millennia to build wooden boats and furniture, while cork has been used for the sole of shoes since Roman times.

Nature's cellular materials are structured at multiple length scales, allowing organisms to adapt to external stimuli for given (even multiple) functions. Hierarchical structuring, which is a simple consequence of (adaptive) growth, is present in many living systems such as wood, bone, and crustacean cuticles. The architecture of these materials is based on the combination of simple building blocks arranged in complex functional structures. Such systems are studied to inspire the development of novel synthetic materials with improved functional and structural performances.

The oldest examples of man-made foams are found in food. Popcorn and bread are two examples of how different processing technologies can be used to make cellular structures. Popcorn represents the closest example on how the principles employed for the formation of cellular structures are used in the modern industry (Figure 1.1). Most types of corn (sweet corn, dent corn, etc.) don't pop when they are cooked, at least not with the same intensity. Popcorn has a

FIGURE 1.1 (a) Popped maize (popcorn), bar = 10 mm; (b) cut surface of endosperm of popcorn showing reticulate foam structure of gelatinized starch, bar = 1 mm; (c) cut surface of popcorn showing outline of an endosperm cell (indicated by dashes) filled with starch foam, and each air bubble is formed from one expanded starch granule, bar = 100 µm. (Adapted from Parker, M.L. et al., *J. Cereal Sci.*, 30, 209–216, 1999.)

couple of things going for it in this respect. First, and foremost, the shell surrounding the kernel (the pericarp) is denser than in other types of corn. The pericarp forms the tough outer surface of the kernel and serves as a moisture barrier. This facilitates the water vapor, trapped inside the kernel, to build up pressure (Hoseney et al. 1983). In the other kinds of corn, the water vapor just passes through the shell. A region of starch granules, the hard endosperm, is the largest feature inside the pericarp. The soft endosperm contains starch and most of the water in the kernel, which popcorn processors carefully control to 13%–14%. The heated kernel forms a solid foam that is produced when the pericarp bursts in response to the vapor pressure of superheated water. As the kernel is heated, water vapor infiltrates the hard endosperm, which softens and becomes gelatinous above 150°C.

The pericarp bursts explosively when the gelatinous starch reaches a temperature of 180°C–190°C. Vapor pressure produces a tiny hole in each starch granule, creating a cell of the foam. Evaporation of the superheated water cools the foam to produce the familiar solid flake. The mechanisms involved during the formation of popcorn have been considered for the teaching and learning of chemistry (Fantini et al. 2006).

Bread can be considered as one of the oldest man-made foams. Soon after mixing, the structure of dough is represented by small gas cells dispersed in a continuous starch–protein matrix. Each discrete gas cell expands in response to CO_2 production during fermentation, and thin membranes separating adjacent cells at the end of proof maintain the foam structure. During baking, starch gelatinization induces a dramatic increase in dough viscosity, resulting in a rapid increase in tensile strength in the membrane. This initiates membrane rupture, converting the foam into a sponge (Gan et al. 1995). The unique position of wheat, compared with other cereals, in bread making is mainly related to the properties of the protein gluten. This protein plays a key role in determining the unique baking quality of wheat by conferring water absorption capacity, cohesivity, viscosity, extensibility, elasticity, resistance to stretch, mixing tolerance, and gas holding capacity (Lazaridou et al. 2007). The increasing demand for high-quality gluten-free bread represents a challenging task for the food technologist due to the low baking quality of gluten-free flours as a consequence of the absent gluten network (Hüttner and Arendt 2010).

In this chapter, we introduce the *building blocks* used in nature to create hierarchical structures in cellular solids. The structures of proteins and polysaccharides as well as minerals and the way they are structured in nano- and micro-composite are briefly discussed. Some examples on how these structures are inspiring researchers to develop bio-inspired cellular structures are included. Bio-based polymers, including biodegradable polymers, used to develop sustainable foams and cellular structures for biomedical applications are also presented. Some of them, such as starch and cellulose, are well known and actively used in products today, while many others are known only in very specific sectors and remain underutilized despite the great potential. The future growth of polymers and composites from renewable resources is reliant on continued research, particularly in the fields of processing techniques, and it is through an understanding of these that they are expected to replace more and more petroleum-based plastics.

1.2 MATERIALS OF NATURE

Polysaccharides, proteins, lipids, and nucleic acid are the four macromolecule building blocks used by nature to make up the majority of life's structures. The *blend* of two or all of these ingredients is necessary to produce all the known complex biological structures present on the earth. Polysaccharide-based materials (e.g., starch, cellulose, alginate, carrageen, chitosan, pectin, and various gums) are hydrophilic and provide strong hydrogen bonding that can be used to bind with functional additives such as flavors, colors, and micronutrients. Due to the ability of adjacent chains to cross-link, these materials have good oxygen but poor moisture barrier properties. Protein-based materials are also hydrophilic (poor moisture barrier) but have the mechanical strength that provides structure or biological activity in plants and animals. The proteins considered here as protein-based biomaterials are those found in greatest quantities: proteins of animal sources (collagen, gelatin, keratin, casein, and whey protein), cereal coproducts (corn zein and wheat gluten), or reserve proteins of grains (sunflower, soybean). Depending on the sequential order of the amino acids (primary structure of the protein), the protein will assume different structures along the polymer chain (secondary structure of the protein), based on van der Waals, hydrogen bonding, electrostatic, hydrophobic, and disulfide cross-link interactions among the amino acid units. The tertiary protein structure reflects how the secondary structures organize relative to each other, based on the same types of interactions, to form overall globular, fibrous, or random protein structure. Finally, quaternary structure occurs when whole proteins interact with each other into associations to provide unique structure or biological activity.

Various physical and chemical agents, including heat, pressure, acids, and alkalis, can modify the secondary, tertiary, and quaternary structure of proteins. This structural modification (denaturation) is often used deliberately in the course of films, fibers, and coating formations. Following denaturation, some proteins will return to their native structures under proper conditions but extreme conditions, such as strong heating, usually cause irreversible change. The long-range ordered molecular secondary structures (e.g., beta-pleated sheets, coiled coils, or triple helices) features promote self-assembly and formation of structural hierarchy and thus prompt their utility as a treasured resource of polymers for biomaterials. Lipid-based materials have good moisture barrier, but low mechanical properties.

The use of composite materials helps to minimize the disadvantages of the individual components while making use of the strength in their properties. For example, the foaming properties of proteins are due to their ability to decrease the interfacial tension at the water–air interface of air bubbles. Polysaccharides also exhibit foam stabilization properties, but these are mainly due to their ability to thicken the aqueous medium. The combination of the foaming properties and stabilization effects of these two classes of biopolymers was expected to enhance the surface properties of the resulting composites. It allows a wide range of naturally occurring polymers for several material applications.

Modern technologies provide powerful tools to elucidate microstructures at different levels, and to understand the relationships between structures and properties. These new levels of understanding bring opportunities to enhance the natural biosynthetic systems and produce new molecules for targeted applications.

Despite the wide ranges of naturally occurring polymers and their impressive properties, naturally occurring polymers represent, for several applications, only an industrial niche area. However, this may shift over time based on the following: (1) changes in consumer values for environmentally safe products, (2) decreased cost and increased performance, and (3) increased global regulatory pressure.

1.2.1 Vegetal Origin

1.2.1.1 Cellulose

Cellulose is the most abundant natural polymer available on this planet. It is found in plant cell walls of all fruits and vegetables, and it is present in the leaves, trunks, and barks of higher plants as a structural component (Collins and Ferrier 1995). Cellulose is also the main component of a variety of natural fibers, such as cotton, bast fibers, and leaf fibers (Gilbert and Kadla 1998; Stevens 1998). In nature, cellulose is generally found in close association with other biomolecules, mainly hemicelluloses, pectin, wax, proteins, lignin, and mineral substances (Gilbert and Kadla 1998; Stevens 1998). The structure of the cellulose is composed by a long-chain polysaccharide made up of 5,000–15,000 glucose monomer units. These molecules align to form microfibrils with a diameter of about 3–4 nm (Fernandes et al. 2011; Hori et al. 2002; Peura et al. 2007). The microfibrils have both crystalline and noncrystalline regions that merge together (Hori et al. 2002). The properties of the cellulose polymers depend mainly on the length of the polymer chain and their degree of polymerization. Although it has a large number of hydroxyl groups, it is insoluble in water and most organic solvents. Cellulose is soluble in strong mineral acids, sodium hydroxide solution, and metal complexes solutions. Under strong acidic conditions, it is hydrolyzed completely to D-glucose, whereas very mild hydrolysis produces hydrocellulose with shorter chains, lower viscosity, and tensile strength. Due to high extents of intramolecular and intermolecular hydrogen bonding, cellulose is not thermoplastic although it decomposes before melting (Yamamoto et al. 1990). Chemical modification of cellulose introduces various chemical groups onto the hydroxyl groups of glucose units, posing steric hindrance to native intermolecular interactions of cellulose. Methyl, carboxymethyl, and hydroxypropyl cellulose ethers provide both nonionic- and anionic-modified derivatives that possess unique functional properties, such as thermal gelation and superior clarity (Gilbert and Kadla 1998; Popa and Hon 1996; Stevens 1998).

Cellulose and its derivatives are widely used in textiles, food, membranes, films, paper, and wood products. Cellulose is also used in medicines (e.g., wound dressing), suspension agents, composites, making rope, mattress, netting, upholstery, sacking, matting, coating, cordage, twine, packaging, linoleum backing gas mantles, muslin, thermal and acoustic insulation, and so on (Popa and Hon 1996; Stevens 1998). Because of its linear structure, nonionic nature, and high degree of solubility, methylcellulose films are strong and clear. It is used as a food-thickening agent, and as an ingredient in certain adhesive, ink, and textile-finishing formulations (Yamamoto et al. 1990). Ultra-light and highly porous cellulose is a new very promising material offering a wide range of potential applications, from biomedical and cosmetics (delivery systems and scaffolds) to insulation and electrochemical (when pyrolyzed). Two different preparation methods were followed for the preparation of these aerogels. The first method requires the formation of the gel by dissolving the cellulose in a suitable solvent, followed by the removal of the solvent using water or alcohol, and then dried in a special way that prevents pores collapse (i.e., either via freeze-drying or under CO_2 supercritical conditions) (Innerlohinger et al. 2006; Liebner et al. 2009). No chemical cross-linking is used to stabilize the cellulose network (Duchemin et al. 2010; Gavillon and Budtova 2008; Sescousse et al. 2010). The resulting aerogels have wide pore size distribution, from tens of nanometers to several microns, and a high specific surface area. The density of aerocellulose depends on the initial cellulose concentration in solution and on the control of drying. Foaming agents can be added to increase aerocellulose porosity (Aaltonen and Jauhiainen 2009; Deng et al. 2009; Jin et al. 2004; Tsioptsias et al. 2008).

The second method requires the use of cellulose nanofibers prepared by either bacterial cellulose or microfibrillated cellulose prepared via native cellulose mechanical disintegration and enzymatic treatment (Figure 1.2). In both cases, the starting material is a network of cellulose nanofibers filled with water that can be extracted using freeze-drying or drying in supercritical carbon dioxide (Liebner et al. 2010). The resulting material is similar to that obtained with the first method but with higher porosities and lower densities because of lower initial cellulose concentrations and absence of contraction during drying due to higher skeleton crystallinity and molecular weight (Maeda et al. 2006; Pääkkö et al. 2008).

1.2.1.2 Hemicelluloses

Next to cellulose, hemicelluloses are the most abundant polysaccharide found on the earth and represent about 20%–30% of wood mass (Kanke et al. 1992). They are a group of polysaccharides—for example, xyloglucans, xylans, mannans and glucomannans, and β-(1→3, 1→4)-glucans—present in the cell walls of all terrestrial plants (Collins and Ferrier 1995; Popa and Hon 1996).

The detailed structure of the hemicelluloses and their abundance vary widely between different species and cell types. The most important biological role of hemicelluloses is their contribution to strengthening the cell wall by interaction with cellulose and, in some walls, with lignin (Gilbert and Kadla 1998). As of today, hemicelluloses have not yet found broad

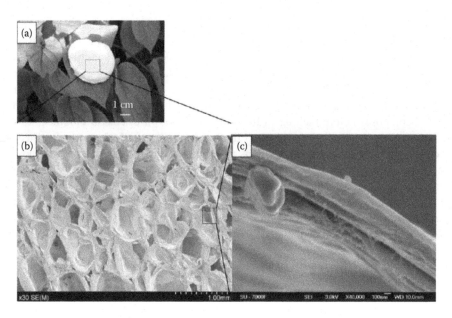

FIGURE 1.2 Various magnification of (a) nano-fibrillated dried cellulose foam, (b) cellulose foam structure, and (c) cell wall morphology. (Reprinted from Cervin, N.T. et al., *Biomacromolecules*, 14, 503–511, 2013.)

applications in structural material production, because of low molecular weight (a degree of polymerization generally below 200), branching, and varying chemical composition (Kanke et al. 1992). Hemicelluloses exist in amorphous form, and unlike cellulose, they are generally very soluble in water. They are hydrolyzed to saccharides (e.g., glucose, mannose, arabinose, glactose, and xylose), acetic acid, and uronic acid. Hemicelluloses and their derivatives are very important in industry due to their useful chemical properties and biological activity. Isolated hemicelluloses are used as food additives, thickeners, emulsifiers, gelling agents, adhesives, and adsorbents (Popa and Hon 1996).

Furfural, produced by dehydration of xylose, is a very useful precursor for the synthesis of tetrahedrofuran, adiponitrile (hence nylon), furfuryl alcohol, tetrahedrofurfuryl alcohol, glutamic acid, polyurethane foams, resins, and plastics (Popa and Hon 1996). Tannin-based rigid foams are known for a long time as natural products resistant to fire, such as synthetic phenolic foams (Meikleham and Pizzi 1994; Pizzi et al. 2008; Tondi et al. 2009). The foaming process involves the mixing of a tannin-formaldehyde resin with furfuryl alcohol, which is used as a heat-generating agent by both the self-polymerization reaction and the reaction with tannin. These tannins are vegetal products obtained by water extraction from the wood of mimosa (*Acacia mearnsii* and *Castanea mollissima*), bark tannin, and quebracho (*Schinopsis balansae* and *Schinopsis lorentzii*) (Tondi et al. 2008). The resulting tannin-based rigid foams are good thermal insulating materials, which do not burn, and which can be used in buildings for interior and exterior applications (Tondi and Pizzi 2009).

Hemicellulose citrate–chitosan cross-linked foams were synthesized by treating hemicellulose citrate and chitosan in an aqueous medium (Salam et al. 2011). The incorporation of carboxylic acid groups into hemicellulose via reaction with citric acid followed by cross-linking with chitosan greatly improved water absorption and strength, and decreased weight loss in water and saline solution compared to hemicellulose–chitosan foam. The hemicellulose citrate–chitosan foams were found to have interesting properties including high elasticity at room temperature, high porosity, and super absorbents properties in saline uptake (Salam et al. 2011).

1.2.1.3 Starch (Amylose/Amylopectin)

Starch is the principal carbohydrate reserve of almost all plants. It is a mixture of the predominantly linear α-(1→4) glucan or amylose, and the highly branched, high molecular weight glucan or amylopectin with α-(1→4) glycosidic linkages containing α-(1→6) branch points (Kaplan 1998; Shogren 1998; Stevens 1998). The relative proportion of these components varies as a function of the starch source (e.g., potato, corn, wheat, and tapioca) and influences the crystallinity and molecular order of the polysaccharide. Starch is isolated from its sources by various physical methods such as steeping, milling, and sedimentation. It is not soluble in water at ambient temperature, but in hot water, its granules gelatinize to form an opalescent dispersion (Shogren 1998).

Upon supercritical drying of the starch gel, an aerogel with a high surface area can be obtained (Mehling et al. 2009). However, water-to-ethanol solvent exchange in starch gels is needed to avoid the collapse of the pore structure or gel particles coalescence upon drying (Glenn et al. 2010). Amylose is the component providing the amorphous character to the resulting starch gel, while the ordered amylopectin exerts a certain control on the local structural ordering of the material. Moreover, higher amylose content, as well as high gelatinization temperature, promotes high generation of mesoporosity in the resulting materials (White et al. 2008). On the contrary, low cooling temperatures are prone to reach higher surface areas in the gel since the number of crystals formed with respect to crystal growth (Hoover et al. 1994). Like cellulose, starch can be modified by esterification, by etherification, and by forming graft copolymers with synthetic materials such as polyacrylic acid. Amylose is used for making films, whereas amylopectin is used for textile sizing and finishing and as a food thickener. The considerable interest in graft copolymers of starch is due to their utility as biodegradable packing materials and agricultural mulches.

Starch-based foams have been studied to develop food, as well as lightweight biodegradable common products (Cotugno et al. 2005). They were produced by using several techniques such as compression molding, extrusion with blowing agents, *in situ* polymerization, solvent casting, and particulate leaching (Gomes et al. 2002; Xu et al. 2005). Starch-based porous structures were produced by using a combination of supercritical carbon dioxide and water as blowing agent (Cho and Rizvi 2008). Continuous extrusion-foaming technologies have been used to produce porous materials. In that case, the steam generated by the moisture present in the sample was used as a blowing agent (Kalambur 2006; Willett and Shogren 2002).

Tensile strength and density were found to increase, while the foam flexibility decreased with increasing starch concentration, molecular weight, and amylase content (Shogren et al. 1998). The authors reported that the tuber starches, such as potato, produce foams with lower densities and higher flexibilities than those from cereals such as corn. However, the main disadvantages of the starch foams are their fragility and their high affinity for water. Natural fibers, such as aspen, jute, softwood fiber, and flax, were added to improve strength and flexibility of these foams (Carr et al. 2006; Glenn et al. 2001; Lawton et al. 2004; Soykeabkaew et al. 2004). The water resistance of baked starch-based foams was improved by addition of hydrophobic materials such as monostearyl citrate, latex, and polycarpolactone (PCL) (Gáspár et al. 2005; Shey et al. 2006; Shogren et al. 2002). The addition of other polysaccharides and proteins, such as cellulose, hemicellulose, casein, gelatin, and zeins, to the formulation, conferred best mechanical properties to the resulting materials (Arvanitoyannis et al. 1996, 1997; Coughlan et al. 2006; Jagannath et al. 2003; Psomiadou et al. 1996; Wongsasulak et al. 2006, 2007).

1.2.1.4 Pectins

Pectins are a class of biopolymers found in the cell walls of higher plants, where they function as a hydrating agent and cementing material for the cellulosic network. They are linear polysaccharides of α-linked anhydrogalacturonic acid with a certain degree of methyl esterification of carboxyl groups (degree of esterification) depending on the polysaccharide quality and source (May 1990).

Although pectin occurs commonly in most of the plant tissues, the number of sources that may be used for its commercial manufacture is very limited. At present, apple pomace and citrus peels are the main sources of commercially acceptable pectins (May 1990). Pectin is soluble in hot water where they form a viscous solution. A number of factors, including pH, presence of other solutes, molecular size, degree of methoxylation, and number and arrangement of side chains, influence the gelation of pectin. After solvent exchange of the gel with ethanol, the supercritical drying of pectin gels avoids the massive shrinkage observed using other drying techniques and yields aerogels with high surface areas and porosity (White et al. 2010).

Films made from natural products are of increasing interest because they are biodegradable and potentially recyclable (Coffin and Fishman 1993; Mangiacapra et al. 2006). Because of its film-forming properties, pectin is useful as a sizing agent for paper and textiles. Blends of pectin and starch can be used to make strong, self-supporting films (Coffin and Fishman 1993).

1.2.1.5 Konjac Glucomannan

Konjac glucomannan is a water-soluble nonionic polysaccharide extracted from tubers of the *Amorphophallus konjac* plant (Kaplan 1998). Different from many other biopolymers, its molecular weight distribution is fairly narrow, and the molecular chains are extending, semi-flexible, linear, little rigid, and without branching. Konjac, like other gums with a linear structure, produces strong films useful for food and pharmaceutical applications (Davé and McCarthy 1997).

A further interesting characteristic of konjac lies in its synergy with other hydrocolloids. According to the literature, konjac and xanthan form an elastic gel, whereas a significant increase in the gel strength is observed for konjac–carrageenan and konjac–agar mixtures. Furthermore, konjac is unique in that it produces a very high viscosity solution also at low concentration (Wen et al. 2009).

The natural acetylation of konjac allows it to be solubilized in water. However, deacetylation of water-cast films renders those films hot- and cold-water stable. This, combined with its high molecular weight and its nonionic nature, makes konjac a good polymer material for edible films (Gao and Nishinari 2006; Ratcliffe et al. 2005).

1.2.1.6 Alginate

Alginate is a natural polysaccharide derived from marine brown algae of the family Phaeophyceae. Its chemical structure consists of a copolymer of 1,4-linked-β-D-mannuronic acid (M) and α-L-guluronic acid (G) of varying composition and sequence (Amsden and Turner 1999; Martinsen et al. 1992). Specific algae that form alginates include the giant kelp, *Macrocystos pyrifera*, as well as other species such as *Ascophyllum nodosum* and *Laminaria digitata*. Different seaweeds give rise to varying ratios of M and G groups within the alginic acid, with seasonal changes in the ratio of M to G groups also seen within species (Siew et al. 2005). Gelation of alginate, with calcium or a bivalent ion, is instantaneous. The G-block responds to calcium cross-linking faster than the M-blocks, because of its three-dimensional (3D) molecular conformation (Mehling et al. 2009).

Alginate hydrogels can be easily prepared by cross-linking at low temperature and in the absence of organic solvents (Puppi et al. 2010). The water-to-ethanol solvent exchange produces a strong reduction in the surface tension of the alginate gel pores followed by a reduction in volume of the gel (shrinkage) (Mehling et al. 2009). The supercritical drying of the alcogels leads to the formation of alginate aerogels with different morphologies: monoliths, beads, and microspheres (Alnaief et al. 2011; Quignard et al. 2008; Robitzer et al. 2008, 2011; Valentin et al. 2006).

Porous alginate scaffolds have been developed for different biomedical applications such as drug delivery, release of cell for revascularization, and repair of cartilage (Freeman et al. 2008; Li et al. 2005; Miralles et al. 2001; Re'em et al. 2010; Sapir et al. 2011). Dry porous alginate scaffolds have

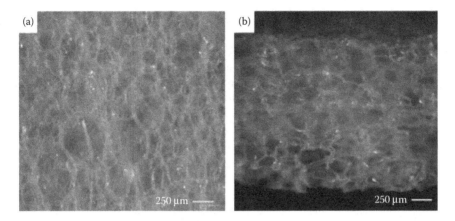

FIGURE 1.3 Light microscope images of dry alginate foams: (a) top surface and (b) cross sections morphology. (Adapted from Andersen, T. et al., *Biomacromolecules*, 13, 3703–3710, 2012.)

the ability to absorb high amounts of wound exudates (Boateng et al. 2008). Alginate foams were prepared by different freezing regimes of ionic and covalent cross-linked hydrogels (Thornton et al. 2004; Zmora et al. 2002). Porogen salts, as well as *in situ* gas formation and emulsion templating methods, are alternative preparation methods used for alginate foams (Barbetta et al. 2009; Leone et al. 2004; Partap et al. 2006).

Different foaming agents such as polysorbate, poloxamers, bovine albumin, hydroxypropyl methylcellulose, sodium dodecylsulfate, tyloxapol, and triton were used for enabling incorporation of air by high shear mixing, and stabilizing the wet foam after it is formed (Barbetta et al. 2009). The alginate foams were proposed as a potential delivery system of active molecules (Hegge et al. 2010, 2011).

One of the methods for preparation of alginate-based foams is based on homogeneous, ionotropic gelation of aerated alginate solutions, followed by air drying (Andersen et al. 2012). The method allows higher flexibility and better control of the pore structure, hydration properties, and mechanical integrity compared to foams prepared by other techniques (Figure 1.3).

1.2.1.7 Carrageenans

Carrageenan is a collective term for polysaccharides extracted from certain species of red seaweed belonging to the family Rhodophyceae. The chemical structure of carrageenans is formed by units of D-galactose and anhydrogalactose joined by glycosidic linkages and containing ester sulfate groups (Collins and Ferrier 1995; Kaplan 1998;). The number and position of the ester sulfate groups as well as the anhydrogalactose content influences the type of carrageenan (kappa, iota, and lambda carrageenans).

Refined carrageenans produce clear solutions and, therefore, clear films. The dried film is only partially soluble in cold water. Lambda carrageenans, being the most soluble ones in cold water and the most negatively charged, produce a slightly weaker film than both kappa and iota carrageenans. The molecular weights of these products range from 100,000 to one million. The gelling, thickening, and stabilizing properties of carrageenans make them useful in several industrial applications. Hot aqueous solution and a cationic salt are needed to obtain a thermo-reversible carrageenan hydrogel (Quignard et al. 2008; Thrimawithana et al. 2010). An important volume reduction in the gel structure takes place during the supercritical drying of carrageenan alcogels. This massive reduction in the volume of the resulting aerogel largely influences the textural properties of the dried material (Rauter et al. 2010).

1.2.1.8 Agar

Agar is extracted from two major commercial sources of red seaweed: *Gelidium* sp. and *Gracilaria* sp. Agar consists of a combination of agaropectin (nongelling fraction) and agarose (gelling fraction) (Stanley 2006). Agarose is a linear polymer with a molecular weight of about 120,000 and is composed of a repeating dimer of D-galactosyl and 3,6-anhydro-L-galactosyl units connected via alternating α-(1→3) and β-(1→4) glycosidic linkages. At temperature above 85°C, agarose exists as a disordered random coil, which upon cooling forms a strong thermally reversible gel. It forms a double helix in solution, which aggregates to form a 3D network (Stanley 2006). The linear, nonionic nature of agar allows hydrated molecules to associate more intimately, forming a network that is stabilized by intensive intermolecular hydrogen bonding during drying of the film.

Agar aerogels were prepared from hot solution with sucrose and consecutive drying using supercritical carbon dioxide and ethanol. The resulting materials presented extensive shrinkage and void formation (Brown et al. 2010). In the literature are also reported the production of agar aerogels as beads by means of an ethanol–water solvent exchange steps (Robitzer et al. 2011).

1.2.1.9 Lignin

Lignin is a highly branched amorphous hetero-polymers formed by three basic substituted phenylpropane repeat units, also known as *monolignols*. The proportion of each of these monomers in lignin varies considerably depending upon the type of plant material under consideration (Reddy and Yang 2005). Due to the abundance of lignins and their structural characteristics, they are under study as a base material for composites and in polymer grafts (Meister et al. 1984; Meister and Patil 1985).

The properties and uses of lignin combined with other polymers have recently been reviewed (Lyubeshkina 1983; Satheesh Kumar et al. 2009). Starch-kraft lignin foams have also been examined. The major applications for starch–lignin foams would be packaging containers for single or short-term use, as biodegradable alternatives to foamed polystyrene (Stevens 2010). The effect of the addition of two kinds of lignins, namely alkaline lignin and sodium lignosulfonate, on the foamability of thermoplastic zein-based bionanocomposites was investigated (Oliviero et al. 2012). In particular, different amounts of alkaline lignin and sodium lignosulfonate were added to zein powder and poly(ethylene glycol) through melt mixing to achieve thermoplastic biopolymers, which were subsequently foamed in a batch process, with a mixture of CO_2 and N_2 as blowing agent, in the temperature range of 50°C–60°C.

1.2.1.10 Soy Protein

Soy protein refers to the protein that is found in soybeans, a legume that consists of about 40% proteins and 20% oil on a dry-weight basis (Montgomery 2003). The most common technique used in the production of soy protein and oil consists of hexane extraction. However, the solvent has some considerable economic, environmental, and safety drawbacks. Aqueous extraction processing is an emerging technology that may be a viable and environmentally compliant alternative to conventional solvent extraction (Kunte et al. 1997; Petruccelli and Anon 1994). Among the plant proteins, soy protein contains relatively high contents of glutamic and aspartic acid residues. These anionic repeat units are responsible for its solubility in alkali and precipitation in acid conditions. Soy proteins are globular, reactive, and often water soluble. Films are formed on the surface of heated soymilk and plasticizers are required to maintain their elasticity (Cho and Rhee 2004). These materials are often used as biodegradable, nonelectrostatic, and nonflammable films and coatings (Wang et al. 1996). Molded plastics made of soy protein alone, or mixed with starch, have adequate mechanical and water resistance properties for several one-time use consumer products such as disposable containers, utensils, toys, and outdoor sporting goods (Cunningham et al. 2000; Rhim et al. 2000).

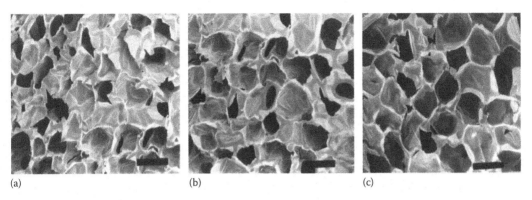

(a)   (b)   (c)

FIGURE 1.4 Scanning electron micrographs of soy protein-based plastic foams containing (a) 30% of defatted soy flour, (b) 30% of soy protein concentrate, and (c) 30% of soy protein isolate. Magnification bars: 500 μm.

Soy protein films have good oxygen barriers and ultraviolet-blocker properties. Biodegradable foams from soy protein are evaluated for their potential to replace synthetic foams for numerous one-time use products such as packaging material and nonflammable thermal insulation (Park et al. 2001; Petruccelli and Anon 1994).

Figure 1.4 shows scanning electron micrographs of defatted soy flour-, soy protein concentrate-, and soy protein isolate-based foams. The density, compressive stress, resilience, and dimensional stability of foams with soy protein isolate, soy protein concentrate, and defatted soy flour were found to increase with the concentration of the soy protein. Defatted soy flour-containing foam had a great expansion with large air gaps and over foam densities than those of foams containing soy protein concentrate and soy protein isolate. This was attributed to the less active hydrogen atoms of the defatted soy flour as compared to that of the soy protein isolate and soy protein concentrate (Park and Hettiarachchy 1999).

1.2.1.11 Zein

Zein proteins are found in amounts of 2%–10% in corn. The greater fraction of zein has a molecular weight of about 40,000. It does not contain lysine or tryptophan (Shukla and Cheryan 2001). Zein is insoluble in water or acetone alone. However, zein is soluble in water/acetone mixtures, aqueous alcohol, the glycols, monoethyl ether of ethylene glycol, tetrahedrofuryl alcohol, and aqueous alkaline solutions (Ghanbarzadeh et al. 2007).

When formed into films, zein is glossy, tough, and greaseproof, with a low water vapor permeability compared to most other agriculturally based protein films. For commercial purposes, it is extracted from gluten meal with dilute isopropanol (Dickey et al. 2001; Wolf and Lawton 1997).

Thermoplastic zein can be obtained by applying heat and shear stresses in the presence of a suitable plasticizer. Thermoplastic zein can be foamed by means of physical blowing agents. The size and number of pores of the resulting foam can be tuned by controlling the nature of the blowing agent composition (N_2 and/or CO_2), the foaming temperature (44°C–140°C), and the pressure drop rate (250–700 bar/s) (Di Maio et al. 2010).

1.2.1.12 Gluten

Gluten is a wheat protein intermixed with the starchy endosperm of the grain. Wheat gluten contains two main groups of proteins, gliadin and glutenin (Payne et al. 1982). Gliadins are proteins with disulfide bonds and low molecular weight, while glutenins have at least 10 times higher molecular weight. Gluten is insoluble in water, but partly soluble in alcohol and dilute acids, and soluble in alkalies (Micard et al. 2000). It is used as an adhesive and as a substitute for flour. Wheat gluten is excellent for forming films but, without plasticizer, the resulting films are brittle (Gontard et al. 1993; Vieira et al. 2011).

Mechanical treatment of gluten leads to disulfide bridge formation created by the amino acid, cysteine, which is relatively abundant in gluten. Gluten films have good oxygen and carbon dioxide barrier properties, but exhibit relatively high water vapor permeation (Debeaufort et al. 1998). Gluten-based films have also been extruded in the presence of glycerol. The resulting films tested either parallel or perpendicular to the direction of extruder flow gave different tensile properties (Hochstetter et al. 2006).

1.2.2 ANIMAL ORIGIN

1.2.2.1 Chitin/Chitosan

Next to cellulose, chitin is the second-most abundant biopolymer on the planet. For example, it is found in a wide range of insects, crustaceans, annelids, mollusks, coelenterata, and most fungi (Arcidiacono and Kaplan 1992; Austin et al. 1981; Cohen-Kupiec and Chet 1998; Hudson and Smith 1998; Shepherd et al. 1997). The deacetylated form of chitin is known as *chitosan*. Chitin is slightly soluble in conventional solvents due to its high crystallinity based on strong hydrogen bonds between chains (Gardner and Blackwell 1975), but can be solubilized by using some special organic solvents. Chitosan is obtained by deacetylation of chitin in a degree beyond 50% and it is easily soluble in dilute acids (Ravi Kumar 2000).

Chitosan is soluble in water and acidic solutions and is slightly soluble in weakly alkaline solutions. High-quality chitin and chitosan are nontoxic, biocompatible, and biodegradable (Kumar et al. 2004). Chitin and chitosan can be processed into beads, gels, powders, sheets, tablets, and sponges. Chitin gels can be obtained in large excess of water (hydrogel) or alcohol (alcogel) (Tamura et al. 2006). Chitin alcogels can then be dried by means of supercritical carbon dioxide to obtain aerogels with high porosity, high surface area, and low density (Tsioptsias et al. 2009). The mesoporous structure of the chitosan aerogels showed an increase in surface area as the chitosan concentration increases and the cross-linker content decreases (Chang et al. 2008; Kadib et al. 2011). An alternative water emulsion-template method was used to produce highly porous structures in hydrophilic poly(vinyl alcohol) and chitosan (Lee et al. 2007). High-strength fibers have been spun from chitin and chitosan acetates (Kumar et al. 2004).

Chitosan foams can be obtained by means of different techniques, such as freeze-drying, rapid prototyping, and internal bubbling process (Di Martino et al. 2005; Ji et al. 2011). Solvent casting/salt leaching and freeze-drying have also been used to generate chitosan materials with high porosity (Madihally and Matthew 1999; Wan et al. 2008; Wu et al. 2008).

A traditional gas-foaming method was proposed to generate a highly interconnected 3D porous structure able to support cell penetration and proliferation (Ji et al. 2011).

1.2.2.2 Casein

Casein is a mixture of related phosphoproteins found in milk and cheese. In bovine milk, casein is found as calcium casemate (up to 3%), whereas in human milk, it is present as potassium caseinate. The major components of casein are α-, β-, γ-, and κ-caseins, which can be distinguished by electrophoresis (Swaisgood 2003). Each of the four protein fractions has unique properties that affect casein ability to form films.

Casein is only sparingly soluble in water and nonpolar solvents but is soluble in aqueous alkaline solution (deWit and Klarenbeek 1984). Casein is used in the manufacture of molded plastics, adhesive, paints, glues, textile finishes, paper coatings, and man-made fibers (Kozempel and Tomasula 2004; Mauer et al. 2000; Sohail et al. 2006).

1.2.2.3 Whey Proteins

Whey protein is the collection of globular proteins isolated from whey, a by-product of cheese manufactured from bovine milk. The protein in bovine milk is 20% whey protein and 80% casein protein. The protein fraction in whey constitutes approximately 10% of the total dry solids in whey. This protein is typically a mixture of β-lactoglobulin (about 65%), α-lactoglobulin (about 25%), bovine serum albumin (about 8%), and immunoglobulins (Haug et al. 2007). Among these proteins, β-lactoglobulin is the one most intensively studied.

This globular protein denatures when heated, leading to small primary aggregates. The small aggregates can further grow into larger aggregates and at a certain concentration self-assemble into gels. In acidic conditions, this protein forms worm-like and flexible fibrils in several water–alcohol mixtures. An increase in temperature resulted in longer and stiffer fibrils. On the contrary, thinner fibrils were obtained when the β-lactoglobulin protein was heated in neutral conditions. The protein interactions that occur between chains determine film network formation and properties.

Recent studies have investigated the processing parameters necessary to obtain transparent, flexible whey protein sheets by means of a twin-screw extruder (Hernandez-Izquierdo et al. 2008). The effect of feed composition, temperature, and screw speed was found to be important to obtain materials with improved mechanical properties (Hernandez-Izquierdo and Krochta 2008).

1.2.2.4 Elastin

The proteinaceous material elastin is the primary structural component underlying the elastomeric mechanical response of compliant tissues in vertebrates and is critically important for appropriate function of tissues including artery, lung, and skin. Elastin is an important load-bearing tissue in the bodies of vertebrates and is used in places where mechanical energy is required to be stored (Daamen et al. 2001; Jia and Kiick 2009). It is highly cross-linked between lysine residues to generate an extensive network of fibers and sheets. Assembly of elastin fibers involves deposition of its precursor (tropoelastin) on a preformed template of microfibrils. Under physiological conditions, elastin undergoes a process of self-assembling called *coacervation*. This specific property has led to the development of a new class of synthetic polypeptides, which mimic elastin in its composition and are therefore also known as *elastin-like polypeptides* (Cappello et al. 1990).

These families of polymers show a wide range of interesting properties that are rarely found together in any other polymeric material, namely, biocompatibility, mechanical properties, stimuli-responsive nature, and self-assembly behavior. In particular, these materials are thermally responsive biopolymers, which can be switched between an extended water-soluble state and a collapsed state, leading to insolubility in water. The transition temperature can be varied over an extended range by changing the chemical of the polymer. In contrast to the majority of fibrous polypeptides, these compliant biomaterials are characterized by high extensibility and low value of the Young's modulus. Therefore, they more closely resemble synthetic polymer gels and elastomers rather than engineering fibers.

A salt-leaching/gas-foaming technique was used to synthesize chemically cross-linked highly porous elastin-like polypeptides and hydrogels with tunable physical properties. The pore size can be controlled by varying the size of the salt particles incorporated during the cross-linking reaction. Using the gas-foaming technique, a highly interconnected porous structure can be formed, which is otherwise difficult to obtain with more conventional techniques (Martín et al. 2009).

A collagen–elastin-like foam-fiber bilayer material was designed to produce improved tissue engineering scaffolds. The incorporation of elastin into the scaffolds improved the uniformity and continuity of the pore network, decreased the pore size and the fibers diameter, and increased their flexibility (Kinikoglu et al. 2011).

1.2.2.5 Collagen and Gelatin

Collagen proteins are particularly present in articular and bone tissues, where they are responsible for important mechanical functions throughout the body. It contains an unusually high level of the cyclic amino acids proline and hydroxyproline (Catchpole 1977). Collagen consists of three helical polypeptide chains wound around each other and connected by intermolecular cross-links (Tanzer 1976). It is used in its native fibrillar form, as well as after denaturation. Denaturation offers the possibility to fabricate several collagen forms, including sheets, tablets, pellets, and sponges. When the triple-helix structure of collagen is broken into single-strand molecules by acid, alkaline, or enzymatic hydrolysis, gelatin is obtained.

Depending on the method in which the collagen is treated, two different types of gelatin can be produced. Gelatin A is obtained after acidic pretreatment of collagen, after which it is thermally broken down. Gelatin B is afforded when an alkaline pretreatment is performed. In that case, glutamine and asparagine residues are converted into glutamic and aspartic acid, resulting in a higher carboxylic acid content (Badii and Howell 2006; Viidik et al. 1982).

Gelatin may be chemically treated to bring significant changes in its physical and chemical properties. Typical reactions include acylation, esterification, deamination, cross-linking, and polymerization, as well as simple reactions with acids and bases (Finch 1983; Fratzl et al. 1998).

The formation of thermo-reversible gels is obtained when aqueous solution of gelatin with a concentration greater than 0.5% is cooled to approximately 35°C–40°C. The rigidity or strength of the gel depends upon gelatin concentration, the structure and molecular mass of the gelatin, pH, temperature, and the presence of any additives (Ottani et al. 2001, 2002).

Thermoplastic gelatin can be foamed by means of a physical blowing agent when it is heated above its glass transition temperature (Salerno et al. 2007a). The resulting foams show a closed pore morphology. Also, it is possible to foam thermoplastic gelatin by blending with PCL. In that case, a selective extraction of the water-soluble gelatin phase permits to develop porous network pathways characterized by multimodal porosities (Salerno et al. 2007b).

Gelatin cryogels, for the production of porous materials, were used as cell carriers for a panel of human cells (Dubruel et al. 2007; Van Vlierberghe et al. 2007). A well-defined *curtain-like* pore architecture was induced by applying a cryogenic treatment on scaffolds containing both gelatin and chondroitin sulfate (Van Vlierberghe et al. 2008).

1.2.2.6 Actin

Actin is a globular multifunctional protein that forms microfilaments. It is found in all eukaryotic species as fibrous proteins (Oda et al. 2009; Reisler 1993). They can be present as either a free monomer, also known as *globular actin* (G-actin), or part of a linear polymer microfilament called *filamentous actin* (F-actin). Both forms of actin are essential for such important cellular functions as the contraction during cell division and the cells mobility (Remedios and Chhabra 2008). Actin was recently proposed as mechanics nanocomponent in nanotransformers able to convert mechanical energy into electrical energy (Xu et al. 1998).

1.2.2.7 Keratin

Keratin is a family of fibrous structural proteins. They form intermediate filaments, which are nonpolarized coiled-coil structures of 10 nm in diameter and several 100 nm in length (Tombolato et al. 2010). Keratin is the key structural material making up the outer layer of human skin. It is also the key structural component of hair and nails. Keratin fibers can assemble *in vitro* in the absence of auxiliary proteins or factors to form filaments, which are tough and insoluble and form strong un-mineralized tissues. The only other biological matter known to approximate the toughness of keratinized tissue is chitin (Omary et al. 2009). The physical properties of keratin depend on their supermolecular aggregation. The α-helix, the β-sheet motifs, and the disulfide bridges play

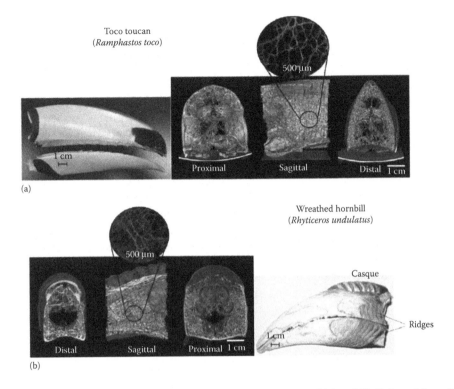

FIGURE 1.5 Structure and morphology of the beaks of (a) toucan and (b) hornbill. (Adapted from Seki, Y. et al., *Acta Biomaterialia*, 6, 331–343, 2010.)

a dominant role in the architecture and aggregation of keratins. Examples of α-keratins are present in the hair and wool. The harder β-keratins can be found in nails and claws of reptiles and in their shells (Kreplak et al. 2004).

The beaks of toucan and hornbill (Figure 1.5) represent interesting examples of sandwich-structured keratin composite. The exterior shell, or rhamphotheca, is made of β-keratin tiles. The internal foam consists of a fibrous network of trabeculae. These two components are separated by the dermis. It was demonstrated that the buckling resistance of the beak is enhanced by the internal cellular core due to the synergism between the two components (Seki et al. 2010).

A range of technical applications in filter, membrane, paper, textile, and leather fabrication are being targeted to employ keratin fibers. Applications of keratin fibers are also being explored in the field of biosensors and in the medical and biomedical sectors, including the use of protein fibers as surgical threads and sutures and for the development of biological membranes and scaffolds to support cell growth and tissue function.

1.2.2.8 Silk Fibroin

Even if the silk fibroins produced by insects, spiders, and worms are often classified as keratins, they represent an important class of natural proteins phylogenetically distinct from the vertebrate keratins (Hardy et al. 2008; Vepari and Kaplan 2007; Vollrath and Porter 2009). Spider silk is typically about 1 to 2 μm thick, compared with about 60 μm for human hair, and more for some mammals. The silk fibroin contains two protein components: fibroin and sericins. Fibroin consists of light (about 26 kDa) and heavy chain (about 390 kDa) polypeptides, which are present in a 1:1 ratio and are connected by a disulfide bridge. The sericins are hydrophilic proteins surrounding the fibroin (Sofia et al. 2001; Van Vlierberghe et al. 2011; Zhou et al. 2000).

The biologically and commercially useful properties of silk fibers depend on the organization of multiple adjacent protein chains into hard, crystalline regions of varying size (β-sheets), alternating with flexible, amorphous regions where the chains are randomly coiled (Jin et al. 2002). Due to the presence of the β-sheets, the silk fibroin becomes insoluble in water. These β-sheets are also responsible for the formation of silk fibroin hydrogels. The hydrogels are biocompatible, have unique mechanical properties, are biodegradable, and display cellular interactions. It was shown that the hydrogels permitted bone healing and stimulated cell proliferation. Furthermore, the bone remodeling process was accelerated, meaning that the bone becomes more mature and wider. These results show that silk fibroin hydrogels are potentially useful as synthetic injectable biomaterials for bone healing applications.

Recent technical applications for protein fibers include their use for patterning on the nanoscale and in the field of biosensors (Jin and Kaplan 2003). Porous materials were obtained from regenerated silk fibroin solutions, both aqueous and solvent, using porogens, gas foaming, and lyophilization (Nazarov et al. 2004). The concentration of silk fibroin solution and the size of porogen affect the stability of the sponge and its stiffness, compressive strength, and modulus (Kim et al. 2005a, c). Solvent-based porous sponges were also prepared by addition of a small amount of organic solvents into aqueous silk fibroin solution before pouring into a mold and freezing (Tamada 2005). Due to their rougher surface morphology and high porosity, aqueous-based silk sponges demonstrated improved cell attachment than the solvent-based porous sponges (Kim et al. 2005a). Porous 3D silk sponges have been utilized in a number of studies with cells to generate various connective tissues (Kim et al. 2005b; Meinel et al. 2004b, 2005). Sponges formed from the blends of poly(vinyl alcohol) and chitosan were also proposed as support for cartilage tissue engineering (Meinel et al. 2004a; Wang et al. 2006; Yeo et al. 2000).

1.2.2.9 Albumin

The albumins are a family of water-soluble proteins, the most common of which is serum albumin. Other types include the storage protein ovalbumin in egg white and different storage albumins in the seeds of some plants (Johnson and Zabik 1981). Within the human body, albumin is an important component of life. It contributes to the regulation of osmosis, helping to transport hormones, drugs, and other substances through the blood. Albumin is a globular protein that in water solution forms a colloid (Damodaran et al. 1998). The ovalbumin is also used for purification, as it tends to trap and store impurities. When heated, albumin begins to unfold, recombining in a new configuration that tends to coagulate.

Ovalbumin is well known in cooking, where it is used to obtain meringues in an oven at elevated temperature. In that case, the combination of mechanical mixing and high temperature partially unfolds the protein structure and entraps air bubble that is stabilized in a foam with a light and fluffy texture. Due to the poor strength, this material cannot be used for industrial applications. However, the cross-link of the protein molecules with formaldehyde or other aldehydes to form stable networks, once heated, was proposed to obtain stable foams (Li et al. 2012b). The resulting foams had a bulk density of 0.29 g/cm^3 and a porosity of around 77%. The addition of plasticizers such as glycerol, polyethylene glycol, and camphor to albumin protein foams yields very flexible soft foams suitable for different applications. The foams so prepared had a good capability of shape recovery under cyclic mechanical loads and they do not show any change of flexibility with aging (Li et al. 2012a).

The egg white protein was applied as binder or foaming agent. The samples fabricated using the protein direct foaming method were stronger and more uniform pore structures in the similar porosity (He et al. 2011). Tuck and Evans (1999) reported the preparation of porous alumina by blending alumina powder with egg white protein solution. Several parameters, such as the ceramic solid loading, the egg white protein–water ratio, the foaming time, and the sintering temperature, were found to affect the foaming processing conditions and then to influence the overall porosity and foam microstructure (Dhara and Bhargava 2003; He et al. 2009; Lemos and Ferreira 2004). Figure 1.6 illustrated

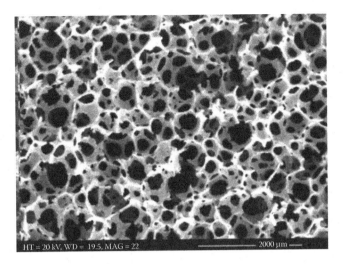

HT = 20 kV, WD = 19.5, MAG = 22   2000 μm

FIGURE 1.6 Example of a cellular structure obtained by protein direct foaming for alumina samples. The sample had a density of 0.42 g cm^{-3} at 89% porosity. (Adapted from He, X. et al., *J. Porous Mater.,* 19, 761–766, 2011.)

a spherical shell-like structure interconnected by small windows on large pore walls. This structure was formed by well-distributed ceramic particles at the air bubble interface, which was generated by the amphiphilic egg white protein during direct foaming.

1.2.3 BACTERIAL AND FUNGAL ORIGIN

1.2.3.1 Cellulose

Many living organisms synthesize cellulose affording a wide range of supramolecular structures, morphologies, and properties. Cellulose is produced by certain types of bacteria such as *Acetobacter, Pseudomonas, Rhizobium, Agrobacterrium,* and *Sarcina* species and can be synthesized as a continuous film by cultivating the bacteria in a glucose medium (Stevens 1998). Marine chordates, such as sea squirts, also secrete cellulose. Biosynthesis of cellulose by enzymatic processes is also known (Gilbert and Kadla 1998; Popa and Hon 1996). Among all cellulosic materials, bacterial cellulose displays the highest mechanical properties. In fact, it has a tensile strength and Young's modulus comparable to Kevlar (Klemm et al. 2006). The outstanding performance of bacterial cellulose stems from its high purity, high crystallinity (75%), and ultra-fine network structure (Klemm et al. 2001).

Cellulose esters such as acetates are used as plastic materials, fibers in textiles, and coatings. Cellulose xanthates are employed in the viscose rayon process (Gilbert and Kadla 1998). Unlike cellulose, cellulose ethers, for example, carboxymethyl cellulose, methylcellulose, and hydroxyl propyl cellulose, are soluble in water and form films for various applications.

1.2.3.2 Pullulan

Pullulan is a water-soluble, extracellular polysaccharide consisting of maltotriose units. Compared to the gum arabic, pullulan does not present re-branching in the structure, and hence, its polymer chains are able to associate and form stronger films. It is produced from a variety of carbon sources such as glucose, sucrose, and starch hydrolyzates by the fungus *Aureobasidium pullulans* (*Kirk-Othmer Encyclopedia of Chemical Technology* 2000). Pullulan molecular weight depends on the production strain and the physiological conditions used during the fermentation (Wiley et al. 1993) (Pollock et al. 1992). Pullulan has some very attractive characteristics: it is highly water soluble, is tasteless, is odorless, has excellent oxygen barriers, and is resistant to oils and greases. They form viscous pseudo-plastic solutions in water, which are stable in the presence of cations, but do not form gels. Pullulan has been identified as suitable polymer for a large number of industrial applications. Due to its unique film-forming and oxygen barrier properties, pullulan has been used in protective and adhesive edible coatings.

Other proposed applications of pullulan are in degradable films and fibers, paper coatings and binders, cosmetics, pharmaceutical tablet coatings, and contact lenses that contain slow release bioactive medicines (Khan and Ahmad 2013). The casting solution produces film with high tensile strength and puncture strength. Its films are clear and may include colors or flavors for food uses (Ball et al. 1992).

1.2.3.3 Scleroglucan/Schizophyllan

Scleroglucan/schizophyllan represents a group of neutral extracellular homo-polysaccharides produced by several fungal species such as *Sclerotium rolfsii*, *Sclerotium glucanicum*, and *Schizophyllum commune*. Both polysaccharides share the chemical structure of the backbone with curdlan. They are known for several things, including their ability to stimulate the immune system, carry metals in water, aid in delivering drugs, and use in some nanofibers (Bluhm et al. 1982). These high molecular weight polymers (about 500,000 Da) form highly viscous solutions with water and exist in a stable triple helix conformation. Because of its peculiar rheological properties and its resistance to hydrolysis, temperature, and electrolytes, scleroglucan has various industrial applications, especially in the oil industry for thickening, for drilling muds, and for enhanced oil recovery (Coviello et al. 2005; Giavasis et al. 2005). Other industrial uses include the preparation of adhesives, water colors, printing inks, and liquid animal feed composition (Schmid et al. 2011).

1.2.3.4 Xanthan Gum

Xanthan gum is an extracellular polysaccharide produced commercially by aerobic submerged fermentation from *Xanthomonas campestris* (Becker et al. 1998; Marchessault 1984). Xanthan gum solutions are highly pseudoplastic, rapidly regaining viscosity on removal of shear stress. Since the viscosity of xanthan gum solutions/dispersions is relatively temperature independent and the polymer is resistant to acids, alkali, and enzymes, it is a popular thickener and stabilizer in the food industry.

Xanthan gum is also used as a stabilizer for a wide variety of suspensions (food, agrochemical pesticides, and sprays), emulsions (foods and thixotropic paints), and foams (beer and firefighting fluids). It is used in petroleum production to increase the viscosity of drilling fluids and extensively in food and pharmaceutical industries to stabilize emulsions and improve texture. Experimentally, a mixture of xanthan–locust bean gum forms thermo-reversible gels at polymer concentrations as low as 0.1%; in contrast, xanthan–guar gum mixtures only exhibit increased viscosity regardless of the concentration (Hassler and Doherty 1990).

Foam as aerogel composites based on xanthan gum and sodium montmorillonite clay has been prepared using friendly freeze-drying process. Agar was used to improve the properties of xanthan gum/clay aerogels. A wide range of microstructures and mechanical properties were obtained with a minimal variation in the density by changing the blended ratio of xanthan gum and agar (Wang et al. 2014).

1.2.3.5 Curdlan

Curdlan gum is an extracellular polysaccharide commercially produced by microbial fermentation of a mutant strain of *Alcaligenes faecalis* (McIntosh et al. 2005). Native curdlan has a granular structure and is water insoluble but can be dissolved in alkaline solution, formic acid, dimethyl sulfoxide, aqueous saturated urea, and thiourea. Curdlan gum can form gels by heating the suspensions or reduce the pH of an alkali solution (Stone 2009). The structure of curdlan (in both hydrated and anhydrous forms), determined from oriented fibers, assumes a right-handed, parallel, sixfold triple-helical conformation. Heat treatment of curdlan suspensions may produce two types of gels, depending on temperature. A thermo-reversible low-set gel is obtained when heated to 55°C–60°C and forms a weak, low-set gel upon cooling. A thermo-irreversible high-set gel requires heating up to 80°C or higher. The gel strength increases with heating temperature (gelation starts at 55°C) but stays constant between 60°C and 80°C. Further increase in temperature, up to 120°C, changes the molecular structure to a triple helix (melting temperatures above 140°C). These gels are very susceptible to shrinkage and syneresis but are resistant to degradation (Gidley and Nishinari 2009).

The property of such gels on acidification can be used as a method for *in situ* gelation in oil fields (Brown et al. 2007). Curdlan is used as an additive to improve the texture of food products such as jellies, jams, puddings, noodles, ice cream, fish paste, and meat products (Sun et al. 2011).

The water-retention, adsorbing, and calcium gel-forming properties of curdlans make them useful as a segregation reducing agent in making super-workable concrete form cement and small stones. It is also used as a binding agent for ceramics, active carbon, and reconstituted tobacco sheets. Since it remains in the form of a gel at high temperatures, it can be used in cultures of microorganisms (Zhan et al. 2012).

1.3 BIO-BASED POLYMERS

1.3.1 Degradable Bio-Based Polymer

Since the last decade, the whole world is facing a sharp increase in energy consumption in conjunction with a fast growth in world population and an increasing demand in emerging economies. The vast majority of energy sources are actually derived from fossil. Chemical industry also uses fossil sources, primarily oil and gas, both for the manufacturing of chemical products and base chemicals and for its internal energy consumption. Many countries have started relevant processes to implement national policies ultimately aimed to use resources with no or minimal environmental impact that are alternatives to fossil ones. Vegetable or animal biomasses can be an option provided they do not compete with food or with feed, because of lack of food availability in developing countries and potential risk of unacceptable agricultural commodities price increase. Therefore, a new kind of chemistry, based on the use of natural renewable resources, like vegetable biomasses, widely available, inexpensive, and with marginal environmental impact, is developing. This so-called bio-based chemistry will contribute to address some of the main concerns of the chemical industry (Cobror and Iannace 2011).

According to the growing demand of low carbon footprint polymeric materials, two families of bio-based polymers are currently under development: (1) biopolymers entirely from new bio-building blocks, typically biodegradable and (2) biopolymers equivalent to the conventional oil-based polymers but obtained starting from renewable resources such as polyesters, polyamide, polyethylene, polyurethane, and epoxy resins.

The first family comprises polyesters obtained, respectively, by fermentation from biomass or from genetically modified plants (e.g., polyhydroxyalkanoate, PHA) and by synthesis from monomers obtained from biomass (e.g., polylactic acid, PLA). The second family comprises polyesters, totally synthesized by the petrochemical process (e.g., polycaprolactone; polyesteramide, PEA; and aliphatic or aromatic copolyesters) (Di Maio and Iannace 2011). In some cases, oil-based monomers are combined with bio-based monomers to develop partially bio-based polyesters. Some examples of the chemical structures of these polymers are given in Figure 1.7.

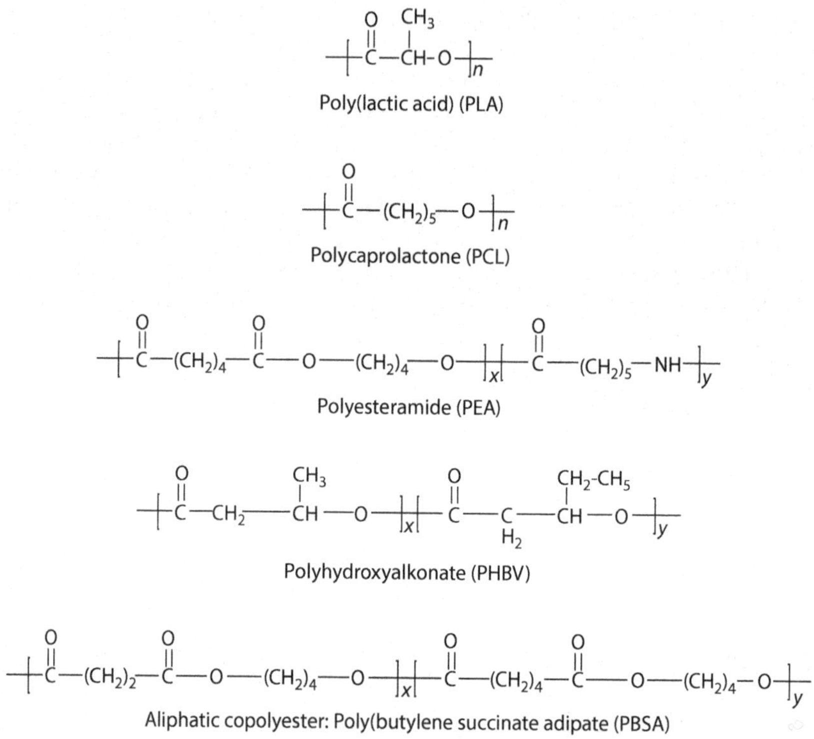

FIGURE 1.7 Chemical structure of different biodegradable polyesters.

Sources, properties, and uses of biodegradable thermoplastics and elastomers have been reviewed by Baillie (2004); Mohanty et al. (2000, 2005); Steinbuchel and Doi (2002); and Steinbüchel (2003).

1.3.1.1 Polylactic Acid

PLA is a biodegradable polyester produced via fermentation of renewable feedstocks. Properties of PLA, such as degree of crystallinity, melting temperature, and glass transition temperature, can be tailored by controlling the monomer composition of the two forms: L and D. A wide spectrum of products can be obtained by using conventional process technologies such as extrusion, injection molding, injection stretch blow molding, casting, blown film thermoforming, foaming, blending, fiber spinning, and compounding (Lim et al. 2008). PLA has been used for a wide range of application areas, such as packaging (cups, bottles, films, and trays), textiles (shirts and furniture), nonwovens (diapers), electronics (mobile phone housing), agriculture (usually blended with thermoplastic starch), and cutlery.

1.3.1.2 Polyhydroxyalkanoates

Like PLA, PHAs are aliphatic polyesters produced via fermentation of renewable feedstocks through accumulation within the microorganism (Holmes 1988). The PHA accumulates as granules within the cytoplasm of cells and serves as a microbial energy reserve material. PHAs have a semi-crystalline structure, the degree of crystallinity ranging from about 40% to around 80% (Abe and Doi 1999).

Depending upon the chemical structure of these bio-based polyesters, homopolymers, copolymers, and terpolymers with a wide range of physical properties can be obtained. Processing technologies such as film blowing, film casting, fiber spinning, and injection molding can be used and require PHAs with specific chemical structures and molecular weights. The injection molded and/or extruded PHA products cover a wide range of applications, such as cutlery, packaging (bags, boxes, and foams), agriculture mulch films, personal care (razors and tooth brush handles), office supplies (pens), golf pins, toys, and various household wares, and fibers (Shen et al. 2009).

1.3.1.3 Poly(ε-Caprolactone)

PCL is obtained by ring opening polymerization of ε-caprolactone (Albertsson and Indra 2002; Chiellini and Solaro 1996; Okada 2002). PCL is widely used as a polyvinyl chloride solid plasticizer or for polyurethane applications. PCL shows a very low T_g (−61°C) and a low melting point (65°C). Therefore, PCL is generally blended (Averous 2000; Bastioli 1998) or modified (e.g., copolymerization and cross-link) (Koenig and Huang 1994).

1.3.1.4 Aliphatic Copolyesters

A large number of biodegradable aliphatic copolyesters are obtained by the combination of diols, such as 1,2-ethanediol, 1,3-propanediol, or 1,4-butadenediol, and dicarboxylic acids, such as adipic, sebacic, and succinic acid. Polybutylene succinate is obtained by polycondensation of 1,4-butanediol and succinic acid, while polybutylene succinate/adipate is obtained by addition of adipic acid. Polybutylene succinate is a white crystalline thermoplastic with a density between 1.2 and 1.3 g cm^{-3} (comparable to PLA) and can be processed by injection molding, extrusion molding, or blown molding using conventional polyolefin equipment. New grades of polybutylene succinate copolymers have recently been produced with a high recrystallization rate and high melt tension, suitable for preparing stretched blown films and highly expanded foams.

1.3.1.5 Aromatic Copolyesters

Compared to totally aliphatic copolyesters, aromatic copolyesters are often based on terephthalic diacid and bio-based monomers such as butylene diols (BDO), succinic acid, and adipic acid. Poly(butylene succinate-terephthalate) is based on BDO and succinic acid, while poly(butylene adipate terephthalate) is based on BDO and adipic acid. Poly(butylene adipate terephthalate) is a very attractive product because it has been successfully used in many blends with bio-based polymers, with PLA possibly being the most prominent example (a blend with PLA has been commercialized by BASF under the name Ecovio®).

1.3.2 NONDEGRADABLE BIO-BASED POLYMERS

Several durable bio-based plastics with varying bio-based content that are potentially interesting for the development of bio-based nondegradable foams include polyesters and polyurethanes.

1.3.2.1 Bio-Based PTT, PBT, PET, and Other Polyesters

Partially or totally bio-based polyesters are usually based on diols and/or diacids that are derived from biomass. In most of the cases, the bio-based diols are ethylene glycol, 1,3-propanediol (trimethylene glycol, 3G, or PDO), and 1,4-butanediol (BDO). These bio-based diols are then polymerized with diacid that can be either bio-based (succinic, adipic acid) or petrochemical based such as dimethyl terephthalate (DMT) or terephthalic acid (PTA) (Cobror and Iannace 2011).

Poly(trimethylene terephthalate) (PTT) is a linear aromatic polyester produced by polycondensation of bio-based PDO with either purified PTA or DMT. PTT combines physical properties similar to polyethylene terephthalate (PET) (strength, stiffness, toughness, and heat resistance) with processing properties of poly(butylene terephthalate) (PBT) (low melt and mold temperatures, rapid crystallization, and faster cycle time than PET).

PBT is a linear aromatic polyester produced by transesterification and polycondensation of DMT with BDO. PBT can also be produced from purified PTA and BDO.

PET can be produced from either DMT or purified PTA, and ethylene glycol. It has the potential to be completely produced from bio-based feedstock. For example, PTA could be produced by using bio-based xylene produced by depolymerization of lignin or it could be replaced by 2,5-furandicarboxylic acid FDCA, a monomer that can be obtained by fructose (Taarning et al. 2008).

1.3.2.2 Polyurethane from Bio-Based Polyols

Polyurethanes are prepared by reacting two components: a polyol and an isocyanate. While so far the isocyanate component has been produced from petrochemical feedstock, there are a number of possibilities for the polyol to be produced from a renewable source. There are three approaches to produce polyols from bio-based resources (Petrovic 2008; Sherman 2007):

1. Producing polyether polyol from sucrose and sorbitol or from PDO.
2. Producing polyester polyol from bio-based diols (ethylene oxide, BDO, and PDO) and dicarboxylic acids (succinic acid or adipic acid).
3. Producing oleochemical polyols from vegetable oils. The commercially available vegetable oil-based polyols can be produced from, for example, soybean oil, castor oil, sunflower oil, and rapeseed oil.

1.4 BIO-INSPIRED CELLULAR MATERIALS

Nature is able to produce a wide variety of materials or structures by combining few types of component materials. Minerals (e.g., calcium carbonate, calcium phosphate, and silica), and a range of polymers, based either on protein (e.g., collagen and silk) or on sugar (cellulose and chitin) are assembled in composite structures with a remarkable range of material properties. The control over shape and structure on many length scales through a hierarchical structuring not only allows materials to be grown in a self-organized manner but also allows the structure to be adapted to needs at each of the different scales. Wood and bone are examples of multilevel cellular architectures found in nature whose structures and functions may lead to new design concepts for bio-inspired cellular materials (Fratzl and Weinkamer 2007).

The hierarchical structure of wood is based on a complex arrangement of cellulose fibrils in a matrix of hemicelluloses and lignin (Dunlop and Fratzl 2013). Crystalline microfibrils are built up of aligned molecules having length of 30–60 nm. These microfibrils represent the reinforcing phase of the composite structure. The orientation of the microfibrils in the composite lamellae determines the rigidity of the three. By simply controlling the microfibril angle, at which these fibrils are wrapped around the lumen of the wood cell (Figure 1.8), it is possible for the tree to vary the mechanical properties, from high stiffness at small angles to more flexibility at larger angles. At larger scale, wood is characterized by a honeycomb-like structure that is stiff and lightweight and can transport fluids through the lumen of the cells themselves. The consequence is that wood exceeds most engineering materials in terms of specific properties (Gibson et al. 2010).

Bone is another example of hierarchical cellular structure found naturally (Figure 1.9). In this case, the constituent building blocks are assembled in a nanocomposite structure, where the reinforcing phase is constituted by hydroxyapatite nanoplatelets, while the composite matrix is based on triple-helical structure of collagen molecules. The nano-sized inorganic phase is organized in an ordered arrangement giving rise to highly mineralized fibers having strength and stiffness able to support the structural loads that bone must carry in vertebrate organisms. Bone can be therefore viewed as a fiber composite of ordered collagen molecules reinforced by plate-like mineral particles.

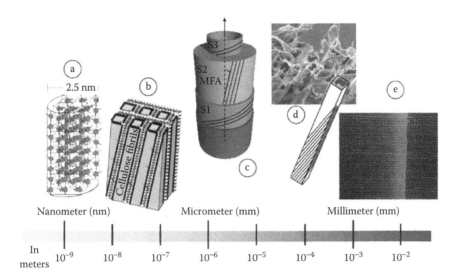

FIGURE 1.8 Hierarchical structure of wood: (a) the crystalline part of a cellulose microfibril, (b) model of the arrangement of the cellulose fibrils in a matrix of hemicelluloses reinforced lignin within, (c) structure of the cell wall of softwood tracheids, (d) broken tracheids within a fracture surface of spruce wood and the schematically drawn wood cell and the cellulose fibrils spiraling around with an inclination against the long cell axis known as the *microfibril angle*, and (e) cross section through the stem. (Data from Weinkamer, R. and Fratzl, P., *Mater. Sci. Eng. C*, 31, 1164–1173, 2011. With permission.)

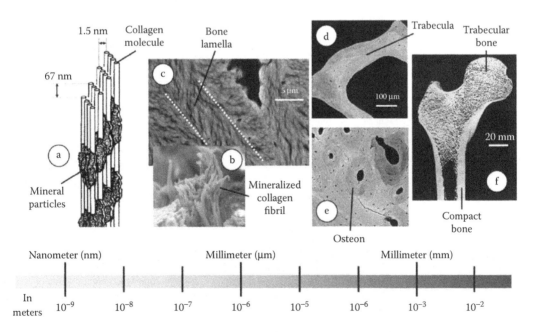

FIGURE 1.9 Hierarchical structure of bone: (a) the nanocomposite of collagen and mineral particles, (b) collagen fibrils sticking out of a fracture surface, (c) lamellar arrangement of bone around an osteocyte lacuna (on top), (d) one trabecula consisting of several bone packets, (e) osteons forming compact bone, and (f) cross section through a proximal femur showing trabecular bone in the upper part and compact bone at the shaft. (Data from Weinkamer, R. and Fratzl, P., *Mater. Sci. Eng. C*, 31, 1164–1173, 2011. With permission.)

These mineralized fibrils are *glued* together by noncollagenous proteins to make up a mineralized fiber bundle. These bundles, in turn, are arranged into lamellae (Dunlop and Fratzl 2013; Weinkamer and Fratzl 2011). At larger scale, in trabecular bone, for example, the bone tissue is characterized by an open cell foam-like structure with a relative density varying from about 0.05 to 0.5 (technically, any bone with a relative density less than 0.7 is classified as *trabecular*).

Even though biological materials are of interest for material scientist as a source of inspiration, the development of hierarchical bio-inspired materials will not be easily applicable immediately to the design of new engineering materials. The reason arises from striking differences between the design strategies common in engineering and those used by nature. These differences are contributed by the different sets of elements used by nature and the engineer—with the engineer having a greater choice of materials and fabrication methods (Fratzl and Weinkamer 2007). In particular, most of the structural materials used by nature are polymers or composites of polymers and ceramic particles. Such materials would not be the first choice of an engineer who intends to build very stiff and long-lived mechanical structures.

Although rules learnt from nature about assemblies and architectures can be applied to the design of synthetic materials, the natural constituents themselves often have performance deficiencies such as thermal and chemical stability, resistance to environment changes, and very simple chemical compositions that limit their applications (Table 1.1).

A large number of studies on nanocomposite and organic–inorganic hybrid materials in the last few decades have stimulated the development of man-made cellular structures that could recall the multilevel structures of natural cellular materials. However, there are several limitations that do not allow engineers and materials scientists to successfully apply the principle and technologies, typically employed in the production of foamed materials, with nanocomposites. For example, when a polymeric matrix is reinforced with inorganic nanofillers, such as SiO_2 nanospheres or clay nanoplatelets, the maximum amount of inorganic phase that can be loaded in the matrix is in

---

TABLE 1.1

Contrast in Properties between Natural and Synthetic Hierarchical Materials

	Natural Hierarchical Materials	Synthetic Hierarchical Materials
Structure		
Composition	Light elements dominate: C, H, O, N, S, P, Ca, Si, and so on	Large variety of elements: C, H, O, N, S, P, Ca, Si, Ti, Zr, Al, Fe, Co, Ni, Cu, Zn, Y, Nb, and so on
Process	Controlled self-assembly	Templating synthesis or spontaneous synthesis
Structure	Very precise and complex	Relatively simple
Length	Molecule, nano-, and micro-levels	Molecule, nano-, and micro-levels
Morphology	Self-changing according to the changes in the environment	Controlled size and shape
Functional Properties		
Functions	Specific functionality	Multiple functionalities design
Lifetime	Self-healing or self-repairing	Few examples of self-repairing
Stability	Very narrow temperature range and pH values range	Thermal and chemical stabilities
Influence of environment	Natural conditions (suited temperature, moisture, pressure, etc.)	Wide condition range

Source: Su, B.-L. et al., *Hierarchically Structured Porous Materials*, Wiley-VCH Verlag GmbH, Weinheim, Germany, 2011.

the range of 5%–10%. At higher loadings, the dispersion or the exfoliation of nanoparticles is difficult to obtain, and they tend to aggregate leading to micro-sized clusters that do not contribute to the improvement of the nanocomposite strength. Even at 5%–10% of nanoparticles, the viscosity of the polymeric melt becomes several order of magnitude higher than the neat polymeric phase. This remarkable increase in viscosity can affect the nucleation and growth of the gas bubbles in a typical gas-foaming process, thus limiting the amount of inorganic phase that can be employed in these structures. In highly mineralized natural cellular materials like bone, the total amount of inorganic phase is around 50%–60% (w/w), which is much higher than any typical man-made polymeric nanocomposites. The high content of inorganic phase and the particular nanostructural arrangement of the organic and inorganic phases results in very high mechanical properties of the cell wall in a trabecular bone. Based on ultrasound and nanoindentation tests, studies reported by several research groups (Gibson et al. 2010) have recently summarized the mechanical properties of a human trabecular bone: 1800 Kg m^{-3} for density, 18 GPa for Young's modulus, 182 MPa for compressive strength, and 113 MPa for tensile strength.

The comparison with nanocomposite foams suggests that another major difference between materials from nature and the engineer is in the way they are made. While the engineer selects a material to fabricate a part according to an exact design and to an available fabrication technology, nature goes the opposite direction and grows both the material and the whole organism (a plant or an animal) using the principles of (biologically controlled) self-assembly (Fratzl and Weinkamer 2007). For example, while in man-made foams, gas bubbles are created in a solid by controlling, using different techniques, the separation mechanisms of a gas or a liquid phase from a solid phase, nature uses a bottom-up approach and builds tubular, honeycomb, or foamed structures at the micron scale by a proper assembling of organic and inorganic building blocks during the growth of the living organism.

Several approaches inspired by the mechanisms of biomineralization in nature and how they can be applied to the synthesis of functional and advanced materials such as bone implants, nanowires, semiconductors, and nanostructured silica have been utilized by scientists. The main biominerals (silica, magnetite, $CaCO_3$, and calcium phosphate) found in nature have been employed with biomimetic strategies to develop organic–inorganic hybrid materials for a variety of functionalities that are inspired by those utilized in living systems such as navigation, mechanical support, photonics, and protection of the soft parts of the body. Nudelman and Sommerdijk (2012) have shown that it is possible to tailor the structure, size, function, and properties of materials from the nanometer to the macrometer level by controlling the self-organization of building block by using methods such as supramolecular template synthesis, template-directed crystal growth, phase separation, and self-assembly. Materials with controllable sizes, composition, morphology, crystallinity, and hierarchical organization can be developed by using organic scaffolds and templates. However, so far such functional materials have only been produced on a laboratory scale by using a bottom-up approach, and their synthesis on an industrial scale with direct technological applications and in a cost-effective way is still very limited. With the current state of self-assembly science, a successful outcome probably cannot yet be obtained by sole reliance on the bottom-up concept. Fusion with well-developed top-down nanofabrication is important for the development of self-assembly science and technology (Ariga et al. 2008).

REFERENCES

Aaltonen, O., and Jauhiainen, O. (2009). The preparation of lignocellulosic aerogels from ionic liquid solutions. *Carbohydrate Polymers*, 75(1), 125–129. doi:10.1016/j.carbpol.2008.07.008.

Abe, H., and Doi, Y. (1999). Structural effects on enzymatic degradabilities for poly[(R)-3-hydroxybutyric acid] and its copolymers. *International Journal of Biological Macromolecules*, 25(1–3), 185–192. doi:10.1016/S0141-8130(99)00033-1.

Albertsson, A.-C., and Indra, K. V. (2002). Aliphatic polyesters: Synthesis, properties and applications. In C. A. Albertsson (Ed.), *Degradable Aliphatic Polyesters* (Vol. 157, pp. 1–40). Berlin, Germany: Springer. doi:10.1007/3-540-45734-8.

Alnaief, M., Alzaitoun, M. A., García-González, C. A., and Smirnova, I. (2011). Preparation of biodegradable nanoporous microspherical aerogel based on alginate. *Carbohydrate Polymers*, *84*(3), 1011–1018. doi:10.1016/j.carbpol.2010.12.060.

Amsden, B., and Turner, N. (1999). Diffusion characteristics of calcium alginate gels. *Biotechnology and Bioengineering*, *65*(5), 605–610. doi:10.1002/(SICI)1097-0290(19991205)65:5<605::AID-BIT14>3.0.CO;2-C.

Andersen, T., Melvik, J. E., Gåserød, O., Alsberg, E., and Christensen, B. E. (2012). Ionically gelled alginate foams: Physical properties controlled by operational and macromolecular parameters. *Biomacromolecules*, *13*(11), 3703–3710. doi:10.1021/bm301194f.

Arcidiacono, S., and Kaplan, D. L. (1992). Molecular weight distribution of chitosan isolated from Mucor rouxii under different culture and processing conditions. *Biotechnology and Bioengineering*, *39*(3), 281–286. doi:10.1002/bit.260390305.

Ariga, K., Hill, J. P., Lee, M. V, Vinu, A., Charvet, R., and Acharya, S. (2008). Challenges and breakthroughs in recent research on self-assembly. *Science and Technology of Advanced Materials*, *9*(1), 014109. doi:10.1088/1468-6996/9/1/014109.

Arvanitoyannis, I., Psomiadou, E., and Nakayama, A. (1996). Edible films made from sodium caseinate, starches, sugars or glycerol. Part 1. *Carbohydrate Polymers*, *31*(4), 179–192. doi:10.1016/S0144-8617(96)00123-3.

Arvanitoyannis, I., Psomiadou, E., Nakayama, A., Aiba, S., and Yamamoto, N. (1997). Edible films made from gelatin, soluble starch and polyols, Part 3. *Food Chemistry*, *60*(4), 593–604. doi:10.1016/S0308-8146(97)00038-1.

Austin, P., Brine, C., Castle, J., and Zikakis, J. (1981). Chitin: New facets of research. *Science*, *212*(4496), 749–753. doi:10.1126/science.7221561.

Averous, L. (2000). Properties of thermoplastic blends: Starch–polycaprolactone. *Polymer*, *41*(11), 4157–4167. doi:10.1016/S0032-3861(99)00636-9.

Badii, F., and Howell, N. (2006). Fish gelatin: Structure, gelling properties and interaction with egg albumen proteins. *Food Hydrocolloids*, *20*(5), 630–640. doi:10.1016/j.foodhyd.2005.06.006.

Baillie, C. (ed.). (2004). Why Green Composites?, Woodhead Publishing Series in Composites Science and Engineering, Woodhead Publishing.

Ball, D. H., Wiley, B. J., and Reese, E. T. (1992). Effect of substitution at C-6 on the susceptibility of pullulan to pullulanases. Enzymatic degradation of modified pullulans. *Canadian Journal of Microbiology*, *38*(4), 324–327. doi:10.1139/m92-054.

Barbetta, A., Barigelli, E., and Dentini, M. (2009). Porous alginate hydrogels: Synthetic methods for tailoring the porous texture. *Biomacromolecules*, *10*(8), 2328–2837. doi:10.1021/bm900517q.

Bastioli, C. (1998). Properties and applications of Mater-Bi starch-based materials. *Polymer Degradation and Stability*, *59*(1–3), 263–272. doi:10.1016/S0141-3910(97)00156-0.

Becker, A., Katzen, F., Pühler, A., and Ielpi, L. (1998). Xanthan gum biosynthesis and application: A biochemical/genetic perspective. *Applied Microbiology and Biotechnology*, *50*(2), 145–52. Retrieved from http://www.ncbi.nlm.nih.gov/pubmed/9763683.

Bluhm, T. L., Deslandes, Y., Marchessault, R. H., Pérez, S., and Rinaudo, M. (1982). Solid-state and solution conformation of scleroglucan. *Carbohydrate Research*, *100*(1), 117–130. doi:10.1016/S0008-6215(00)81030-7.

Boateng, J. S., Matthews, K. H., Stevens, H. N. E., and Eccleston, G. M. (2008). Wound healing dressings and drug delivery systems: A review. *Journal of Pharmaceutical Sciences*, *97*(8), 2892–2923. doi:10.1002/jps.21210.

Brown, J., O'Callaghan, C. A., Marshall, A. S. J., Gilbert, R. J. C., Siebold, C., Gordon, S., Brown, G. D., and Jones, E. Y. (2007). Structure of the fungal beta-glucan-binding immune receptor dectin-1: Implications for function. *Protein Science: A Publication of the Protein Society*, *16*(6), 1042–1052. doi:10.1110/ps.072791207.

Brown, Z. K., Fryer, P. J., Norton, I. T., and Bridson, R. H. (2010). Drying of agar gels using supercritical carbon dioxide. *The Journal of Supercritical Fluids*, *54*(1), 89–95. doi:10.1016/j.supflu.2010.03.008.

Cappello, J., Crissman, J., Dorman, M., Mikolajczak, M., Textor, G., Marquet, M., and Ferrari, F. (1990). Genetic engineering of structural protein polymers. *Biotechnology Progress*, *6*(3), 198–202. doi:10.1021/bp00003a006.

Carr, L. G., Parra, D. F., Ponce, P., Lugão, A. B., and Buchler, P. M. (2006). Influence of fibers on the mechanical properties of cassava starch foams. *Journal of Polymers and the Environment*, *14*(2), 179–183. doi:10.1007/s10924-006-0008-5.

Catchpole, H. R. (1977). Biochemistry of collagen. *JAMA: The Journal of the American Medical Association*, *237*(26), 2856. doi:10.1001/jama.1977.03270530064034.

Cervin, N. T., Andersson, L., Ng, J. B. S., Olin, P., Bergström, L., and Wågberg, L. (2013). Lightweight and strong cellulose materials made from aqueous foams stabilized by nanofibrillated cellulose. *Biomacromolecules*, *14*(2), 503–11. doi:10.1021/bm301755u.

Chang, X., Chen, D., and Jiao, X. (2008). Chitosan-based aerogels with high adsorption performance. *The Journal of Physical Chemistry. B*, *112*(26), 7721–7725. doi:10.1021/jp8011359.

Chiellini, E., and Solaro, R. (1996). Biodegradable polymeric materials. *Advanced Materials*, *8*(4), 305–313. doi:10.1002/adma.19960080406.

Cho, K. Y., and Rizvi, S. S. H. (2008). The time-delayed expansion profile of supercritical fluid extrudates. *Food Research International*, *41*(1), 31–42. doi:10.1016/j.foodres.2007.09.002.

Cho, S. Y., and Rhee, C. (2004). Mechanical properties and water vapor permeability of edible films made from fractionated soy proteins with ultrafiltration. *LWT—Food Science and Technology*, *37*(8), 833–839. doi:10.1016/j.lwt.2004.03.009.

Cobror, S., and Iannace, S. (2011). Bio-based chemistry for polymeric matrices. In L. Nicolais (Ed.), *Wiley Encyclopedia of Composites* (2nd ed., pp. 79–84). Hoboken, NJ: John Wiley & Sons. doi:10.1002/9781118097298.

Coffin, D. R., and Fishman, M. L. (1993). Viscoelastic properties of pectin/starch blends. *Journal of Agricultural and Food Chemistry*, *41*(8), 1192–1197. doi:10.1021/jf00032a005.

Cohen-Kupiec, R., and Chet, I. (1998). The molecular biology of chitin digestion. *Current Opinion in Biotechnology*, *9*(3), 270–277. doi:10.1016/S0958-1669(98)80058-X.

Collins, P. C. and Ferrier, R. J. (1995). *Monosaccharides: Their Chemistry and Their Roles in Natural Products* (p. 594). New York: John Wiley & Sons.

Cotugno, S., Di Maio, E., Iannace, S., Mensitieri, G., and Nicolais, L. (2005). Biodegradabile foams. In B. Narasimhan and S. K. Mallapragada (Eds.), *Handbook of Biodegradable Polymeric Materials and their Applications*. Ames, IA: American Scientific Publishers.

Coughlan, K., Shaw, N. B., Kerry, J. F., and Kerry, J. P. (2006). Combined effects of proteins and polysaccharides on physical properties of whey protein concentrate-based edible films. *Journal of Food Science*, *69*(6), E271–E275. doi:10.1111/j.1365-2621.2004.tb10997.x.

Coviello, T., Palleschi, A., Grassi, M., Matricardi, P., Bocchinfuso, G., and Alhaique, F. (2005). Scleroglucan: A versatile polysaccharide for modified drug delivery. *Molecules (Basel, Switzerland)*, *10*(1), 6–33. Retrieved from http://www.ncbi.nlm.nih.gov/pubmed/18007275.

Cunningham, P., Ogale, A. A., Dawson, P. L., and Acton, J. C. (2000). Tensile properties of soy protein isolate films produced by a thermal compaction technique. *Journal of Food Science*, *65*(4), 668–671. doi:10.1111/j.1365-2621.2000.tb16070.x.

Daamen, W., Hafmans, T., Veerkamp, J., and van Kuppevelt, T. (2001). Comparison of five procedures for the purification of insoluble elastin. *Biomaterials*, *22*(14), 1997–2005. doi:10.1016/S0142-9612(00)00383-5.

Damodaran, S., Anand, K., and Razumovsky, L. (1998). Competitive adsorption of egg white proteins at the air–water interface: Direct evidence for electrostatic complex formation between lysozyme and other egg proteins at the interface. *Journal of Agricultural and Food Chemistry*, *46*(3), 872–876. doi:10.1021/jf970722i.

Davé, V., and McCarthy, S. P. (1997). Review of konjac glucomannan. *Journal of Environmental Polymer Degradation*, *5*(4), 237–241. Retrieved from http://www.scopus.com/inward/record.url?eid=2-s2.0-0031253273&partnerID=40&md5=742d97793df23896448348bab3a2a8ff.

Debeaufort, F., Quezada-Gallo, J. A., and Voilley, A. (1998). Edible films and coatings: Tomorrow's packagings—A review. *Critical Reviews in Food Science and Nutrition*, *38*(4), 299–313. doi:10.1080/10408699891274219.

Deng, M., Zhou, Q., Du, A., van Kasteren, J., and Wang, Y. (2009). Preparation of nanoporous cellulose foams from cellulose-ionic liquid solutions. *Materials Letters*, *63*(21), 1851–1854. doi:10.1016/j.matlet.2009.05.064.

DeWit, J. N., and Klarenbeek, G. (1984). Effects of various heat treatments on structure and solubility of whey proteins. *Journal of Dairy Science*, *67*(11), 2701–2710. doi:10.3168/jds.S0022-0302(84)81628-8.

Dhara, S., and Bhargava, P. (2003). A simple direct casting route to ceramic foams. *Journal of the American Ceramic Society*, *86*(10), 1645–1650. doi:10.1111/j.1151-2916.2003.tb03534.x.

Dickey, L., Parris, N., Craig, J., and Kurantz, M. (2001). Ethanolic extraction of zein from maize. *Industrial Crops and Products*, *13*(1), 67–76. doi:10.1016/S0926-6690(00)00054-6.

Di Maio, E., and Iannace, S. (2011). Biodegradable composites. In L. Nicolais (Ed.), *Wiley Encyclopedia of Composites* (2nd ed., pp. 84–101). Hoboken, NJ: John Wiley & Sons. doi:10.1002/9781118097298.

Di Maio, E., Mali, R., and Iannace, S. (2010). Investigation of thermoplasticity of zein and kafirin proteins: Mixing process and mechanical properties. *Journal of Polymers and the Environment, 18*(4), 626–633. doi:10.1007/s10924-010-0224-x.

Di Martino, A., Sittinger, M., and Risbud, M. V. (2005). Chitosan: A versatile biopolymer for orthopaedic tissue-engineering. *Biomaterials, 26*(30), 5983–5990. doi:10.1016/j.biomaterials.2005.03.016.

Dubruel, P., Unger, R., Van Vlierberghe, S., Cnudde, V., Jacobs, P. J. S., Schacht, E., and Kirkpatrick, C. J. (2007). Porous gelatin hydrogels: 2. In vitro cell interaction study. *Biomacromolecules, 8*(2), 338–344. doi:10.1021/bm0606869.

Duchemin, B. J. C., Staiger, M. P., Tucker, N., and Newman, R. H. (2010). Aerocellulose based on all-cellulose composites. *Journal of Applied Polymer Science, 115*(1), 216–221. doi:10.1002/app.31111.

Dunlop, J. W. C., and Fratzl, P. (2013). Multilevel architectures in natural materials. *Scripta Materialia, 68*(1), 8–12. doi:10.1016/j.scriptamat.2012.05.045.

Fantini, J. L., Fuson, M. M., and Evans, T. A. (2006). Popping popcorn kernels: Expanding relevance with linear thinking. *Journal of Chemical Education, 83*(3), 414. doi:10.1021/ed083p414.

Fernandes, A. N., Thomas, L. H., Altaner, C. M., Callow, P., Forsyth, V. T., Apperley, D. C., Kennedy, C.J.,and Jarvis, M. C. (2011). Nanostructure of cellulose microfibrils in spruce wood. *Proceedings of the National Academy of Sciences of the United States of America, 108*(47), E1195–E1203. doi:10.1073/pnas.1108942108.

Finch, C. A. (1983). Chemical modification and some cross-linking reactions of water-soluble polymers. In C. A. Finch (Ed.), *Chemistry and Technology of Water-Soluble Polymers* (pp. 81–111). Boston, MA: Springer. doi:10.1007/978-1-4757-9661-2.

Fratzl, P., Misof, K., Zizak, I., Rapp, G., Amenitsch, H., and Bernstorff, S. (1998). Fibrillar structure and mechanical properties of collagen. *Journal of Structural Biology, 122*(1–2), 119–122. doi:10.1006/jsbi.1998.3966.

Fratzl, P., and Weinkamer, R. (2007). Nature's hierarchical materials. *Progress in Materials Science, 52*(8), 1263–1334. doi:10.1016/j.pmatsci.2007.06.001.

Freeman, I., Kedem, A., and Cohen, S. (2008). The effect of sulfation of alginate hydrogels on the specific binding and controlled release of heparin-binding proteins. *Biomaterials, 29*(22), 3260–3268. doi:10.1016/j.biomaterials.2008.04.025.

Gan, Z., Ellis, P. R., and Schofield, J. D. (1995). Gas cell stabilisation and gas retention in wheat bread dough. *Journal of Cereal Science, 21*(3), 215–230. doi:10.1006/jcrs.1995.0025.

Gao, S., and Nishinari, K. (2006). Effect of degree of acetylation on gelation of konjac glucomannan. *Biomacromolecules, 5*(1), 175–185. doi:10.1021/bm034302f.

Gardner, K. H., and Blackwell, J. (1975). Refinement of the structure of beta-chitin. *Biopolymers, 14*(8), 1581–1595. doi:10.1002/bip.1975.360140804.

Gáspár, M., Benkő, Z., Dogossy, G., Réczey, K., and Czigány, T. (2005). Reducing water absorption in compostable starch-based plastics. *Polymer Degradation and Stability, 90*(3), 563–569. doi:10.1016/j.polymdegradstab.2005.03.012.

Gavillon, R., and Budtova, T. (2008). Aerocellulose: New highly porous cellulose prepared from cellulose-NaOH aqueous solutions. *Biomacromolecules, 9*(1), 269–277. doi:10.1021/bm700972k.

Ghanbarzadeh, B., Musavi, M., Oromiehie, A. R., Rezayi, K., Razmi Rad, E., and Milani, J. (2007). Effect of plasticizing sugars on water vapor permeability, surface energy and microstructure properties of zein films. *LWT—Food Science and Technology, 40*(7), 1191–1197. doi:10.1016/j.lwt.2006.07.008.

Giavasis, I., Harvey, L. M., and McNeil, B. (2005). Scleroglucan. In *Biopolymers Online*. Wiley-VCH Verlag GmbH. doi:10.1002/3527600035.bpol6002.

Gibson, L. J., Ashby, M. F., and Harley, B. A. (2010). *Cellular Materials in Nature and Medicine* (p. 320). Cambridge: Cambridge International Science Publishing.

Gidley, M. J., and Nishinari, K. (2009). Chapter 2.2—Physico-chemistry of (1,3)-β-Glucans. In A. Bacic, G. B. Fincher, and B. A. Stone (Eds.), *Biochemistry, and Biology of 1-3 Beta Glucans and Related Polysaccharides* (pp. 47–118). San Diego, CA: Academic Press. doi:http://dx.doi.org/10.1016/B978-0-12-373971-1.00003-0.

Gilbert, R. D., and Kadla, J. F. (1998). Polysaccharides—Cellulose. In D. Kaplan (Ed.), *Biopolymers from Renewable Resources SE—3* (pp. 47–95). Berlin, Germany: Springer. doi:10.1007/978-3-662-03680-8_3.

Glenn, G., Orts, W., and Nobes, G. A. (2001). Starch, fiber and CaCO3 effects on the physical properties of foams made by a baking process. *Industrial Crops and Products, 14*(3), 201–212. doi:10.1016/S0926-6690(01)00085-1.

Glenn, G. M., Klamczynski, A. P., Woods, D. F., Chiou, B., Orts, W. J., and Imam, S. H. (2010). Encapsulation of plant oils in porous starch microspheres. *Journal of Agricultural and Food Chemistry*, *58*(7), 4180–4184. doi:10.1021/jf9037826.

Gomes, M., Godinho, J., Tchalamov, D., Cunha, A., and Reis, R. (2002). Alternative tissue engineering scaffolds based on starch: Processing methodologies, morphology, degradation and mechanical properties. *Materials Science and Engineering: C*, *20*(1–2), 19–26. doi:10.1016/S0928-4931(02)00008-5.

Gontard, N., Guilbert, S., and Cuq, J.-L. (1993). Water and glycerol as plasticizers affect mechanical and water vapor barrier properties of an edible wheat gluten film. *Journal of Food Science*, *58*(1), 206–211. doi:10.1111/j.1365-2621.1993.tb03246.x.

Hardy, J. G., Römer, L. M., and Scheibel, T. R. (2008). Polymeric materials based on silk proteins. *Polymer*, *49*(20), 4309–4327. doi:10.1016/j.polymer.2008.08.006.

Hassler, R. A., and Doherty, D. H. (1990). Genetic engineering of polysaccharide structure: Production of variants of xanthan gum in *Xanthomonas campestris*. *Biotechnology Progress*, *6*(3), 182–187. doi:10.1021/bp00003a003.

Haug, A., Høstmark, A. T., and Harstad, O. M. (2007). Bovine milk in human nutrition—A review. *Lipids in Health and Disease*, *6*, 25. doi:10.1186/1476-511X-6-25.

He, X., Su, B., Tang, Z., Zhao, B., Wang, X., Yang, G., Qiu, H., Zhang, H., and Yang, J. (2011). The comparison of macroporous ceramics fabricated through the protein direct foaming and sponge replica methods. *Journal of Porous Materials*, *19*(5), 761–766. doi:10.1007/s10934-011-9528-z.

He, X., Zhou, X., and Su, B. (2009). 3D interconnective porous alumina ceramics via direct protein foaming. *Materials Letters*, *63*(11), 830–832. doi:10.1016/j.matlet.2008.12.021.

Hegge, A. B., Andersen, T., Melvik, J. E., Bruzell, E., Kristensen, S., and Tønnesen, H. H. (2011). Formulation and bacterial phototoxicity of curcumin loaded alginate foams for wound treatment applications: Studies on curcumin and curcuminoides XLII. *Journal of Pharmaceutical Sciences*, *100*(1), 174–185. doi:10.1002/jps.22263.

Hegge, A. B., Andersen, T., Melvik, J. E., Kristensen, S., and Tønnesen, H. H. (2010). Evaluation of novel alginate foams as drug delivery systems in antimicrobial photodynamic therapy (aPDT) of infected wounds—An in vitro study: Studies on curcumin and curcuminoides XL. *Journal of Pharmaceutical Sciences*, *99*(8), 3499–3513. doi:10.1002/jps.22119.

Hernandez-Izquierdo, V. M., and Krochta, J. M. (2008). Thermoplastic processing of proteins for film formation—A review. *Journal of Food Science*, *73*(2), R30–R39. doi:10.1111/j.1750-3841.2007.00636.x.

Hernandez-Izquierdo, V. M., Reid, D. S., McHugh, T. H., Berrios, J. D. J., and Krochta, J. M. (2008). Thermal transitions and extrusion of glycerol-plasticized whey protein mixtures. *Journal of Food Science*, *73*(4), E169–E175. doi:10.1111/j.1750-3841.2008.00735.x.

Hochstetter, A., Talja, R. A., Helén, H. J., Hyvönen, L., and Jouppila, K. (2006). Properties of gluten-based sheet produced by twin-screw extruder. *LWT—Food Science and Technology*, *39*(8), 893–901. doi:10.1016/j.lwt.2005.06.013.

Holmes, P. A. (1988). Biologically produced (R)-3-hydroxyalkanoate polymers and copolymers. In D. C. Bassett (Ed.), *Developments in Crystalline Polymers* (pp. 1–65). Dordrecht, the Netherlands: Springer. doi:10.1007/978-94-009-1341-7.

Hoover, R., Vasanthan, T., Senanayake, N. J., and Martin, A. M. (1994). The effects of defatting and heat-moisture treatment on the retrogradation of starch gels from wheat, oat, potato, and lentil. *Carbohydrate Research*, *261*(1), 13–24. doi:10.1016/0008-6215(94)80002-2.

Hori, R., Müller, M., Watanabe, U., Lichtenegger, H. C., Fratzl, P., and Sugiyama, J. (2002). The importance of seasonal differences in the cellulose microfibril angle in softwoods in determining acoustic properties. *Journal of Materials Science*, *37*(20), 4279–4284. doi:10.1023/A:1020688132345.

Hoseney, R. C., Zeleznak, K., and Abdelrahman, A. (1983). Mechanism of popcorn popping. *Journal of Cereal Science*, *1*(1), 43–52. doi:10.1016/S0733-5210(83)80007-1.

Hudson, S. M., and Smith, C. (1998). Polysaccharides: Chitin and chitosan: Chemistry and technology of their use as structural materials. In D. Kaplan (Ed.), *Biopolymers from Renewable Resources SE—4* (pp. 96–118). Berlin, Germany: Springer. doi:10.1007/978-3-662-03680-8_4.

Hüttner, E. K., and Arendt, E. K. (2010). Recent advances in gluten-free baking and the current status of oats. *Trends in Food Science & Technology*, *21*(6), 303–312. doi:10.1016/j.tifs.2010.03.005.

Innerlohinger, J., Weber, H. K., and Kraft, G. (2006). Aerocellulose: Aerogels and aerogel-like materials made from cellulose. *Macromolecular Symposia*, *244*(1), 126–135. doi:10.1002/masy.200651212.

Jagannath, J. H., Nanjappa, C., Das Gupta, D. K., and Bawa, A. S. (2003). Mechanical and barrier properties of edible starch-protein-based films. *Journal of Applied Polymer Science*, *88*(1), 64–71. doi:10.1002/app.11602.

Ji, C., Annabi, N., Khademhosseini, A., and Dehghani, F. (2011). Fabrication of porous chitosan scaffolds for soft tissue engineering using dense gas CO_2. *Acta Biomaterialia, 7*(4), 1653–1664. doi:10.1016/j.actbio.2010.11.043.

Jia, X., and Kiick, K. L. (2009). Hybrid multicomponent hydrogels for tissue engineering. *Macromolecular Bioscience, 9*(2), 140–156. doi:10.1002/mabi.200800284.

Jin, H., Nishiyama, Y., Wada, M., and Kuga, S. (2004). Nanofibrillar cellulose aerogels. *Colloids and Surfaces A: Physicochemical and Engineering Aspects, 240*(1–3), 63–67. doi:10.1016/j.colsurfa.2004.03.007.

Jin, H.-J., Fridrikh, S. V., Rutledge, G. C., and Kaplan, D. L. (2002). Electrospinning Bombyx mori silk with poly(ethylene oxide). *Biomacromolecules, 3*(6), 1233–1239. Retrieved from http://www.ncbi.nlm.nih.gov/pubmed/12425660.

Jin, H.-J., and Kaplan, D. L. (2003). Mechanism of silk processing in insects and spiders. *Nature, 424*(6952), 1057–1061. doi:10.1038/nature01809.

Johnson, T. M., and Zabik, M. E. (1981). Egg albumen proteins interactions in an angel food cake system. *Journal of Food Science, 46*(4), 1231–1236. doi:10.1111/j.1365-2621.1981.tb03029.x.

Kadib, A. El, Molvinger, K., Cacciaguerra, T., Bousmina, M., and Brunel, D. (2011). Chitosan templated synthesis of porous metal oxide microspheres with filamentary nanostructures. *Microporous and Mesoporous Materials, 142*(1), 301–307. doi:10.1016/j.micromeso.2010.12.012.

Kalambur, S. (2006). An overview of starch-based plastic blends from reactive extrusion. *Journal of Plastic Film and Sheeting, 22*(1), 39–58. doi:10.1177/8756087906062729.

Kanke, M., Koda, K., Koda, Y., and Katayama, H. (1992). Application of curdlan to controlled drug delivery. I. The preparation and evaluation of theophylline-containing curdlan tablets. *Pharmaceutical Research, 9*(3), 414–418. doi:10.1023/A:1015811523426.

Kaplan, D. L. (1998). Introduction to biopolymers from renewable resources. In D. Kaplan (Ed.), *Biopolymers from Renewable Resources SE—1* (pp. 1–29). Berlin, Germany: Springer. doi:10.1007/978-3-662-03680-8_1.

Khan, F., and Ahmad, S. R. (2013). Polysaccharides and their derivatives for versatile tissue engineering application. *Macromolecular Bioscience, 13*(4), 395–421. doi:10.1002/mabi.201200409.

Kim, H. J., Kim, H. S., Matsumoto, A., Chin, I.-J., Jin, H.-J., and Kaplan, D. L. (2005a). Processing windows for forming silk fibroin biomaterials into a 3D porous matrix. *Australian Journal of Chemistry, 58*(10), 716. doi:10.1071/CH05170.

Kim, H. J., Kim, U.-J., Vunjak-Novakovic, G., Min, B.-H., and Kaplan, D. L. (2005b). Influence of macroporous protein scaffolds on bone tissue engineering from bone marrow stem cells. *Biomaterials, 26*(21), 4442–4452. doi:10.1016/j.biomaterials.2004.11.013.

Kim, U.-J., Park, J., Kim, H. J., Wada, M., and Kaplan, D. L. (2005c). Three-dimensional aqueous-derived biomaterial scaffolds from silk fibroin. *Biomaterials, 26*(15), 2775–2785. doi:10.1016/j.biomaterials.2004.07.044

Kinikoglu, B., Rodríguez-Cabello, J. C., Damour, O., and Hasirci, V. (2011). A smart bilayer scaffold of elastin-like recombinamer and collagen for soft tissue engineering. *Journal of Materials Science: Materials in Medicine, 22*(6), 1541–1554. doi:10.1007/s10856-011-4315-6.

Kirk-Othmer. (2000). *Kirk-Othmer Encyclopedia of Chemical Technology*. Hoboken, NJ: John Wiley & Sons. doi:10.1002/0471238961.

Klemm, D., Schumann, D., Kramer, F., Heßler, N., Hornung, M., Schmauder, H.-P., and Marsch, S. (2006). Nanocelluloses as innovative polymers in research and application. In D. Klemm (Ed.), *Polysaccharides II SE—97* (Vol. 205, pp. 49–96). Berlin, Germany: Springer. doi:10.1007/12_097.

Klemm, D., Schumann, D., Udhardt, U., and Marsch, S. (2001). Bacterial synthesized cellulose—artificial blood vessels for microsurgery. *Progress in Polymer Science, 26*(9), 1561–1603. doi:10.1016/S0079-6700(01)00021-1.

Koenig, M. F., and Huang, S. J. (1994). Evaluation of crosslinked poly(caprolactone) as a biodegradable, hydrophobic coating. *Polymer Degradation and Stability, 45*(1), 139–144. doi:10.1016/0141-3910(94)90189-9.

Kozempel, M., and Tomasula, P. M. (2004). Development of a continuous process to make casein films. *Journal of Agricultural and Food Chemistry, 52*(5), 1190–1195. doi:10.1021/jf0304573.

Kreplak, L., Doucet, J., Dumas, P., and Briki, F. (2004). New aspects of the alpha-helix to beta-sheet transition in stretched hard alpha-keratin fibers. *Biophysical Journal, 87*(1), 640–647. doi:10.1529/biophysj.103.036749.

Kumar, M. N. V. R., Muzzarelli, R. A. A., Muzzarelli, C., Sashiwa, H., and Domb, A. J. (2004). Chitosan chemistry and pharmaceutical perspectives. *Chemical Reviews, 104*(12), 6017–6084. doi:10.1021/cr030441b.

Kunte, L. A., Gennadios, A., Cuppett, S. L., Hanna, M. A., and Weller, C. L. (1997). Cast films from soy protein isolates and fractions 1. *Cereal Chemistry, 74*(2), 115–118. doi:10.1094/CCHEM.1997.74.2.115.

Lawton, J., Shogren, R., and Tiefenbacher, K. (2004). Aspen fiber addition improves the mechanical properties of baked cornstarch foams. *Industrial Crops and Products*, *19*(1), 41–48. doi:10.1016/S0926-6690(03)00079-7.

Lazaridou, A., Duta, D., Papageorgiou, M., Belc, N., and Biliaderis, C. G. (2007). Effects of hydrocolloids on dough rheology and bread quality parameters in gluten-free formulations. *Journal of Food Engineering*, *79*(3), 1033–1047. doi:10.1016/j.jfoodeng.2006.03.032.

Lee, J.-Y., Tan, B., and Cooper, A. I. (2007). CO_2-in-water emulsion-templated poly(vinyl alcohol) hydrogels using poly(vinyl acetate)-based surfactants. *Macromolecules*, *40*(6), 1955–1961. doi:10.1021/ma0625197.

Lemos, A. F., and Ferreira, J. M. F. (2004). The valences of egg white for designing smart porous bioceramics: As foaming and consolidation agent. *Key Engineering Materials*, *254–256*, 1045–1050. doi:10.4028/www.scientific.net/KEM.254-256.1045.

Leone, G., Barbucci, R., Borzacchiello, A., Ambrosio, L., Netti, P., and Migliaresi, C. (2004). Preparation and physico-chemical characterisation of microporous polysaccharidic hydrogels. *Journal of Materials Science: Materials in Medicine*, *15*(4), 463–467. doi:10.1023/B:JMSM.0000021121.91449.f4.

Li, X., Pizzi, A., Cangemi, M., Fierro, V., and Celzard, A. (2012a). Flexible natural tannin-based and protein-based biosourced foams. *Industrial Crops and Products*, *37*(1), 389–393. doi:10.1016/j.indcrop.2011.12.037.

Li, X., Pizzi, A., Cangemi, M., Navarrete, P., Segovia, C., Fierro, V., and Celzard, A. (2012b). Insulation rigid and elastic foams based on albumin. *Industrial Crops and Products*, *37*(1), 149–154. doi:10.1016/j.indcrop.2011.11.030.

Li, Z., Ramay, H. R., Hauch, K. D., Xiao, D., and Zhang, M. (2005). Chitosan-alginate hybrid scaffolds for bone tissue engineering. *Biomaterials*, *26*(18), 3919–3928. doi:10.1016/j.biomaterials.2004.09.062.

Liebner, F., Haimer, E., Potthast, A., Loidl, D., Tschegg, S., Neouze, M.-A., Wendland, M., and Rosenau, T. (2009). Cellulosic aerogels as ultra-lightweight materials. Part 2: Synthesis and properties 2nd ICC 2007, Tokyo, Japan, October 25–29, 2007. *Holzforschung*, *63*(1), 3–11. doi:10.1515/HF.2009.002.

Liebner, F., Haimer, E., Wendland, M., Neouze, M.-A., Schlufter, K., Miethe, P., Heinze, T., Potthast, A., and Rosenau, T. (2010). Aerogels from unaltered bacterial cellulose: Application of scCO$_2$ drying for the preparation of shaped, ultra-lightweight cellulosic aerogels. *Macromolecular Bioscience*, *10*(4), 349–352. doi:10.1002/mabi.200900371.

Lim, L.-T., Auras, R., and Rubino, M. (2008). Processing technologies for poly(lactic acid). *Progress in Polymer Science*, *33*(8), 820–852. doi:10.1016/j.progpolymsci.2008.05.004.

Lyubeshkina, E. G. (1983). Lignins as components of polymeric composite materials. *Russian Chemical Reviews*, *52*(7), 675–692. doi:10.1070/RC1983v052n07ABEH002873.

Madihally, S. V., and Matthew, H. W. T. (1999). Porous chitosan scaffolds for tissue engineering. *Biomaterials*, *20*(12), 1133–1142. doi:10.1016/S0142-9612(99)00011-3.

Maeda, H., Nakajima, M., Hagiwara, T., Sawaguchi, T., and Yano, S. (2006). Preparation and properties of bacterial cellulose aerogel. *Kobunshi Ronbunshu*, *63*(2), 135–137. doi:10.1295/koron.63.135.

Mangiacapra, P., Gorrasi, G., Sorrentino, A., and Vittoria, V. (2006). Biodegradable nanocomposites obtained by ball milling of pectin and montmorillonites. *Carbohydrate Polymers*, *64*(4), 516–523. doi:10.1016/j.carbpol.2005.11.003.

Marchessault, R. H. (1984). Carbohydrate polymers: Nature's high performance materials. In E. J. Vandenberg (Ed.), *Contemporary Topics in Polymer Science SE—3* (pp. 15–53). New York and London: Springer. doi:10.1007/978-1-4613-2759-2_3.

Martín, L., Alonso, M., Girotti, A., Arias, F. J., and Rodríguez-Cabello, J. C. (2009). Synthesis and characterization of macroporous thermosensitive hydrogels from recombinant elastin-like polymers. *Biomacromolecules*, *10*(11), 3015–3022. doi:10.1021/bm900560a.

Martinsen, A., Storrø, I., and Skjårk-Braek, G. (1992). Alginate as immobilization material: III. Diffusional properties. *Biotechnology and Bioengineering*, *39*(2), 186–194. doi:10.1002/bit.260390210.

Mauer, L., Smith, D., and Labuza, T. (2000). Water vapor permeability, mechanical, and structural properties of edible β-casein films. *International Dairy Journal*, *10*(5–6), 353–358. doi:10.1016/S0958-6946(00)00061-3.

May, C. D. (1990). Industrial pectins: Sources, production and applications. *Carbohydrate Polymers*, *12*(1), 79–99. doi:10.1016/0144-8617(90)90105-2.

McIntosh, M., Stone, B. A., and Stanisich, V. A. (2005). Curdlan and other bacterial (1-->3)-beta-D-glucans. *Applied Microbiology and Biotechnology*, *68*(2), 163–173. doi:10.1007/s00253-005-1959-5.

Mehling, T., Smirnova, I., Guenther, U., and Neubert, R. H. H. (2009). Polysaccharide-based aerogels as drug carriers. *Journal of Non-Crystalline Solids*, *355*(50–51), 2472–2479. doi:10.1016/j.jnoncrysol.2009.08.038.

Meikleham, N. E., and Pizzi, A. (1994). Acid- and alkali-catalyzed tannin-based rigid foams. *Journal of Applied Polymer Science*, *53*(11), 1547–1556. doi:10.1002/app.1994.070531117.

Meinel, L., Fajardo, R., Hofmann, S., Langer, R., Chen, J., Snyder, B., Vunjaknovakovic, G., and Kaplan, D. (2005). Silk implants for the healing of critical size bone defects. *Bone*, *37*(5), 688–698. doi:10.1016/j. bone.2005.06.010.

Meinel, L., Hofmann, S., Karageorgiou, V., Zichner, L., Langer, R., Kaplan, D., and Vunjak-Novakovic, G. (2004a). Engineering cartilage-like tissue using human mesenchymal stem cells and silk protein scaffolds. *Biotechnology and Bioengineering*, *88*(3), 379–391. doi:10.1002/bit.20252.

Meinel, L., Karageorgiou, V., Fajardo, R., Snyder, B., Shinde-Patil, V., Zichner, L., Kaplan, D., Langer, R., and Vunjak-Novakovic, G. (2004b). Bone tissue engineering using human mesenchymal stem cells: Effects of scaffold material and medium flow. *Annals of Biomedical Engineering*, *32*(1), 112–122. doi:10.1023/B:ABME.0000007796.48329.b4.

Meister, J. J., and Patil, D. R. (1985). Solvent effects and initiation mechanisms for graft polymerization on pine lignin. *Macromolecules*, *18*(8), 1559–1564. doi:10.1021/ma00150a006.

Meister, J. J., Patil, D. R., and Channell, H. (1984). Properties and applications of lignin–acrylamide graft copolymer. *Journal of Applied Polymer Science*, *29*(11), 3457–3477. doi:10.1002/app.1984.070291122.

Micard, V., Belamri, R., Morel, M.-H., and Guilbert, S. (2000). Properties of chemically and physically treated wheat gluten films. *Journal of Agricultural and Food Chemistry*, *48*(7), 2948–2953. doi:10.1021/jf0001785.

Miralles, G., Baudoin, R., Dumas, D., Baptiste, D., Hubert, P., Stoltz, J. F., Dellacherie, E., Mainard, D., Netter, P., and Payan, E. (2001). Sodium alginate sponges with or without sodium hyaluronate: In vitro engineering of cartilage. *Journal of Biomedical Materials Research*, *57*(2), 268–278. doi:10.1002/1097-4636(200111)57:2<268::AID-JBM1167>3.0.CO;2-L.

Mohanty, A. K., Misra, M., Drzal, L. T., Selke, S. E., Harte, B. R., and Hinrichsen, G. (2005). *Natural Fibers, Biopolymers, and Biocomposites* (p. 896). CRC Press.

Mohanty, A. K., Misra, M., and Hinrichsen, G. (2000). Biofibres, biodegradable polymers and biocomposites: An overview. *Macromolecular Materials and Engineering*, *276–277*(1), 1–24. doi:10.1002/(SICI)1439-2054(20000301)276:1<1::AID-MAME1>3.0.CO;2-W.

Montgomery, K. S. (2003). Soy protein. *The Journal of Perinatal Education*, *12*(3), 42–45. doi:10.1624/105812403X106946.

Nazarov, R., Jin, H.-J., and Kaplan, D. L. (2004). Porous 3-D scaffolds from regenerated silk fibroin. *Biomacromolecules*, *5*(3), 718–26. doi:10.1021/bm034327e.

Nudelman, F., and Sommerdijk, N. A. J. M. (2012). Biomineralization as an inspiration for materials chemistry. *Angewandte Chemie (International Ed. in English)*, *51*(27), 6582–6596. doi:10.1002/anie.201106715.

Oda, T., Iwasa, M., Aihara, T., Maéda, Y., and Narita, A. (2009). The nature of the globular- to fibrous-actin transition. *Nature*, *457*(7228), 441–445. doi:10.1038/nature07685.

Okada, M. (2002). Chemical syntheses of biodegradable polymers. *Progress in Polymer Science*, *27*(1), 87–133. doi:10.1016/S0079-6700(01)00039-9.

Oliviero, M., Verdolotti, L., Nedi, I., Docimo, F., Di Maio, E., and Iannace, S. (2012). Effect of two kinds of lignins, alkaline lignin and sodium lignosulfonate, on the foamability of thermoplastic zein-based bionanocomposites. *Journal of Cellular Plastics*, *48*(6), 516–525. doi:10.1177/0021955X12460043.

Omary, M. B., Ku, N.-O., Strnad, P., and Hanada, S. (2009). Toward unraveling the complexity of simple epithelial keratins in human disease. *The Journal of Clinical Investigation*, *119*(7), 1794–1805. doi:10.1172/JCI37762.

Ottani, V., Martini, D., Franchi, M., Ruggeri, A., and Raspanti, M. (2002). Hierarchical structures in fibrillar collagens. *Micron*, *33*(7–8), 587–596. doi:10.1016/S0968-4328(02)00033-1.

Ottani, V., Raspanti, M., and Ruggeri, A. (2001). Collagen structure and functional implications. *Micron*, *32*(3), 251–260. doi:10.1016/S0968-4328(00)00042-1.

Pääkkö, M., Vapaavuori, J., Silvennoinen, R., Kosonen, H., Ankerfors, M., Lindström, T., Berglund, L. A., and Ikkala, O. (2008). Long and entangled native cellulose I nanofibers allow flexible aerogels and hierarchically porous templates for functionalities. *Soft Matter*, *4*(12), 2492. doi:10.1039/b810371b.

Park, S., and Hettiarachchy, N. S. (1999). Physical and mechanical properties of soy protein-based plastic foams. *Journal of the American Oil Chemists' Society*, *76*(10), 1201–1205. doi:10.1007/s11746-999-0094-3.

Park, S. K., Rhee, C. O., Bae, D. H., and Hettiarachchy, N. S. (2001). Mechanical properties and water-vapor permeability of soy-protein films affected by calcium salts and glucono-δ-lactone. *Journal of Agricultural and Food Chemistry*, *49*(5), 2308–2312. doi:10.1021/jf0007479.

Parker, M. L., Grant, A., Rigby, N. M., Belton, P. S., and Taylor, J. R. N. (1999). Effects of popping on the endosperm cell walls of sorghum and maize. *Journal of Cereal Science*, *30*(3), 209–216. doi:10.1006/jcrs.1999.0281.

Partap, S., Rehman, I., Jones, J. R., and Darr, J. A. (2006). Supercritical carbon dioxide in water emulsion-templated synthesis of porous calcium alginate hydrogels. *Advanced Materials*, *18*(4), 501–504. doi:10.1002/adma.200501423.

Payne, P. I., Holt, L. M., Lawrence, G. J., and Law, C. N. (1982). The genetics of gliadin and glutenin, the major storage proteins of the wheat endosperm. *Qualitas Plantarum Plant Foods for Human Nutrition*, *31*(3), 229–241. doi:10.1007/BF01108632.

Petrovic, Z. (2008). Polyurethanes from vegetable oils. *Polymer Reviews*, *48*(1), 109–155. doi:10.1080/15583720701834224.

Petruccelli, S., and Anon, M. C. (1994). Relationship between the method of obtention and the structural and functional properties of soy proteins isolates. 1. Structural and hydration properties. *Journal of Agricultural and Food Chemistry*, *42*(10), 2161–2169. doi:10.1021/jf00046a017.

Peura, M., Müller, M., Vainio, U., Sarén, M.-P., Saranpää, P., and Serimaa, R. (2007). X-ray microdiffraction reveals the orientation of cellulose microfibrils and the size of cellulose crystallites in single Norway spruce tracheids. *Trees*, *22*(1), 49–61. doi:10.1007/s00468-007-0168-5.

Pizzi, A., Tondi, G., Pasch, H., and Celzard, A. (2008). Matrix-assisted laser desorption/ionization time-of-flight structure determination of complex thermoset networks: Polyflavonoid tannin-furanic rigid foams. *Journal of Applied Polymer Science*, *110*(3), 1451–1456. doi:10.1002/app.28545.

Pollock, T. J., Thorne, L., and Armentrout, R. W. (1992). Isolation of new aureobasidium strains that produce high-molecular-weight pullulan with reduced pigmentation. *Applied and Environmental Microbiology*, *58*(3), 877–883. Retrieved from http://aem.asm.org/cgi/content/long/58/3/877.

Popa, V. I. (1996). Hemicelluloses. In S. Dumitriu (Ed.), *Polysaccharides in Medicinal Applications* (p. 107). Chapter 5. New York: Marcel Dekker.

Psomiadou, E., Arvanitoyannis, I., and Yamamoto, N. (1996). Edible films made from natural resources; microcrystalline cellulose (MCC), methylcellulose (MC) and corn starch and polyols—Part 2. *Carbohydrate Polymers*, *31*(4), 193–204. doi:10.1016/S0144-8617(96)00077-X.

Puppi, D., Chiellini, F., Piras, A. M., and Chiellini, E. (2010). Polymeric materials for bone and cartilage repair. *Progress in Polymer Science*, *35*(4), 403–440. doi:10.1016/j.progpolymsci.2010.01.006.

Quignard, F., Valentin, R., and Di Renzo, F. (2008). Aerogel materials from marine polysaccharides. *New Journal of Chemistry*, *32*(8), 1300. doi:10.1039/b808218a.

Ratcliffe, I., Williams, P. A., Viebke, C., and Meadows, J. (2005). Physicochemical characterization of konjac glucomannan. *Biomacromolecules*, *6*(4), 1977–1986. doi:10.1021/bm0492226

Rauter, A. P., Vogel, P., and Queneau, Y. (Eds.). (2010). *Carbohydrates in Sustainable Development I* (Vol. 294). Berlin, Germany: Springer. doi:10.1007/978-3-642-14837-8.

Ravi Kumar, M. N. (2000). A review of chitin and chitosan applications. *Reactive and Functional Polymers*, *46*(1), 1–27. doi:10.1016/S1381-5148(00)00038-9.

Reddy, N., and Yang, Y. (2005). Biofibers from agricultural byproducts for industrial applications. *Trends in Biotechnology*, *23*(1), 22–27. doi:10.1016/j.tibtech.2004.11.002.

Re'em, T., Tsur-Gang, O., and Cohen, S. (2010). The effect of immobilized RGD peptide in macroporous alginate scaffolds on TGFbeta1-induced chondrogenesis of human mesenchymal stem cells. *Biomaterials*, *31*(26), 6746–6755. doi:10.1016/j.biomaterials.2010.05.025.

Reisler, E. (1993). Actin molecular structure and function. *Current Opinion in Cell Biology*, *5*(1), 41–47. Retrieved from http://www.ncbi.nlm.nih.gov/pubmed/8448029.

Remedios, C. G., and Chhabra, D. (Eds.). (2008). *Actin-Binding Proteins and Disease*. New York: Springer. doi:10.1007/978-0-387-71749-4.

Rhim, J. W., Gennadios, A., Handa, A., Weller, C. L., and Hanna, M. A. (2000). Solubility, tensile, and color properties of modified soy protein isolate films. *Journal of Agricultural and Food Chemistry*, *48*(10), 4937–4941. doi:10.1021/jf0005418.

Robitzer, M., David, L., Rochas, C., Di Renzo, F., and Quignard, F. (2008). Nanostructure of calcium alginate aerogels obtained from multistep solvent exchange route. *Langmuir: The ACS Journal of Surfaces and Colloids*, *24*(21), 12547–12552. doi:10.1021/la802103t.

Robitzer, M., Renzo, F. D., and Quignard, F. (2011). Natural materials with high surface area. Physisorption methods for the characterization of the texture and surface of polysaccharide aerogels. *Microporous and Mesoporous Materials*, *140*(1–3), 9–16. doi:10.1016/j.micromeso.2010.10.006.

Salam, A., Venditti, R. A., Pawlak, J. J., and El-Tahlawy, K. (2011). Crosslinked hemicellulose citrate–chitosan aerogel foams. *Carbohydrate Polymers*, *84*(4), 1221–1229. doi:10.1016/j.carbpol.2011.01.008.

Salerno, A., Oliviero, M., Di Maio, E., and Iannace, S. (2007a). Thermoplastic foams from zein and gelatin. *International Polymer Processing*, *22*(5), 480–488. doi:10.3139/217.2065

Salerno, A., Oliviero, M., Di Maio, E., Iannace, S., and Netti, P. A. (2007b). Design and preparation of µ-bimodal porous scaffold for tissue engineering. *Journal of Applied Polymer Science, 106*(5), 3335–3342. doi:10.1002/app.26881.

Sapir, Y., Kryukov, O., and Cohen, S. (2011). Integration of multiple cell-matrix interactions into alginate scaffolds for promoting cardiac tissue regeneration. *Biomaterials, 32*(7), 1838–1847. doi:10.1016/j.biomaterials.2010.11.008.

Satheesh Kumar, M. N., Mohanty, A. K., Erickson, L., and Misra, M. (2009). Lignin and its applications with polymers. *Journal of Biobased Materials and Bioenergy, 3*(1), 1–24. doi:10.1166/jbmb.2009.1001

Schmid, J., Meyer, V., and Sieber, V. (2011). Scleroglucan: Biosynthesis, production and application of a versatile hydrocolloid. *Applied Microbiology and Biotechnology, 91*(4), 937–47. doi:10.1007/s00253-011-3438-5.

Seki, Y., Bodde, S. G., and Meyers, M. A. (2010). Toucan and hornbill beaks: A comparative study. *Acta Biomaterialia, 6*(2), 331–343. doi:10.1016/j.actbio.2009.08.026.

Sescousse, R., Le, K. A., Ries, M. E., and Budtova, T. (2010). Viscosity of cellulose-imidazolium-based ionic liquid solutions. *The Journal of Physical Chemistry. B, 114*(21), 7222–7228. doi:10.1021/jp1024203.

Shen, L., Haufe, J., and Patel, M. K. (2009). *Product Overview and Market Projection of Emerging Bio-Based Plastics PRO-BIP 2009* (p. 243). Utrecht, the Netherlands. Retrieved from http://www.plastice.org/fileadmin/files/PROBIP2009_Final_June_2009.pdf.

Shepherd, R., Reader, S., and Falshaw, A. (1997). Chitosan functional properties. *Glycoconjugate Journal, 14*(4), 535–542. Retrieved from http://www.ncbi.nlm.nih.gov/pubmed/9249156.

Sherman, L. M. (2007). Polyurethanes: Bio-based materials capture attention. *Plastics Technology*. Retrieved from http://www.ptonline.com/articles/polyurethanes-bio-based-materials-capture-attention.

Shey, J., Imam, S. H., Glenn, G. M., and Orts, W. J. (2006). Properties of baked starch foam with natural rubber latex. *Industrial Crops and Products, 24*(1), 34–40. doi:10.1016/j.indcrop.2005.12.001.

Shogren, R., Lawton, J., Doane, W., and Tiefenbacher, K. (1998). Structure and morphology of baked starch foams. *Polymer, 39*(25), 6649–6655. doi:10.1016/S0032-3861(97)10303-2.

Shogren, R., Lawton, J., and Tiefenbacher, K. (2002). Baked starch foams: Starch modifications and additives improve process parameters, structure and properties. *Industrial Crops and Products, 16*(1), 69–79. doi:10.1016/S0926-6690(02)00010-9.

Shogren, R. L. (1998). Starch: Properties and materials applications. In D. Kaplan (Ed.), *Biopolymers from Renewable Resources SE—2* (pp. 30–46). Berlin, Germany: Springer. doi:10.1007/978-3-662-03680-8_2.

Shukla, R., and Cheryan, M. (2001). Zein: The industrial protein from corn. *Industrial Crops and Products, 13*(3), 171–192. doi:10.1016/S0926-6690(00)00064-9.

Siew, C. K., Williams, P. A., and Young, N. W. G. (2005). New insights into the mechanism of gelation of alginate and pectin: Charge annihilation and reversal mechanism. *Biomacromolecules, 6*(2), 963–969. doi:10.1021/bm0493411.

Sofia, S., McCarthy, M. B., Gronowicz, G., and Kaplan, D. L. (2001). Functionalized silk-based biomaterials for bone formation. *Journal of Biomedical Materials Research, 54*(1), 139–148. Retrieved from http://www.ncbi.nlm.nih.gov/pubmed/11077413.

Sohail, S. S., Wang, B., Biswas, M. A. S., and Oh, J.-H. (2006). Physical, morphological, and barrier properties of edible casein films with wax applications. *Journal of Food Science, 71*(4), Boca Raton, FL: C255–C259. doi:10.1111/j.1750-3841.2006.00006.x.

Soykeabkaew, N., Supaphol, P., and Rujiravanit, R. (2004). Preparation and characterization of jute- and flax-reinforced starch-based composite foams. *Carbohydrate Polymers, 58*(1), 53–63. doi:10.1016/j.carbpol.2004.06.037.

Stanley, N. (2006). Agars. In *Food Polysaccharides and Their Applications* (pp. 217–238). Boca Raton, FL: CRC Press. doi:10.1201/9781420015164.ch7

Steinbüchel, A. (2003). *Biopolymers: General Aspects and Special Applications* (p. 526). Weinheim, Germany: Wiley-Blackwell.

Steinbuchel, A., and Doi, Y. (2002). *Polyesters III—Applications and Commercial Products* (p. 385). Weinheim, Germany: Wiley-Blackwell.

Stevens, E. S. (2010). Starch-lignin foams. *eXPRESS Polymer Letters, 4*(5), 311–320. doi:10.3144/expresspolymlett.2010.39.

Stevens, M. P. (1998). *Polymer Chemistry: An Introduction* (p. 576). Oxford, UK: Oxford University Press.

Stone, B. A. (2009). Chapter 2.1—Chemistry of β-Glucans. In A. Bacic, G. B. Fincher, and B. A. Stone (Ed.), *Biochemistry, and Biology of 1-3 Beta Glucans and Related Polysaccharides* (pp. 5–46). San Diego, CA: Academic Press. doi:http://dx.doi.org/10.1016/B978-0-12-373971-1.00002-9.

Su, B.-L., Sanchez, C., and Yang, X.-Y. (Eds.). (2011). *Hierarchically Structured Porous Materials*. Weinheim, Germany: Wiley-VCH Verlag GmbH. doi:10.1002/9783527639588.

Sun, Y., Liu, Y., Li, Y., Lv, M., Li, P., Xu, H., and Wang, L. (2011). Preparation and characterization of novel curdlan/chitosan blending membranes for antibacterial applications. *Carbohydrate Polymers*, *84*(3), 952–959. doi:10.1016/j.carbpol.2010.12.055.

Swaisgood, H. E. (2003). Chemistry of the caseins. In P. F. Fox and P. L. H. McSweeney (Eds.), *Advanced Dairy Chemistry—1 Proteins SE—3* (pp. 139–201). New York: Springer. doi:10.1007/978-1-4419-8602-3_3.

Taarning, E., Nielsen, I. S., Egeblad, K., Madsen, R., and Christensen, C. H. (2008). Chemicals from renewables: Aerobic oxidation of furfural and hydroxymethylfurfural over gold catalysts. *ChemSusChem*, *1*(1–2), 75–78. doi:10.1002/cssc.200700033.

Tamada, Y. (2005). New process to form a silk fibroin porous 3-D structure. *Biomacromolecules*, *6*(6), 3100–3106. doi:10.1021/bm050431f.

Tamura, H., Nagahama, H., and Tokura, S. (2006). Preparation of chitin hydrogel under mild conditions. *Cellulose*, *13*(4), 357–364. doi:10.1007/s10570-006-9058-z.

Tanzer, M. (1976). Cross-linking. In G. N. Ramachandran and A. H. Reddi (Eds.), *Biochemistry of Collagen SE—4* (pp. 137–162). New York: Springer. doi:10.1007/978-1-4757-4602-0_4.

Thornton, A. J., Alsberg, E., Albertelli, M., and Mooney, D. J. (2004). Shape-defining scaffolds for minimally invasive tissue engineering. *Transplantation*, *77*(12), 1798–1803. doi:10.1097/01.TP.0000131152.71117.0E.

Thrimawithana, T. R., Young, S., Dunstan, D. E., and Alany, R. G. (2010). Texture and rheological characterization of kappa and iota carrageenan in the presence of counter ions. *Carbohydrate Polymers*, *82*(1), 69–77. doi:10.1016/j.carbpol.2010.04.024.

Tombolato, L., Novitskaya, E. E., Chen, P.-Y., Sheppard, F. A., and McKittrick, J. (2010). Microstructure, elastic properties and deformation mechanisms of horn keratin. *Acta Biomaterialia*, *6*(2), 319–330. doi:10.1016/j.actbio.2009.06.033.

Tondi, G., Fierro, V., Pizzi, A., and Celzard, A. (2009). Tannin-based carbon foams. *Carbon*, *47*(6), 1480–1492. doi:10.1016/j.carbon.2009.01.041.

Tondi, G., and Pizzi, A. (2009). Tannin-based rigid foams: Characterization and modification. *Industrial Crops and Products*, *29*(2–3), 356–363. doi:10.1016/j.indcrop.2008.07.003.

Tondi, G., Pizzi, A., Pasch, H., and Celzard, A. (2008). Structure degradation, conservation and rearrangement in the carbonisation of polyflavonoid tannin/furanic rigid foams—A MALDI-TOF investigation. *Polymer Degradation and Stability*, *93*(5), 968–975. doi:10.1016/j.polymdegradstab.2008.01.024.

Tsioptsias, C., Michailof, C., Stauropoulos, G., and Panayiotou, C. (2009). Chitin and carbon aerogels from chitin alcogels. *Carbohydrate Polymers*, *76*(4), 535–540. doi:10.1016/j.carbpol.2008.11.018.

Tsioptsias, C., Stefopoulos, A., Kokkinomalis, I., Papadopoulou, L., and Panayiotou, C. (2008). Development of micro- and nano-porous composite materials by processing cellulose with ionic liquids and supercritical CO_2. *Green Chemistry*, *10*(9), 965. doi:10.1039/b803869d.

Tuck, C., and Evans, J. R. G. (1999). Porous ceramics prepared from aqueous foams. *Journal of Materials Science Letters*, *18*(13), 1003–1005. doi:10.1023/A:1006665829967.

Valentin, R., Horga, R., Bonelli, B., Garrone, E., Di Renzo, F., and Quignard, F. (2006). FTIR spectroscopy of NH3 on acidic and ionotropic alginate aerogels. *Biomacromolecules*, *7*(3), 877–882. doi:10.1021/bm050559x.

Van Vlierberghe, S., Cnudde, V., Dubruel, P., Masschaele, B., Cosijns, A., Paepe, I. D., Jacobs, P. J., Hoorebeke, L. V., Remon, J. P., and Schacht, E. (2007). Porous gelatin hydrogels: 1. Cryogenic formation and structure analysis. *Biomacromolecules*, *8*(2), 331–337. doi:10.1021/bm060684o.

Van Vlierberghe, S., Dubruel, P., Lippens, E., Masschaele, B., Van Hoorebeke, L., Cornelissen, M., Unger, R., Kirkpatrick, C. J., and Schacht, E. (2008). Toward modulating the architecture of hydrogel scaffolds: Curtains versus channels. *Journal of Materials Science: Materials in Medicine*, *19*(4), 1459–1466. doi:10.1007/s10856-008-3375-8.

Van Vlierberghe, S., Dubruel, P., and Schacht, E. (2011). Biopolymer-based hydrogels as scaffolds for tissue engineering applications: A review. *Biomacromolecules*, *12*(5), 1387–1408. doi:10.1021/bm200083n.

Vepari, C., and Kaplan, D. L. (2007). Silk as a biomaterial. *Progress in Polymer Science*, *32*(8–9), 991–1007. doi:10.1016/j.progpolymsci.2007.05.013.

Vieira, M. G. A., da Silva, M. A., dos Santos, L. O., and Beppu, M. M. (2011). Natural-based plasticizers and biopolymer films: A review. *European Polymer Journal*, *47*(3), 254–263. doi:10.1016/j.eurpolymj.2010.12.011.

Viidik, A., Danielson, C. C., and Oxlund, H. (1982). On fundamental and phenomenological models, structure and mechanical properties of collagen, elastin and glycosaminoglycan complexes. *Biorheology, 19*(3), 437–51. Retrieved from http://www.ncbi.nlm.nih.gov/pubmed/6286009.

Vollrath, F., and Porter, D. (2009). Silks as ancient models for modern polymers. *Polymer, 50*(24), 5623–5632. doi:10.1016/j.polymer.2009.09.068.

Wan, Y., Wu, H., Cao, X., and Dalai, S. (2008). Compressive mechanical properties and biodegradability of porous poly(caprolectone)/chitosan scaffolds. *Polymer Degradation and Stability, 93*(10), 1736–1741. doi:10.1016/j.polymdegradstab.2008.08.001.

Wang, L., Schiraldi, D. A., and Sanchez-Soto, M. (2014). Foam-like xanthan gum/clay aerogel composites and tailoring properties by blending with agar. *Industrial & Engineering Chemistry Research*, 140409072312001. doi:10.1021/ie500490n.

Wang, S., Sue, H.-J., and Jane, J. (1996). Effects of polyhydric alcohols on the mechanical properties of soy protein plastics. *Journal of Macromolecular Science, Part A, 33*(5), 557–569. doi:10.1080/10601329608010878.

Wang, Y., Blasioli, D. J., Kim, H.-J., Kim, H. S., and Kaplan, D. L. (2006). Cartilage tissue engineering with silk scaffolds and human articular chondrocytes. *Biomaterials, 27*(25), 4434–42. doi:10.1016/j.biomaterials.2006.03.050.

Weinkamer, R., and Fratzl, P. (2011). Mechanical adaptation of biological materials—The examples of bone and wood. *Materials Science and Engineering: C, 31*(6), 1164–1173. doi:10.1016/j.msec.2010.12.002.

Wen, X., Cao, X., Yin, Z., Wang, T., and Zhao, C. (2009). Preparation and characterization of konjac glucomannan–poly(acrylic acid) IPN hydrogels for controlled release. *Carbohydrate Polymers, 78*(2), 193–198. doi:10.1016/j.carbpol.2009.04.001.

White, R. J., Budarin, V. L., and Clark, J. H. (2008). Tuneable mesoporous materials from alpha-D-polysaccharides. *ChemSusChem, 1*(5), 408–411. doi:10.1002/cssc.200800012.

White, R. J., Budarin, V. L., and Clark, J. H. (2010). Pectin-derived porous materials. *Chemistry (Weinheim an Der Bergstrasse, Germany), 16*(4), 1326–1335. doi:10.1002/chem.200901879.

Wiley, B. J., Ball, D. H., Arcidiacono, S. M., Sousa, S., Mayer, J. M., and Kaplan, D. L. (1993). Control of molecular weight distribution of the biopolymer pullulan produced byAureobasidium pullulans. *Journal of Environmental Polymer Degradation, 1*(1), 3–9. doi:10.1007/BF01457648.

Willett, J., and Shogren, R. (2002). Processing and properties of extruded starch/polymer foams. *Polymer, 43*(22), 5935–5947. doi:10.1016/S0032-3861(02)00497-4.

Wolf, W. J., and Lawton, J. W. (1997). Isolation and characterization of zein from corn distillers' grains and related fractions. *Cereal Chemistry, 74*(5), 530–536. doi:10.1094/CCHEM.1997.74.5.530.

Wongsasulak, S., Yoovidhya, T., Bhumiratana, S., and Hongsprabhas, P. (2007). Physical properties of egg albumen and cassava starch composite network formed by a salt-induced gelation method. *Food Research International, 40*(2), 249–256. doi:10.1016/j.foodres.2006.03.011.

Wongsasulak, S., Yoovidhya, T., Bhumiratana, S., Hongsprabhas, P., McClements, D. J., and Weiss, J. (2006). Thermo-mechanical properties of egg albumen–cassava starch composite films containing sunflower-oil droplets as influenced by moisture content. *Food Research International, 39*(3), 277–284. doi:10.1016/j.foodres.2005.07.014.

Wu, H., Wan, Y., Cao, X., and Wu, Q. (2008). Proliferation of chondrocytes on porous poly(DL-lactide)/chitosan scaffolds. *Acta Biomaterialia, 4*(1), 76–87. doi:10.1016/j.actbio.2007.06.010.

Xu, J., Schwarz, W. H., Käs, J. A., Stossel, T. P., Janmey, P. A., and Pollard, T. D. (1998). Mechanical properties of actin filament networks depend on preparation, polymerization conditions, and storage of actin monomers. *Biophysical Journal, 74*(5), 2731–2740. doi:10.1016/S0006-3495(98)77979-2.

Xu, Y. X., Kim, K. M., Hanna, M. A., and Nag, D. (2005). Chitosan–starch composite film: Preparation and characterization. *Industrial Crops and Products, 21*(2), 185–192. doi:10.1016/j.indcrop.2004.03.002.

Yamamoto, I., Takayama, K., Honma, K., Gonda, T., Matsuzaki, K., Hatanaka, K., Uryu, T. et al. (1990). Synthesis, structure and antiviral activity of sulfates of cellulose and its branched derivatives. *Carbohydrate Polymers, 14*(1), 53–63. doi:http://dx.doi.org/10.1016/0144-8617(90)90006-E.

Yeo, J. H., Lee, K. G., Kim, H. C., Oh, H.Y.L., Kim, A. J., and Kim, S. Y. (2000). The effects of Pva/chitosan/fibroin (PCF)-blended spongy sheets on wound healing in rats. *Biological & Pharmaceutical Bulletin, 23*(10), 1220–1223. Retrieved from http://www.ncbi.nlm.nih.gov/pubmed/11041255.

Zhan, X.-B., Lin, C.-C., and Zhang, H.-T. (2012). Recent advances in curdlan biosynthesis, biotechnological production, and applications. *Applied Microbiology and Biotechnology, 93*(2), 525–531. doi:10.1007/s00253-011-3740-2.

Zhou, C. Z., Confalonieri, F., Medina, N., Zivanovic, Y., Esnault, C., Yang, T., Jacquet, M. et al. (2000). Fine organization of Bombyx mori fibroin heavy chain gene. *Nucleic Acids Research*, *28*(12), 2413–9. Retrieved from http://www.pubmedcentral.nih.gov/articlerender.fcgi?artid=102737&tool=pmcentrez& rendertype=abstract.

Zmora, S., Glicklis, R., and Cohen, S. (2002). Tailoring the pore architecture in 3-D alginate scaffolds by controlling the freezing regime during fabrication. *Biomaterials*, *23*(20), 4087–4094. doi:10.1016/S0142-9612(02)00146-1.

2 Organic–Inorganic Bio-Hybrid Materials by SolGel Processing

Marino Lavorgna, Letizia Verdolotti, and Leno Mascia

CONTENTS

2.1 INTRODUCTION

Organic–inorganic (O–I) hybrid materials are systems in which the organic and inorganic components are intimately assembled at nano-sized scale length. The properties of these materials strictly depend upon the interactions established between the phases, which can be tailored to have new and multifunctional materials exhibiting the properties of both the inorganic material (e.g., rigidity, thermal stability) and the organic phase (e.g., flexibility, ductility, and processability). The intimate bonding between organic and inorganic phase is evident in many natural materials such as bone and nacre. Over the last few years, the understanding of the formation of these natural materials has significantly increased the interest in research on O–I hybrid materials. Knowledge of self-assembly and biomineralization mechanisms is a major challenge of modern materials science, which is increasingly adopting a multidisciplinary approach aimed to produce new materials for applications in many fields, such as optics, packaging, environment, and energy and medicine. The growing interest in these areas has broadened the meaning of the term *hybrid* to include all cases where organic or inorganic nano-sized domains are intimately assembled into homogeneous morphology. However, the term *hybrid material* is more widely used to denote O–I nanocomposites produced by the solgel method.

In this chapter, the types of O–I hybrids derived by solgel reactions of alkoxysilane compounds as precursors of the polysiloxane domains (i.e., low density network silica-type structures), as well as aspects of the preparation of compact and porous hybrid materials, and the way that the procedure can be adapted to produce polymer-based O–I materials, giving particular attention to systems obtained from biopolymers (i.e., polymers from renewable sources such as biomass or biocompatible polymers), are discussed.

2.2 POLYMER-BASED O–I HYBRIDS BY SOLGEL PROCESS

Contrary to traditional nanocomposites, obtained by mixing preformed nanoparticles with a polymer as the matrix, O–I hybrid materials are produced by dispersing the precursors, usually in solution, from which at least the inorganic component is produced *in situ* (Kickelbick 2007). For silica O–I hybrids, the inorganic phase is produced by the solgel process using tetraethoxysilane (TEOS) with minor amounts of a reactive organo-trialkoxysilane as coupling and compatibilizing agent to promote the formation of siloxane nanodomains as continuous phase within a polymer matrix (Mascia 1995). These materials have been known also as ORMOSILs (ORganically MOdified SILicates), CERAMERs (CERAmics and polyMERs), or ORMOCER (ORganically MOdified CERamics). The main features of these systems are the large polymer-inorganic interfacial area and the wide range of interactions that can be achieved between the two phases. In the category that has been called *class I type hybrid materials*, there are only weak physical interactions between the two phases, such as van der Waals types or hydrogen bonding. In class II type systems, the interfacial interactions are strong chemical types, that is, covalent or ionic bonds, which also assist the formation of nanodimensioned co-continuous morphologies (Kickelbick 2007; Mascia and Lavorgna 2012).

2.2.1 SolGel Chemistry for the Production of O–I Hybrids

Polymeric O–I hybrid materials are produced by solgel chemistry, using alkoxysilane precursors to generate inorganic-like siloxane domains. The several steps involved in the solgel process are (1) hydrolysis of the precursors, (2) condensation reactions between the resulting monomers to produce highly branched oligomers that evolve into *sol* particles, and (3) self-assembling and aggregation of the *sol* particles to form an irreversible *gel*.

The overall process revolves around chemical reactions in the precursor solution, which are acid or base catalyzed. Examples of hydrolysis and condensation reactions involved for systems using alkoxysilane precursors are shown below

$$-\text{Si-OR} + \text{H}_2\text{O} \xrightarrow{\text{H}^+ \text{or OH}^-} -\text{Si-OH} + \text{ROH} \tag{2.1}$$

with R an alkyl group, and

$$-\text{Si-OH} + \text{HO-Si-} \xrightarrow{\text{H}^+ \text{or OH}^-} -\text{Si-O-Si-} + \text{H}_2\text{O} \tag{2.2}$$

and/or

$$-\text{Si-OH} + \text{RO-Si-} \xrightarrow{\text{H}^+ \text{or OH}^-} -\text{Si-O-Si-} + \text{ROH} \tag{2.3}$$

The Si-O-Si bonds, which are the backbone of the silica or polysiloxane structure, are formed during the condensation reactions. (Note that the term *siloxane* is used to indicate that the condensation reactions have not progressed to completion). The most common precursors for the production of inorganic three-dimensional network are TEOS and tetramethoxysilane (TMOS). Since neither TEOS nor TMOS dissolves totally in water, it is necessary to add an organic solvent to enhance the solubility. However, for the production of bio-hybrids, the organic solvent is generally omitted to avoid a possible denaturating effect and/or a decrease in the solubility of the biopolymer used as the matrix component. For the latter systems, tetrakis(2-hydroxyethyl)orthosilicate, THEOS is often used as an alternative to TEOS and TMOS as it is completely water soluble (Shchipunov 2003). In this respect, it should be noted that the silica/siloxane phase can be produced also from sodium silicate (Cornelissen et al. 2003) and silicic acid as precursor (Laugel et al. 2007).

Alkoxysilane-containing polymers have also been used in the solgel process as precursors for the organic phase and compatibilizing agents to achieve a co-continuous morphology in the resulting O–I hybrid. They are obtained by reacting polymers or oligomers with functional alkoxysilanes (i.e., conventional silane coupling agents used to increase the interfacial bonding in composites) to improve the miscibility of the organic component. The general structure of the coupling agents can be represented by the formula $RSi(OX)_3$, where the OX represents the hydrolyzable groups, typically ethoxy or methoxy groups. The organo R group can have a variety of functionalities to react with the polymer. The coupling agents can also be used alone for the formation of three-dimensional networks through the formation of the so-called silsesquioxanes (SSQO)—general formula $R-(SiO_{1.5})$—materials. The most widely used coupling agent in the compatibilization of polymers and polysiloxane domains are 3-aminopropyltriethoxysilane (APTES), $H_2N(CH_2)_3Si(OC_2H5)_3$; 3-isocyanatopropyltriethoxysilane (ICPTES), $OCN(CH_2)_3Si(OC_2H_5)_3$; mercaptopropyl triethoxysilane (MPTS), $SH(CH_2)_3Si(OC_2H_5)_3$; (3-acryloxypropyl)trimethoxysilane (APTMS), $CH_2=CHCOO(CH_2)_3Si(OCH_3)_3$; and 3-glycidoxypropyltrimethoxysilane (GOTMS), $CH_2(O)CHCH_2O(CH_2)_3Si(OCH_3)_3$.

Many factors influence the kinetics of the hydrolysis and condensation reactions in the solgel process, which include the water–silane ratio, catalyst, temperature, and the nature of solvent. Alcohols (ethanol and isopropanol) or tetrahydrofuran are often used as solvents.

It has generally been acknowledged that the hydrolysis of TEOS is very fast compared to the condensation reaction in an acidic environment and that the resulting inorganic phase consists of open and highly ramified chemical structure. In contrast, slower hydrolysis and faster condensation reactions take place under alkaline conditions, which result in the formation of heavily cross-linked and dense colloidal particles (from 100 nm to about 1 micron). For this reason, O–I hybrids are usually produced either under slightly acidic conditions or even in neutral conditions where the condensation reactions are catalyzed with the usual condensation catalysts, such as organo-tin compounds.

A schematic illustration of the formation of silica nanoparticles from hydrolysis and condensation of TEOS and the subsequent evolution of the morphology is shown in Figure 2.1.

At some stage, depending on the pH of the reaction medium, the *sol* particles start to react or strongly interact, by the formation of hydrogen bonds, through the hydroxyl groups generated

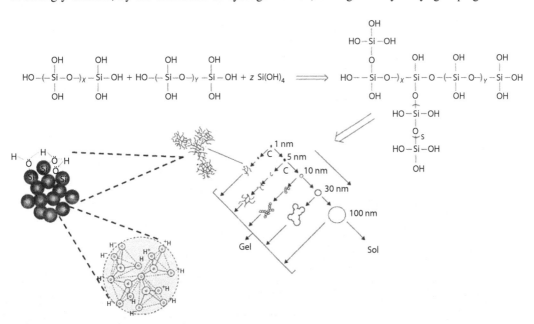

FIGURE 2.1 Mechanism of the formation of silica from pre-hydrolyzed tetraethoxysilane by the SolGel method (Data from Mascia, L. and Lavorgna, M., Nanostructured polymer composites by sol-gel method. In *Wiley Encyclopedia of Composites*, 1922–1941, VCH Publishers, New York, 2012), formation of siloxane aggregate through hydrogen bonding among silanol groups (Data from Jana, S.C. and Sachin, J., *Polymer*, 42, 6897–6905, 2001), and chemical structure of silica nanoparticles at room temperature. (Data from Jones, J.R., *Acta Biomater.*, 9, 4457–4486, 2013.)

during hydrolysis (Zou et al. 2008). As a result, the linked nanoparticles form a *gel* phase consisting of a three-dimensional network of interconnected particles.

2.2.2 PREPARATION ROUTES FOR THE OBTAINMENT OF POLYMER–SILOXANE HYBRIDS

In the preparation of polymeric-inorganic hybrids, the functionalized organic component can be a monomer, an oligomer, or a polymer (in molten state, in solution, or as an emulsion) or a cross-linkable system (for thermoset and elastomeric matrices).

There are basically the following three different approaches to produce polymeric-inorganic hybrid materials by the solgel method: (1) blending, (2) *in situ* structuring of the inorganic phase, and (3) simultaneous structuring of the organic and inorganic phases.

The blending method consists of mixing the siloxane component, previously produced by solgel method, with the polymer in solution or liquid organic oligomers, followed by the casting of films or coating of substrates and the subsequent removal of solvent leading to gelation and vitrification of the final product, which can be post-cured by thermal treatments. In this context, infusion of organic monomers in preformed porous inorganic structures or vice versa is also an example of hybrid systems produced by blending approach.

The method by *in situ structuring of the inorganic phase* consists of dissolving the polymer into an inorganic solgel precursor solution, which contains the silane alkoxide and often also a silane coupling agent. The silane species in the produced solution are hydrolyzed and then allowed to undergo condensation reactions to form interpenetrating polysiloxane domains within the swollen polymer, from which the final O–I hybrid coating or film is obtained through the evaporation of the solvent. When biopolymers (such as polysaccharides or proteins) are used as organic component, the method requires the polymer to be functionalized with a silane coupling agent, avoiding the use of alcoholic solvent to avoid any adverse effect on the morphological structure through premature aggregation of *sol* particles or lack of total solubilization of the precursors.

In such cases, it is preferable to use THEOS as the source of siloxane domain or to produce the required O–I hybrid via the prior formation of microemulsions of siloxane domains in the polymer solution. (Shchipunov 2003; Zalzberg and Avnir 2008).

In the *simultaneous structuring of organic and inorganic phases* method, both the organic and inorganic phases can be formed through the simultaneous chain extension reactions within liquid oligomers or the polymeric precursor solutions and the events leading to the structuring of the inorganic domains, such as the condensation reactions and the percolation of sol particles. The silane coupling agent ensures that the organic and inorganic components will co-react to form interphases that will stabilize the morphological structure by preventing agglomerations of nucleated particles in the two separate phases. A variant of this method is represented by the possibility to perform a preliminary hydrolysis and condensation of the inorganic precursors before mixing with the organic component (Matejka et al. 2000; Matejka and Dukh 2001; Mascia et al. 2005, 2006).

A schematic representation of the several approaches for the preparation of polymeric-inorganic hybrid materials is reported in the Figure 2.2.

It should be noted that the formation of co-continuous phase morphology creates the conditions to achieve the most efficient mechanism for the transfer of external excitations through the two main phases, thereby maximizing the contribution of both components (organic and inorganic) to the overall properties of the polymeric hybrid material. The diagram on the left of Figure 2.3 shows a schematic representation of the co-continuous domains, while the transmission electron micrograph illustrates the actual morphology of an epoxy–silica hybrid displaying the existence of *graded* interphases, which consist of siloxane units derived from reactions of the silane coupling agent with both the organic component and the TEOS (Mascia 2010).

In order to obtain morphology exhibiting co-continuity of phases, the preparation conditions are normally imposed to allow the condensation reactions for the formation of the inorganic domains

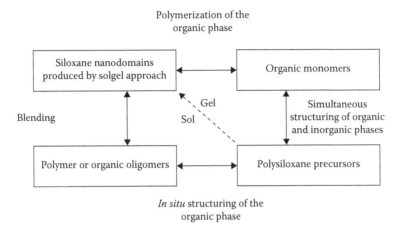

Polymerization of the
organic phase

FIGURE 2.2   General approaches to prepare polymeric-inorganic hybrid materials by solgel approach.

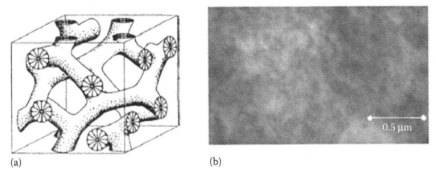

(a)   (b)

FIGURE 2.3   Illustration of phase co-continuity in polymeric-inorganic hybrid materials. (a) Sketch of the inorganic three-dimensional skeleton and (b) transmission electron micrograph of an epoxy–silica hybrid. (Data from Mascia, L., Fillers generated in-situ: Bicontinuous nanocomposites. In *Functional Fillers for Plastics*, 459–468, Wiley-VCH, Weinheim, Germany, 2010.)

to take place quickly both in the *in situ structuring of the inorganic phase* method and in the *simultaneous structuring of organic and inorganic phases* method. Thus, small particles can nucleate and grow to form very small (nanometric) particles dispersed into the surrounding liquid medium constituted by the polymer or the precursors of the organic phase. At this stage, the system is still in a fluid state and constitutes a *sol* system. The inorganic particles can grow bigger and/or percolate to form either macroscopic aggregates or a tridimensional arrangement of interconnected particles to produce a *gel* system.

Examples of widely studied polymeric O–I hybrid systems exhibiting a co-continuous morphology are described below:

1. Siloxane hybridization of Nafion® membranes (Lavorgna et al. 2007, 2009). These have been produced as class I ionomeric O–I hybrid materials for fuel cells applications through two routes: (1) the infusion of siloxane precursors into an existing Nafion membrane and (2) casting of the films starting from a dispersion of ionomeric polymer and siloxane precursors. In both cases, the siloxane domains were formed from a mixture of TEOS and mercapto-proprylsilane $HS(CH_2)_3Si(OC_2H_5)_3$. The mercapto groups in the siloxane domains of the membrane were oxidized with hydrogen peroxide to produce strong-SO_3H acid groups. Both procedures produced membranes exhibiting a morphology consisting of

inorganic domains assembled in the Nafion ionomeric clusters. Water vapor absorption and ionic conductivity of the hybridized membranes were higher than the values recorded for neat Nafion membrane, particularly at high temperatures (80°C–120°C). These improvements were ascribed to the presence of additional water species absorbed on specific sites of hydrophilic polysiloxane domains (i.e., SiOSi siloxane backbone bonds and —SiOH and —SO₃H groups).

2. Particularly promising are O–I hybrid materials based on epoxy and polyimide in view of the wide range of possibilities for coupling the organic and inorganic phase and consequently improving the properties of conventional epoxy or polyimides resins for coating, adhesives, or composite applications (Abbate et al. 2004; Musto et al. 2004, 2005; Mascia et al. 2005; Prezzi and Mascia EP 2005055218). Cold-cured epoxy–siloxane hybrid systems have been produced to overcome the main limitations of conventional cold-cured epoxy resins, which require long curing times and the T_g of the final products can only reach values of about 10°C–20°C higher than the temperature used for curing (Lionetto et al. 2013). It has been shown that cold-cured epoxy–silica hybrids, based on epoxy resins functionalized with mercapto γ-propyltrimethoxysilane as coupling agent and hydrolyzed mixture of TEOS and γ-glycidoxypropyltrimethoxysilane (GOTMS), can exhibit a higher T_g than the cured original resin/hardener mixture. It was shown that the moisture from the atmosphere is often sufficient to induce further hydrolysis of the alkoxysilane species for the formation of silica/siloxane domains, which improve further the mechanical properties and solvent resistance (Mascia et al. 2006).

2.3 BIOPOLYMER-BASED O–I HYBRID MATERIALS

It is well recognized that the presence of the polymeric or oligomeric phase modifies the chemistry of solgel process. This is more significant with polymers from renewable source or biopolymers. In fact, biopolymers such as peptides produce a shortening of the interval time for gelation (Kröger et al. 1999). The polysaccharides, such as cellulose (Perry et al. 1992), alginate (Coradin et al. 2001), and carrageenans (Shchipunov 2003), influence the morphology of the generated silica, even though they are chemically involved in the solgel process. Polysaccharide macromolecules have been found to promote the mineralization process, that is, nucleation and growth of siloxane domains, associated with the large number of hydroxyl groups in the system, which may undergo condensation reactions with the silanol groups in the siloxane domains, in addition to (Yoshinaga and Katayama 1996) the formation of hydrogen bondings. Linked siloxanes and biopolymer molecules can further hydrolyze and also participate in the condensation reactions of the whole solgel process, which can lead to the formation of a silica shell and initiate the mineralization of the dispersed biopolymer macromolecules.

In this section, the most widely studied hybrid systems involving biopolymers for potential application as films, membrane, and coatings are reviewed. Particular attention is given to chemical aspects related to the compatibilization of the polymeric and the polysiloxane phases by mimicking the mechanism operating in natural O–I materials.

2.3.1 BIO-HYBRIDS OF CLASS I TYPE

Class I type hybrid materials can be readily produced using polymers or oligomers bearing functional groups that form hydrogen bonds with the silanol groups in the siloxane domains. Many kinds of polymers such as poly(2-hydroxyethyl methacrylate), PHEMA (Hajji et al. 1999; Huang et al. 2005); polyvinylalcohol, PVA (Nakane et al. 1999); and polysaccharide (Shchipunov and Karpenko 2004) have been successfully used to prepare class I hybrid materials.

Hajji et al. (1999) have prepared PHEMA-based hybrid materials by using two different synthesis routes: (1) the blending method, consisting of a bulk free radical polymerization of 2-hydroxyethyl methacrylate (HEMA) in the presence of HEMA-functionalized preformed silica nanoparticles and (2) the method of *simultaneous structuring of organic and inorganic phases*, derived from the *in situ* generation of the silica phase during the free radical polymerization of HEMA. The morphology of these systems consists of an open mass fractal silicate structure as illustrated in Figure 2.4. At temperatures below the glass transition temperature of the polymer, the silica exhibits a moderate reinforcing effect that is independent of the method used for the synthesis, whereas in the rubbery plateau, the systems obtained through simultaneous polymerization exhibit a much greater reinforcing efficiency, which is ascribed to the presence of co-continuous silica domains. In the TEOS/PHEMA hybrid systems, the damping peak associated with the glassy transition is almost totally depressed when the silica content is increased to 30 wt%, which can be considered a criterion for the validation of the formation of a *true* O–I hybrid.

A detailed comparison of the reactivity of many polysaccharides toward THEOS was made by Shchipunov and Karpenko (2004) (Shchipunov 2003). The data show that neutral polymers, such as hydroxyethyl cellulose, are prone to induce syneresis in the resulting gel, whereas stable monoliths were obtained with cationic (chitosan) or anionic (alginate, carrageenan) polysaccharides. Since THEOS is completely soluble in water, the solgel reactions for these systems were performed in aqueous solution, without the addition of an organic solvent or catalyst. The formation of siloxane domains was accelerated by the presence of the polysaccharide through a catalytic effect. Gelation was observed within a few minutes after mixing the polysaccharides with the inorganic precursor,

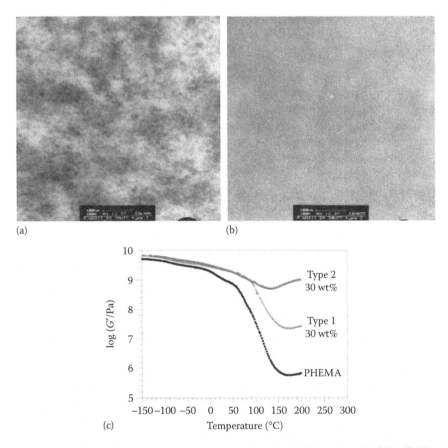

(a)   (b)

(c)

FIGURE 2.4 Transmission electron microscopy micrographs of hybrid systems with 30 wt% silica obtained through (a) blending method, (b) simultaneous structuring of organic and inorganic phases, and (c) Temperature dependence of storage modulus for unfilled PHEMA, type 1 is the system obtained through blending method and type 2 is the system obtained through simultaneous polymerization. (Data from Hajji, P. et al., *J. Polym. Sci. Part B Poly. Phys.*, 37, 3172–3187, 1999.)

whereas in the presence of an acid catalyst, gelation took place after a few hours. The reactions or strong associations at the interface lead to the formation of a silica shell around the macromolecules giving rise to fibril-like species, resulting from self-organization of the products of the reaction. This mechanism is thought to simulate formation of silica domains in living species, where glycoproteins, proteoglycans, and polysaccharides act as the main nucleating and templating agents present in living cells. Figure 2.5 shows scanning electron micrographs of a gel obtained from a cationic hydroxyethyl cellulose solution mixed with the inorganic precursor solution.

Shchipunov and coworkers have derived the following conclusions: (1) polysaccharides promote the biomineralization of silica by the formation of hydrogen bonds between hydroxyl groups of macromolecules and silanols generated by the hydrolysis of precursor, (2) the polysaccharides serve as a template for *in situ* generation of silica, and (3) the structure and properties of synthesized biomaterials are controlled by the nature of the polysaccharide used and the strength and concentration of charges on the macromolecule.

These authors (Shchipunova et al. 2004) also showed that polysaccharide-based hybrids obtained by using THEOS as precursor of the inorganic phase are effective agents for the immobilization of enzymes, which was evidenced by the sharp increase in the enzyme lifetime (more than 100 times) after their immobilization. The efficient entrapment of enzymes was ascribed to the THEOS presence, which allows the following: (1) the removal of organic solvents and catalysts and catalytic effect of polysaccharides on the solgel reactions; (2) the entrapment of enzymes at any pH, which helps to maintain their structural integrity and functionality; and (3) the obtainment of a gel at very low concentrations of THEOS (1%–2%) in the initial solution.

Chitosan-based hybrid materials with a co-continuous morphology were found to be useful as pH sensitive materials for the drug delivery applications (Park et al. 2001), using TEOS to obtain a siloxane gel that entraps the chitosan. The proposed mechanism is based on the ability of chitosan to swell at pH 2.5 (the isoelectric point of the siloxane gel) and to shrink at pH 7.5. Since the swelling is completely reversible, an increase in pH from 2.5 to 7.5 induces the agglomeration of chitosan in the siloxane domains and causes an internal pressure to start the release of the drug molecules. These types of O–I hybrid materials were also prepared from natural chitosan and THEOS by solgel process without using organic solvents and catalyst (Wang and Zhang 2006). A potential application of chitosan O–I coatings is in packaging of fresh fruit as it would extend the useful life producing a barrier to the infusion of oxygen from the atmosphere (Dhanasingh et al. 2011; Shengyou et al. 2013).

Cellulose is the most abundant natural polymer available and has been used in numerous applications, ranging from papermaking and plastics to biomedicine. Value-added cellulosic

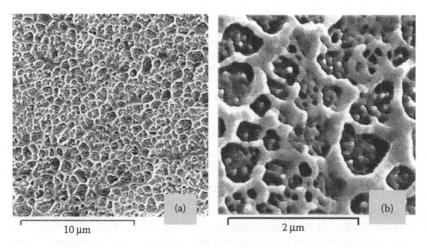

FIGURE 2.5   SEM images of cationic cellulose coated by silica shell (a) 10 μm and (b) 2 μm. (Data from Shchipunov, Y.A. et al., *J. Colloid Interf. Sci.*, 285, 574–580, 2005.)

materials are obtained through appropriate chemical modification of the natural polymer, such as etherification, esterification, and silylation, as well as via the formation of O–I hybrids by the solgel method. Cellulose acetate and hydroxypropyl cellulose have been used for the latter case (Yano 1994; Zoppi et al. 2002; Mo et al. 2008).

Innovative hybrid materials based on cellulose fibers (Sequeira et al. 2007; Gamelas et al. 2012) have been prepared at room temperature by hydrolyzing TEOS in the presence of $H_3PW_{12}O_{40}$ and other Keggin-type polyoxometalates catalysts. Heteropolyacids have shown to have a higher catalytic efficiency than that of conventional mineral acids (HCl and HNO_3). The synthesis conditions have a significant effect on the structure of inorganic phase and has been suggested that an increase in H_2O/TEOS molar ratio up to 4.4 enhances the formation of gels, whereas high catalyst concentrations (i.e., higher than 3×10^{-4} mol L^{-1}) inhibit the generation of silica in favor of the formation of linear siloxane fragments. Hybrid materials showed considerably higher hydrophobicity (about four times) and thermal stability when compared to the starting fibrous material (bleached kraft pulp).

For coatings on cellulose fibers, the siloxane precursor was functionalized with polyoxometalates via electrostatic interactions with protonated propylamino groups introduced with APTES aminosilanes (see Figure 2.6). Despite the high silica content, the resulting cellulose–silica hybrids retained fibrous structure and porosity without increasing the density of the fibers.

O–I hybrids based on bacterial cellulose were prepared by mimicking the synthesis of natural biocomposites (Maeda et al. 2006). It is well known that in plants, such as rice and sugarcane,

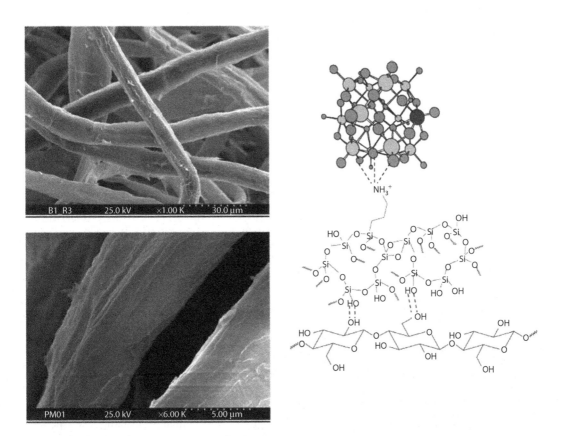

FIGURE 2.6 Typical SEM images of cellulose–silica hybrids (left side) and schematic representation of the chemical interactions that occur in the novel functionalized cellulose–silica hybrids (right side). (Data from Gamelas, J.A.F. et al., *RSC Adv.*, 2, 831–839, 2012.)

biomineralization takes place via the transfer of amorphous silica particles from soil to become encapsulated in polysaccharide matrix. To simulate this process, O–I hybrids were produced by immersing a bacterial cellulose hydrogel in an aqueous solution of pre-hydrolyzed TEOS, and the products were subsequently post-cured under pressure at 120°C to obtain products with excellent mechanical properties. These properties were attributed to the formation of a silica network chemically bound to the surface of cellulose microfibrils, which having hydroxyl groups may undergo condensation reactions with silanol groups, which in turn becomes fixed to microfibril. In this case, the formation of silica takes place *in situ*, similar to the biomineralization that takes place in rice plants.

As indicated previously, the silica domains can also be obtained starting from an alkaline sodium silicate solution. The chemistry that leads to the formation of silica particles takes place essentially through condensation of silanol groups to form a three-dimensional network of —SiOSi— bonds. This approach has been successfully used to upgrade cellulose waste from paper manufacturing (De Pisapia et al. 2012; Verdolotti et al. WO2013105040 A1 and PCT/IB2013/050199). The sodium silicate solution was chosen for its ability to swell the cellulose crystalline structure, through the formation of hydrogen bonds between the silanol groups of polysilicate domains and the hydroxyl groups of the cellulose. These cellulose–silicate hybrids have shown to be useful for the replacement of traditional polyols in the reaction with the isocyanate groups to produce polyurethane hybrid foams exhibiting improved mechanical properties and thermal stability.

2.3.2 BIO-HYBRIDS OF CLASS II TYPE

As stated previously, the most widely used way of involving a coupling agent in the solgel process is via a prior functionalization reaction with the polymer or reactive oligomer precursor for the inorganic component. Accordingly, Silvestri et al. (2009) have prepared a silane functionalized monomer by the Michael reaction of 2-hydroxyethylmethacrylate (HEMA) with APTES. The corresponding O–I hybrid materials were prepared via a prior solgel reaction of TEOS and the silane functionalized monomer, followed by an *in situ* free-radical polymerization of an unmodified methacrylate monomer. In the synthesis, it was demonstrated that the silanol groups inside the O–I hybrid were able to stimulate the deposition of apatite layers in contact with simulated body fluids.

Chitosan–siloxane class II type hybrids have also been investigated in several studies aimed to produce covalently bonded siloxane–chitosan hybrids using different functionalized silane coupling agents, such as 3-glycidyloxy-propyltrimethoxysilane (Liu et al. 2004; Shirosaki et al. 2009), alkyloxosilane-3-(trimethoxysilyl) propyl methacrylate (Fei et al. 2006), and 3-isocyanatopropyltriethoxysilane (Silva et al. 2005). In this synthesis, the use of 3-glycidyloxy-propyltrimethoxysilane (GOTMS or GPTMS) as coupling agent, it is expected that a certain amount of GOTMS is linked to the amino group of chitosan, through reactions with the epoxy group. These hybrids were found to exhibit a reduction in water uptake by increasing the GOTMS content. The same approach was used by Zhu et al. (2012) to produce a chitosan bio-hybrid for use as an interlayer between TiO_2-fluorinated nanoparticles deposited as coatings on a polylactic acid substrate. The cross-linked hybrid layer improved the water resistance and showed a higher dimensional stability than pristine chitosan coating.

The surface modification of cellulose with pre-hydrolyzed and partially condensed alkoxysilanes with various functional groups (e.g., propylamine and alkyl) has been found to alter the hydrophilicity/hydrophobicity balance of the resulting hybrid materials (Abdelmouleh et al. 2005; Chen et al. 2010). A significant level of hydrophobization of cellulose has been obtained using alkyl-substituted silanes, whereas the modification of silica with triazine derivatives has been found to improve the target fixation of reactive dyes for cellulose fibers. The cellulose substrate modified via a solgel process using a combination of TEOS and γ-aminopropyltriethoxylsilane (APTES) with 2,4,6-tri[(2-epihydrin-3-bimethyl-ammonium)propyl]-1,3,5-triazine chloride (Tri-EBAC) as a cross-linking agent have been used for the modification of the surface of cellulose fibers in order to improve the surface wettability by traditional dyes (Aiqin et al. 2009; Kongliang et al. 2009).

The preparation of biodegradable polymer-based hybrids for coatings on biopolymer substrate films has been investigated for potential application in biodegradable food packaging films (Iotti et al. 2009). These studies include the preparation of biodegradable PLA-based O–I hybrid materials as gas barrier coatings for PLA films (Gree and Seong 2012). These hybrid materials were synthesized using TEOS as the source of silica and 3-isocyanatopropyltriethoxysilane (IPTES) as a coupling agent. The barrier properties were improved by increasing the content of silica, resulting in maximum at around 65 wt%, than decrease due to some inherent porosity arising from internal shrinkage.

Iotti et al. (2009) have used 3-isocyanatopropyltriethoxysilane (ICPTES) as coupling agent for the silane functionalization of poly(caprolactone) (PCL). The resulting triethoxysilane-terminated PCL was mixed with TEOS to produce an O–I hybrid coating for PLA films. A similar approach has also been used to produce shape-memory PCL materials. Recovery times from water immersion tests were less than one s, depending on temperature, and were related to cross-linking density and molecular weight of PCL used as the organic precursors (Paderni et al. 2012). End-capped PCL's were also reacted with pre-hydrolyzed TEOS to prepare hybrid materials as a bioactive bone substitute (Yoo and Rhee 2004; Rhee 2003).

Proteins represent the main constituents of soft and hard tissues in animals and provide structural and biological activities in plants. The interest toward the fabrication of inorganic materials under the control of proteins has grown considerably in recent years, which arise from the realization that in diatoms and sponges, some specific proteins (e.g., silicatein and silaffin) direct the formation of silica in highly organized structure at multiple length scales. This confirms that proteins have a templating role and a catalytic effect on the solgel reactions. In this respect, Hecky et al. (1973) have indicated that amino acids have considerable influence on the solgel process in view of their ability to covalently bond the siloxane oligomers through hydroxyl and silanol condensation reactions. Furthermore, Coradin et al. (2003 and 2004) have shown that gelatin solutions are able to control the formation of silica nanoparticles from sodium silicate. In particular, the size of these particles depends on the concentration of the silicate in the precursor, whereas their packing density appears to depend on the protein content. It was shown that the formation of silica nanoparticles is equivalent to the growth of colloids on the surface of the gelatin films. The formation of micrometric size particles was observed, on the other hand, when gelatin chain brushes were formed at the film/water interface (Coradin et al. 2005).

Recently, the use of several agro-based proteins, as in the case of zein from corn, has been proposed for the preparation of bio-based thermoplastic films or foams for packaging application. Plasticization of zein can be brought about using water-soluble compounds, such as glycerol and polyethylene glycols, which can be carried out in practice by the combined effect of heat and shear to induce the denaturation of the hierarchical structure of zein protein and the development of molecular entanglements (Di Maio et al. 2010). These materials exhibit a relatively lower glass transition temperature, due to the presence of both hydrophilic amino acids and plasticizers, which enables the system to absorb large amount of water in humid environments. These limitations can be overcome by hybridization reactions associated with the solgel process. Recently, bio-hybrid films have been prepared using a thermoplastic zein (TPZ) and GOTMS in a two-step procedure. This includes the silane functionalization of zein macromolecules by reactive melting and the subsequent hydrolysis and condensation of the anchored GOTMS units, which result in the formation of SSQO nanostructures (Verdolotti et al. 2015). These nanostructures bind the zein macromolecules to each other through the reaction between the epoxy groups and the zein amine groups to produce an O–I network. The results confirm that solgel approach has only a slight effect on the secondary structure of zein proteins (Figure 2.7). The obtained hybrids show an interesting combination of improved characteristics, such better mechanical properties, enhanced resistance to water swelling, as well as increased durability in hostile biodegradation environments.

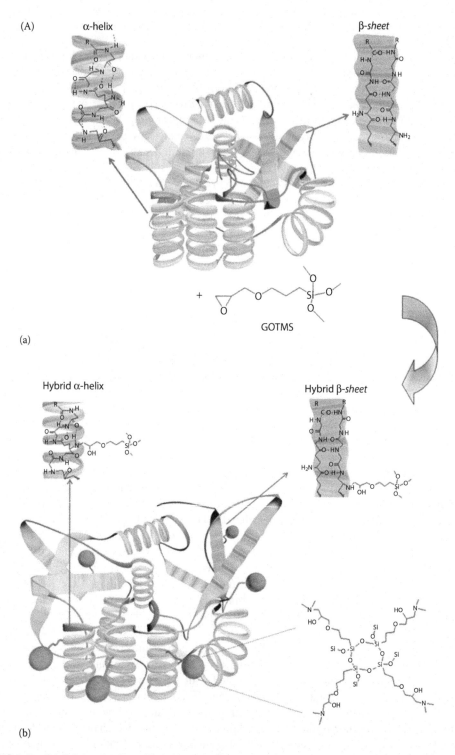

FIGURE 2.7 (A) Schematization of hybrid production: zein structure (a) before and (b) after reaction with GOTMS and formation of silsesquioxane domains. *(Continued)*

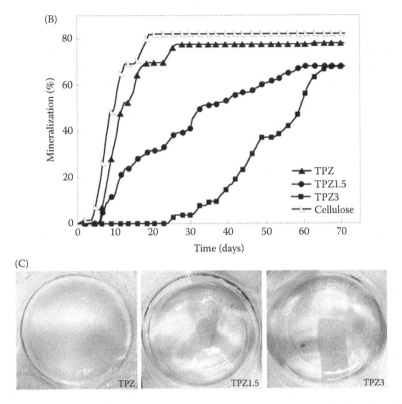

FIGURE 2.7 (Continued) (B) Swelling test in ethanol for neat zein (TPZ) and hybrids with increasing silica content; (C) evolution of the biodegradation of cellulose, zein, and hybrid films in compost. (Data from Verdolotti, L. et al., *Cell. Plast.* Special Issue 2015.)

A recent review by Heinemann et al. (2013) is particularly interesting insofar as the authors propose silica–collagen materials as a possible basis for understanding the correlation between the self-assembly of the polymer molecules and the mechanism for the condensation reactions to produce the inorganic domains in the solgel process. In this respect, type I collagen is considered to be an ideal component for the production of hybrid xerogel or scaffold materials for cell adhesion and proliferation, which are important requirements for wound healing and tissue regeneration. Strong interactions between silica species and collagen in diluted solutions result in a well-controlled biomineral interface at a molecular level, offering the possibility to modify the silica-induced collagen fibrillogenesis and/or formation of collagen-templated silica. This becomes particularly interesting when GOTMS (GPTMS) is used to cross-link the protein macromolecules (Chen et al. 2012). However, these materials exhibit poor mechanical properties, which can be improved only by increasing the collagen concentration and/or by speeding up the mineralization rate. Heinemann et al. (2009) have demonstrated that hydrogels, obtained from highly concentrated mixtures of silicic acid and a fibrillar collagen suspension, can be dried at ambient temperature and/or high relative humidity to obtain monolithic xerogels with excellent mechanical properties.

2.4 POROUS O–I BIO-HYBRIDS

Porous O–I hybrids in the form of foams, xerogels, and aerogels are attractive materials for applications in a variety of sectors, ranging from energy to packaging and above all as biomaterials for the regenerative medicine. New possible applications are proposed in the field of catalysis and water or air cleaning.

There are several methods for the production of porous hybrid systems, which resemble the known procedures for the production of polymeric or ceramic foams. The more feasible methods would be through any of the following routes: (1) producing a xerogel from an O–I hybrid gel, (2) carrying out a polymerization-induced phase separation in oxide solgels, (3) foaming by using foam stabilizers and accelerators of gelation, and (4) foaming of preformed hybrid materials by gas solubilization and expansion.

Porous xerogels are readily obtained by removing the solvents from the solid gel obtained from an O–I hybrid precursor solution. Removal of solvents is carried by gently heating the gel in order to avoid the shrinkage and collapse of the inorganic structure. Alternatively, the removal of solvents can be carried out through freeze-drying or supercritical drying methods. The size of the obtained cells is in the region of a few tens of nanometers.

Polymerization-induced phase separation, especially by spinodal decomposition, has been extensively used to obtain hierarchical porous hybrid structure. In these solgel systems, phase separation takes place in concomitance with the formation of heterogeneous structures in both the gel phase and the fluid phase. After gelation, the fluid phase can be removed to form pores in the length scale of micrometers. Generally, polymers or surfactants that do not interact with silanols tend to be located predominantly in the fluid phase, while the repartition fraction controls the volume fractions of macropores formed when the fluid phase is removed.

Conversely, polymers or surfactants interacting with silanols, such as polyethylene oxide, are located also in the gel phase. In this case, the volume of the fluid phase depends on the solvent amount, whereas the size of gel phase is determined by the polymeric additive (Nakanishi 1997).

Porous O–I hybrids produced using foam stabilizers require suitable *sol* solution composition. Foaming is induced by vigorous agitation after the addition of a surfactant as foam stabilizer (see Figure 2.8). An appropriate catalyst, such as hydrofluoric acid, is usually added to accelerate the gelation of the hybrid *sol* and to trap the air bubbles within the material during the solgel transition. The foaming gels are cast into molds, aged and dried. The type and concentration of the surfactant is critical to achieve the required surface tension of the liquid to stabilize the foamed sol prior to gelation. Drying can be carried out by heating, freeze-drying, or supercritical carbon dioxide (SC-CO_2) after partially removing the water through solvent exchange with an alcohol.

Using conventional gas foaming technology, it would be necessary to start with a polymeric-inorganic hybrid stock in a predetermined geometric shape, such as a sheet, which is saturated with a suitable gas, such as CO_2, N_2, or their mixture, at high pressure (see Figure 2.8). After solubilization of the gas, the stock is cooled to the desired foaming temperature, and subsequently, the pressure is released to ambient conditions to allow expansion of the cells and the formation of the foam. To stabilize the cellular structure, foams are immediately cooled at ambient conditions.

Other foaming processes include the use of polymeric spherical particles in the form of microemulsions as support first for the formation of an O–I hybrid structure, which is then removed by dissolution in a suitable solvent. For inorganic foams of high-temperature polymer-based hybrids, such as polyimides, the dispersed polymer particles can be burned out. Alternatively, microemulsion-templated approach may be also used to have hybrid foams. Polymerization of the continuous phase and removal of the droplets of the dispersed one, used as soft template, conducts to solid microcellular foams. These materials are known as porous polymers synthesized within high internal phase emulsions (polyHIPE). Particularly interesting are the silica foams of the type Si-polyHIPE produced by solgel process by using an organo-silane to stabilize microemulsion after hydrolyzing the TEOS (Brun et al. 2011).

2.4.1 Porous Bio-Hybrids for Biological Applications

Biological applications, such as encapsulation systems for biosensors or bioreactors and also as scaffolds for the regenerative medicine, require materials exhibiting a variation of mechanical properties, morphological structure, and chemical characteristics. For these applications, the solgel method is particularly suitable in view of the natural porosity of xerogels, which provides suitable

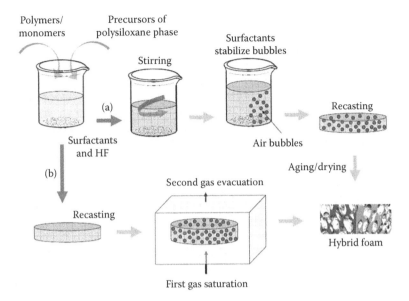

FIGURE 2.8 Schematic representation of hybrid foam formation by (a) foam stabilizers and (b) gas solubilization and expansion.

sites for the encapsulation of biological entities and drugs. At the same time, the porosity in hybrid foams or aerogel makes them suitable for the preparation of bioactive three-dimensional interconnected porous scaffold for the regeneration of soft and hard tissue. The regeneration of tissues may be induced with the use of a scaffold that acts as a temporary three-dimensional template. For these systems, the scaffold should be biocompatible and bioactive and should have an interconnected porous structure to allow the required body processes to take place, such as fluid flow, cell migration, bone ingrowths, and vascularization. It should also degrade at a specified rate.

In the case of bone repair, the scaffold should be able to link to the human bone and share the load during cycle applications. Bioactive glass scaffolds or polymeric hybrid scaffolds can be both obtained by solgel process, which makes it possible to tailor the morphology of the materials by mimicking the hierarchical structure of the human bone, consisting of collagen and carbonated hydroxyapatite. In this respect, it is obvious that bioactive glass scaffolds obtained by melt-quenching or by the solgel process cannot meet all the desired requirements due to the intrinsic brittleness for the majority of cases. The polymeric component of a hybrid system would not only overcome the inherent deficiency of bio-glass but also assist the initiation of the bioactive action and the congruent degradation of both the bioactive inorganic and the polymer components. Porous O–I hybrids allow bone cells to come into contact with both phases at one time, and the material to degrade at a single rate (Jones 2013).

In the production of porous O–I hybrids, the chain-like structure of the silicate phase intermingles with the polymer chains up to the foaming and gelation stages. The O–I hybrid foam obtained after drying is biocompatible due to the presence of a large amount of silanol groups. It should be noted that the bioactivity characteristics hinge on the possibility to induce nucleation and deposition of layers of hydroxyapatite in the presence of simulated body fluids and also in the presence of calcium ions in the chemical structure. It has been demonstrated, however, that the introduction of calcium ions as calcium nitrate in the hybrid precursor solution does not provide an effective system, insofar as the calcium ions deposited on the surface of the silica nanoparticles do not enter into the structure of the hybrids unless a thermal treatment is carried out at high temperatures. In this respect, some promising results have been obtained by Yu et al. (2011), showing that calcium enters the silica network at room temperature if used as calcium methoxyethoxide.

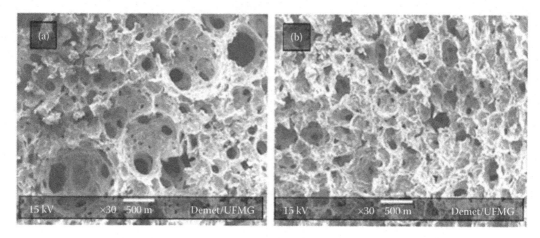

FIGURE 2.9 SEM images of hybrid foams with PVA content (a) 20 wt% and (b) 60 wt%. (Data from De Oliveira, A.A.R. et al., *J. Sol-Gel Sci. Tech.*, 47, 335–346, 2008.)

From the above discussion, it is evident that the porous hybrids can be produced both as class I and as class II materials. In class I hybrids, the polymer is weakly entrapped in the silica network during condensation through hydrogen bonding to the surface silanol (Si–OH) groups. This interaction is very weak and the porous hybrid is prone to be dissolved during treatment with human fluids. The principle was illustrated by producing bioactive glass/poly(vinyl alcohol) (PVA) nanocomposite scaffolds (de Oliveira et al. 2009). The hybrids were prepared with an inorganic phase composition consisting of 70% SiO_2–30% CaO and with PVA fractions in the range 20%–60% using two different routes for the solgel reactions starting from mixtures of TEOS and $CaCl_2$.

In the studied systems, the porosity was found to be in the region of 65% at 60 wt% PVA content and 80% with systems with the lower PVA content (20 wt%). The pore size was in the range 100–500 µm. Micrographs of the foam are shown in Figure 2.9.

Class II hybrids are more attractive owing to their strong resistance to body fluids. As already mentioned, the most used coupling agent is the glycidoxypropyltrimethoxysilane (GOTMS) (Valliant and Jones 2011). Class II type silica–gelatin hybrid scaffolds have been produced using GOTMS to bond the silica network, derived from hydrolyzed TEOS, to the gelatin molecules (Mahony et al. 2010). The porous hybrids were obtained using foam stabilizers and hydrofluoric acid and removing the water by using freeze-drying in order to create an interconnected pore network. Porous scaffolds were produced with up to 90% porosity. The compressive strength, as well as the degradation rate (i.e., rate of polymer dissolution) of hybrids, depends on the extent of interfacial covalent bonding. The higher the level of covalent bond coupling introduced at the interface, the better the mechanical properties and the lower the degradation rate.

Porous hybrid scaffolds, produced by a solgel foaming process, have been obtained through reactions between glycidoxypropyltrimethoxysilane with poly(γ-glutamic acid) in the presence of hydrolyzed silica (Poologasundarampillai et al. 2010, 2012). The epoxy ring in the GOTMS is thought to react with —COOH groups in the polypeptide chain, producing a functionalized polymer with Si—OCH_3 groups. When added to the sol, the siloxane is hydrolyzed to form Si—OH, which can form O—Si—O bonds with the silica network through condensation reactions. For these systems, calcium ions were added as $CaCl_2$ dissolved in the precursor solution.

Freeze-drying is an alternative method to foaming and is particularly suitable for polysaccharides such as chitosan or chitosan hybrids (Shirosaki et al. 2008). Increasing the GOTMS content brings about a significant reduction of the breaking stress with a consequent increment of the Young's modulus. When freeze-drying was used to produce scaffolds, the pore size was strongly dependent on the freezing temperature. In detail, the lower freezing temperature produced smaller

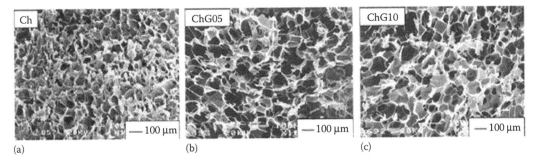

FIGURE 2.10 SEM photographs of the fracture surface of porous scaffolds of chitosan (Ch) (a) and hybrids with different content of coupling agents (ChG05 and ChG10) (b and c). (Data from Shirosaki, Y. et al., *Chem. Eng. J.*, 137, 122–128, 2008.)

pores, which were found to be in the region of 110 μm for the hybrids frozen at −20°C and about 50 μm for those treated at −85°C. The pore size was found to depend to some extent also on the GOTMS content. This indicates, therefore, that the porous microstructure can be altered through a combination of freezing temperature and composition parameters. Furthermore biocompatibility tests showed that the hybrid had a high capability to promote the cell attachment and proliferation. Figure 2.10 shows the effect of the composition on the pore size and demonstrates that the hybrids exhibit a porosity with higher void dimensions.

2.4.2 Aerogels Based on Bio-Hybrids

The main characteristics of aerogels are the very low density (0.004–0.500 g cm⁻³), the large internal surface area, and the open porosity. The use of totally inorganic aerogels, however, has been hampered by their poor mechanical properties and would, therefore, benefit from any modifications that would increase the toughness. In this respect, hybridization of the inorganic glasses with organic polymers constitutes a viable solution, and Leventis et al. (2010) have demonstrated the feasibility of achieving these objectives in their studies on hybrid silica aerogels using polymers such as polyurea, polyurethane, poly(methyl methacrylate), polyacrylonitrile, and polystyrene (Meador et al. 2007; Randall et al. 2011). Other candidates for the reinforcement of inorganic aerogels are insoluble polysaccharides or proteins, which are abundantly available in different structures and properties. In particular, regenerated cellulose gel prepared from aqueous alkali–urea solution has shown promising results as possible scaffold/template for the preparation of hybrid cellulose–silica aerogels (Cai et al. 2012). These materials exhibited a thermal conductivity in the range of 0.025–0.045 Wm K⁻¹. Since the cellulose and the related aerogels do not soften or decompose at temperatures up to 300°C, they can be used at much higher temperatures than common synthetic polymers and are, therefore, potentially useful as heat insulating materials.

Shchipunov and Shipunova (2008) have demonstrated that common proteins such as albumin, caseins, and gelatin can modulate the formation of silica from THEOS by the solgel method. Aerogel hybrid materials were prepared by drying the initially synthesized hydrogels under supercritical conditions. The results showed that proteins offer an additional opportunity for manipulating the silica morphology and architecture at the nanoscale level through the modification of the secondary and tertiary structures brought about by changing the pH of aqueous solution or the processing temperature. Since the pH of the medium can change the morphology of proteins from a fibrillar to the globular type, a similar role can also be expected with respect to the morphology of the hybrids produced from different proteins. The morphological peculiarities of aerogels observed at high magnification are shown in Figure 2.11.

(a) 100 µm   (b) 100 µm   (c) 100 µm

FIGURE 2.11 SEM micrographs of some hybrid aerogels: (a) gelatin, pH 3.3; (b) albumin, pH 4.3; and (c) casein, pH 5.6. (Data from Shchipunov, Y.A. and Shipunova, N., *Colloids Surf. B Biointerf.*, 63, 7–11, 2008.)

Fibrillar nanoparticles have been found in gelatin-containing aerogel, whereas a spherical structure was observed in most of the samples produced with albumin and caseins. It is worth mentioning that the morphological variations are related to the tertiary structure of the protein. However, the process of biomineralization in living organisms is more complex and several biopolymers, such as polysaccharides or proteins, act concordantly in regulating the silica morphology. Many studies have addressed the effect of the concomitant presence of polysaccharides and proteins in natural materials.

2.4.3 OTHER POROUS MATERIALS BASED ON BIO-HYBRIDS

Porous hybrid materials can be potentially obtained by gas foaming or supercritical drying of films or other products based on polymeric hybrid materials. An important factor in these systems is the cross-linking in the organic domains. The higher the cross-linking density, the lower will be the

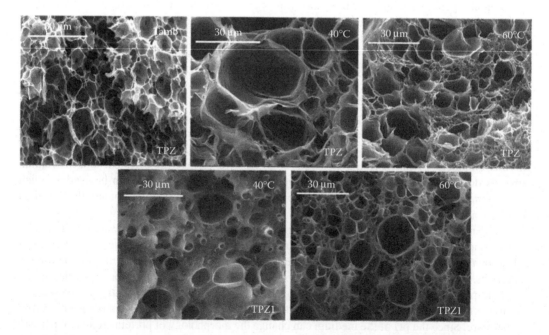

FIGURE 2.12 SEM images of foams obtained from neat zein (TPZ) (top) and hybrid (TPZ1) (bottom) with different ethanol swelling pretreatments temperatures. (Data from Verdolotti, L. et al., *Cell. Plast.* Special Issue 2015.)

capability of the hybrid to absorb gases and expand by foaming. Zein hybrid films were produced by the functionalization of the protein with GOTMS and subsequent film forming by pressure casting. Hybrid microcellular foams with homogeneous cellular structures have been obtained by gas foaming or supercritical drying. However, a bimodal structure with non-collapsed bubbles, varying from several microns to a few nanometers in size, was obtained by foaming hybrid materials with scCO$_2$ drying (see Figure 2.12).

REFERENCES

Abbate, M., Musto, P., Ragosta, G., Scarinzi, G., and Mascia, L. 2004. Polyimide-silica hybrids: Spectroscopy, morphology and mechanical properties. *Macromol Symp* 218: 211–220.

Abdelmouleh, M., Boufi, S., Belgacem, M.N., Dufresne, A., and Gandini, A. 2005. Modification of cellulose fibers with functionalized silanes: Effect of the fiber treatment on the mechanical performances of cellulose–thermoset composites. *J Appl Polym Sci* 98: 974–984.

Aiqin, H., Yaqi, S., and Yanhong, Y. 2009. Preparation of the cellulose/silica hybrid containing cationic group by sol-gel crosslinking process and its dyeing properties. 77: 201–205.

Brun, N., Ungureanu, S., Deleuze, H., and Backov, R. 2011. Hybrid foams, colloids and beyond: From design to applications. *Chem Soc Rev* 40: 771–788.

Cai, J., Liu, S., Feng, J. et al. 2012. Cellulose–silica nanocomposite aerogels by in situ formation of silica in cellulose gel. *Angew Chem Int Ed* 51: 2076–2079.

Chen, S., Chinnathambi, S., Shi, X., Osaka, A., Zhu, Y., and Hanagata, N. 2012. Fabrication of novel collagen-silica hybrid membranes with tailored biodegradation and strong cell contact guidance ability. *J Mater Chem* 22: 21885–21892.

Chen, X., Liu, Y., Lu, H., Yang, H., Zhou, X., and Xin, J.H. 2010. In-situ growth of silica nanoparticles on cellulose and application of hierarchical structure in biomimetic hydrophobicity. *Cellulose* 17: 1103–1113.

Coradin, T., Bah, S., and Livage, J. 2004. Gelatine/silicate interactions: From nanoparticles to composite gels. *Colloids Surf B* 35: 53–58.

Coradin, T., Coupe, A., and Livage, J. 2003. Interactions of bovine serum albumin and lysozyme with sodium silicate solutions. *Colloids Surf B* 29: 189.

Coradin, T., Marchal, A., Abdoul-Aribi, N., and Livage, J. 2005. Gelatine thin films as biomimetic surfaces for silica particles formation. *Colloids Surf B Biointerf* 44: 191–196.

Coradin, T., Mercey, E., Lisnarda, L., and Livage, J. 2001. Design of silica-coated microcapsules for bioencapsulation. *Chem Commun* 23: 2496–2497.

Cornelissen, J.J.L.M., Connor, E.F., Kim, H.-C. et al. 2003. Versatile synthesis of nanometer sized hollow silica spheres. *Chem Commun* 8: 1010–1011.

De Oliveira, A.A.R., Ciminelli, V., Dantas, M.S.S., Mansur, H.S., and Pereira, M.M. 2008. Acid character control of bioactive glass/polyvinyl alcohol hybrid foams produced by sol-gel. *J Sol-Gel Sci Tech* 47: 335–346.

De Oliveira, A.A.R., Gomide, V.S., de Fátima Leite, M., Mansur, H.S., and de Magalhães Pereira, M. 2009. Effect of polyvinyl alcohol content and after synthesis neutralization on structure mechanical properties and cytotoxicity of sol-gel derived hybrid foams. *Mater Res* 12: 239–244.

De Pisapia, L., Verdolotti, L., Di Mauro, E., Di Maio, E., Lavorgna, M., and Iannace, S. 2012. Cellulose based hybrid hydroxylated adducts for polyurethane foams. *AIP Conference Proceedings*, 1459: 123–125. doi:10.1063/1.4738418.

Dhanasingh, S., Mesha, L., and Hiriyannaiah, J.J. 2011. Preparation, characterization and antimicrobial studies of chitosan/silica hybrid polymer. *Biointerf Res Appl Chem* 1: 48–56.

Di Maio, E., Mali, R., and Iannace, S. 2010. Investigation of thermoplasticity of Zein and Kafirin proteins: Mixing process and mechanical properties. *J Appl Polym Sci* 18: 626–633.

Fei, B., Lu, H.F., and Xin, J.H. 2006. One-step preparation of organosilica-chitosan crosslinked nanospheres. *Polymer* 47: 947–950.

Gamelas, J.A.F., Evtyugina, M.G., Portugalc, I., and Evtuguin, D.V. 2012. New polyoxometalate-functionalized cellulosic fibre/silica hybrids for environmental applications. *RSC Adv* 2: 831–839.

Gree, B., and Seong, W.K. 2012. Biodegradable poly(lactic acid)-based hybrid coating materials for food packaging films with gas barrier properties. *J Ind Eng Chem* 18: 1063–1068.

Hajji, P., David, L., Gerard, J.F., Pascault, J.P., and Vigier, G. 1999. Synthesis, structure, and morphology of polymer–silica hybrid nanocomposites based on hydroxyethyl methacrylate. *J Polym Sci Part B Poly Phys* 37: 3172–3187.

Hecky, R.E., Mopper, K., Kilham, P., and Degens, E.T. 1973. The amino acid and sugar composition of diatom cell walls. *Marine Biology* 19: 323–331.

Heinemann, S., Coradin, T., and De Simone, M.F. 2013. Bio-inspired silica–collagen materials: Applications and perspectives in the medical field. *Biomater Sci* 1: 688–702.

Heinemann, S., Heinemann, C., and Bernhardt, R. 2009. Bioactive silica-collagen composite xerogels modified by calcium phosphate phases with adjustable mechanical properties for bone replacement. *Acta Biomater* 5: 1979–1990.

Huang, S.L., Chin, W.K., and Yang, W.P. 2005. Structural characteristics and properties of silica/poly(2-hydroxyethyl methacrylate) (PHEMA) nanocomposites prepared by mixing colloidal silica or tetraethyloxysilane (TEOS) with PHEMA. *Polymer* 46: 1865–1877.

Iotti, M., Fabbri, P., Pilati, M.M.F., and Fava, P. 2009. Organic–inorganic hybrid coatings for the modification of barrier properties of poly(lactic acid) films for food packaging applications. *J Polym Environ* 17: 10–19.

Jana, S.C., and Sachin, J. 2001. Dispersion of nanofillers in high performance polymers using reactive solvents as processing aids. *Polymer* 42: 6897–6905.

Jones, J.R. 2013. Review of bioactive glass: From Hench to hybrids. *Acta Biomater* 9: 4457–4486.

Kickelbick, G. 2007. Introduction to hybrid materials. In *Hybrid Materials. Synthesis, Characterization, and Applications*, 1–48. Weinheim, Germany: Wiley-VCH.

Kongliang, X., Yanhong, Y., and Shi, Y. 2009. Synthesis and characterization of cellulose/silica hybrid materials with chemical crosslinking. *Carbohyd Polym* 78: 799–805.

Kröger, N., Deutzmann, R., and Sumper, M. 1999. Polycationic peptides from diatom biosilica that direct silica nanosphere formation. *Science* 286: 1129–1132.

Laugel, N., Hemmerlé, J., and Porcel. C. 2007. Nanocomposite silica/polyamine films prepared by a reactive layer-by-layer deposition. *Langmuir* 23: 3706–3711.

Lavorgna, M., Gilbert, M., Mascia, L. et al. 2009. Hybridization Nafion membranes with an acid function-alised polysiloxane: Effect of morphology on water sorption and proton conductivity. *J Membr Sci* 330: 214–226.

Lavorgna, M., Mascia, L., Mensitieri, G. et al. 2007. Hybridization of Nafion membranes by the infusion of functionalized siloxane precursors. *J Membr Sci* 294: 159–168.

Leventis, N., Sadekar, A., Chandrasekaran, N., and Leventis, C.S. 2010. Click synthesis of monolithic silicon carbide aerogels from polyacrylonitrile-coated 3D silica networks. *Chem Mater* 22: 2790–2803.

Lionetto, F., Mascia, L., and Frigione, M. 2013. Evolution of transient-states and properties of an epoxy-silica hybrid cured at ambient temperature. *Eur Polym J* 49: 1298–1313.

Liu, Y.L., Su, Y.H., and Lai, J.Y. 2004. In situ crosslinking of chitosan and formation of chitosan–silica hybrid membranes with using 3-glycidoxypropyltrimethoxysilane as a crosslinking agent. *Polymer* 45: 6831–6837.

Maeda, H., Nakajima, M., Hagiwara, T., Sawaguchi, T., and Yano, S. 2006. Bacterial cellulose/silica hybrid fabricated by mimicking biocomposites. *J Mater Sci* 41: 5646–5656.

Mahony, O., Tsigkou, O., Ionescu, C., Minelli, C., Hanly, R., Ling, L. et al. 2010. Silica-gelatin hybrids with tailorable degradation and mechanical properties for tissue regeneration. *Adv Funct Mater* 20: 3835–3845.

Mascia, L. 1995. Current developments in organic-inorganic hybrid materials: Ceramers. *Trends Polym Sci* 3(1): 61.

Mascia, L. 2010. Fillers generated in-situ: Bicontinuous nanocomposites. In *Functional Fillers for Plastics*, 459–468. Weinheim, Germany: Wiley-VCH.

Mascia, L., and Lavorgna, M. 2012. Nanostructured polymer composites by sol-gel method. In *Wiley Encyclopedia of Composites*, 1922–194. New York: VCH Publishers.

Mascia, L., Prezzi, L., and Lavorgna, M. 2005. Peculiarities in the solvent absorption characteristics of epoxy-siloxane hybrids. *Poly Eng Sci* 45: 1039–1048.

Mascia, L., Prezzi, L., Wilcox, G.D. et al. 2006. Molybdate-doping of networks in epoxy-silica hybrids: Domain structuring and corrosion inhibition. *Prog Org Coat* 56: 13–22.

Matejka, L. and Dukh, O. 2001. Organic-inorganic hybrid networks. *Macromol Symp* 171: 181–188.

Matejka, L., Dukh, O., and Kolarik, J. 2000. Reinforcement of crosslinked rubbery epoxies by in-situ formed silica. *Polymer* 41: 1449–1459.

Meador, M.A.B., Capadona, L.A., McCorkle, L., Papadopoulos, D.S., and Leventis, N. 2007. Structure-property relationships in porous 3D nanostructures as a function of preparation conditions: Isocyanate cross-linked silica aerogels. *Chem Mater* 19: 2247–2260.

Mo, Z., Zhao, Z., Chen, H., and Niu, G. 2008. Heterogeneous preparation and properties of nano-SiO$_2$/cellulose composites. *Fuhe Cailiao Xuebao/Acta Materiae Compositae Sinica* 4: 24–28.

Musto, P., Mascia, L., Mensitieri, G., and Ragosta, G. 2005. Diffusion of water and ammonia through polyimide-silica bicontinuous nanocomposites: Interactions and reactions. *Polymer* 46: 4492–4503.

Musto, P., Ragosta, G., Scarinzi, G., and Mascia, L. 2004. Polyimide-silica nanocomposites: Spectroscopic, morphological and mechanical investigations. *Polymer* 45: 1697–1706.

Nakane, K., Yamashita, T., Iwakura, K., and Suzuki, F.J. 1999. Properties and structure of poly(vinyl alcohol)/silica composites. *J Appl Polym Sci* 74: 133–138.

Nakanishi, K. 1997. Pore structure control of silica gels based on phase separation. *J Porous Mater* 4: 64–112.

Paderni, K., Pandini, S., Passera, S., Pilati, F., Toselli, M., and Messori, M. 2012. Shape-memory polymer networks from sol-gel cross-linked alkoxysilane-terminated poly(e-caprolactone). *J Mater Sci* 47: 4354–4362.

Park, S.-B., You, J.O., Park, H.Y., Haam, S.J., and Kim, W.S. 2001. A novel pH-sensitive membrane from chitosan-TEOS IPN; preparation and its drug permeation characteristics. *Biomaterials* 22: 323–330.

Perry, C.C., and Lu, Y. 1992. Preparation of silicas from silicon complexes: Role of cellulose in polymerisation and aggregation control. *Chem Soc Faraday Trans* 88: 2915–2921.

Poologasundarampillai, G., Ionescu, C., Tsigkou, O. et al. 2010. Synthesis of bioactive class II poly(gamma-glutamic acid)/silica hybrids for bone regeneration. *J Mater Chem* 20: 8952–8961.

Poologasundarampillai, G., Yu, B., Tsigkou, O. et al. 2012. Bioactive silica–poly(c-glutamic acid) hybrids for bone regeneration: Effect of covalent coupling on dissolution and mechanical properties and fabrication of porous scaffolds. *Soft Matter* 8: 4822–4832.

Prezzi, L., and Mascia, L. Polymeric compositions with modified siloxane networks, corresponding production and uses thereof. European Patent EP 2005055218.

Randall, J.P., Meador, M.A.B., and Jana, S.C. 2011. Tailoring mechanical properties of aerogels for aerospace. *ACS Appl Mater Interf* 3: 613–626.

Rhee, S. 2003. Effect if MW of polycaprolactone on interpenetrating network structure, apatite-forming ability and degradability of polycaprolactone/silica nano-hybrid material. *Biomaterials* 24: 1721–1727.

Sequeira, S., Evtuguin, D.V., Portugal, I., and Esculcas, A.P. 2007. Synthesis and characterisation of cellulose/silica hybrids obtained by heteropoly acid catalysed sol-gel process. *Mater Sci Eng C* 27: 172–179.

Shchipunov, Y.A. 2003. Sol–gel-derived biomaterials of silica and carrageenans. *J Colloid Interf Sci* 268: 68–76.

Shchipunov, Y.A., and Karpenko, Y. 2004. Hybrid polysaccharide-silica nanocomposites prepared by the sol-gel technique. *Langmuir* 20: 3882–3887.

Shchipunov, Y.A., Karpenko, T.Yu., Bakunin, I.Yu., Burtseva, V.Y., and Zvyagintsev, N.T. 2004. A new precursor for the immobilization of enzymes inside sol-gel-derived hybrid silica nanocomposites containing polysaccharides. *J Biochem Biophys Methods* 58: 25–38.

Shchipunov, Y.A., Kojima, A., and Imae T. 2005. Polysaccharides as a template for silicate generated by sol-gel processes. *J Colloid Interf Sci* 285: 574–580.

Shchipunov, Y.A., and Shipunova, N. 2008. Regulation of silica morphology by proteins serving as a template for mineralization. *Colloids Surf B Biointerf* 63: 7–11.

Shengyou, S., Wei, W., Liqin, L., Shijia, W., Yongzan, W., and Weicai, L. 2013. Effect of chitosan/nano-silica coating on the physicochemical characteristics of longan fruit under ambient temperature. *J Food Eng* 118: 125–131.

Shirosaki, Y., Okayama, T., Tsuru, K., Hayakawa, S., and Osaka, A. 2008. Synthesis and cytocompatibility of porous chitosan-silicate hybrids for tissue engineering scaffold application. *Chem Eng J* 137: 122–128.

Shirosaki, Y., Tsuru, K., Hayakawa, S. et al. 2009. Physical, chemical and in vitro biological profile of chitosan hybrid membrane as a function of organosiloxane concentration. *Acta Biomater* 5: 346–355.

Silva, S.S., Ferreira, R.A.S., Fu, L. et al. 2005. Functional nanostructured chitosan-siloxane hybrids. *J Mater Chem* 15: 3952–3961.

Silvestri, B., Luciani, G., Costantini, A. et al. 2008. In-situ sol-gel synthesis and characterization of bioactive pHEMA/SiO2 blend hybrids. *J Biomed Mater Res Part B Appl Biomater* 369–378.

Valliant, E.M., and Jones, J.R. 2011. Softening bioactive glass for bone regeneration: Sol-gel hybrid materials. *Soft Matter* 7: 5083–5095.

Verdolotti, L., Lavorgna, M., Di Maio, E., and Iannace, S. Process for destructuring a cellulose pulp, resulting product and uses of the product. Patent WO2013105040 A1 and PCT/IB2013/050199.

Verdolotti, L., Lavorgna, M., Oliviero, M. et al. 2013. Functional zein–siloxane bio-hybrids. *ACS Sust Chem* doi:10.1021/sc400295w.

Verdolotti, L., Oliviero, M., Lavorgna, M., Iozzino, V., Larobina, D., and Iannace, S. 2014. Bio-hybrid foams by silsequioxanes cross-linked thermoplastic zein films. *Cell Plast* Special Issue (Proceedings of Biofoams 2013).

Wang, G.-H., and Zhang, L.M. 2006. Using novel polysaccharide-silica hybrid material to construct an ampero-metric biosensor for hydrogen peroxide. *J Phys Chem B* 110: 24864–24868.

Yano, S. 1994. Preparation and characterization of hydroxypropyl cellulose/silica micro-hybrids. *Polymer* 35: 5565–5570.

Yoo, J.J., and Rhee, S.-H. 2004. Evaluations of bioactivity and mechanical properties of poly (ε-caprolactone)/silica nanocomposite following heat treatment. *J Biomed Mater Res Part A* 68: 401–410.

Yoshinaga, I., and Katayama, S. 1996. Synthesis of inorganic-organic hybrid by incorporation of inorganic component into organic polymer using metal alkoxides. *J Sol-Gel Sci Technol* 6: 151–154.

Yu, B., Poologasundarampillai, G., Turdean-Ionescu, C., Smith, M.E., and Jones, J.R. 2011. A new calcium source for bioactive sol-gel hybrids. *Bioceramics Dev Appl* 1: 1–3.

Zalzberg, L., and Avnir, D. 2008. Biocompatible hybrid particles of poly(l-lactic acid)-silica. *J Sol-Gel Sci Tech* 48: 47–50.

Zhu, Y., Piscitelli, F., Buonocore, G.G., Lavorgna, M., Amendola, E., and Ambrosio, L. 2012. Effect of surface fluorination of tio$_2$ particles on photocatalitytic activity of a hybrid multilayer coating obtained by sol-gel method. *ACS Appl Mater Interf* 4: 150–157.

Zoppi, R.A., and Goncalves, M.C. 2002. Hybrids of cellulose acetate and sol-gel silica: Morphology, thermomechanical properties, water permeability, and biodegradation evaluation. *J Appl Polym Sci* 84: 2196–2205.

Zou, H., Wu, S., and Shen, J. 2008. Polymer/Silica nanocomposites: Preparation, characterization, properties, and applications. *Chem Rev* 108: 3893–3957.

3 Equation-of-State Approach in Polymer Solution and Polymer Foaming Thermodynamics

Ioannis Tsivintzelis and Costas Panayiotou

CONTENTS

ABSTRACT: In this chapter, the challenging issues related to the production of porous polymers are reviewed in an effort to pinpoint at the requirements that should be met by prospect thermodynamic models to be used for describing and, ultimately, designing such processes. The role of external conditions of temperature, pressure, and composition of solvent or antisolvent is first discussed and the polymer-foaming processes are categorized accordingly. The role of the glassy state or the (semi)crystalline state for stabilizing the final porous polymer structure is also discussed. It becomes clear that an appropriate model for handling such systems and processes should be applicable to gases, vapors, liquids, or glasses; to low molecular weight solvents as well as to high polymers; to homogeneous as well as to heterogeneous systems and should handle vapor–liquid, liquid–liquid, solid–liquid, glass-to-rubber transitions, and interfacial properties. The nonrandom with hydrogen bonding (NRHB) equation-of-state model is such a model, and its essential working formalism is reviewed and applied to a variety of systems and thermodynamic properties including systems of polymers and pharmaceuticals. The very same model is finally applied for describing and providing with designing elements of the polymer foaming process with supercritical fluids.

3.1   INTRODUCTION

Porous polymeric structures can be produced by a large variety of methods, which include the use of bio-plotters or bio-printers for development of highly ordered three-dimensional porous structures, particulate leaching, electrospinning and other spinning processes for the production of nanofiber mats, and microphase separation techniques (Hentze and Antonietti 2002; Ma 2004; Quirk et al. 2004; Tsivintzelis and Panayiotou 2013). The latter category includes a large variety of methods, but most of them are based on a common idea: A homogeneous multicomponent mixture is led to a supersaturated or thermodynamically unstable state, which triggers phase separation. Usually, the thermodynamic instability is caused by a rapid change in temperature or pressure, addition of a nonsolvent, or transformation of at least one mixture component by chemical reaction (Ma 2004; Quirk et al. 2004). Consequently, a homogeneous polymer solution separates in two phases, the polymer-rich phase and the polymer-lean phase, or rather in a network of polymer-lean microphases dispersed typically in a polymer-rich matrix phase. After the removal of solvent, under specific conditions, the polymer-rich phase can solidify retaining pores or voids in the places of the solvent pools in the polymer-lean microphases.

From the above plain exposition, it becomes obvious that the knowledge of thermodynamic properties of such polymer systems is of particular interest in understanding, designing, and optimizing the physicochemical processes for the production of porous polymeric materials. This will become clearer by describing, in the subsequent paragraphs, the most common phase separation techniques from a thermodynamics point of view.

3.1.1   TEMPERATURE-INDUCED PHASE SEPARATION

The temperature-induced phase separation, TIPS, technique is a fairly common method for the formation of porous membranes. It is based on the fact that the ability of a solvent to dissolve a polymer changes with temperature. Subsequently, in a homogeneous binary polymer–solvent mixture, a rapid change in temperature induces thermodynamic instability and subsequently the solution is separated in a polymer-lean and a polymer-rich phase. After demixing, the solvent is removed by extraction, evaporation, or freeze-drying (Van de Witte et al. 1996a; Nam and Park 1999; Hentze and Antonietti 2002).

This method is mainly used in polymer–solvent systems that present upper critical solution temperature, in which the solvent quality decreases with temperature decrease. A schematic phase diagram of a binary polymer solution is shown in Figure 3.1. Such a system presents a liquid–liquid demixing gap at relatively low temperatures. Depending on the solution composition, the liquid–liquid demixing proceeds with different mechanisms (Van de Witte et al. 1996a). Compositions located between the binodal and the spinodal curve are metastable, and consequently, the dimixing requires large enough fluctuations. In this area, liquid–liquid phase separation will take place by nucleation and growth of droplets inside a continuous phase. More specifically, sufficiently low polymer concentrations will result in the nucleation and growth of polymer-rich phase droplets inside a continuous polymer-lean phase and, consequently, cannot result in porous structures, but could lead to the formation of polymer particles (case A). On the other hand, a sufficiently high polymer concentration leads to the nucleation and growth of polymer-lean phase droplets inside a continuous polymer-rich phase (case C), which could result in the formation of a porous polymer structure.

A sufficiently high fraction of the minor phase could lead to the phase separation inside the spinodal region, which results in the formation of bicontinuous structures (case B). However, as the droplets evolve, the requirement for minimization of the interfacial free energy will cause the coalescence of neighboring droplets (Van de Witte et al. 1996a). Subsequently, if the obtained structure will not be stabilized by other means, the structures of Figure 3.1 (a, b, and c) will grow and coarsen in time, until a fully demixed structure consisting of two fully separated layers will

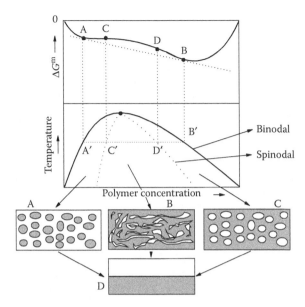

FIGURE 3.1 Liquid–liquid phase separation of a polymer–solvent system. The upper part shows a typical plot of the Gibbs free energy of mixing against the composition of the mixture, while the lower part shows schematically the liquid–liquid phase separation. The demixing routes are nucleation of a polymer-rich phase inside the polymer-lean phase (A), nucleation of a polymer-lean phase inside the polymer-rich phase (C), and spinodal decomposition (B). Ultimately, they may lead to full phase separation (D).

occur (case D). This is prevented and the initial liquid–liquid demixing microstructure is *frozen* when the removal of solvent from the polymer-rich phase and/or the decrease of temperature induce polymer crystallization (solid–liquid demixing), vitrification, or gelation (Van de Witte et al. 1996a, b; Mulder 2000; Hentze and Antonietti 2002). Figure 3.2 shows an equilibrium phase diagram for a system showing both liquid–liquid and solid–liquid demixing. Consequently, it is clear that the thermodynamic behavior of the system plays an important role in understanding the process. However, it should be kept in mind that kinetics is also important, especially for the crystallization of polymer phases (Van de Witte et al. 1996a).

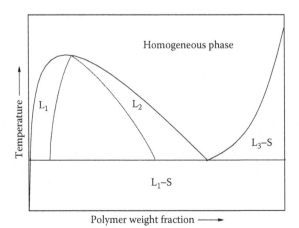

FIGURE 3.2 Combination of liquid–liquid demixing with crystallization.

3.1.2 ADDITION OF NONSOLVENT

3.1.2.1 Liquid Solvent—Nonsolvent Systems

In a homogenous mixture at constant pressure and temperature conditions, thermodynamic instability can be caused by the addition of another component. In processes relevant to the production of biofoams, such a component would be miscible with the solvent of the initial solution, but immiscible with the polymer (nonsolvent). Depending on the nature of the components and the experimental conditions, the subsequent precipitation of the polymer can lead to the formation of porous structures (Van de Witte et al. 1996a). Thermodynamics and, especially, knowledge of the phase behavior of such systems are of vital importance in understanding the process.

Figure 3.3a presents the equilibrium phase diagram of a three-component system exhibiting liquid–liquid demixing phase behavior. The arrows do not represent the composition change over time, but only connect the initial and the final compositions. Similar to the TIPS method, demixing of a mixture with low polymer concentration inside the metastable area (arrow 1) results in the nucleation and growth of a polymer-rich phase inside the polymer-lean (solvent) phase. Demixing inside the unstable area results in the production of bicontinuous phases (arrow 2), while demixing of a dense polymer solution in the area between the binodal and the spinodal curve (metastable area) leads to the nucleation and growth of a polymer-lean phase droplets inside the polymer-rich phase (arrow 3).

The collapse of the porous structure is prevented when the demixing results in polymer crystallization (solid–liquid demixing), vitrification, or gelation. An equilibrium phase diagram of a system exhibiting both liquid–liquid demixing and solid (crystalline)–liquid demixing is shown in Figure 3.3b. However, kinetics is crucial and usually the process proceeds away from equilibrium (Mulder 2000). For such cases, nonequilibrium phase diagrams have been suggested (Van de Witte et al. 1996b). Usually, the nucleation and growth of crystallites proceeds slowly, while the liquid–liquid phase separation proceeds relatively faster. Consequently, the latter phenomenon dominates even at conditions where the crystallization (solid–liquid demixing) is thermodynamically favored (Van de Witte et al. 1996a).

3.1.2.2 Addition of Supercritical Fluids—Pressure, an Extra Variable

Using the immersion precipitation method with the addition of a liquid nonsolvent described in the previous paragraph, porous structures with desired properties can be produced by fine-tuning the dissolution ability (solvent power) of the solvent/nonsolvent system, which is mainly accomplished by the appropriate selection of solvents and temperature conditions (Van de Witte et al. 1996a, b;

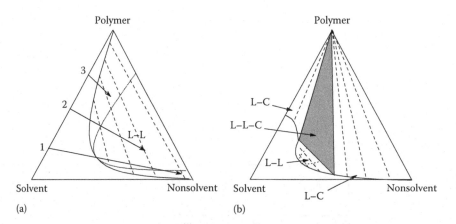

FIGURE 3.3 Equilibrium phase diagrams of polymer–solvent–nonsolvent mixtures exhibiting (a) liquid–liquid (L–L) and (b) liquid–crystalline (L–C) phase separation.

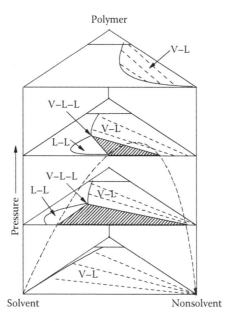

FIGURE 3.4 Polymer–solvent–nonsolvent phase behavior at various pressures, showing vapor–liquid (V–L), liquid–liquid (L–L) and vapor–liquid–liquid (V–L–L) equilibria.

Mulder 2000). On the other hand, since the dissolution ability of liquid solvents does not significantly depend on pressure, the process is usually performed at ambient pressure.

However, the selection of a liquid solvent or nonsolvent for the polymer is not the only option, since supercritical fluids can be also used for this purpose. Usually, they are used as nonsolvents, since few polymers present significant solubility in the common supercritical fluids used in such processes (Kirby and McHugh 1999). In such cases, the pressure is an additional process variable that can be tuned in order to alter and control the properties of the final porous polymer structure. Consequently, the phase diagrams that describe the phase equilibria of such systems have an extra dimension, as shown in Figure 3.4.

CO_2 is the most common fluid used as nonsolvent at supercritical conditions (Matsuyama et al. 2001; Reverchon and Cardea 2005; Tsivintzelis et al. 2007b; Duarte et al. 2012), and many experimental studies report data for the phase behavior of polymer–solvent–CO_2 systems (Kirby and McHugh 1999). However, very rarely the experimental data cover a wide temperature, pressure, and composition range in order to allow for the construction of phase diagrams similar to Figure 3.4. In all cases, complementary to the experimental data, an appropriate thermodynamic model is needed in order to predict the phase behavior (Shim and Johnston 1989).

3.1.3 FOAMING OF POLYMERS USING GASES OR SUPERCRITICAL FLUIDS

Porous polymers can be produced through the gas foaming method (Martini-Vvedensky et al. 1984; Kumar and Suh 1990; Goel and Beckman 1994a, b; Baldwin et al. 1996; Arora et al. 1998; Rodeheaver and Colton 2001; Reverchon and Cardea 2007; Tsivintzelis et al. 2007a). According to this technique, the polymer is saturated with a gas or supercritical fluid (usually CO_2 or N_2) at constant temperature and pressure. Then, the system is led to a supersaturated state either by reducing pressure (pressure-induced phase separation [PIPS]) or by increasing temperature (TIPS) resulting in the nucleation and growth of gas bubbles inside a supersaturated polymer-blowing agent matrix. The growth of pores continues until the polymer vitrifies (Goel and Beckman 1994a; Arora et al. 1998; Tsivintzelis et al. 2007a) or until the viscosity of the polymer matrix is high enough to cause the retractive force restricting cell growth to become unexcelled. Further diffusion of the blowing agent outside the matrix leaves pores and voids in places where gas bubbles existed. Furthermore, foaming of polymers with gases or supercritical fluids has a strong advantage, especially in the processing of polymers for biomedical applications. There is no need for use of harmful organic solvents that in most cases are not easily removed from the final product material (Quirk et al. 2004).

The most important phenomena that control the final porous structure are the solubility of the blowing agent in the polymer matrix, the interfacial energy between the gas nuclei and the surrounding polymer matrix, and the plasticization profile of the polymer–gas system (Tsivintzelis et al. 2007a). However, considering amorphous polymers, the plasticization profile of the polymer matrix is of particular interest, since it defines the route of the foaming procedure (Tsivintzelis and Panayiotou 2013).

Gas molecules that are dissolved inside a glassy polymer matrix force the polymer chains to rearrange, increasing locally the chain distance and, subsequently, decreasing the inter- or intra-chain interactions and increasing the chain mobility. This effect resembles the glass-to-rubber transition and, indeed, a decrease in the glass transition temperature is experimentally observed (Condo et al. 1992). According to Condo et al. (1992), there are four types of plasticization profiles, which are shown in Figure 3.5. Such profiles were predicted by a thermodynamic lattice model, which is similar to, but simpler than the model described below in this chapter. It is worth mentioning that the types II and III were first predicted by Condo et al. (1992) and then were experimentally observed (Condo et al. 1994).

From the point of view of the foaming methods, the most interesting plasticization profile is type IV of Figure 3.5, which is called retrograde vitrification. Such polymer–fluid systems present

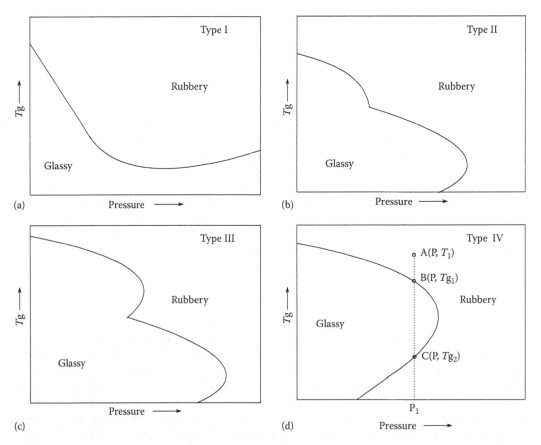

FIGURE 3.5 Four types of plasticization behavior for polymer–supercritical fluid systems. (a) Type I, (b) Type II, (c) Type III, and (d) Type IV.

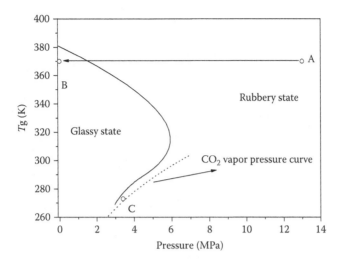

FIGURE 3.6 Plasticization behavior of the PMMA–CO_2 system and foaming routes.

two glass transition temperatures at the same pressure. A polymer–fluid system of Figure 3.5, which presents type IV behavior, is initially in the rubbery state at conditions referred to point A. When the temperature decreases at constant pressure, the system enters in the glassy region (point B). However, as the temperature continues to decrease, the system again enters the rubbery region (point C). As the temperature decreases, two competitive phenomena take place. First, the decrease of temperature induces a decrease of the chain mobility and the subsequent transition of the system from the liquid (rubbery) to the glassy state. At the same time, however, the decrease of temperature increases the solubility of the fluid inside the polymer matrix and, thus, increases the chain mobility, an effect that favors the transition of the system from the glassy to the rubbery state (Condo et al. 1992).

Usually, when the PIPS is used in amorphous polymers, the saturated with gas polymer exists in the rubbery state, at the equilibrium conditions. The rapid reduction of pressure induces the phase separation and the production of gas nuclei, which will be grown to pores dispersed in the polymer matrix. The growth of such pores continues until the polymer vitrifies. Such a foaming route for a system that exhibits type IV plasticization profile (retrograde vitrification) is illustrated as the AB route in Figure 3.6. However, such mixtures can be saturated and exist in the rubbery state at temperature and pressure conditions much lower than the usual conditions employed in the PIPS method. Handa and Zhang (2000) used this idea and successfully prepared poly(methyl-methacrylate) (PMMA) porous structures by saturating the polymer at much lower pressure and temperature conditions (−0.2°C and 34 atm, illustrated by point C in Figure 3.6) and heating the polymers after the depressurization step (up to 24°C–90°C).

It should be mentioned here that equilibrium thermodynamics can provide with a state diagram or a map for the phases and the thermodynamically favored transitions. However, nonequilibrium thermodynamics can also serve as a valuable tool in describing such processes, since the rapid changes in temperature and pressure conditions or in the mixture's composition may render a significant part of the process at nonequilibrium. In addition, the kinetics of phase separation determines whether the thermodynamically favored separation will take place in the process time length.

3.1.4 Need for Thermodynamic Models

Overall, knowledge of thermodynamic properties is of particular importance in understanding and designing such processes. Since the experimental data are scarce and do not cover wide ranges of temperature–pressure-composition conditions, there is much need for describing or predicting thermodynamic properties using appropriate models. As discussed in the previous paragraphs, such models should be able to describe various system properties, such as the phase behavior, the interfacial properties, and the plasticization profile in polymer–solvent mixtures over a wide range of external conditions. Consequently, they should be able to describe (properties of) the following:

- Polymer solutions
- Fluid systems at high pressures
- Liquids, gases, or supercritical fluids
- The retrograde vitrification
- Systems with specific interactions, that is, hydrogen bonding interactions

From the above, it is clear that equation-of-state models able to describe polymer solutions over a broad range of conditions are the ideally suited models for our purpose.

Over the last years, advanced thermodynamic equation-of-state models were developed based on statistical thermodynamics (Sanchez and Panayiotou 1994; Economou 2002; Kontogeorgis and Folas 2010). These models are more complex than traditional approaches (e.g., cubic equations of state and activity coefficient models), but result in significantly more accurate predictions for systems that exhibit nonideal behavior, such as polymer solutions and hydrogen bonding fluid mixtures (Kontogeorgis and Folas 2010). Two of the most successful and widely used families of such models are based on Wertheim's first order thermodynamic perturbation theory (Wertheim 1984a, b, 1986a, b) and lattice fluid theory (Sanchez and Lacombe 1976a, b, 1978; Panayiotou and Sanchez 1991). The first family contains all the variants of the SAFT (statistical associating fluid theory) equation of state (Huang and Radosz 1990, 1991). On the other hand, lattice models have been used in industry and academia since the 1970s. Advanced models of both families are able to successfully describe polymer solutions, mixtures of hydrogen bonding fluids, and systems at high pressures, in many cases showing similar performance and limitations (Grenner et al. 2008; Tsivintzelis et al. 2007c, 2008). However, only the lattice models have been applied to describe the plasticization profile of polymer–gas or supercritical fluid mixtures, which is of particular interest in the foaming of amorphous polymers (Condo et al. 1992; Tsivintzelis et al. 2007a).

3.2 NRHB MODEL

3.2.1 OVERVIEW

The NRHB theory is a lattice model that has mainly two versions that were recently developed (Panayiotou et al. 2004, 2007) as an extension of previous successful models (Panayiotou and Vera 1982; Panayiotou and Sanchez 1991). Both account explicitly for hydrogen bonding interactions; however, the simplified one (mentioned here as simplified NRHB) accounts for the nonrandom distribution of empty sites and molecular sites without taking into account the differences in molecular sites occupied by different molecules (Panayiotou et al. 2004; Tsivintzelis et al. 2007a), while the most recent version (Panayiotou et al. 2007; Tsivintzelis et al. 2007c, 2008, 2009a, b; Grenner et al. 2008; Tsivintzelis and Kontogeorgis 2009; Tsioptsias et al. 2010) (mentioned here solely as NRHB) accounts both for the nonrandom distribution of molecular segments of different type molecules as well as for the free volume. This is the model that we will, primarily, focus on in this chapter.

According to the NRHB theory, the molecules are distributed in a three-dimensional quasi lattice of N_r cells, N_0 of which are empty. Every cell (segment) has z neighboring cells. In the general case of a system with $N_1, N_2, \ldots, N_t$ molecules of $1, 2, \ldots, t$ components, in temperature T and pressure P, every molecule of the ith component occupies r_i cells with volume equal to v_i^*. Consequently, the total number of lattice sites in the system is given by the following equation:

$$N_r = N_1 r_1 + N_2 r_2 + \ldots + N_t r_t + N_0 = rN + N_0 = N\left(x_1 r_1 + x_2 r_2 + \ldots + x_t r_t\right) + N_0 \tag{3.1}$$

where:

N is the total number of molecules

x_i the mol fraction of component i

The average per segment interaction energy is

$$\varepsilon_i^* = \left(\frac{z}{2}\right)\varepsilon_{ii} \tag{3.2}$$

where:

ε_{ii} is the interaction energy for every i–i contact

The average intersegmental energy is connected with the traditional scaling parameters of the lattice fluid equation of state (Sanchez and Lacombe 1976a, b) P^* and T^*, by the following equation:

$$\varepsilon^* = RT^* = P^*v^* \tag{3.3}$$

The reduced temperature and pressure are defined by the following relations:

$$\tilde{T} = \frac{T}{T^*}, \quad \tilde{P} = \frac{P}{P^*} \tag{3.4}$$

The number of external contacts of an i type molecule is

$$zq_i = zr_is_i \tag{3.5}$$

where s_i is the characteristic molecular surface-to-volume ratio or

$$s_i = \frac{q_i}{r_i} \tag{3.6}$$

Mixtures are modeled using the one-fluid approach, and the parameters r, q, and s are calculated through the following mixing rules:

$$r = \sum_i x_ir_i \tag{3.7}$$

$$q = \sum_i x_iq_i \tag{3.8}$$

$$s = \frac{q}{r} \tag{3.9}$$

The segment fractions, ϕ_i, and the contact (surface) fractions, θ_i, of component i are defined as

$$\phi_i = \frac{r_iN_i}{rN} = \frac{x_ir_i}{r} \quad i = 1,2,...,t \tag{3.10}$$

$$\theta_i = \frac{q_iN_i}{\sum_k q_kN_k} = \frac{q_iN_i}{qN} = \frac{\phi_is_i}{\sum_k \phi_ks_k} = \frac{\phi_is_i}{s} \tag{3.11}$$

The total number of external contacts in the system, zN_q, accounting also for the empty cells is

$$zN_q = zqN + zN_0 \tag{3.12}$$

For the calculation of the total system volume, it is assumed that every empty cell has the same segmental volume, v^*, with the occupied cells, while two neighboring empty cells remain discrete. Thus, the total volume is

$$V = N\ r\ v^* + N_0 v^* = N_r v^* = V^* + N_0 v^* \tag{3.13}$$

The pivotal point of the theory is the formulation of the partition function, Q, which is assumed that it can be factored into three contributions:

$$Q = Q_R Q_{NR} Q_{HB} \tag{3.14}$$

The first term, Q_R, refers to the random distribution of molecular and empty segments in the quasi lattice; the second term, Q_{NR}, is the correction due to the nonrandom distribution of molecular and empty segments, while the third term, Q_{HB}, accounts for specific interactions, such as hydrogen bonding or Lewis acid–base interactions. By ignoring for the moment contributions due to hydrogen bonding, the isothermal-isobaric partition function can be written as

$$Q(N,P,T) = \Omega_R \Omega_{NR} \exp\left(-\frac{E + PV}{RT}\right) \tag{3.15}$$

where the combinatorial term Ω_R is given by the Staverman approximation (Staverman 1950; Panayiotou et al. 2004, 2007):

$$\Omega_R = \prod_i \omega_i^{N_i} \frac{N_r! \prod_i N_r^{l_i N_i}}{\prod_i N_i!} \left(\frac{N_q!}{N_r!}\right)^{z/2} \tag{3.16}$$

where:

$$l_i = \frac{z}{2}(r_i - q_i) - (r_i - 1) \tag{3.17}$$

and ω_i is a characteristic parameter, which accounts for the symmetry and the flexibility of molecule i. This quantity cancels out in all phase equilibrium calculations.

For the nonrandom correction, the quasi-chemical theory of Guggenheim (1952) is used (Panayiotou and Vera 1981):

$$\Omega_{NR} = \frac{\prod_{i=0}^{t}(N_{ii}^0) \prod_{j>i}^{t}\left[(N_{ij}^0/2)!\right]^2}{\prod_{i=0}^{t}(N_{ii}) \prod_{j>i}^{t}\left[(N_{ij}/2)!\right]^2} \tag{3.18}$$

where:

N_{ij} is the number of external i–j intersegmental contacts

In this equation, i runs from zero (empty site) to the number of components, t, while the superscript 0 refers to the case of random distribution of empty and molecular sites.

As already mentioned, two alternative approaches have appeared in the literature, resulting in two different versions of the model. Initially, Panayiotou et al. (2004) suggested that only the nonrandom distribution of empty sites around a central empty site or a central molecular site should be accounted for. Consequently, the above relation (Equation 3.18) can be simplified as follows:

$$\Omega_{NR} = \frac{N_{rr}^0! N_{00}^0! \left[\left(N_{r0}^0/2 \right)! \right]^2}{N_{rr}! N_{00}! \left[\left(N_{r0}/2 \right)! \right]^2} \tag{3.19}$$

where:

N_{rr} is the number of external contacts between segments belonging to molecules

N_{00} is the number of contacts between empty sites

In the second case (Panayiotou et al. 2007), the full equation (Equation 3.18) is used and the nonrandomness of every i–j contact ($i = 0$ to t and $j \geq i$) is accounted for. These two approaches resulted in two different versions of the theory. In this chapter by NRHB, we refer to the second approach (Panayiotou et al. 2007; Grenner et al. 2008; Tsivintzelis et al. 2007c, 2008, 2009a, b; Tsivintzelis and Kontogeorgis 2009; Tsioptsias et al. 2010).

In order to derive the equations for estimating the number of contacts N_{ij} and N_{00}, we will introduce some new quantities. The site fractions f_0 and f_i, for the empty and the molecular fractions, respectively, are calculated from the following relations:

$$f_0 = \frac{N_0}{N_r} = \frac{N_r - \sum_i r_i N_i}{N_r} = 1 - \sum_i f_i \tag{3.20}$$

The reduced volume is defined as

$$\tilde{v} = \frac{V}{V^*} = \frac{1}{\tilde{\rho}} = \frac{1}{\sum_i f_i} \tag{3.21}$$

while the contact fractions, θ_r and θ_0, are given by

$$\theta_r = 1 - \theta_0 = \frac{q/r}{q/r + \tilde{v} - 1} \tag{3.22}$$

Among contact fractions, the following holds true:

$$\theta_0 + \sum_{i=1}^{t} \theta_i \theta_r = 1 \tag{3.23}$$

In the random case, the contact balance results in the following equations ($i,j > 0$):

$$N_{00}^0 = \frac{1}{2} N_0 z \frac{N_0}{N_q} = \frac{z}{2} N_0 \theta_0 \tag{3.24}$$

$$N_{i0}^0 = zq_i N_i \frac{N_0}{N_q} = zN_0 \frac{q_i N_i}{N_q} = zq_i N_i \theta_0 = zN_0 \theta_0 \theta_r \tag{3.25}$$

$$N_{ii}^0 = \frac{z}{2} q_i N_i \frac{q_i N_i}{N_q} = \frac{z}{2} q_i N_i \theta_i \theta_r \tag{3.26}$$

$$N_{ij}^0 = zq_i N_i \frac{q_j N_j}{N_q} = zq_i N_i \theta_j \theta_r = zq_j N_j \theta_i \theta_r \quad i \neq j \tag{3.27}$$

$$N_{r0}^0 = zqN \frac{N_0}{N_q} = zN_0 \frac{qN}{N_q} = zqN \theta_0 = zN_0 \theta_r \tag{3.28}$$

$$N_{rr}^0 = \sum_i N_{ii}^0 + \sum_i \sum_{j>i} N_{ij}^0 = \frac{1}{2}zqN\frac{qN}{N_0 + qN} = \frac{z}{2}qN\theta_r \qquad (3.29)$$

In order to calculate nonrandom distribution of molecular segments and empty sites, appropriate nonrandom factors Γ are introduced, as explained below:

$$N_{ii} = N_{ii}^0\Gamma_{ii} \quad i = 1,\ldots,t$$

$$N_{ij} = N_{ij}^0\Gamma_{ij} \quad t \geq j > i$$

$$N_{00} = N_{00}^0\Gamma_{00} \qquad\qquad\qquad\qquad (3.30)$$

$$N_{i0} = N_{i0}^0\Gamma_{i0} \quad i = 1,\ldots,t$$

In the case of random distribution, all nonrandom factors, Γ, are equal to one, while in all cases they should obey the following material (contact) balance expressions:

$$\sum_{i=0}^{t}\Theta_i\Gamma_{ij} = 1 \quad j = 0,1,\ldots,t \qquad (3.31)$$

where:
$$\Theta_0 = \theta_0$$
$$\Theta_i = \theta_i\theta_r$$

The number of contacts, N_{ij}, or, equivalently, the nonrandom factors Γ_{ij} are calculated from the following set of minimization conditions:

$$\left(\frac{\partial G}{\partial N_{ij}}\right)_{T,P,N,\tilde{\rho}} = 0 \quad i = 0,1,\ldots,t \quad \text{and} \quad j = i+1,\ldots,t \qquad (3.32)$$

which leads to the following set of equations:

$$\frac{\Gamma_{ii}\Gamma_{jj}}{\Gamma_{ij}^2} = \exp\left(\frac{\Delta\varepsilon_{ij}}{RT}\right) \quad i = 0,1,\ldots,t \quad \text{and} \quad j = i+1,\ldots,t \qquad (3.33)$$

where:

$$\Delta\varepsilon_{ij} = \varepsilon_i + \varepsilon_j - 2(1 - k_{ij})\sqrt{\varepsilon_i\varepsilon_j} \qquad (3.34)$$

and $\varepsilon_0 = 0$. Equations 3.31 and 3.33 form a system of nonlinear algebraic equations, which is solved analytically for pure fluids and numerically for the case of multicomponent mixtures (Abusleme and Vera 1985). However, for the simplified NRHB model (Panayiotou et al. 2004; Tsivintzelis et al. 2007a), the Γ balance $t + 1$ Equations 3.31 is simplified to the following system of two equations:

$$\theta_0\Gamma_{00} + \theta_r\Gamma_{r0} = 1$$

$$\theta_r\Gamma_{rr} + \theta_0\Gamma_{r0} = 1 \qquad\qquad\qquad (3.35)$$

and Equation 3.33 is simplified to the following equation:

$$\frac{4N_{rr}N_{00}}{N_{r0}^2} = \frac{\Gamma_{rr}\Gamma_{00}}{\Gamma_{r0}^2} = \exp\left(\frac{\varepsilon_{rr}}{RT}\right) = \exp\left(\frac{2\varepsilon^*/z}{RT}\right) = A \qquad (3.36)$$

Consequently, for the simplified NRHB model, this system of equations can be analytically solved also in the case of mixtures. Thus, from the quadratic equation (3.36), after some algebra, we obtain for the Γ_{r0} (root with physical meaning):

$$\Gamma_{r0} = \frac{2}{1+\left[1-4\theta_0\theta_r\left(1-A\right)\right]^{1/2}} \tag{3.37}$$

Since the partition function of the system is known, all thermodynamic properties can be calculated. The Gibbs free energy of the system is calculated from the following relation:

$$G = -RT\ln\left[Q\left(N,P,T\right)\right] = -RT\ln\left[\Omega_R\Omega_{NR}\exp\left(-\frac{E+PV}{RT}\right)\right] \tag{3.38}$$

At equilibrium, the density of the system can be calculated by minimization of the Gibbs free energy:

$$\left(\frac{\partial G}{\partial\tilde\rho}\right)_{T,P,N,N_{r0}} = 0 \tag{3.39}$$

which leads to the equation of state for the model:

$$\tilde P + \tilde T\left[\ln\left(1-\tilde\rho\right)-\tilde\rho\sum_i\phi_i\frac{l_i}{r_i}-\frac{z}{2}\ln\left(1-\tilde\rho+\frac{q}{r}\tilde\rho\right)+\frac{z}{2}\ln\Gamma_{00}\right] = 0 \tag{3.40}$$

The chemical potential of component i is obtained from the classical equation:

$$\mu_i = \left(\frac{\partial G}{\partial N_i}\right)_{T,P,N_j,N_{r0},\tilde\rho} \tag{3.41}$$

which, in the case of the simplified NRHB (Panayiotou et al. 2004; Tsivintzelis et al. 2007a), leads to the following equation:

$$\frac{\mu_i}{RT} = \ln\frac{\phi_i}{\omega_i r_i} - r_i\sum_j\frac{\phi_j l_j}{r_j} + \ln\tilde\rho + r_i(\tilde v - 1)\ln(1-\tilde\rho) - \frac{z}{2}r_i\left[\tilde v - 1 + \frac{q_i}{r_i}\right]\ln\left[1-\tilde\rho+\frac{q}{r}\tilde\rho\right]$$

$$+ \frac{zq_i}{2}\left[\ln\Gamma_{rr}+\frac{r_i}{q_i}(\tilde v - 1)\ln\Gamma_{00}\right] + r_i\frac{\tilde P\tilde v}{\tilde T} - \frac{q_i}{\tilde T} + \frac{\tilde P\tilde v}{\tilde T}\left[\frac{rN}{v^*}\frac{\partial v^*}{\partial N_i}\right] - \frac{\Theta_r\Gamma_{rr}}{\tilde T}\left[\frac{qN}{\varepsilon^*}\frac{\partial\varepsilon^*}{\partial N_i}\right] \tag{3.42}$$

In the simplified NRHB, the average intersegmental energy, ε^*, and the segment volume, v^*, of mixtures are calculated using simple combining and mixing rules. The partial derivatives of Equation 3.42 depend on the selected mixing rules and the most common of them are presented in Table 3.1.

In the case of the NRHB model (Panayiotou et al. 2007; Grenner et al. 2008; Tsivintzelis et al. 2007c, 2008, 2009a, b; Tsivintzelis and Kontogeorgis 2009; Tsioptsias et al. 2010), the average segment volume, v^*, is assumed to have a constant value for all fluids (Panayiotou and Vera 1982) equal to 9.75 cm³ mol⁻¹ and, consequently, no mixing rule is needed for mixtures. However, mixing and combining rules are needed for the intersegmental energy, ε^*, which are

$$\varepsilon^* = \sum_{i=1}^t\sum_{j=1}^t\theta_i\theta_j\varepsilon_{ij}^* \tag{3.43}$$

TABLE 3.1

Mixing and Combining Rules for v^* and ε^* of the Simplified NRHB Model

Mixing Rule	Combining Rule	Partial Derivative (Equation 3.42)
$v^* = \sum_i \phi_i v_i^*$		$\dfrac{rN}{v^*}\dfrac{\partial v^*}{\partial N_i} = r_i\left[\dfrac{v_i^*}{v^*} - 1\right]$
$v^* = \sum_i \sum_j \phi_i \phi_j v_{ij}^*$	$v_{ij}^* = \left[\dfrac{v_i^{*(1/3)} + v_j^{*(1/3)}}{2}\right]^3$	$\dfrac{rN}{v^*}\dfrac{\partial v^*}{\partial N_i} = 2r_i\left[\dfrac{\sum_j \phi_j v_{ij}^*}{v^*} - 1\right]$
$\varepsilon^* = \sum_i \sum_j \theta_i \theta_j \varepsilon_{ij}^*$	$\varepsilon_{ij}^* = (1-k_{ij})\sqrt{\varepsilon_i^* \varepsilon_j^*}$	$\dfrac{qN}{\varepsilon^*}\dfrac{\partial \varepsilon^*}{\partial N_i} = 2q_i\left[\dfrac{\sum_j \theta_j \varepsilon_{ij}^*}{\varepsilon^*} - 1\right]$
$\varepsilon^* v^* = \sum_i \sum_j \theta_i \phi_j \varepsilon_{ij}^* v_{ij}^*$	$\varepsilon_{ij}^* = (1-k_{ij})\sqrt{\varepsilon_i^* \varepsilon_j^*}$	$\dfrac{qN}{\varepsilon^*}\dfrac{\partial \varepsilon^*}{\partial N_i}$
		$= q_i\left[\dfrac{\sum_j \phi_j \varepsilon_{ij}^* v_{ij}^*}{\varepsilon^* v^*} - 1\right] + \dfrac{q}{r}r_i\left[\dfrac{\sum_j \theta_j \varepsilon_{ij}^* v_{ij}^*}{\varepsilon^* v^*}\right]$
		$-\dfrac{q}{r}\left[\dfrac{rN}{v^*}\dfrac{\partial v^*}{\partial N_i}\right]$

and

$$\varepsilon_{ij}^* = \sqrt{\varepsilon_i^* \varepsilon_j^*}\left(1-k_{ij}\right) \tag{3.44}$$

where:

k_{ij} is a binary interaction parameter between species i and j and is, usually, fitted to binary experimental data

The equation for the chemical potential of the NRHB model is (Panayiotou et al. 2007):

$$\frac{\mu_i}{RT} = \ln\frac{\phi_i}{\omega_i r_i} - r_i\sum_j \frac{\phi_j l_j}{r_j} + \ln\tilde{\rho} + r_i(\tilde{v}-1)\ln(1-\tilde{\rho})$$

$$-\frac{z}{2}r_i\left[\tilde{v}-1+\frac{q_i}{r_i}\right]\ln\left[1-\tilde{\rho}+\frac{q}{r}\tilde{\rho}\right] \tag{3.45}$$

$$+\frac{zq_i}{2}\left[\ln\Gamma_{ii} + \frac{r_i}{q_i}(\tilde{v}-1)\ln\Gamma_{00}\right] + r_i\frac{\tilde{P}\tilde{v}}{\tilde{T}} - \frac{q_i}{\tilde{T}_i}$$

The expression for the chemical potential of a pure component, μ_i^o, can be obtained from Equations 3.42 or 3.45, for the simplified NRHB and the NRHB, respectively, by setting $x_i = \varphi_i = \theta_i = 1$ and the number of components in the summations equal to 1.

3.2.2 Association Term

The model can account also for specific interactions using a formalism analogous to that of the lattice fluid hydrogen bonding theory (Panayiotou and Sanchez 1991). For a given system with

m proton donor types and n proton acceptor types, let us assume that d_α^k is the number of donor groups of type α in each molecule of type k and c_β^k the number of acceptor groups of type β in each molecule of type k. Furthermore, the total number of hydrogen bonds between a donor of type α and an acceptor of type β in the system is $N_{\alpha\beta}^{\text{hb}}$. The average per molecular segment number of hydrogen bonds in the system, v_{hb}, is

$$v_{\text{hb}} = \sum_\alpha^m \sum_\beta^n v_{\alpha\beta} = \sum_\alpha^m \sum_\beta^n \frac{N_{\alpha\beta}^{\text{hb}}}{rN} = \frac{N^{\text{hb}}}{rN} \tag{3.46}$$

The number of donors of type α per molecular segment, v_d^α, is

$$v_d^\alpha = \frac{N_d^\alpha}{rN} = \frac{\sum_k d_\alpha^k N_k}{rN} \tag{3.47}$$

and the number of acceptors of type β per molecular segment, v_c^β, is

$$v_c^\beta = \frac{N_\alpha^\beta}{rN} = \frac{\sum_k c_\beta^k N_k}{rN} \tag{3.48}$$

The number of nonbonded donors of type α per molecular segment, $v_{\alpha 0}$, is

$$v_{\alpha 0} = v_d^\alpha - \sum_{\beta=1}^n v_{\alpha\beta} \tag{3.49}$$

and similarly the number of nonbonded acceptors of type β per molecular segment, $v_{0\beta}$,

$$v_{0\beta} = v_c^\beta - \sum_{\alpha=1}^m v_{\alpha\beta} \tag{3.50}$$

The free enthalpy of formation of the hydrogen bond of type $\alpha - \beta$, $G_{\alpha\beta}^{\text{hb}}$, is given in terms of hydrogen bonding energy (E), volume (V), and entropy (S) by the equation

$$G_{\alpha\beta}^{\text{hb}} = E_{\alpha\beta}^{\text{hb}} + PV_{\alpha\beta}^{\text{hb}} - TS_{\alpha\beta}^{\text{hb}} \tag{3.51}$$

Following the lattice fluid hydrogen bonding procedure (Panayiotou and Sanchez 1991), the contribution of the hydrogen bonding interactions to the partition function (Equation 3.14) is

$$Q_{\text{HB}} = \Omega_{\text{HB}} \left(\frac{\tilde{\rho}}{rN} \right)^{N^{\text{hb}}} \exp\left(\frac{-\sum N_{\alpha\beta} G_{\alpha\beta}^{\text{hb}}}{RT} \right) \tag{3.52}$$

where:
   Ω_{HB} is the number of different ways of distributing the hydrogen bonds in the system without requiring that donor and acceptor groups be neighbors

The latter requirement is taken into account by the second term of the above equation, which reflects the possibility that two sites are in close proximity. The hydrogen bonding contribution to the Gibbs free energy of the system is

$$G_{\text{HB}} = -RT \ln Q_{\text{HB}} \tag{3.53}$$

The minimization of Equation 3.53 with respect to the reduced density, $\tilde{\rho}$, and to the number of the i type molecules N_i results in hydrogen bonding contribution on the equation of state and the chemical potential, respectively. Consequently, the full equation of state of the model becomes

$$\tilde{P} + \tilde{T}\left[\ln\left(1 - \tilde{\rho}\right) - \tilde{\rho}\left(\sum_i \phi_i \frac{l_i}{r_i} - v_H\right) - \frac{z}{2}\ln\left(1 - \tilde{\rho} + \frac{q}{r}\tilde{\rho}\right) + \frac{z}{2}\ln\Gamma_{00}\right] = 0 \qquad (3.54)$$

while the hydrogen bonding contribution to the chemical potential of component i is given by the expression

$$\frac{\mu_{i,H}}{RT} = r_i v_H - \sum_{\alpha=1}^{m} d_\alpha^i \ln\frac{v_d^\alpha}{v_{\alpha 0}} - \sum_{\beta=1}^{n} c_\beta^i \ln\frac{v_c^\beta}{v_{0\beta}} \qquad (3.55)$$

Finally, the minimization of Equation 3.53 with respect to the number of each type of hydrogen bonds results in the following general equations:

$$\frac{v_{\alpha\beta}}{v_{\alpha 0}v_{0\beta}} = \tilde{\rho}\exp\left(\frac{-G_{\alpha\beta}^{hb}}{RT}\right) \text{ for all }\left(\alpha,\beta\right) \qquad (3.56)$$

The formalism presented in this section is general and allows for the modeling of intermolecular hydrogen bonding in systems with any number of proton donors and proton acceptors.

3.2.3 GLASS TRANSITION TEMPERATURE

The glass transition temperature is calculated using the Gibbs–DiMarzio criterion, according to which the entropy of the system is zero at the glass transition temperature (Panayiotou et al. 2004). When the partition function of the system is known, the entropy can be subsequently estimated. For the simplified NRHB model, the entropy of a nonhydrogen bonding mixture is calculated through the following relation:

$$\frac{S}{rNR} = \sum_i \frac{\varphi_i}{r_i}\ln\delta_i + \left(1 - \tilde{v}\right)\ln\left(1 - \tilde{\rho}\right) + \frac{l + \ln(r\tilde{v}) - \sum_i x_i \ln x_i}{r}$$

$$+ \frac{z}{2}\left[\tilde{v} - 1 + \frac{q}{r}\right]\ln\left[1 - \tilde{\rho} + \frac{q}{r}\tilde{\rho}\right] + \frac{zq}{2r}\left[\theta_0\Gamma_{r0}\frac{2\varepsilon^*}{zRT} - \ln\Gamma_{rr}\right] - \frac{z}{2}\left(\tilde{v} - 1\right)\ln\Gamma_{00} \qquad (3.57)$$

where

$$l = \sum_i x_i l_i \qquad (3.58)$$

and δ_i is the number of different conformations of the polymer chain and is calculated through the following equation (Panayiotou et al. 2007):

$$\ln\delta_i = \ln\left(\frac{Z}{2}\right) - \left(r_i - 2\right)\ln\left(1 - f_i\right) + f_i\left(r_i - 2\right)\frac{u_i}{RT} \qquad (3.59)$$

where:
   Z is the number of discrete conformations available to each bond
   f_i is the fraction of the $r - 2$ bonds of the macromolecule that are in high-energy level and is
      calculated by the equation:

$$f_i = \frac{(Z-2)\exp(-u_i/RT)}{1+(Z-2)\exp(-u_i/RT)} \tag{3.60}$$

Parameter u_i represents the increase in the intramolecular energy due to the *flexing* of the bond in the chain molecule and is a characteristic parameter of each fluid that is calculated from zeroing the entropy of the pure polymer in the glass transition temperature at ambient pressure. Once u_i is estimated, the glass transition of the pure polymer in other pressures or the glass transition temperature of polymer–gas (supercritical fluids) systems can be predicted. The corresponding equations for NRHB are somewhat more complex but do not offer any substantial improvement over Equations 3.57 through 3.60 of the simplified NRHB and are not reported here.

3.2.4 SOLUBILITY PARAMETERS

In the NRHB model, polar and dispersive interactions are treated together and are characterized as physical interactions, in contrast to the hydrogen bonding or other specific interactions. Thus, it is not possible to estimate separately the polar and the dispersive partial solubility parameters. For this task, a more general version of the model was developed, which will not be used here, but the reader is referred to the corresponding publication (Stefanis et al. 2006). However, the partial hydrogen bonding and the total solubility parameter can be directly calculated using the NRHB model described here.

The cohesive energy of a compound is the sum of the potential energy due to physical (dispersive and polar interactions), E_{ph}, and the energy contribution due to hydrogen bonding interactions, E_{hb}:

$$E_{coh} = E_{ph} + E_{hb} \tag{3.61}$$

For pure fluids, these quantities are calculated from the following equations:

$$E_{ph} = \sum_i N_{ii}\varepsilon_{ii} = \Gamma_{rr}qN\theta_r\varepsilon^* \tag{3.62}$$

$$E_{hb} = -\sum_\alpha \sum_\beta N_{\alpha\beta}^{hb}E_{\alpha\beta}^{hb} = -rN\sum_\alpha \sum_\beta \nu_{\alpha\beta}E_{\alpha\beta}^{hb} \tag{3.63}$$

The volume of the system is given by

$$V = rN \, \tilde{v} \, v^* + \sum_\alpha \sum_\beta N_{\alpha\beta}^{hb}V_{\alpha\beta}^{hb} \tag{3.64}$$

For simplicity, the volume change upon formation of a hydrogen bond, $V_{\alpha\beta}^{hb}$, is set equal to zero. With the above definitions, the total solubility parameter and the partial hydrogen bonding solubility parameter are given by Equations 3.65 and 3.66, respectively:

$$\delta = \sqrt{\delta_d + \delta_p + \delta_{hb}} = \sqrt{\frac{E_{coh}}{V}} = \sqrt{\frac{\Gamma_{11}q\Theta_r\varepsilon^* - r\sum_\alpha \sum_\beta \nu_{\alpha\beta}E_{\alpha\beta}^{hb}}{r\tilde{v}\,v^*}} \tag{3.65}$$

and

$$\delta_{hb} = \sqrt{\frac{E_{hb}}{V}} = \sqrt{\frac{-r\sum_\alpha \sum_\beta \nu_{\alpha\beta}E_{\alpha\beta}^{hb}}{r\tilde{v}v^*}} \tag{3.66}$$

3.2.5 SCALING CONSTANTS

In order to model real systems with the NRHB theory or other similar equation-of-state models, the knowledge of characteristic pure fluid parameters is required. For low molecular weight compounds, such pure fluid parameters are estimated by fitting the predictions of the model to some pure fluid experimental data, usually vapor pressures and liquid densities. However, since high polymers do not exhibit any detectable vapor pressure, extended pressure-volume-temperature (PVT) data are used instead.

NRHB requires the knowledge of three scaling constants for each pure fluid, while in cases of associating compounds, it also requires knowledge of two additional parameters that characterize the specific association, namely, the association energy, $E_{\alpha\beta}^{hb}$, and the association entropy, $S_{\alpha\beta}^{hb}$ (as mentioned, for simplicity, the association volume, $V_{\alpha\beta}^{hb}$, is set equal to zero).

For the simplified NRHB model (Panayiotou et al. 2004; Tsivintzelis et al. 2007a), the three scaling constants are the mean interaction energy per segment, ε_i^*; the hardcore volume per segment, v_i^*; and the hardcore density, $\rho_i^* = 1/v_{sp,i}^*$. On the other hand, in the NRHB model (Panayiotou et al. 2007; Grenner et al. 2008; Tsivintzelis et al. 2007c, 2008, 2009a, b; Tsivintzelis and Kontogeorgis 2009; Tsioptsias et al. 2010), the hardcore volume per segment, v^*, is constant and equal to 9.75 cm³ mol⁻¹ for all fluids. The first two scaling parameters, ε_h^* and ε_s^*, are used for the calculation of the mean interaction energy per molecular segment, ε^*, according to the following equation:

$$\varepsilon^* = \varepsilon_h^* + \left(T - 298.15\right)\varepsilon_s^* \tag{3.67}$$

The third scaling parameter, $v_{sp,0}^*$, is used for the calculation of the close-packed density, $\rho^* = 1/v_{sp}^*$, as described by the following equation:

$$v_{sp}^* = v_{sp,0}^* + \left(T - 298.15\right)v_{sp,1}^* \tag{3.68}$$

Parameter $v_{sp,1}^*$ in Equation 3.68 is treated as a constant for a given homologous series. It is equal to $-0.412 \cdot 10^{-3}$ cm³ g⁻¹ K⁻¹ for nonaromatic hydrocarbons, $-0.310 \cdot 10^{-3}$ cm³ g⁻¹ K⁻¹ for alcohols, $-0.240 \cdot 10^{-3}$ cm³ g⁻¹ K⁻¹ for acetates, $-0.300 \cdot 10^{-3}$ cm³ g⁻¹ K⁻¹ for water, and $0.150 \cdot 10^{-3}$ cm³ g⁻¹ K⁻¹ for all the other fluids (Tsivintzelis and Kontogeorgis 2009).

For both versions of the models, the following relation holds: $r = MW\, v_{sp}^*/v^*$, where MW stands for molecular weight. Furthermore, the geometric factor s_i, which is defined as the ratio of molecular surface to molecular volume, $s = zq/zr = q/r$, is calculated from UNIFAC group contribution method (Panayiotou et al. 2004).

3.2.6 MIXTURE PARAMETERS

Having the scaling parameters for every fluid of interest, the model can predict all basic thermodynamic properties of pure fluids and mixtures. As already mentioned, parameters for mixtures are calculated through mixing and combining rules (see Table 3.1 and Equations 3.43 and 3.44 for the scaling parameters of the two versions of the model, respectively). However, combining rules are also used for estimating the cross-hydrogen bonding parameters in cases where no experimental or predicted (using, for example, ab initio calculations) data exist. More specifically, the following combining rules were used for the cross interaction between two self-associating groups:

$$E_{\alpha\beta}^{hb} = \frac{E_{\alpha\alpha}^{hb} + E_{\beta\beta}^{hb}}{2}, \quad S_{\alpha\beta}^{hb} = \left(\frac{S_{\alpha\alpha}^{hb\,1/3} + S_{\beta\beta}^{hb\,1/3}}{2}\right)^3 \tag{3.69}$$

while for the cross interaction between one self-associating and one nonself-associating group the combining rules were

$$E_{\alpha\beta}^{hb} = \frac{E_{\alpha\alpha}^{hb}}{2}, \quad S_{\alpha\beta}^{hb} = \frac{S_{\alpha\alpha}^{hb}}{2} \tag{3.70}$$

3.3 APPLICATIONS TO SYSTEMS RELEVANT TO THE PRODUCTION OF POROUS POLYMERS

As mentioned in the introductory Section 3.1, many methods for preparing porous polymer structures are based on phase separation phenomena. Thus, the phase behavior of polymer–fluid systems is of particular interest in designing such processes. This section is divided into three parts, in which representative applications of the model are presented. In the first one and in view of the importance of solubility parameters in the thermodynamic treatments of these systems, predictions of the NRHB model for the solubility parameters of low and high molecular weight compounds and for the PVT properties of some biocompatible polymers will be shown. In the second part, some characteristic applications of the model for the phase behavior of systems that contain polymers, liquid solvents, pharmaceutical compounds, and gases or supercritical fluids will be reported. Finally, in the third part, a characteristic application of the model to the foaming of polymers with supercritical fluids will be described. The NRHB model (Panayiotou et al. 2007; Grenner et al. 2008; Tsivintzelis et al. 2007c, 2008, 2009a, b; Tsivintzelis and Kontogeorgis 2009; Tsioptsias et al. 2010) is used for the applications of the first two parts, while the simplified NRHB model (Panayiotou et al. 2004; Tsivintzelis et al. 2007a) is used for the third part since it captures the essentials with much simplicity.

All pure fluid and association parameters that were used for the calculations of the first two parts are shown in Tables 3.2 through 3.4, while the parameters for the simplified NRHB model of the third part were adapted from Tsivintzelis et al. (2007a).

TABLE 3.2
NRHB Parameters for the Fluids That Were Used in This Study

Fluid	ε_h^* (J/mol)	ε_s^* (J/molK)	$v_{sp,0}^*$ (cm³/g)	s	Reference
Ethylene glycol (EG)	5,800.9	0.86530	0.8424	0.933	Tsivintzelis and Kontogeorgis (2009)
1,2 Fropylene glycol (PG)	5,088.9	1.05260	0.9084	0.903	Tsivintzelis and Kontogeorgis (2009)
Diethylene glycol (DEG)	5,372.9	1.91910	0.8384	0.892	Tsivintzelis and Kontogeorgis (2009)
Methanol	4,202.3	1.52690	1.15899	0.941	Grenner et al. (2008)
Ethanol	4,378.5	0.75100	1.15867	0.903	Grenner et al. (2008)
1-Propanol	4,425.6	0.87240	1.13923	0.881	Grenner et al. (2008)
1-Butanol	4,463.1	1.19110	1.13403	0.867	Grenner et al. (2008)
2-Butanol	4,125.1	1.45711	1.11477	0.867	Tsivintzelis and Kontogeorgis (2009)
1-Hexanol	4,522.7	1.64571	1.11545	0.850	This work

(Continued)

TABLE 3.2 (Continued)
NRHB Parameters for the Fluids That Were Used in This Study

Fluid	ε_h^* (J/mol)	ε_s^* (J/molK)	$v_{sp,0}^*$ (cm³/g)	s	Reference
1-Octanol	4,532.1	1.86863	1.12094	0.839	This work
1-Propylamine	4,275.0	0.06561	1.18720	0.891	Grenner et al. (2008)
Acetone	4,909.0	−1.15000	1.14300	0.908	Grenner et al. (2008)
Methyl ethyl ketone	4,809.6	−0.58530	1.11659	0.885	Tsivintzelis and Kontogeorgis (2009)
Propyl acetate	4,491.3	0.68180	1.00634	0.880	Grenner et al. (2008)
Water	5,336.5	−6.5057	0.97034	0.861	Grenner et al. (2008)
Carbon dioxide	3,468.4	−4.5855	0.79641	0.909	Tsivintzelis et al. (2009b)
n-Octane	4,105.3	1.8889	1.23687	0.844	This work
4-Heptanone	4,605.0	0.54282	1.0787	0.853	This work
Ibuprofen	4,837.1	−7.6536	0.8236	0.784	Tsivintzelis et al. (2009b)
Poly(ethylene glycol) $200 < MW < 100,000$ g mol⁻¹	6,273.0	0.71110	0.86620— Mw·7.98·10⁻⁰⁸	0.829	Tsivintzelis and Kontogeorgis (2009)
Poly(ethylene glycol) $MW > 100,000$ g mol⁻¹	6,273.0	0.71110	0.85820	0.829	Tsivintzelis and Kontogeorgis (2009)
Poly(vinyl acetate)	5,970.4	2.59194	0.80924	0.825	Tsivintzelis and Kontogeorgis (2009)
Linear polyethylene	7,434.9	−7.94296	1.20994	0.800	This work
Poly(methyl methacrylate)	4,594.4	11.18424	0.78411	0.843	This work
Poly(n-butyl methacrylate)	3,967.7	10.6393	0.87151	0.828	This work

TABLE 3.3
Association Parameters

HB Groups	E^{hb} (J/mol)	S^{hb} (J/molK)	Reference
–OH....OH– (alkanols)	−24,000	−27.5	Grenner et al. (2008)
–OH....OH– (methanol)	−25100	−26.5	Grenner et al. (2008)
–OH....OH– (glycols)	−22500	−27.5	Grenner et al. (2008)
–NH₂...–NH₂	−11100	−20.6	Tsivintzelis and Kontogeorgis (2009)
HOH...HOH	−16100	−14.7	Grenner et al. (2008)
HOH.... –O–	−17000	−27.0	Tsivintzelis and Kontogeorgis (2009)
–OH....–O– (glycols)	−14900	−27.8	Tsivintzelis and Kontogeorgis (2009)
HOH....–OH–	Combining rule (Equation 3.69)	Combining rule (Equation 3.69)	Tsivintzelis and Kontogeorgis (2009)
–OH....O=C–O–	−22,100	−27.3	Tsivintzelis et al. (2009b)
–COOH....HOOC– (ibuprofen)	−16,380	−20.3	Tsivintzelis et al. (2009b)

TABLE 3.4
Association Sites on Hydrogen Bonding Groups

	HB Sites	
Group	Proton Donors	Proton Acceptors
–OH	1	1
–O–	0	2
HOH	2	2
–O–C=O	0	1
–NH$_2$	2	1
–COOH (ibuprofen)	1	1

3.3.1 PURE FLUIDS PROPERTIES—SOLUBILITY PARAMETERS

In many cases, especially in applications relevant to the preparation of porous membranes with the immersion precipitation method (addition of a nonsolvent), the screening of various solvent/nonsolvent systems is performed using the solubility parameter approach. For estimating the solubility parameters, some solubility data of the investigated compound in various solvents are usually needed (Hansen 1999). However, such parameters can be predicted using thermodynamic models (Stefanis et al. 2006). Table 3.5

TABLE 3.5
NRHB Predictions and Literature Values for the Partial Hydrogen Bonding and the Total Solubility Parameters of Some Characteristic Fluids (NRHB Pure Fluid Parameters from Tables 3.2 and 3.3)

	δ^{hb} (MPa$^{1/2}$)		δ_{total} (MPa$^{1/2}$)	
Fluid	Literature (Experimental or Predicted)	NRHB	Literature (Experimental or Predicted)	NRHB
Water	42.3	40.3	47.8	45.5
Ethylene glycol	26.0	26.6	33.0	35.4
1,2 Propylene glycol	23.3	23.4	30.2	31.0
Diethylene glycol	23.3	23.4	30.2	31.0
Methanol	22.3	23.6	29.6	29.5
Ethanol	19.4	19.0	26.1	26.1
1-Propanol	17.4	16.7	24.4	24.4
1-Butanol	15.8	15.0	23.2	23.3
2-Butanol	14.5	14.9	22.2	22.4
n-Propylamine	8.6	8.8	19.6	19.3
Acetone	6.95	–	19.9	19.2
Methyl ethyl ketone	5.1	–	19.0	18.7
Ibuprofen	7.2	7.2	19.5	19.4
Polyethylene		–		24.0
Poly(methyl-methacrylate)		–	17.8–22.8	18.0
Poly(butyl-methacrylate)		–	17.8	16.0

Sources: Hansen, C.M., *Hansen Solubility Parameters, A User's Handbook*, CRC Press, Boca Raton, FL, 1999; Bustamante, P. et al., *Int. J. Pharm.*, 194, 117–124, 2000.

presents the NRHB predictions and literature values for the partial hydrogen bonding and the total solubility parameter of some characteristic solvents and polymers.

Usually, the literature experimental and calculated solubility parameter values refer to 298 K and ambient pressure. However, many polymer processes, such as the production of porous polymer membranes via the immersion precipitation method, are performed at various temperatures, since temperature is a key operational variable that can be tuned in order to obtain the required final product properties. In addition, elevated pressures are needed in various supercritical fluid applications, such as the immersion precipitation using a supercritical fluid as a nonsolvent. In such cases, it is very useful to have the solubility parameters of the process fluids predicted by appropriate thermodynamic models. Figure 3.7a presents the NRHB predictions for the hydrogen bonding component and the total solubility parameter of water, while Figure 3.7b presents the variation of the total solubility parameter for supercritical CO_2 with pressure for five isotherms.

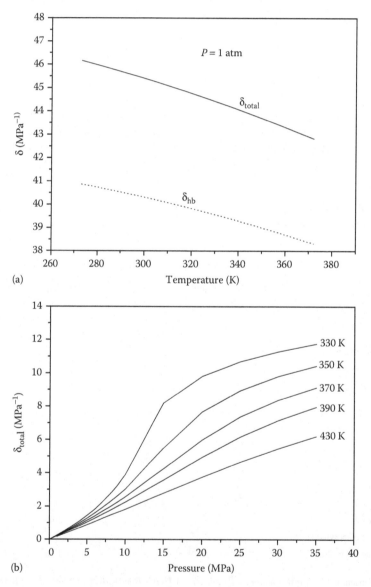

FIGURE 3.7 NRHB predictions for the solubility parameters for (a) water (total and hydrogen bonding component) and (b) supercritical CO_2 (total).

Solubility parameters find numerous applications, especially in the pharmaceutical industry for selecting solvents, characterizing surfaces, and estimating polymer compatibility (Stefanis et al. 2006). However, despite their extended use, which is mainly based on the simplicity of the corresponding approach, the thermodynamic framework behind them is based on the rather arbitrary principle that *like dissolve like*. Very recently, the whole concept of solubility parameter has been reconsidered and its key limitations have been removed (Panayiotou 2012a, b, c, 2013). An alternative approach, the partial solvation parameter, has been proposed that expands significantly the capacity of the widely used partial or Hansen solubility parameter approach. For space limitations, however, this new partial solvation parameter approach is not reviewed here.

Besides solubility parameters, the NRHB model can describe all basic thermodynamic properties of pure fluids. Figure 3.8 presents the description by the NRHB model of the specific volume of two

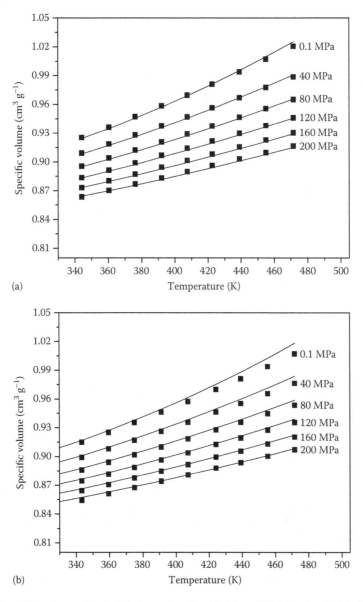

FIGURE 3.8   Specific volume of poly(ethylene glycol) with $M_n = 600$ (a) and poly(ethylene oxide) with $M_n = 100,000$ (b). Experimental data (symbols) and NRHB calculations (lines).

characteristic high molecular weight compounds often used in bioengineering applications. The first plot refers to a hydrogen bonding compound—poly(ethylene glycol) with an average molecular weight equal to 600—while the second one to a polymer of the same family—poly(ethyl oxide)— with high molecular weight. Model predictions were obtained using NRHB pure fluid parameters of Table 3.2, while for PEG-600 the inter-chain hydrogen bonding interactions were explicitly accounted for using the parameters of Tables 3.3 and 3.4.

3.3.2 PROPERTIES OF MIXTURES

The aforementioned applications refer to thermodynamic properties of pure fluids. However, the strong advantage of the equation-of-state thermodynamic models is their ability to predict properties of mixtures over extended ranges of external conditions. Predictions are feasible if the pure fluid parameters, mentioned in the previous sections, are known. Next in this section, some representative calculations for mixtures will be reported. In all cases, unless otherwise mentioned, the calculations were performed using the pure fluid and hydrogen bonding parameters presented in Tables 3.2 and 3.3.

Figure 3.9 presents the NRHB predictions for the sorption of various solvent vapors in poly(vinyl acetate). The model satisfactorily describes vapor–liquid equilibrium of this system without the use of any parameters adjusted to the binary experimental data (pure predictions). The predictions are accurate for, both, nonassociating systems (systems with ketones or propyl-acetate) and the 2-butanol-poly(vinyl acetate) (PVAC) mixture, where self-association (—OH...HO—between 2-butanol molecules) and cross-association interactions (—OH...O=C—O—between 2-butanol molecules and PVAC chain groups) occur (Tsivintzelis and Kontogeorgis 2009). Model predictions for vapor–liquid equilibrium are accurate even for systems with more complex hydrogen bonding behavior as the polyethylene glycol–water mixture shown in Figure 3.10. By explicitly accounting for all self- and cross-association interactions (Tsivintzelis and Kontogeorgis 2009), the NRHB model very accurately predicts the sorption of water in two polymer samples of different molecular weight.

However, for most systems and for the accurate description of more complex phase behavior, such as the liquid–liquid demixing, typically one binary interaction parameter (k_{ij}) is adjusted using experimental data for the binary system. Figure 3.11 presents the liquid–liquid equilibrium of two binary polymer systems containing PMMA or poly(butyl-methacrylate) (PBMA) and low

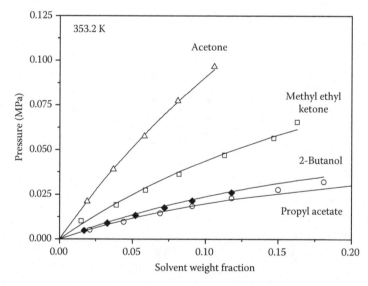

FIGURE 3.9 PVAC–solvent VLE. Experimental data (symbols) and NRHB predictions ($k_{ij} = 0$, lines) for the sorption of acetone, methyl ethyl ketone, 2-butanol, and propyl acetate in PVAC.

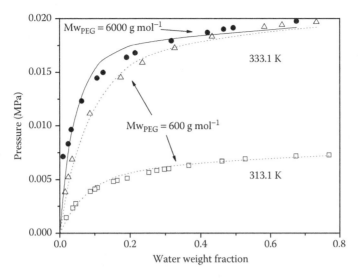

FIGURE 3.10 PEG–water VLE. Experimental data (symbols) and NRHB predictions ($k_{ij} = 0$, lines) for the sorption of water in two PEG samples of different average molecular weight.

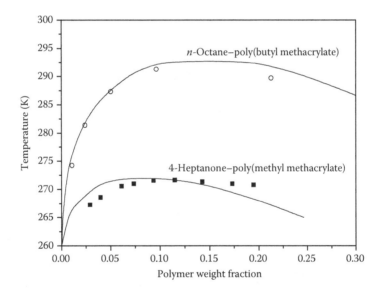

FIGURE 3.11 4-Heptanone–PMMA and n-octane–PBMA LLE. Experimental data (symbols) and NRHB correlations ($k_{ij} = -0.014$ for 4-heptanone–PMMA and $k_{ij} = 0.018$ for n-octane–PBMA).

molecular weight solvents. No hydrogen bonding interactions occur in these systems, and the use of one temperature independent adjustable parameter (the binary interaction parameter, k_{ij}) is sufficient for the accurate description of the experimental data. The liquid–liquid equilibrium of high density linear polyethylene mixtures with n-hexanol and n-octanol is presented in Figure 3.12. Such systems are also accurately described using one temperature independent binary interaction parameter (k_{ij}).

Recently, numerous novel processes involving supercritical fluids were proposed for the encapsulation of drugs and pharmaceuticals in polymer matrices (i.e., polymer nanoparticles or tissue engineering porous polymer scaffolds) for controlled release applications. In this direction, the impregnation of polymers with drugs is based on the balance between the drug solubility in the supercritical fluid and its affinity to the polymer matrix. Figure 3.13a presents the

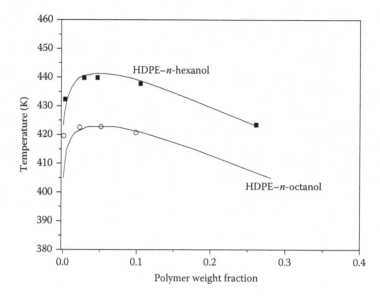

FIGURE 3.12 HDPE–*n*-hexanol and HDPE–*n*-octanol LLE. Experimental data (points) and NRHB calculations (lines) using one binary interaction parameter ($k_{ij} = -0.0230$ for *n*-hexanol and $k_{ij} = -0.0248$ for *n*-octanol).

solubility of the optical isomer, S(+), and the racemic mixture of ibuprofen in supercritical CO_2, using the approach of Tsivintzelis et al. (2009a), according to which both ibuprofen samples were modeled using the same pure fluid parameters and only the values (shown in Table 3.6) of the fusion properties are different, since optical isomers and their racemic mixture usually present different crystalline structures. Using one binary interaction parameter, the model can describe such solid–liquid equilibrium (SLE) in a rather satisfactory way, considering the nonideality of the mixture.

In many drug micronization or polymer impregnation processes, the pressure required to melt the solute and produce a saturated liquid solution is important. Such pharmaceutical–supercritical fluid systems exhibit solid–gas (fluid) equilibrium at low pressures, but as the pressure increases, there is a point where three phase equilibria (solid–liquid–fluid) is observed. The *P–T* projection of the solid–liquid–fluid curve is often called the melting point depression curve. Tsivintzelis et al. (2009a) used the binary interaction parameter for CO_2–ibuprofen, which was obtained from the solubility (SLE) data and applied the model to predict the melting point depression curve. The NRHB predictions for S(+)-ibuprofen are presented in Figure 3.13b, where it can be seen that the model rather satisfactorily predicts the melting point behavior of this system. It is worth mentioning that one point on this curve corresponds to a four-phase solid–liquid 1–liquid 2–gas (SL_1L_2G) equilibrium (De Loos 2006). The NRHB prediction for this point of four phase equilibria for the CO_2–S(+)-ibuprofen system is shown in Figure 3.13b, while in Figure 3.14 the characteristic diagram of the Gibbs free energy of mixing (ΔG^m) against the mixture composition is presented. The tangent line, which defines the points that correspond to the compositions of the four phases in equilibrium, is also shown in the same figure.

3.3.3 Case Study: Foaming of Polymers with CO_2 as Blowing Agent

In the previous paragraphs, some characteristic applications of the NRHB model to systems relevant to polymer and pharmaceutical applications were shown. Next, an application of the model to systems relevant to the foaming of polymers with supercritical CO_2 will be presented.

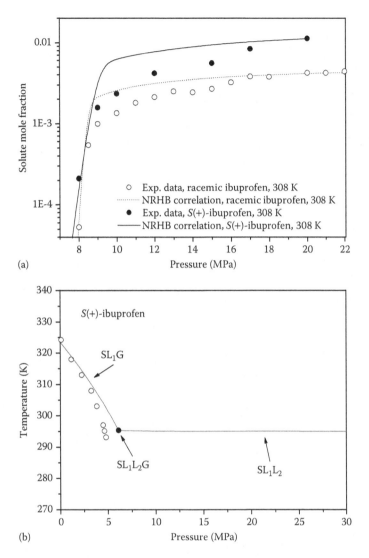

FIGURE 3.13 $S(+)$-ibuprofen–CO_2 phase behavior. Solid–gas equilibrium (a) and pressure–temperature projection of the SLG equilibrium line (b). Experimental data (symbols) and NRHB predictions (lines) ($k_{ij} = 0.1080$ optimized in solubility data).

TABLE 3.6

Melting Temperature (T_m), Enthalpy of Fusion (ΔH_f), Difference between the Heat Capacities in Liquid and Solid State (ΔC_p) and Solid Molar Volume (V_{solid}) for Ibuprofen

Pharmaceutical	T_m (K)	ΔH_f (J/mol)	ΔC_p (J/molK)	V_{solid} (cm^3/mol)
Racemic ibuprofen	347.2[a]	23,100[a]	50.3[b]	175[a]
$S(+)$-ibuprofen	323.5[a]	15,400[a]	50.3[c]	175[a]

Sources: [a]Perlovich, G.L. et al., *J. Pharm. Sci.*, 93, 654–665, 2004.

[b] Gracin, S. and Rasmuson, A.C., *J. Chem. Eng. Data*, 47, 1379–1383, 2002.

[c] The value for the racemic compound was used.

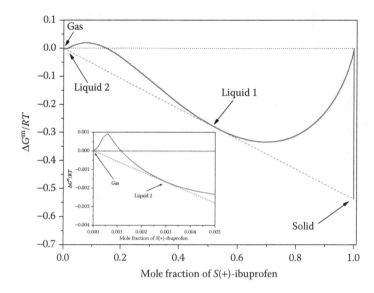

FIGURE 3.14 Gibbs energy of mixing (ΔG^m) for CO_2–$S(+)$-ibuprofen at the SL_1L_2G equilibrium point.

Tsivintzelis et al. (2007a) used the NRHB theory in order to model the foaming of polystyrene using CO_2 as blowing agent and finally to correlate the observed pore population density of the porous polymer matrix against the pressure and temperature saturation conditions, assuming very rapid depressurization rate. The simplified NRHB model was used. As already mentioned, no major differences are expected when using the newer NRHB version. Here, we will present calculations for a more expanded range of temperature and pressure conditions. All calculations were obtained with the pure fluid parameters reported by Tsivintzelis et al. (2007a).

The first step was the combination of the classical nucleation theory with the NRHB model. According to the former theory and assuming homogenous nucleation, the free energy change related to the formation of a nuclei of the new phase is a function of the energy gain due to the thermodynamically favored phase separation and the energy loss due to the introduction of the interface (Gabbard 2002):

$$\Delta G = -V\,\Delta P + A\gamma = -\frac{4\pi\,r^3}{3}\Delta P + 4\pi\,r^2\gamma \tag{3.71}$$

where:

V and A are the volume and surface, respectively, of the new phase nuclei
r is the radius of this spherical cluster
γ is the interfacial tension
ΔP is the pressure difference between the two sides of the interface

The energy barrier for nucleation, ΔG^*_{hom}, and the critical radius of a stable nuclei, r_c, correspond to the maximum of the ΔG function (Colton and Suh 1987; Goel and Beckman 1994a; Gabbard 2002):

$$\frac{d\Delta G}{dr} = 0 \tag{3.72}$$

from which we have

$$r_c = \frac{2\gamma}{\Delta P} \tag{3.73}$$

$$\Delta G *_{\text{hom}} = \frac{16\pi\gamma^3}{3\Delta P^2}$$ (3.74)

The rate of nucleation is described by the following equation (Goel and Beckman 1994a):

$$N_o = C_o f_o \exp\left(\frac{-\Delta G *_{\text{hom}}}{kT}\right)$$ (3.75)

where:

C_o is the concentration of the dissolved fluid in the polymer

T is the temperature

k is the Boltzmann factor

f_o is a characteristic frequency factor (Gabbard 2002)

The total number of stable nuclei that are formed in the system is obtained through the following integration (Goel and Beckman 1994a):

$$N_{\text{total}} = \int_0^{t,\text{vitr}} N_o dt = \int_{P,\text{sat}}^{P,\text{vitr}} N_o \frac{dP}{dP/dt}$$ (3.76)

The integration starts at $t = 0$ (starting point of the depressurization) and ends at the point where the porous structure is stabilized, which for amorphous polymers below the neat polymer glass transition temperature is the moment that the polymer–CO_2 system vitrifies (t_{vitr}). In terms of pressure, the corresponding integration limits are the saturation pressure (P_{sat}) and the vitrification pressure (P_{vitr}). The term dP/dt corresponds to the depressurization rate and usually for theoretical calculations is kept constant, despite the corresponding experimental difficulties on performing a really rapid depressurization with constant rate.

In order to calculate the energy barrier for nucleation, the critical nuclei radius and the number of stable nuclei that are formed in the system, Tsivintzelis et al. (2007a) used the simplified NRHB model to estimate the concentration of the gas inside the polymer matrix as a function of pressure and temperature, the vitrification pressure–temperature profile of the PS–CO_2 system and the surface tension of the neat polymer as a function of temperature (Panayiotou 2003). The surface tension of the polymer—CO_2 mixture was calculated, for simplicity, using the following empirical equation (Reid et al. 1988):

$$\gamma_{\text{mix}}^{1/r} = (1 - w_{CO_2})\gamma_{\text{pol}}^{1/r}$$ (3.77)

where:

γ is the surface tension

w_{CO_2} is the weight fraction of the dissolved CO_2 in the polymer matrix

Subscripts mix and pol denote mixture and neat polymer, respectively

The parameter r in Equation 3.77 was adjusted using the available literature experimental data (Park et al. 2006) and it was estimated to be equal to 13.

Using this approach with pure fluid the parameters from Tsivintzelis et al. (2007a) and a binary interaction parameter equal to −0.074, the NRHB predictions for the sorption of CO_2 in polystyrene is shown in Figure 3.15, while Figure 3.16 shows the plasticization behavior of this binary mixture. The model accurately describes the sorption of CO_2 in the polymer matrix and, at the same time, satisfactorily predicts the plasticization behavior (of type II according to the classification of Condo et al. 1992; see Figure 3.5) of this system. In Figure 3.16 are shown the experimental data (solid symbols) and the model predictions (solid line) for the glass transition temperature. However, on the same plot, the values predicted by the model sorption isopleths (dot lines) and the

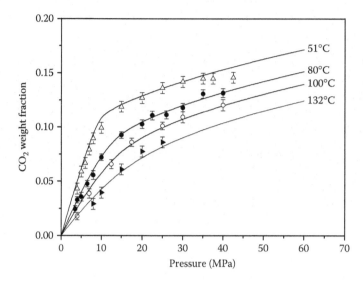

FIGURE 3.15 Sorption of CO_2 in polystyrene. Experimental data (points) and NRHB calculations (lines) using a k_{ij} equal to −0.074.

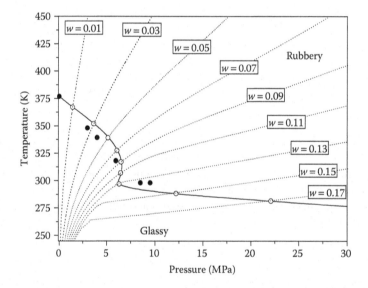

FIGURE 3.16 The plasticization behavior of CO_2–polystyrene system. Experimental data (solid symbols) and NRHB predictions (solid line) using a k_{ij} equal to −0.074 (open symbols show the glass transition point on every calculated sorption isopleth (dot lines), w is the weigh fraction of CO_2 on every sorption isopleth).

points on those isopleths, where the model predicts the glass-to-rubber transition (open symbols), are also shown. The model predictions for the glass transition temperature are based on the Gibbs–DiMarzio criterion, according to which the entropy of the system is zero at the glass transition temperature.

Using the model predictions for the equilibrium weight fraction of CO_2 in the polystyrene matrix as input in Equation 3.77, one can calculate the surface tension of this binary polymer–sorbed gas binary system for all pressure and temperature conditions of interest to the foaming process. Figure 3.17 shows the predictions for the interfacial tension, where an interesting phenomenon is revealed. At low pressures, the interfacial tension decreases with increase of temperature, as expected.

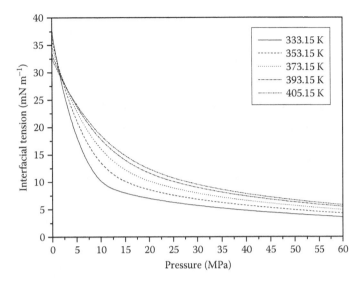

FIGURE 3.17 Surface tension of a polystyrene matrix with dissolved CO_2.

However, for relatively higher pressures, the surface tension increases with increase of temperature indicating that the reduction of CO_2 sorption due to the temperature increase has here a dominant effect.

Using the calculations of Figures 3.15 through 3.17 as input for Equation 3.74, the energy barrier for homogeneous nucleation can be estimated and the results are presented in Figure 3.18. At constant temperature, the energy barrier for nucleation, ΔG^*_{hom}, decreases exponentially with increase of pressure. In other words, stable nuclei are more easily generated inside the metastable polymer matrix as the depressurization of the system starts from a higher saturation pressure. This can explain the experimentally observed increase of cell population density with increase of the saturation pressure (Goel and Beckman 1994a; Tsivintzelis et al. 2007a). The energy barrier is very high for relatively low saturation pressures, which can explain the difficulty in producing homogeneous porous polymer structures at such conditions (Goel and Beckman 1994a, b). On the other hand, at

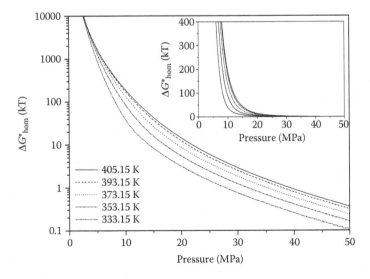

FIGURE 3.18 Energy barrier for nucleation in the polystyrene–CO_2 system.

relatively high saturation pressures, the energy barrier seems to reach a plateau and this can explain the leveling of the cell population density that is experimentally observed (Goel and Beckman 1994a; Tsivintzelis et al. 2007a).

Furthermore, when the pressure remains constant, the energy barrier for nucleation increases with increase of temperature, which means that the generation of stable nuclei becomes more difficult. This is a consequence of the interfacial tension increase with temperature increase, due to the reduction of the CO_2 solubility in the polymer matrix (Figure 3.17). Consequently, fewer nuclei are generated in the polymer matrix, resulting in the decrease of the cell population density with increase of the foaming temperature and this is what is experimentally observed (Goel and Beckman 1994a; Tsivintzelis et al. 2007a). This indicates that such an analysis with the simplified NRHB model can be used for describing the physicochemical phenomena occurring during the production of porous polymer structures and could contribute to the optimization and the rational design of the polymer foaming processes.

3.4 CONCLUSIONS

From the above exposition, it is clear that the foaming of polymers by using supercritical fluids as blowing agents is a much challenging process but it could be rationally designed and controlled by applying versatile and efficient equation of state models like the NRHB model. This is particularly useful for the selection of the appropriate polymer matrices as well as the optimum operational conditions.

Apart from its usefulness for the foaming process design, NRHB model is also useful for predicting or correlating a variety of thermodynamic properties and phase behavior of complex polymer–solvent systems. Its capacity to apply to homogeneous as well as to heterogeneous systems, to simple solvents as well as to macromolecules, to vapors or liquids or glasses make it one of the most versatile equation-of-state models currently available for handling polymer systems.

The applications to drugs and pharmaceuticals are most demanding ones, and it is most rewarding that NRHB can meet these challenges and describe quite satisfactorily technologically important phenomena such as the drug melting point depression upon processing it with supercritical fluids. From the technological point of view, it is worth mentioning its capacity to handle systems relevant to drug encapsulation processes with polymers for controlled drug delivery. By combining these, one may use NRHB model for designing a porous polymer structure (e.g., a tissue engineering scaffold) impregnated with an appropriate drug or a growth factor and other agents of importance in biomedical/biopharmaceutical or nutraceutical applications.

REFERENCES

Abusleme, J.A., and J.H. Vera. 1985. A generalized solution method for the quasi-chemical local composition equations. *Can. J. Chem. Eng.* 63: 845–849.

Arora, K.A., A.J. Lesser, and J.T. McCarthy. 1998. Preparation and characterization of microcellular polystyrene foams processed in supercritical carbon dioxide. *Macromolecules* 31: 4614–4620.

Baldwin, D.F., C.B. Park, and N.P. Suh. 1996. A microcellular processing study of poly(ethylene terephthalate) in amorphous and semicrystalline states. Part I: Microcell nucleation. *Polym. Eng. Sci.* 36: 1437–1445.

Bustamante, P., M.A. Pena, and J. Barra. 2000. The modified extended Hansen method to determine partial solubility parameters of drugs containing a single hydrogen bonding group and their sodium derivatives: Benzoic acid/Na and ibuprofen/Na. *Int. J. Pharm.* 194: 117–124.

Colton, J.S., and N.P. Suh. 1987. Nucleation of microcellular foam: Theory and practice. *Pol. Eng. Sci.* 27: 500–503.

Condo, P.D., R.D. Paul, and P.K. Johnston. 1994. Glass transition of polymers with compressed fluid diluents: Type II and III behavior. *Macromolecules* 27: 365–371.

Condo, P.D., C.I. Sanchez, G.C. Panayiotou, and P.K. Johnston. 1992. Glass transition behavior including retrograde vitrification of polymers with compressed fluid diluents. *Macromolecules* 25: 6119–6127.

De Loos, T.W. 2006. On the phase behaviour of asymmetric systems: The three phase curve solid-liquid-gas. *J. Supercrit. Fluids* 39: 154–159.

Duarte, A.R.C., J.F. Mano, and R.L. Reis. 2012. The role of organic solvent on the preparation of chitosan scaffolds by supercritical assisted phase inversion. *J. Supercritical Fluids* 72: 326–332.

Economou, I.G. 2002. Statistical association fluid theory: A successful model for the calculation of thermodynamic and phase equilibrium properties of complex fluid mixtures. *Ind. Eng. Chem. Res.* 41: 953–962.

Gabbard, R.G. 2002. *The Development of a Homogeneous Nucleation Rate Model for the Thermoplastic Foams Based on a Molecular Partition Function and Fickian Diffusion.* PhD thesis, New Jersey Institute of Technology, Department of Chemical Engineering, Newark, NJ.

Goel, S.K., and E.J. Beckman. 1994a. Generation of microcellular polymeric foams using supercritical carbon dioxide. I: Effect of pressure and temperature on nucleation. *Polym. Eng. Sci.* 34: 1137–1147.

Goel, S.K., and E.J. Beckman. 1994b. Generation of microcellular polymeric foams using supercritical carbon dioxide. II: Cell growth and skin formation. *Polym. Eng. Sci.* 34: 1148–1156.

Gracin, S., and Å.C. Rasmuson. 2002. Solubility of phenylacetic acid, p-hydroxyphenylacetic acid, p-aminophenylacetic acid, p-hydroxybenzoic acid, and ibuprofen in pure solvents. *J. Chem. Eng. Data* 47: 1379–1383.

Grenner, A., I. Tsivintzelis, G.M. Kontogeorgis, I.G. Economou, and C. Panayiotou. 2008. Evaluation of the non-random hydrogen bonding (NRHB) theory and the simplified perturbed-chain-statistical associating fluid theory (sPC-SAFT). 1. Vapor-liquid equilibria. *Ind. Eng. Chem. Res.* 47: 5636–5650.

Guggenheim, E.A. 1952. *Mixtures.* Oxford: Oxford University Press.

Handa, Y.P., and Z. Zhang. 2000. A new technique for measuring retrograde vitrification in polymer-gas systems and for making ultramicrocellular foams from the retrograde phase. *J. Polym. Sci. Part B: Polym. Phys.* 38: 716–725.

Hansen, C.M. 1999. *Hansen Solubility Parameters, A User's Handbook.* Boca Raton, FL: CRC Press.

Hentze, H.-P., and M. Antonietti. 2002. Porous polymers and resins for biotechnological and biomedical applications. *Rev. Mol. Biotechnol.* 90: 27–53.

Huang, H.S., and M. Radosz. 1990. Equation of state for small, large, polydisperse and associating molecules. *Ind. Eng. Chem. Res.* 29: 2284–2294.

Huang, H.S., and M. Radosz. 1991. Equation of state for small, large, polydisperse and associating molecules: Extension to fluid mixtures. *Ind. Eng. Chem. Res.* 30: 1994–2005.

Kirby, F.C., and A.M. McHugh. 1999. Phase behavior of polymers in supercritical fluid solvents. *Chem. Rev.* 99: 565–602.

Kontogeorgis, G.M., and G.K. Folas. 2010. *Thermodynamic Models for Industrial Applications. From Classical and Advanced Mixing Rules to Association Theories.* Chichester: John Wiley & Sons.

Kumar, V., and N.P. Suh. 1990. A process for making microcellular thermoplastic parts. *Polym. Eng. Sci.* 30: 1323–1329.

Ma, P.X. 2004. Scaffolds for tissue fabrication. *Mater. Today* 7(5): 30–40.

Martini-Vvedensky, J.E., N.P. Suh, and F.A. Waldman. 1984. Microcellular closed cell foams and their method of manufacture. US Patent No. 4,473,665.

Matsuyama, H., H. Yano, T. Maki, M. Teramoto, K. Mishima, and K. Matsuyama. 2001. Formation of porous flat membrane by phase separation with supercritical CO2. *J. Memb. Sci.* 194: 157–163.

Mulder, M. 2000. Phase inversion membranes, in *Encyclopedia of Separation Science*, Eds. M. Cooke, C.F. Poole, and I.D. Wilson. New York: Academic Press.

Nam, Y.S., and T.G. Park. 1999. Biodegradable polymeric microcellular foams by modified thermally induced phase separation method. *Biomaterials* 20: 1783–1790.

Panayiotou, C. 2003. The QCHB model of fluids and their mixtures. *J. Chem. Thermodyn.* 35: 349–381.

Panayiotou, C. 2012a. Redefining solubility parameters: The partial solvation parameters. *Phys. Chem. Chem. Phys.* 14: 3882–3908.

Panayiotou, C. 2012b. Partial solvation parameters and LSER molecular descriptors. *J. Chem. Thermodyn.* 51: 172–189.

Panayiotou, C. 2012c. Partial solvation parameters and mixture thermodynamics. *J. Phys. Chem. B* 116: 7302–7321.

Panayiotou, C. 2013. Polymer-polymer miscibility and partial solvation parameters. *Polymer* 54: 1621–1638.

Panayiotou, C., and I.C. Sanchez. 1991. Hydrogen bonding in fluids: An equation of state approach. *J. Phys. Chem.* 95: 10090–10097.

Panayiotou, C., E. Stefanis, I. Tsivintzelis, M. Pantoula, and I. Economou. 2004. Nonrandom hydrogen-bonding model of fluids and their mixtures 1. Pure fluids. *Ind. Eng. Chem. Res.* 43: 6952–6606.

Panayiotou, C., I. Tsivintzelis, and I.G. Economou. 2007. Nonrandom hydrogen-bonding model of fluids and their mixtures. 2. Multicomponent mixtures. *Ind. Eng. Chem. Res.* 46: 2628–2636.

Panayiotou, C., and J.H. Vera. 1982. Statistical thermodynamics of r-mer fluids and their mixtures. *Polym. J.* 14: 681–694.

Park, H., C.B. Park, C. Tzoganakis, K.H. Tan, and P. Chen. 2006. Surface tension measurement of polystyrene melts in supercritical carbon dioxide. *Ind. Eng. Chem. Res.* 45: 1650–1658.

Perlovich, G.L., S.V. Kurkov, L.Kr. Hansen, and A. Bauer-Brandl. 2004. Thermodynamics of sublimation, crystal lattice energies, and crystal structures of racemates and enantiomers: (+)- and (±)-ibuprofen. *J. Pharm. Sci.* 93: 654–665.

Quirk, R.A., R.M. France, K.M. Shakesheff, and S.M. Howdle. 2004. Supercritical fluid technologies and tissue engineering scaffolds. *Curr. Opin. Solid St. Mater. Sci.* 8: 313–321.

Reid, R., J.M. Prausnitz, and B.E. Poling. 1988. The properties of gases and liquids. 4th edition, Singapore: McGraw-Hill.

Reverchon, E., and S. Cardea. 2005. Formation of polysulfone membranes by supercritical CO2. *J. Membrane Sci.* 35: 140–146.

Reverchon, E., and S. Cardea. 2007. Production of controlled polymeric foams by supercritical CO2. *J. Supercrit. Fluids* 40: 144–152.

Rodeheaver, B.A., and J.S. Colton. 2001. Open-celled microcellular thermoplastic foam. *Polym. Eng. Sci.* 41: 380–400.

Sanchez, I.C., and R.H. Lacombe. 1976a. Statistical thermodynamics of fluid mixtures. *J. Phys. Chem.* 80: 2568–2580.

Sanchez, I.C., and R.H. Lacombe. 1976b. An elementary molecular theory of classical fluids. Pure fluids. *J. Phys. Chem.* 80: 2352–2362.

Sanchez, I.C., and R.H. Lacombe. 1978. Statistical thermodynamics of polymer solutions. *Macromolecules* 11: 1145–1156.

Sanchez, I.C., and C. Panayiotou. 1994. Equations of state thermodynamics of polymer and related solutions, in *Models for Thermodynamic and Phase Equilibria Calculations*, ed. S. Sandler. New York: Dekker.

Shim, J.-J., and K.P. Johnston. 1989. Adjustable solute distribution between polymers and supercritical fluids. *AIChE J.* 35: 1097–1106.

Staverman, A.J. 1950. The entropy of high polymer solutions. Generalizations of formulae. *Recl. Trav. Chim. Pays-Bas* 69: 163–174.

Stefanis, E., I. Tsivintzelis, and C. Panayiotou. 2006. The partial solubility parameters: An equation of state approach. *Fluid Phase Equil.* 240: 144–154.

Tsioptsias, C., I. Tsivintzelis, and C. Panayiotou. 2010. Equation-of-state modeling of mixtures with ionic liquids. *Phys. Chem. Chem. Phys.* 12: 4843–4851.

Tsivintzelis, I., A.G. Angelopoulou, and C. Panayiotou. 2007a. Foaming of polymers with supercritical CO2: An experimental and theoretical study. *Polymer* 48: 5928–5939.

Tsivintzelis, I., I.G. Economou, and G.M. Kontogeorgis. 2009a. Modeling the solid-liquid equilibrium in pharmaceutical-solvent mixtures: Systems with complex hydrogen bonding behavior. *AIChE J.* 55: 756–770.

Tsivintzelis, I., I.G. Economou, and G.M. Kontogeorgis. 2009b. Modeling the phase behavior in mixtures of pharmaceuticals with liquid or supercritical solvents. *J. Phys. Chem. B* 113: 6446–6458.

Tsivintzelis, I., A. Grenner, I.G. Economou, and G.M. Kontogeorgis. 2008. Evaluation of the non-random hydrogen bonding (NRHB) theory and the simplified perturbed-chain-statistical associating fluid theory (sPC-SAFT) 2. Liquid—liquid equilibria and prediction of monomer fraction in hydrogen bonding systems. *Ind. Eng. Chem. Res.* 47: 5651–5659.

Tsivintzelis, I., and G.M. Kontogeorgis. 2009. Modeling the vapor-liquid equilibria of polymer-solvent mixtures: Systems with complex hydrogen bonding behavior. *Fluid Phase Equil.* 280: 100–109.

Tsivintzelis, I., and C. Panayiotou. 2013. Designing issues in polymer foaming with supercritical fluids. *Macromol. Symp.* 331–332: 109–114.

Tsivintzelis, I., E. Pavlidou, and C. Panayiotou. 2007b. Porous scaffolds prepared by phase inversion using supercritical CO2 as antisolvent: Poly(L-lactic acid). *J. Supercrit. Fluids* 40: 317–322.

Tsivintzelis, I., T. Spyriouni, and I.G. Economou. 2007c. Modeling of fluid phase equilibria with two thermodynamic theories: Non-random hydrogen bonding (NRHB) and Atatistical association fluid theory (SAFT). *Fluid Phase Equil.* 253: 19–28.

Van de Witte, P., P.J. Dijkstra, J.W.A. van der Berg, and J. Feijen. 1996a. Phase separation process in polymer solutions in relation to membrane formation. *J. Memb Sci.* 117: 1–31.

Van de Witte, P., H. Esselbrugge, P.J. Dijkstra, J.W.A. van der Berg, and J. Feijen. 1996b. Phase transitions during membrane formation of polylactides. I. A morphological study of membranes obtained from the system polylactide-chloroform-methanol. *J. Memb. Sci.* 113: 223–236.

Wertheim, M.S. 1984a. Fluids with highly directional attractive forces. 1. Statistical thermodynamics. *J. Stat. Phys.* 35: 19–34.

Wertheim, M.S. 1984b. Fluids with highly directional attractive forces. 2. Thermodynamic perturbation-theory and integral-equations. *J. Stat. Phys.* 35: 35–47.

Wertheim, M.S. 1986a. Fluids with highly directional attractive forces. 3. Multiple attraction sites. *J. Stat. Phys.* 42: 459–476.

Wertheim, M.S. 1986b. Fluids with highly directional attractive forces. 4. Equilibrium polymerization. *J. Stat. Phys.* 42: 477–492.

4 Relevant Properties for the Formation of Porous and Cellular Structures

Ernesto Di Maio, Giuseppe Mensitieri, Maria Giovanna Pastore Carbone, and Giuseppe Scherillo

CONTENTS

4.1 INTRODUCTION

Proper qualitative and quantitative description of polymer foaming process is a multiphysics problem involving mass and heat transport and momentum transfer as well as sorption thermodynamics, in a high-pressure polymer–penetrant system. There are several material and thermodynamic properties playing a major role, ranging from volumetric, rheological, and mass transport behavior to interfacial properties of the binary mixtures of polymer and foaming agent.

In particular, in the context of this book, this chapter addresses the theoretical and experimental aspects of gas solubility and diffusivity as well as of the effect of gas concentration on interfacial properties and viscosity of the gas–molten polymer mixture. These are properties that, with reference to the general process of gas foaming, rule the whole process of bubble nucleation and growth.

To this aim, the theoretical models that can be effectively used to provide a reliable framework to interpret material behavior; to predict, where possible, the relevant properties of the system; and to supply expressions to correlate experimental data and extrapolate them for temperature, pressure, and composition conditions where data are not available, are presented first here. In the second part, the experimental techniques more suitable to evaluate these properties are reviewed. These experimental approaches need the proper setup to be effective in tackling the challenging technical problem associated with the meaningful measurement of mixture properties at high pressure/temperature conditions.

4.2 THEORETICAL ASPECTS

4.2.1 Mass Transport of Low Molecular Weight Compounds in Rubbery and Molten Polymers

4.2.1.1 Mass Balance of Mixture Components and Constitutive Expressions for Diffusive Flux

The description of the mass transport phenomena in binary systems consisting of a rubbery/molten polymer and a low molecular weight penetrant involves writing the continuity equation and the mass balance equation for one of the two components of the mixture (penetrant and polymer, in the case at hand). Alternatively, one could write the two mass balances for the two components. In the general case, one needs to couple these two equations with energy and momentum balance equations for the overall mixture.

All these equations are strongly coupled and solving mass transport problems could be rather complex. In fact, as we will discuss later, the mass flux of each component, which appears in the mass balance of each component, can be split into two contributions, one related to a *bulk* velocity of the mixture and the other related to the diffusive flux. The *bulk* velocity is determined by solving the momentum balance equation. In addition, one of the driving forces for the diffusive flux, that is, the gradient of chemical potential of the component itself, depends upon the state of stress that, again, appears also in the momentum balance. In the momentum balance equation, terms related to natural convection (which is promoted by differences in temperature and in concentration of the two components, if these have different molecular mass) are present, which depend upon the profiles of temperature and composition within the mixture. Moreover, material properties, which are present in the constitutive expressions for mass, momentum, and energy fluxes, strongly depend both on the values of local concentration and of local temperature, the profile of the latter being evaluated by solving the energy balance. In turn, in the energy balance, the total mass fluxes of each component (in the term related to the work of external force fields) and the diffusive fluxes of each component (in fact, each component carries in and out of the balance volume of its own enthalpy) are present. More on this is discussed in the following section dealing with the relevant balance equations.

We focus, for the moment, on the mass balance equation for one of the components of the binary mixture. To this end, proper constitutive equations for mass flux have to be introduced. In the following, we will refer with subscripts 1 and 2, respectively, to the penetrant and to the polymer.

The mass balance equation for the penetrant species reads in three-dimensional form:

$$\frac{\partial C_1}{\partial t} = -\underline{\nabla} \cdot \underline{j}_1 = -\underline{\nabla} \cdot C_1 \underline{u}_1 \tag{4.1}$$

where:

C_1 is the molar concentration of the penetrant

$\underline{j}_1$ is the molar flux vector of the penetrant in the lab-fixed frame

$\underline{u}_1$ is the local average velocity vector of component 1

In the following, we refer to a one-dimensional geometry, without loss of generality of the illustrated arguments, indicating with X the position coordinate in the lab-fixed frame. For mixtures where the concentration of penetrant is nonnegligible, it proves convenient to transform the coordinate of the mass balance equations from lab-fixed to polymer-fixed coordinate (x), employing the following variable length scale:

$$x = \int_{X^*}^{X} C_2 \bar{V}_2 dz \qquad (4.2)$$

with $\bar{V}_2$ being the partial molar volume of polymer and X^* being a reference position where the polymer mass flux vanishes (e.g., $X^* = 0$). If the volume change on mixing is neglected, it can be easily shown that

$$j_1^2 = \frac{j_1}{\phi_2} \qquad (4.3)$$

thus following that

$$\frac{\partial C_1}{\partial t} = -\frac{\partial j_1^2}{\partial x} \qquad (4.4)$$

where $j_1^2 = C_1(u_1 - u_2)$ stands for the molar flux of penetrant relative to the polymer.

In fact, since

$$j_1 = C_1 u_1$$

one obtains

$$\frac{j_1^2}{j_1} = \frac{u_1 - u_2}{u_1}$$

If it is further assumed that there is no change in volume at mixing, that is, $\phi_1 u_1 = -\phi_2 u_2$, it follows that

$$\frac{u_2}{u_1} = \frac{(\phi_2 - 1)}{\phi_2}$$

from which Equation 4.3 is finally obtained.

Due to the assumption of no change of volume on mixing, Equation 4.4 can be, equivalently, expressed in terms of volume fractions as follows:

$$\frac{\partial \phi_1}{\partial t} = -\frac{\partial J_1^2}{\partial x} \qquad (4.5)$$

where:

ϕ_1 is the volume fraction of component 1 in the mixture

J_1^2 is the volumetric flux of penetrant

Specifying the flux is a key step in constructing an adequate theory for mutual diffusion. Based on the analysis made by Bearman (1961), it is possible to deduce the following expression relating to the flux of the component 1 to the gradient of its chemical potential. (In this discussion, it is assumed that only the gradient of chemical potential is the driving force for diffusion. This is a simplifying assumption: as will be discussed later, there are other driving forces playing a role.) In fact, shortly after a concentration gradient is established, a quasi-stationary regime is reached, in which the driving force for diffusion of component 1 becomes sensibly equal in magnitude, but opposite in sign, to the frictional force.

$$\frac{d\mu_1}{dX} = -C_2\zeta_{12}(u_1 - u_2) \tag{4.6}$$

In Equation 4.6, ζ_{12} is the friction coefficient, which takes into account all the lack of knowledge about the interaction resistance of the two components during their mixing. From Equation 4.6, we can derive an expression for j_1^2:

$$j_1^2 \equiv C_1(u_1 - u_2) = -\frac{C_1}{C_2\zeta_{12}}\frac{d\mu_1}{dX} = -\frac{\bar{V}_2 C_1}{\zeta_{12}}\frac{d\mu_1}{dx} \tag{4.7}$$

where

$$\zeta_{12} [=] \frac{\text{Joule } s\, m}{\text{mol } i^2}$$

It is also possible to introduce the expression of the diffusive molar flux of component 1 relative to a volume-fixed frame of reference (i.e., a frame moving with the average volume velocity of the mixture), j_1^v, defined as follows:

$$j_1^v \equiv C_1(u_1 - u_0) \tag{4.8}$$

Here u_0 represents the volume-averaged velocity of the mixture (i.e., $u_0 = C_1\bar{V}_1 u_1 + C_2\bar{V}_2 u_2$, with $\bar{V}_i$ the molar volume of each component). We can put this molar flux in terms of the product of mutual diffusion coefficient multiplied by the gradient of penetrant concentration:

$$j_1^v \equiv C_1(u_1 - u_0) = -D_{12}^v \cdot \frac{\partial C_1}{\partial X} \tag{4.9}$$

This is one of the forms of the so-called Fick's I law, where D_{12}^v represents the binary mutual diffusion coefficient in the volume-fixed frame of reference. An analogous expression holds for the polymer:

$$j_2^v \equiv C_2(u_2 - u_0) = -D_{12}^v \cdot \frac{\partial C_2}{\partial X} \tag{4.10}$$

If the constitutive expression (4.9) for the flux of component 1 relative to a volume-fixed frame of reference is substituted in the mass balance, Equation 4.1, one obtains

$$\frac{\partial C_1}{\partial t} = \frac{\partial}{\partial X}\left(-D_{12}^v \cdot \frac{\partial C_1}{\partial X} + C_1 u_0\right) \tag{4.11}$$

In this last equation, it is evident how the expression of the flux of component 1 can be split into two contributions. The first contribution (*diffusive* contribution) represents the relative movement of component 1 as referred to the chosen frame of reference, in this case the volume-fixed one moving locally with a velocity u_0. The second contribution (*bulk* contribution) represents the movement of component 1 along with the whole mixture that is moving with the local volume-averaged velocity. Obviously, analogous expressions would be obtained for the mass balance of component 1 if different frame of reference for the diffusive flux were chosen and hence if the motion of component 1 were decomposed in a different way. In fact, in the case the flux of component 1 was expressed with reference to a mass-fixed or to a molar-fixed frame of reference, the *diffusive* contribution would contain different binary mutual diffusion coefficients, while the second contribution would represent the motion of the component 1 along with the mixture at the *bulk* mass or molar velocity.

If the assumption of no change of volume on mixing is reasonable, u_0 is zero and then

$$j_1^v = C_1 u_1 = j_1 \tag{4.12}$$

from which it follows that

$$j_1^2 = \frac{j_1^v}{\phi_2} \tag{4.13}$$

and

$$j_1^2 = -\frac{D_{12}^v}{\phi_2} \cdot \frac{\partial C_1}{\partial X} = -D_{12}^v \frac{\partial C_1}{\partial x} = -\frac{D_{12}^v}{\overline{V}_1} \frac{\partial \phi_1}{\partial x} \tag{4.14}$$

Now, by comparing Equations 4.7 and 4.14, D_{12}^v can be expressed as (Bearman 1961)

$$D_{12}^v = \frac{\overline{V}_2}{\zeta_{12}} C_1 \left(\frac{\partial \mu_1}{\partial C_1}\right)_{T,P} + \frac{\overline{V}_2}{\zeta_{12}} C_1 \left(\frac{\partial \mu_1}{\partial p}\right)_{T,C_1} \frac{dp}{dC_1} \tag{4.15}$$

where:
μ_1 represents the chemical potential of component 1

We can also consistently derive an expression in terms of volume fraction for J_1^2 for a one-dimensional process:

$$J_1^2 = \frac{\phi_1}{C_1} j_1^2 = -\overline{V}_1 \frac{D_{12}^v}{\overline{V}_1} \frac{\partial \phi_1}{\partial x} \tag{4.16}$$

It has been demonstrated above that, when there is no change of volume on mixing, the total flux of both components in a binary mixture can be simply expressed, respectively, as

$$j_1^v = -D_{12}^v \cdot \frac{\partial C_1}{\partial X} \quad \text{and} \quad j_2^v = -D_{12}^v \cdot \frac{\partial C_2}{\partial X}$$

Hence, the diffusion behavior of a two-component system can be described by using only a single (mutual) diffusion coefficient expressing the diffusive flux of each component as referred to a frame of reference fixed with respect to the total volume of the mixture. In such a case, the mass balance of component 1, expressed in terms of a lab-fixed coordinate, X, is

$$\frac{\partial C_1}{\partial t} = \frac{\partial}{\partial X}\left(-D_{12}^v \cdot \frac{\partial C_1}{\partial X}\right)$$

(4.17)

that is often referred to as *Fick's II law*. However, when the volumes on mixing are not additive, and consequently $u_0 \neq 0$, the definition of the side of the section on which the volume is to remain constant must be chosen arbitrarily, and the diffusion coefficient becomes equally arbitrary (Crank 1975).

Another possible alternative is that of defining the constitutive equation for mass flux in one-dimensional diffusion in a binary mixture by referring to a frame of reference moving with the center of mass of the polymer–penetrant system. In such a case, the total mass flux is expressed as

$$\underline{n}_1 = \underline{j}_1^M + \rho_1 \underline{u}_M = -D_{12}\rho \cdot \frac{\partial \omega_1}{\partial X} + \rho_1 \underline{u}_M$$

(4.18)

where:

ρ is the mass density of the mixture

ρ_1 is the mass of component 1 per volume of mixture (mass concentration of component 1)

ω_1 is the mass fraction of component 1 and

$$u_M = \omega_1 u_1 + \omega_2 u_2$$

(4.19)

Consequently, the differential three-dimensional mass balance of component 1 becomes

$$\frac{\partial \rho_1}{\partial t} = -\nabla \cdot \left(-D_{12}\rho \cdot \underline{\nabla}\omega_1 + \rho_1 \underline{u}_M\right)$$

(4.20)

that, in its one-dimensional form, reads

$$\frac{\partial \rho_1}{\partial t} = -\frac{\partial}{\partial X} \cdot \left(-D_{12}\rho \cdot \frac{\partial \omega_1}{\partial X} + \rho_1 u_M\right)$$

(4.21)

It is possible to recast this mass balance to obtain an expression formally similar to Equation 4.17 by introducing a modified scale of length, ξ, such that equal increments of ξ include equal increments of total mass. Based on this definition, we have that

$$d\xi = V_0(\rho_1 + \rho_2)dX$$

(4.22)

where V_0 is an arbitrary constant-specific volume that is used to recover a dimension of length for ξ. Also the measure of concentration should consistently change. The expression for the differential mass balance of component 1 is then

$$V_0\frac{\partial \omega_1}{\partial t} = -\frac{\partial}{\partial \xi}\left(-D_{12}V_0\frac{\partial \omega_1}{\partial \xi}\right)$$

(4.23)

4.2.1.2   Evaluation of Concentration Profiles in Polymer–Penetrant Mixtures

As anticipated, determination of concentration profiles of penetrant in a molten polymer, in the general case, can be performed by solving a system of equations consisting of the mass balances of the two components (or, alternatively, of the mass balance of one of the components and of the continuity equation) and of the momentum and energy balances of the mixture. In addition, one has to use the proper constitutive equations for the fluxes (of mass, energy, and momentum) as well as for the material properties appearing in the equations (e.g., density, diffusivities, and viscosity).

Before proceeding with the discussion of balance equations, it is worth noticing here that in Section 4.2.1.1, we have introduced the constitutive expressions for mass flux of each component, limiting our attention to ordinary diffusion phenomena, that is, those phenomena where the only relevant driving force for the diffusion of a component is just the gradient of its chemical potential or the gradient of composition.

Actually, in writing the constitutive expression for diffusive flux, one should account for other driving forces that could as well promote a diffusive flux. Subsequently, before introducing all the relevant balance equations for the molten polymer–gas mixture, we will mention more complex expressions for the diffusive mass flux as determined on the grounds of classical irreversible thermodynamics.

If we consider the diffusive flux as referred to the weight average velocity of the mixture, that is, j_1^M in Equation 4.18, it can be demonstrated that (Bird et al. 1960) the diffusive flux can be written as

$$j_1^M = -j_2^M = j_1^{M,\omega} + j_1^{M,P} + j_1^{M,F} + j_1^{M,T} \tag{4.24}$$

where the first three terms on the right-hand side of Equation 4.24 are associated with the *mechanical* driving forces, while the last term is associated with the *thermal* driving force. In detail, the expressions of these four contributions are as follows:

$$j_1^{M,\omega} = -\rho D_{12} \underline{\nabla}\omega_1 = \left[\left(-\frac{c^2}{\rho}\right)M_1 M_2\right] D_{12} \nabla x_1 = \left[\left(-\frac{c^2}{\rho}\right)M_1 M_2\right]\frac{x_1}{RT} D_{12,T}\left(\frac{\partial \mu_1}{\partial x_1}\underline{\nabla}x_1\right) \tag{4.25}$$

$$j_1^{M,P} = \left[\left(-\frac{c^2}{\rho}\right)M_1 M_2\right]\frac{x_1}{RT} D_{12,T}\left[\left(\overline{V}_1 - \frac{M_1}{\rho}\right)\underline{\nabla}P\right] \tag{4.26}$$

$$j_1^{M,\overline{F}} = \left[\left(-\frac{c^2}{\rho}\right)M_1 M_2\right]\frac{x_1}{RT} D_{12,T}\left[M_1 \frac{\rho_2}{\rho}\left(\underline{F}_1 - \underline{F}_2\right)\right] \tag{4.27}$$

$$j_1^{M,T} = -D_{12,T}\underline{\nabla}\ln T \tag{4.28}$$

where:

c represents the moles of mixture components per unit volume
M_i represents the molecular weights of the components
x_i is the molar fraction of each component in the mixture
$\overline{V}_i$ is the molar volume of each component
$D_{12,T}$ is the mutual *thermodynamic diffusivity*
$\underline{F}_i$ represents the body force acting on component i

Since the main focus of this chapter is on mass transport, analogous details on constitutive equations for conductive energy flux and on momentum flux will not be discussed.

In order to determine the concentration profile of polymer and penetrant, one needs to solve the system of equations consisting of the two mass balances of the two components (Equation 4.29), of the momentum balance (Equation 4.30) and of the energy balance (Equation 4.26), reported below:

$$\frac{\partial \rho_i}{\partial t} = -\underline{\nabla} \cdot \underline{n}_i = -\left(\rho_i \underline{u}_M + \underline{j}_i^M\right) \tag{4.29}$$

$$\frac{\partial (\rho \underline{u}_M)}{\partial t} = -\nabla \cdot \left(\rho \underline{u}_M \underline{u}_M + \underline{T}\right) + \sum_{i=1}^{2} \rho_i \underline{F}_i \tag{4.30}$$

$$\frac{\partial \left\{\rho \left[\hat{U} + (1/2) u_M^2\right]\right\}}{\partial t} = -\underline{\nabla} \cdot \left[\left(\hat{U} + \frac{1}{2} u_M^2\right) \underline{u}_M + \underline{q} + \left(\underline{T} \cdot \underline{u}_M\right)\right] + \sum_{i=1}^{2} \underline{n}_i \underline{F}_i \tag{4.31}$$

In Equation 4.31, the contribution of radiant heat flux has been neglected. Here $\hat{U}$ represents the internal energy per unit mass of the mixture. Proper constitutive equations should be provided for the stress tensor, $\underline{T}$, and for the conductive energy flux relative to the average mass velocity, $\underline{q}$. It is worth mentioning that $\underline{q}$ accounts for a contribution associated with temperature gradient, a contribution associated with the amount of enthalpy that each component carries in or out of the balance volume by diffusive flux, and, finally, a contribution associated with the so-called Dufour effect, that is, the contribution associated with the composition gradient.

If we neglect the contribution to the diffusive flux of each component (referred to the mass average velocity) related to thermal driving forces and to mechanical driving forces other than the composition gradient, Equation 4.31, written for the penetrant, reads

$$\frac{\partial \rho_1}{\partial t} = -\underline{\nabla} \cdot \left(-D_{12} \rho \cdot \underline{\nabla} \omega_1 + \rho_1 \underline{u}_M\right) = -\underline{\nabla} \cdot \left(\rho_1 \underline{u}_M\right) + \underline{\nabla} \cdot \left(D_{12} \rho \cdot \underline{\nabla} \omega_1\right)$$

If, in addition, ρ is a constant, it follows, from the mass balance on the whole mixture, that $\underline{\nabla} \cdot \underline{u}_M$ and, in turn, one has

$$\frac{\partial \rho_1}{\partial t} + \underline{u}_M \cdot \nabla \rho_1 = \underline{\nabla} \cdot \left(D_{12} \underline{\nabla} \rho_1\right) \tag{4.32}$$

or, equivalently:

$$\frac{\partial C_1}{\partial t} + \underline{u}_M \cdot \underline{\nabla} C_1 = \underline{\nabla} \cdot \left(D_{12} \underline{\nabla} C_1\right) \tag{4.33}$$

Despite the relevant simplification, the determination of concentration profile of the penetrant still requires the solution of the whole system of equations including the polymer mass balance as well as the momentum and energy balance for the mixture. However, if one considers the case of a *stagnant liquid mixture* (as is the case for a molten polymer–gas mixture), we have $\underline{u}_M = 0$. As a consequence, Equations 4.32 and 4.33 take the following form:

$$\frac{\partial \rho_1}{\partial t} = \underline{\nabla} \cdot \left(D_{12} \underline{\nabla} \rho_1\right) \tag{4.34}$$

$$\frac{\partial C_1}{\partial t} = \underline{\nabla} \cdot \left(D_{12} \underline{\nabla} C_1 \right) \qquad (4.35)$$

It is evident from the previous discussion that, for the case of a stagnant liquid mixture, when one can assume that there is no relevant temperature profile (that, in turn, would determine a dependence of diffusivity on the spatial coordinates), one can determine the concentration profile of the penetrant solving just the differential equation expressing the penetrant mass balance. Moreover, in the case D_{12} is independent of composition of the mixture, these expressions can be further simplified obtaining

$$\frac{\partial \rho_1}{\partial t} = D_{12} \nabla^2 \rho_1 \qquad (4.36)$$

$$\frac{\partial c_1}{\partial t} = D_{12} \nabla^2 C_1 \qquad (4.37)$$

Equation 4.37 is known as *II Fick's law*. In the classical *Crank* book (Crank 1975), mathematical solutions are available for Equations 4.36 and 4.37 in the case of several geometries as well as of different boundary and initial conditions. Mathematical solutions of Equations 4.34 and 4.35 are also available in the same book in the case of concentration-dependent mutual diffusivity, considering different functional forms of this dependence. By comparing the experimental results obtained from tests, conducted at well-defined boundary conditions and initial conditions, with analytical solutions, one can obtain, from fitting procedures, the value of mutual diffusion coefficient.

4.2.1.3 Theories for Diffusion Coefficients

In this section, a short review of the most relevant theoretical approaches aimed at supplying expressions for the mutual diffusivity rooted on firm physical grounds is provided. In the following paragraphs, we will always refer to a frame of reference moving with the mass velocity of the whole mixture and will use the symbol D for D_{12} and the symbol j_1 for j_1^M.

As by the experimental evidence, penetrants diffusivity in polymer melts is weakly dependent upon molecular weight of the polymer, depending primarily upon penetrant size and geometry, penetrant concentration, and on temperature. In particular, changes in concentration and temperature can change in diffusivity of several orders of magnitude.

Actually, molecular rooted theories supply expressions for the intra-diffusion coefficients for the solvent and polymer, respectively, D_1 and D_2, which represents the intrinsic mobility of the components. (Note: these coefficients are often referred in the literature as self-diffusion coefficients, generating some confusion. In fact, in the present context, we will use the term self-diffusion coefficient of a substance when referring to the intrinsic mobility of a molecule of that substance in the pure state (more on this later.) Hence, to the aim of constructing physically sound theories providing expressions for mutual diffusivities, it is necessary to supply an equation relating D, or, equivalently, any other diffusion coefficient defined relatively to another frame of reference, to intra-diffusion coefficients. These three diffusion coefficients have all been related (Bearman 1961) to the molecular friction coefficients using statistical mechanics arguments. As we will see in Equations 4.38 through 4.40, relationships between mutual D and one of the intra-diffusion coefficients D_1 or D_2, for a binary mixture, involve necessarily the knowledge of one of the friction coefficients.

Theories for diffusion, in general, have been developed to describe the displacement of a component in a spatially uniform mixture, without the presence of a driving force due to the gradient of concentration or chemical potential. In other words, theories address the issue of the intrinsic mobility of each species. In the case of mono-component systems, the diffusion process is then described by the so-called self-diffusion coefficient, while, in the case of spatially uniform mixtures, the diffusivity of each of the components is described in terms of the *intra-diffusion coefficient* (Caruthers et al. 1998). The self-diffusion coefficient is, hence, a special case of intra-diffusion coefficient.

Binary mutual diffusion coefficient appearing in Equations 4.18, 4.20, 4.21, and 4.23 is related to intra-diffusion coefficient of the solvent, D_1, and of the polymer, D_2, through relationships involving solvent–solvent, polymer–polymer, and solvent–polymer friction coefficients, respectively, ζ_{11}, ζ_{22}, and ζ_{12} (Bearman 1961):

$$D = \frac{M_2 \cdot \omega_1 \cdot \hat{V}_2}{N_A^2 \cdot \zeta_{12}} \left(\frac{\partial \mu_1}{\partial \omega_1} \right)_{T,P} \tag{4.38}$$

$$D_1 = \frac{RT}{N_A^2 \cdot \left(\dfrac{\omega_1 \cdot \zeta_{11}}{M_1} + \dfrac{\omega_2 \cdot \zeta_{12}}{M_2} \right)} \tag{4.39}$$

$$D_2 = \frac{RT}{N_A^2 \cdot \left(\dfrac{\omega_2 \cdot \zeta_{22}}{M_2} + \dfrac{\omega_1 \cdot \zeta_{12}}{M_1} \right)} \tag{4.40}$$

where N_A is the Avogadro number. Expressions for intra-diffusion coefficients D_1 and D_2 can be independently obtained from free volume theories that will be discussed later. However, since three friction coefficients appear in Equations 4.38 through 4.40, in general it is not possible to obtain an expression for the mutual diffusivity, D, only in terms of D_1 and D_2, containing none of the friction coefficients. Only in particular cases (e.g., introducing an expression relating the three friction coefficients, or considering the limit of trace amount of solvent or when $D_1 \gg D_2$), it is possible to evaluate D only from D_1 and D_2.

Free volume theories used to describe penetrant diffusivity above glass transition temperature of the system are rooted on the hypothesis that the value of diffusion coefficient is determined by the available free volume of the system. These theories provide expressions for intra-diffusion coefficients of polymer and solvent from which, under prescribed conditions and hypotheses, an expression for D can be obtained. *Configuration entropy* models are instead based on the picture that penetrant molecules move under the action of a deterministic drag force combined with a stochastic force originated by the random impact of neighboring molecules. *Molecular models* are based on another approach in that they are based on a molecular picture to describe motions of penetrants and macromolecular chains accounting for relevant intermolecular forces, and directly provide expressions for the mutual binary diffusion coefficient, D. We will discuss in the following paragraphs, in some detail, only the *free volume* theories and the *molecular* models.

4.2.1.3.1 Free Volume Models

The so-called free volume theories for diffusion are based on the model proposed by Cohen and Turnbull (1959), originally formulated for diffusion in a hard-sphere liquid. The basic assumption is that diffusion is not a thermally activated process but rather results from a redistribution, due to random fluctuations in local density, of discontinuous free volume voids. According to this physical picture, these voids are just a part of the total free volume in a system, the other contribution being that arising from molecular vibration. This latter contribution, which can only be redistributed as a consequence of a large energy change, does not play a role in a diffusion process. According to this approach, the probability of a diffusive jump corresponds to the probability that local density fluctuations determine the formation of a hole of sufficient size. No energy is involved in the redistribution of free volume and, hence, in the formation of a hole. The self-diffusion coefficient for a simple liquid has been proposed to be given by

$$D_S = g \cdot a^* \cdot u \cdot \exp\left(-\frac{\gamma \cdot V^*}{V_f}\right) \tag{4.41}$$

where:
  g is a geometric factor
  a^* is the molecular diameter
  u is the kinetic velocity
  V^* denotes the critical free volume required for a jumping of a single hard sphere and corresponds to the volume of an hard sphere
  V_f is the average free volume per sphere
  γ is an overlap factor of free volume elements comprised between 0.5 and 1

This simple model constitutes the framework upon which the subsequent free volume models for binary polymer–penetrant systems were constructed.

One of the first free volume models successfully used to describe the thermodynamic binary diffusion coefficient in rubbery-molten polymers is that proposed by Fujita et al. (1960; Fujita 1961) for which the thermodynamic binary mutual diffusion coefficient, D_T, can be expressed as

$$D_T = R \cdot T \cdot A_d \cdot \exp\left(\frac{B_d}{v_f}\right) \tag{4.42}$$

The thermodynamic binary mutual diffusion coefficient, D_T, is defined in such a way that the diffusive mass flux is expressed as a function of the gradient of chemical potential, that is,

$$j_1 = \frac{D_T \cdot \rho_1}{RT} \frac{\partial \mu_1}{\partial X} \tag{4.43}$$

As a consequence, D_T is related to D, in the case of constant total density of the system, by

$$D = \frac{D_T \cdot \rho_1}{RT} \cdot \left(\frac{\partial \mu_1}{\partial \rho_1}\right)_{T,P} \tag{4.44}$$

In Equation 4.42, A_d and B_d are model parameters, which depend on the nature of polymer and penetrant, while v_f is the fractional free volume and, at low penetrant concentrations, is given by

$$v_f(T,\phi_1) = v_f(T,0) + \beta(T) \cdot \phi_1 \tag{4.45}$$

where $v_f(T,0)$ is the average fractional free volume of pure polymer at temperature T and $\beta(T)$ measures the effectiveness of the penetrant in increasing the free volume of the system. The value of D_T approaches the value of Fickian diffusivity, D, as ϕ approaches 0 value.

The fractional free volume was assumed to depend linearly upon temperature:

$$v_f(T,0) = v_f(T_R,0) + \alpha(T - T_R) \tag{4.46}$$

where:

T_R is a properly chosen reference temperature

α is a material parameter for pure polymer

Fujita's model is adequate for describing diffusivity of small molecules in rubbery polymers, although it fails in the case of small penetrant able to establish specific interactions. An extended version of this model has been proposed by Stern et al. (1972, 1983) to deal with the case of diffusion of light gases in rubbery polymers.

More recently, Vrentas and Duda (1977a, b, 1986; Vrentas et al. 1982) proposed a new model for the solvent intra-diffusion coefficient based on the free volume concept, which was successfully used to describe diffusion in concentrated polymer solutions. The model is based on the Cohen–Turnbull formalism that, for the case of a single component fluid, provides an expression relating the self-diffusion coefficient of the fluid to its free volume. Such an expression, once extended to derive a relation between the intra-diffusion coefficient of one species and the system free volume in a binary mixture, reads

$$D_1 = D_{01} \cdot \exp\left(-\gamma \frac{\overline{V}_1^*}{\overline{V}_{FH}}\right)$$ (4.47)

where $\overline{V}_1^*$ is the critical molar free volume required for a jumping unit of species 1 to migrate, $\overline{V}_{FH}$ is the hole free volume per mole (see the following for the meaning of this term) of all individual jumping units in the mixture, and D_{01} is a temperature-independent constant. The physical picture on which this theoretical approach is rooted is based on the assumption that there are two types of free volume: *interstitial free volume* ($\overline{V}_{FI}$) and *hole free volume* ($\overline{V}_{FH}$). The first is related to the increase with temperature of the amplitude of anharmonic vibrations, is uniformly distributed, is permanently located in the proximity of a given molecule or molecular segment, and has a high redistribution energy. The second is associated with holes or vacancies discontinuously distributed in the material at any instant of time and has no redistribution energy. It is hypothesized that it is only the hole free volume that controls the diffusion process.

A relevant difference with the original Cohen–Turnbull model is that the Vrentas–Duda model is not applied to fluids made of hard spheres, but to solutions of real molecules, including macromolecules. Migration of diffusing molecules is regarded as resulting from elementary jumps of small segments of the molecules, the so-called jumping units. The molecular weight of each jumping units is indicated by the symbol M_{ij}, where, for the case of a penetrant–polymer binary mixture, $i = 1$ for the penetrant and $i = 2$ for the polymer. In fact, the penetrant or polymer molecule does not jump as a whole into holes accommodating the entire molecule but rather by diffusional jumps involving the coordinated motion between several parts of the molecules (Duda and Zielinski 1996). If the molecular weight of the penetrant is low, M_{1j} can be taken as equal to the value of molecular weight of penetrant itself.

Vrentas and Duda generalized the theory of Cohen and Turnbull (Equation 4.47) obtaining for the specific case of a low molecular weight solvent in a polymer solution the following relationship for the intra-diffusion coefficient of penetrant (component 1):

$$D_1 = D_{01} \exp\left[\frac{-\gamma\left(\omega_1 \hat{V}_1^* + \omega_2 \xi \hat{V}_2^*\right)}{\hat{V}_{FH}}\right]$$ (4.48)

Here $\hat{V}_i^*$ is the specific hole free volume of component i that is required for a diffusive step, $\xi = \hat{V}_{1j}^*/\hat{V}_{2j}^*$, and $\hat{V}_{FH}$ is the specific hole free volume of a two components fluid with a weight fraction

ω_i for each of the components characterized, respectively, by a jumping unit molecular weight equal to M_{ij}. $\bar{V}_{FH}$ is related to $\hat{V}_{FH}$ by the following relationship:

$$\bar{V}_{FH} = \frac{\hat{V}_{FH}}{\omega_1/M_{1j} + \omega_2/M_{2j}} \tag{4.49}$$

where:

$\left(\omega_1/M_{1j} + \omega_2/M_{2j}\right)$ represents the total number of moles of jumping units per gram of system

A critical information to implement the described free volume approach is the quantification of $\hat{V}_{FH}$, which represents, as anticipated, that fraction of specific free volume that is redistributed effortlessly and, as such, is the one actually involved in molecular transport. This quantity cannot be simply determined as the difference between the specific volume of the system and the specific occupied volume of the system itself (generally taken as the specific volume of the system at equilibrium at 0 K), which would represent the total specific free volume. To this aim, Vrentas and Duda (1977b; Duda and Zielinski 1996) developed the following relationship that allows the evaluation of $\hat{V}_{FH}$ from volumetric characteristics of the pure components of the system:

$$\hat{V}_{FH} = \omega_1 K_{11}\left(K_{21} - T_{g1} + T\right) + \omega_2 K_{12}\left(K_{22} - T_{g2} + T\right) \tag{4.50}$$

In the previous expression, K_{11} and K_{21} denote free volume solvent parameters, while K_{12} and K_{22} denote free volume polymer parameters. For a discussion on these parameters, see Vrentas and Duda (1977b) and Vrentas and Vrentas (1996). The terms T_{g1} and T_{g2} represent the glass transition temperatures of a pure solvent and of a pure polymer, respectively. The mentioned free volume parameters appearing in Equation 4.50 can be calculated from viscosity data as a function of temperature of the pure components forming the mixture and are related to the parameters appearing in the well-known Williams–Landel–Ferry (WLF) equation (Williams et al. 1955).

When the availability of free volume limits the molecular mobility—as is the case in concentrated polymer solutions—the dependence on temperature of the pre-exponential factor, D_{01}, appearing in Equation 4.48 can be safely neglected as compared to the temperature dependence of the free volume exponential term. In the case, one has to account for the attractive forces with surrounding molecules that a jumping unit has to overcome prior to a diffusive step, Equation 4.48 can be properly modified including a temperature-dependent pre-exponential term, which includes the activation energy, E, that a jumping unit needs to acquire to jump in a contiguous free volume void. Thus, Equation 4.48 should be generalized by introducing this dependence obtaining the following expression for the solvent intra-diffusion coefficient of a solvent in a polymer mixture (Vrentas and Duda 1977a).

$$D_1 = D_1^0 \exp\left(-\frac{E_1}{RT}\right)\exp\left[\frac{-\gamma\left(\omega_1 \hat{V}_1^* + \omega_2 \xi \hat{V}_2^*\right)}{\hat{V}_{FH}}\right] \tag{4.51}$$

An analogous expression can be derived describing the concentration and temperature dependence of the intra-diffusion coefficient of a polymer molecule in the binary mixture:

$$D_2 = D_2^0 \exp\left(-\frac{E_2}{RT}\right)\exp\left[\frac{-\gamma\left(\omega_1 \hat{V}_1^* + \omega_2 \xi \hat{V}_2^*\right)}{\hat{V}_{FH}}\right] \tag{4.52}$$

E_1 and E_2 are expected to be a function of penetrant concentration, since the neighbor molecular environment significantly changes with concentration. To keep the adjustable parameters to a reasonable number, this dependence is often neglected in the presented formulation of the free volume model.

Once the expressions for intra-diffusion coefficients, based on free volume concepts, are available from Equations. 4.51 and 4.52, it is possible, at least in principle, to calculate the binary mutual diffusion coefficient D, that is, the parameter of relevance in engineering applications. In fact, Equations 4.51 and 4.52 provide theoretical expressions for D_1 and D_2 that, coupled with the Equations 4.38 through 4.40, form a set of equations for the three unknown friction coefficients: it is, at the end, a set of two equations in three unknowns, which does not allow the determination of the values of the friction coefficients. Hence, an explicit expression relating D only to D_1 and D_2 can be obtained if some additive information is provided (i.e., an expression relating the three friction coefficients) or if one makes some simplifying assumption or in the case of limiting conditions. As an example, it can be assumed that

$$\zeta_{12} = \left(\zeta_{11}\zeta_{22}\right)^{1/2} \tag{4.53}$$

There is no physically sound basis for such an expression that is used in the absence of a substantiated relationship. It is worth noticing that the properties of *regular* solutions, in the meaning given to this definition by Bearman (1961), are sufficient but not necessary for the validity of Equation 4.53.

By combining Equations 4.38 through 4.40, with Equation 4.53, one obtains

$$D = \frac{\left(D_1 x_1 + D_2 x_2\right)}{RT}\left(\frac{\partial \mu_1}{\partial \ln x_1}\right)_{T,P} \tag{4.54}$$

An explicit expression for D in terms of D_1 can also be obtained in the limit of trace amounts of solvent:

$$D = D_1 = D_{01}\exp\left(-\frac{\gamma\xi\hat{V}_2^*}{\hat{V}_{FH}}\right) \tag{4.55}$$

In the case of a *reasonable* behavior of ζ_{11} and ζ_{12} (i.e., the two friction coefficients always have a finite value), it can be demonstrated that the following expression holds not only in the limit of vanishingly small solvent concentrations but also at higher concentrations, over a major portion of the concentration interval (Vrentas and Duda 1977a, b):

$$D = \frac{D_1 \cdot \rho_2 \cdot \hat{V}_2 \cdot \rho_1}{RT}\left(\frac{\partial \mu_1}{\partial \rho_1}\right)_{T,P} \tag{4.56}$$

To extend the range of validity of Equation 4.56 to higher values of solvent concentration, Vrentas and Vrentas (1993) have proposed an alternative relationship involving also D_2 whose value, however, is seldom experimentally available:

$$D = \frac{Q \cdot D_1}{\alpha\phi_1^2 + \left(1-\phi_1\right)\cdot\left(1+2\phi_1\right)} \tag{4.57}$$

where:

$$Q \equiv \frac{\rho_1\left(\partial \mu_1/\partial \rho_1\right)_{T,P}}{RT} \tag{4.58}$$

and

$$\alpha = \frac{M_1}{M_2} \frac{\hat{V}_1^0}{\hat{V}_{20}} \left(\frac{D_1}{D_2} \right)_{\phi_1=1} \tag{4.59}$$

With $\hat{V}_1^0$ representing the specific volume of pure solvent, and $\hat{V}_{20}$, the partial specific volume of the polymer at $\omega_1 = 1$.

The resulting expression for D contains several parameters, which can be estimated a priori, without using any experimental data for diffusivity. In particular, the key variables are K_{11}/γ, $K_{21} - T_{g1}$, K_{12}/γ, $K_{22} - T_{g2}$, $V_1^*, V_2^*, D_0, E, \xi$ and the parameters of the thermodynamic model chosen to formulate the expression for the chemical potential of the solvent in the polymer mixture.

The model provides a reasonable prediction of dependence of mutual diffusion coefficients of small molecular weight penetrants in rubbery polymers on concentration and temperature. Accurate prediction is particularly dependent on the knowledge of the parameters, D_0 and ξ. In fact, for a better prediction, experimental data for diffusivity are needed to estimate D_0, thus enabling the use of the model to correlate existing data and to extrapolate these data for values of temperature and concentration where they are not available. Moreover, it is important to provide a good estimate of the jumping unit molar volumes for both the penetrant and the polymer (for a discussion on this point, see Duda and Zielinski 1996).

Useful qualitative information can be gathered from the free volume theory for the intra-diffusion and the mutual diffusion coefficients. The penetrant intra-diffusion coefficient, in general, increases with its concentration since the free volume associated with the penetrant is commonly considerably greater than the free volume of pure polymer. By contrast, since the expression of mutual diffusion coefficient is contributed by a mobility-related term, which generally increases with concentration, and by a thermodynamic term, $\left(\partial \mu_1 / \partial \rho_1 \right)_{T,P}$, which, instead, decreases, the mutual diffusion coefficient displays a maximum as a function of penetrant concentration.

The higher is the penetrant concentration and the higher is the temperature above T_g, the lower is the concentration dependence of the mutual binary diffusion coefficient, since the overall free volume increases. In polymers well above their T_g, as is the case of molten polymers, mutual diffusion coefficients are large and relatively independent of temperature and concentration. The temperature and concentration dependence of mutual diffusivity increases with penetrant size. In the case of diffusion in polymers of permanent gases, characterized by low molecular weight molecules, mutual binary diffusion coefficients are virtually independent of concentration and, in view of the low apparent activation energy, are also independent of temperature. Molecular weight of polymer does influence the mutual binary diffusion coefficient at low polymer concentration, while it has a very weak effect in concentrated polymer solutions.

There are relevant differences between the Vrentas–Duda model as compared to Fujita's model, the latter being only able to correlate data with no extrapolative or predictive capability. In the evaluation of the mutual thermodynamic diffusivity, D_T, from intra-diffusion coefficients of penetrant and polymer, it is not considered a polymer-fixed frame of reference (that would imply the equivalence of D_T and the penetrant self-diffusion coefficient). Moreover, the polymer chain sections involved in a diffusive jump have not necessarily, in the Vrentas–Duda approach, the same molecular weight as the diffusing penetrant molecule. Moreover, there is an important conceptual difference between the two approaches related to the free volume concept: while the Fujita's theory considers the free volume per unit volume of the solution, the Vrentas–Duda theory uses the free volume per jumping unit. In fact, the two theories provide identical results only if (Duda and Zielinski 1996):

$$\frac{\xi \hat{V}_2^*}{\hat{V}_2^0} = \frac{\hat{V}_1^*}{\hat{V}_1^0} \tag{4.60}$$

4.2.1.3.2   Molecular Models

The physical picture on which *molecular models* are rooted is that micro-cavities dynamically form and disappear in a polymer above T_g. At equilibrium, a definite time-averaged size distribution establishes within the system. A penetrant molecule solubilized within the polymer completes successfully a diffusive elementary jump when a sequence of events occurs: a hole of sufficient size opens up close to its position, the penetrant molecule possesses enough energy to jump into the cavity, and the surrounding macromolecules rearrange their position thus preventing the penetrant to jump back. In an isotropic medium, these jumps occur with equal probability in any direction, and a net diffusive flux of penetrant in a preferred direction eventually occurs if a driving force is present in the form of chemical potential gradient of penetrant.

The various molecular models proposed differ for the physical picture of the illustrated elementary jumps providing, on this basis, an expression for the apparent experimental activation energy, E_{app}, that rules the temperature dependence of the mutual diffusion coefficient:

$$D = D_0 \exp\left(-\frac{E_{app}}{RT}\right) \tag{4.61}$$

In its pioneering work, Meares (1954) proposed that the activation energy predicted theoretically, E_D, is the amount of energy needed to create a cylindrical void connecting the starting and final position of the jumping penetrant molecule obtaining

$$E_D = \frac{\pi}{4}\sigma^2 \cdot N_A \cdot \lambda \cdot \left(CED\right) \tag{4.62}$$

where:
   σ is the collision diameter of the penetrant
   λ is the jump length
   N_A is the Avogadro's number
   CED represents the cohesive energy density of the polymer

This idea was further developed by Brandt (1959) that assumed the activation energy for diffusion as consisting of the sum of the energy needed to bend chains to create the passage for the penetrant molecule to jump and of an intramolecular contribution that is related to the resistance of a chain to bending. In contrast to the Meares' model, activation energy is not simply related to CED, since chain flexibility is also of importance. According to experimental data, this model indicates that in the case of sufficiently small penetrants (σ^2 comprised between 5 and 8Å), no activation energy is required for diffusion.

One of the problems with these early molecular theories is that they fail to predict the downward curvature of E_{app} or E_D as a function of σ^2. To overcome this limitation, DiBenedetto and Paul (1964) proposed a theory based on the idea that a penetrant molecule diffusing in a polymer can be described as an harmonic oscillator confined in a void formed by a bundle of four parallel polymer segments. The activation energy is the energy needed to create a cylindrical void by breaking Van der Waals interactions between four polymer segments and does not include the interaction energy of the penetrant with surrounding molecules as well as the energy needed to compress the surrounding chains. The fitting parameter of the model is the number of center segments per each of the four polymer chain forming the cylindrical void involved in a single diffusive jump.

Both in the model proposed by Brandt and in the model proposed by DiBenedetto and Paul, E_D is assumed to be equivalent to E_{app}.

Pace and Datyner (1979a, b, c, d, 1980a, b) proposed a model that includes the most relevant features of the models by Brandt and DiBenedetto and Paul. The assumption is that diffusion of penetrants involves two different types of motions of the penetrant molecule: one, which is not associated with any activation energy, occurs along the axis of a tube formed by four parallel bundles, while the other occurs in a direction that is orthogonal to the axis of the tube and is allowed by bending of two of the chains of the bundle. This second motion is associated with an activation energy needed to separate the polymer chains. For diffusion processes exhibiting significant activation energies, the two processes are expected to occur in series with the second acting as rate limiting step. The expression for the diffusion coefficient is

$$D = \frac{1}{6}\lambda^2 \cdot v \qquad (4.63)$$

where λ^2 is the mean-square jump displacement of the penetrant molecule, v is the opening frequency of the polymer chains that is related to the activation energy by the following expression:

$$v \equiv 0.0546 \cdot \left(\frac{1}{\lambda^2}\right)\left(\frac{\varepsilon^*}{\rho^*}\right)^{5/4}\left(\frac{\sqrt{\beta}}{m^*}\right)^{1/2} \cdot \frac{\sigma'}{\partial E_D/\partial\sigma} \exp\left(-\frac{E_D}{RT}\right) \qquad (4.64)$$

where:
σ is the diameter of penetrant molecule
m^* is the mass of the polymer backbone element
β is the average single chain-bending modulus per unit length
ρ^* is the average Lennard-Jones distance parameter
ε^* is the average Lennard-Jones energy parameter
$\sigma' = \sigma + \rho^* - \rho$, with ρ representing the equilibrium chain separation

Finally, the expression of E_D is as follows:

$$E_D = 5.23 \cdot \beta^{1/4}\left(\frac{\varepsilon^* \cdot \rho^*}{L^2}\right)^{3/4}(\sigma')^{-1/4} \cdot$$

$$\left\{0.077\left[\left(\frac{\rho^*}{\rho}\right)^{11}(\rho - 10\sigma') - \rho^*\left(\frac{\rho}{\rho^* + \sigma}\right)^{10}\right] - 0.58\left[\left(\frac{\rho^*}{\rho}\right)^5(\rho - 4\sigma') - \rho^*\left(\frac{\rho}{\rho^* + \sigma}\right)^4\right]\right\}^{3/4} \qquad (4.65)$$

where L is the mean backbone element separation along chain axis. In the case of the model proposed by Pace and Datyner, E_D is not considered to be equivalent to E_{app}, the relationship between the two being

$$E_{app} = \frac{d(E_D/T)}{d(1/T)} \qquad (4.66)$$

The dependence of E_{app} on σ is in reasonable agreement with experimental results. The theory has been further modified by Kloczkowski and Mark (1989), by introducing an analytical expression for the interaction potential of polymer chains and by correcting some inconsistencies of the original model in imposed boundary conditions. As a result, slightly lower values of activation energy were calculated but values for mean-square jump lengths obtained by fitting diffusion data with the model were still unrealistically large as for the case of the Pace–Datyner model.

4.2.2 Rheological Properties of Polymer–Gas Solution

The rheological properties of the expanding matter, both in continuous and in batch processing, are of great importance, for the design of extruder geometries and for minimizing consumption and because rheological properties affect to a great extent the whole foaming process to define the final foam morphology. In this context, the expanding matter is a polymer–blowing agent solution at pressures and temperatures suitable to guarantee the preservation of a single phase within the equipment. A common feature of the experimental data on the viscosity of molten polymers plasticized with gaseous or volatile compounds is that the observed concentration-dependent viscosity curves of these systems are similar in shape. Hence, classical viscoelastic scaling methods (e.g., WLF theory) have been used by several authors (Han and Ma 1983; Elkovitch et al. 1999; Areerat et al. 2002), and this scaling has been performed by defining a concentration dependent viscosity reduction factor, a_c, as the ratio of the viscosity of the molten polymer–plasticizer mixture and of the neat molten polymer. On the basis of the experimental observations, according to this scaling approach, it was established that a_c was very weakly dependent upon shear rate and temperature and, consequently, that the following simple relation can be used (Han and Ma 1983):

$$\eta(\dot{\gamma},c,T) = a_c \eta^0(\dot{\gamma},T) \tag{4.67}$$

where $\eta(\dot{\gamma},c,T)$ is the viscosity of the polymer–plasticizer solution at shear rate $\dot{\gamma}$, temperature T, and polymer concentration c, $\eta^0(\dot{\gamma},T)$ is the viscosity of the pure polymer at the same temperature and shear rate, and a_c is the concentration-dependent shift factor. Several attempts have been made to model those experimental observations to predict the viscosity of molten polymer–plasticizer solutions (namely, the value of a_c), dating back to the early works of Kelley and Bueche (1961) and Berry and Fox (1968), which related the effect of the diluent to the relative increase in free volume. More recently, Richards and Prud'homme (1986) described the concentration dependence of the Newtonian viscosity of the solution in the limit of zero shear rate, η_0, by the following relationship, valid for high molecular weight polymers:

$$\eta_0 = K c^n \bar{M}_W^{3.4} \zeta(c,T) \tag{4.68}$$

where:

K is a constant for a given polymer
n is the concentration exponent
$\bar{M}_W$ is the weight average molecular weight of the polymer
ζ is the segmental friction fraction

For ζ, the Doolittle experimental expression, based on free volume, has been proposed (Doolittle and Doolittle 1957):

$$\zeta = \zeta_0 \exp(\lambda/f) \tag{4.69}$$

in which ζ_0 and λ are constants and f is the fractional free volume, defined as the ratio of the free volume to the occupied volume.

The shifting parameter, $a_{c,0}$, for scaling the viscosity of the pure polymer to that of the polymer–diluent mixtures, both in the limit of zero shear rate, can be obtained from Equations 4.67 through 4.69, once the concentration exponent, n, and the free volume fraction, f, are known:

$$a_{c,0} = \frac{\eta_0(c,T)}{\eta_0^0(T)} = c^n \exp\left(\frac{1}{f_m} - \frac{1}{f_p}\right) \tag{4.70}$$

where f_p and f_m are the fractional free volumes of the pure polymer and of the polymer–plasticizer mixture, respectively.

A further development of this approach was described by Gerhardt el al. (1997) and Kwag et al. (1999), which used the following expression for a_c, for the case of a generic $\dot{\gamma}$ value:

$$a_c = \frac{\eta(c,\dot{\gamma},T)}{\eta^0(\dot{\gamma}/a_c,T)} = (1-\omega_c)^n \left(\frac{\hat{V}_p}{\hat{V}_m}\right)^n \exp\left(\frac{1}{f_m} - \frac{1}{f_p}\right),$$

(4.71)

where ω_c is the weight fraction of the plasticizer and $\hat{V}_p$ and $\hat{V}_m$ are the specific volumes of the pure polymer and of the polymer–plasticizer mixture, respectively. It is worth noting that they used the same shift factor also to scale the shear rate, to account for the wider range of Newtonian behavior of the polymeric solution as compared to the pure polymer.

In order to model and/or to predict the viscosity of polymer–plasticizer solutions by this scaling approach, one needs to evaluate the following:

ω_c: The processing variable that is often imposed by the experimental conditions. In fact, it is determined by solubility of the plasticizer in batch experiments or by the polymer–gas mass flux ratio in continuous (in-line extrusion) experiments.

n: From experimental, theoretical, and semi-empirical analyses (Kwag et al. 2001; Royer et al. 2000), it was observed that the exponent of the concentration term ranges from 3 to 4 in several polymer–diluent systems.

$\hat{V}_p$: Experimental *PVT* data are available on a wide number of polymers and can be interpolated using proper equations of state (EoS)—for example, Tait (Tait 1888), Sanchez and Lacombe (SL) (Sanchez and Lacombe 1978), and Simha and Somcynsky (SS) (Simha and Somcynsky 1969) EoS.

$\hat{V}_m$: PVT data on polymeric solutions, mainly for polymer–gas mixtures, are not widely available, and hence, the specific volume of the mixture has to be evaluated using thermodynamic models. Statistical thermodynamic EoS of multicomponent fluid, such as the SL and the SS, allow the evaluation of $\hat{V}_m$ by knowing some characteristic parameters of the pure components and by adopting proper mixing rules. For example, Gerhardt et al. (1998) and Royer et al. (2000) used the SL-EoS. We have to point out, however, that the scarcity of experimental data on the swelling of molten polymers plasticized with gaseous or volatile compound has not allowed an experimental validation of the $\hat{V}_m$ evaluation through the mentioned EoS.

f_p and f_m: The different approaches to evaluate their values determine the predictive capability of the model. Gerhardt et al. (1998), for example, based on the concept of *occupied volume* (i.e., the portion of the total specific volume inaccessible to the chain motion), defined f_p and f_m as follows:

$$f_p = \frac{\hat{V}_p - \hat{V}_{0,p}}{\hat{V}_p}, \qquad f_m = \frac{\hat{V}_m - \hat{V}_{0,m}}{\hat{V}_m}$$

(4.72)

where $\hat{V}_{0,p}$ and $\hat{V}_{0,m}$ are the specific occupied volumes; $\hat{V}_{0,m}$ was calculated from the occupied volumes of the pure components and using a linear mixing rule. The authors used extrapolated crystalline densities by Bondi (1964) for carbon dioxide, while they evaluated the specific occupied volume of the polymer, $\hat{V}_{0,p}$, by fitting viscosity data obtained at different temperatures on the pure polymer using Equation 4.71. In this way, they defined a free volume targeted to the rheological measurements. However, some ambiguity with respect to the classical thermodynamic definition of free volume remains. This procedure seems difficult to apply; moreover, it does not account for the pressure effects. Royer et al. (2000) described the effect of the gas concentration on free volume by T_g shifts, allowing the gas concentration and pressure to be manipulated in a similar manner to temperature with respect to viscoelastic effects. This approach allowed the authors to simply relate the pressure effect on the free volume and, hence, on the viscosity by using the Ferry and Stratton description (Ferry and Stratton 1960).

4.2.3 Sorption Thermodynamics of Low Molecular Weight Compounds in Molten Polymers: Modeling of the Solubility and of the Specific Volume of the Mixture at Equilibrium

Capability of modeling equilibrium sorption thermodynamics of gaseous low molecular weight compounds (penetrants) in molten polymers plays a key role to the aim of predicting gas foaming process in polymers. In the following paragraphs, this topic is reviewed with particular focus on approaches able to supply workable equations for a quantitative evaluation at the equilibrium of the amount of penetrant absorbed in a polymer melt exposed to gas or vapor environments and of the equilibrium specific volume of the corresponding polymer–penetrant mixture. The case of molten amorphous polymers is examined accounting also for the possible presence of polymer–polymer, penetrant–penetrant, and polymer–penetrant-specific interactions. The models we discuss here have all been developed to treat the equilibrium thermodynamic of multicomponent multiphase amorphous systems, which are sufficiently far away from the critical point and all require the knowledge of characteristic parameters of the pure components involved. In particular, the polymer characteristic parameters of a given model can be obtained by best fitting of its volumetric data at different pressures and temperatures (PVT data) in correspondence of equilibrium molten state.

Theoretical efforts devoted to modeling sorption thermodynamics of low molecular weight compounds in rubbery and molten polymers date back to the pioneering work of Flory (1953) based on uncompressible lattice theory of macromolecular solutions to evolve to the more recent EoS approaches based on statistical thermodynamics, referred to as *semi-theoretical EoS* (Tsivintzelis et al. 2007a).

In fact, the Flory–Huggins theory (FH) (Flory 1953) represents a seminal model in the development of all the lattice fluid (LF) mean field theories that take into account the presence of polymeric species. The theory is based on the assumption that a multicomponent fluid system can be described by using a regular LF arrangement, on the grounds of the intrinsic order at low range distance scale observed in liquids and of the fact that the most significant interactions occur at the same scale of distances. Starting from this assumption, each molecular component is divided into elemental mer units, each occupying a single cell of the lattice, that is characterized by a coordination number, Z, defining the total number of contact allowed for each cell with the adjacent ones. The volume of the mixture is assumed to be additive. This last assumption makes the model intrinsically not suited to describe systems in which significant change of volume can takes place, such as the case of polymer–penetrant mixture in equilibrium with gas at high pressure. This is a typical situation of interest in foaming process, where a significant swelling of the pure polymer matrix can take place as a consequence of penetrant sorption promoted by contact with a high-pressure pure penetrant gas phase.

In order to overcome the intrinsic limitations of the FH approach, new theories have then been developed grounded on statistical thermodynamics. In particular, EoS theories provide an effective framework to model thermodynamic properties and describe phase equilibria of mixtures of polymers and low molecular weight compounds. Two of the main classes of semi-theoretical EoS models proposed are those based on compressible LF theory (LF-EoS) (Guggenheim 1952; Flory et al. 1964) and those based on perturbation theory (Reed and Gubbins 1973). These theories have been widely employed to successfully interpret and predict solubility of gases and vapors in amorphous rubbery polymer systems in the cases in which specific interactions can be safely neglected.

Among LF-EoS are worthy of mention the random mixing compressible LF theory of Sanchez and Lacombe (SL) (1976a, b, 1978) and the random mixing hole theory of Simha and Somcinsky (SS) (1969).

In particular, in the SL theory, holes (vacant cells) are introduced in a regular lattice to provide the additional disorder inherent in the liquid state. The cell volume or lattice size is fixed, so changes in the volume of the system are governed by changes in the number of holes. Starting from this basic assumptions, with a classical statistical mechanics procedure, the Gibbs free energy of a generic multicomponent mixture can be calculated. The equilibrium EoS, which relates the equilibrium density of the system to pressure, temperature, and composition, is then obtained by minimizing the Gibbs free energy as a function of the number of holes (or equivalently of the density of the mixture) at fixed pressure, temperature, and composition. Finally, all the other thermodynamic properties of a mixture can be readily obtained from classic thermodynamic expressions, including the chemical potential of the penetrant which is obtained by calculating the partial derivative of the expression of equilibrium Gibbs free energy as a function of number of moles of penetrant under consideration. At a fixed temperature and pressure, predictions of solubility of low molecular weight penetrants in amorphous rubbery polymers and of the density of the polymer–penetrant mixture are obtained by solving at the same time a set of equations, which establishes the equivalence of chemical potential of each low molecular weight compound in the gas/vapor phase and in the polymer phase (Equation 4.73) and the EoS of the mixture (Equation 4.74) and of the pure penetrant phase (i.e., Equation 4.74 written for pure penetrant density) in contact with polymer mixture (assuming, no solubility of polymer in the penetrant reach phase). These equations, in the case of a binary system, are

$$
\left[-\frac{\tilde{\rho}_1}{\tilde{T}_1} + \frac{\tilde{P}_1}{\tilde{T}_1 \tilde{\rho}_1} + \frac{(1-\tilde{\rho}_1)\ln(1-\tilde{\rho}_1)}{\tilde{\rho}_1} + \frac{\ln\tilde{\rho}_1}{r_1^0} \right]
$$
$$
= \ln\phi_1 + 1 - \phi_1 + \tilde{\rho}r_1^0\chi_1(1-\phi_1)^2 + \left[-\frac{\tilde{\rho}}{\tilde{T}_1} + \frac{\tilde{P}_1}{\tilde{T}_1\tilde{\rho}} + \frac{(1-\tilde{\rho})\ln(1-\tilde{\rho})}{\tilde{\rho}} + \frac{\ln\tilde{\rho}}{r_1^0} \right] r_1^0
\tag{4.73}
$$

and

$$
\tilde{\rho} = 1 - \exp\left[-\frac{\tilde{\rho}^2}{\tilde{T}} - \frac{\tilde{P}}{\tilde{T}} - \left(1 - \frac{\phi_1}{r_1}\right)\tilde{\rho} \right]
\tag{4.74}
$$

where:
$\tilde{\rho}$ represents the reduced density of the mixture
$\tilde{\rho}_1$ the reduced density of the pure penetrant in the external phase
$\tilde{P}$ the reduced pressure
$\tilde{P}_1$ the reduced penetrant pressure
$\tilde{T}$ the reduced temperature
$\tilde{T}_1$ the reduced penetrant temperature
r_1^0 represents, for the pure penetrant, the number of lattice sites occupied per molecule of penetrant
ϕ_1 the volumetric fraction of the penetrant in the mixture
χ_1 the interaction parameter

The reduced variables for the pure components are obtained by scaling the corresponding dimensional variables by means of LF quantities, that is, for the case of penetrant:

$$
\tilde{\rho}_1 = \rho / \rho_1^* = \rho r^0 v^* / M_w
$$
$$
\tilde{P}_1 = P / P_1^* = P v_1^* / \varepsilon_1^*
\tag{4.75}
$$
$$
\tilde{T}_1 = T / T_1^* = RT / \varepsilon_1^*
$$

where:

M_W represents the molecular weight

R the ideal gas constant

P the pressure

T the temperature

ρ the density

r^0 the number of sites occupied by a molecule

v_1^* the close volume packed of a mer, corresponding to the volume of a lattice cell

ε_1^* the interaction energy per mer

Analogous expressions can be written for the polymer. The scaling parameters can be easily obtained for the pure components by fitting experimental data (typically PVT data in the case of pure polymers, vapor pressures, and density data for low molecular weight compounds) with EoS (Equation 4.74) written for pure components. In the case of mixtures, similar reduced variables are introduced, but the scaling variables are averages of the corresponding values of the pure components calculated by proper mixing rules (see for details Sanchez and Lacombe 1978). It is evident that, since this model does not assume volume additivity, as opposed to FH theory, it is intrinsically better suited to describe the behavior of systems in which significant change of the volume is associated with the formation of the mixture.

A further step in the development of mean field theories is provided by the SS model (Simha and Somcynsky 1969). Again solubility of low molecular weight penetrants in rubbery polymers can be interpreted based on the coupling of the expression for chemical potential with EoS. The structure of the model is rather similar to SL, however, unlike the classical LF models described above, in this case the unit cell volume is not a constant, being a function of pressure. In this way, changes in the system volume are accounted for by changes in both the number of holes and of size of unit cell.

All the theories described above are based on a LF framework. Recently, a new class of models have been developed to describe the thermodynamic of multicomponent fluids at equilibrium, which is known as *statistical associating fluid theories (SAFTs)* (Chapman et al. 1988, 1989, 1990; Huang and Radosz 1990, 1991) consisting of a workable set of equations allowing the calculation of the Helmholtz free energy of pure components and of multicomponent fluid mixtures These approaches are based on the associating theory (or thermodynamic perturbation theory) developed by Wertheim in his seminal papers (1984a, b, 1986a, b, c, 1987) grounded on a rigorous statistical mechanical framework.

On the basis of Wertheim's perturbation approach, several versions of SAFT have been proposed, developed in terms of residual Helmholtz free energy, which is factorized in different contributions. Number and type of factorization terms can vary depending on the specific model.

The essence of the SAFT approach is that of considering the molecules of components as chains made of identical hard tangent spherical segments, in a square well potential to take into account the dispersion forces, being able of establishing covalent bonds and association interactions. Each pure fluid is described by a set of parameters, which determine its basic physical features, such as the size of the elementary segment, which comprise the molecules; the number of mers constituting the molecules; and the characteristic parameters, which define the interactional potential (such as the well depth of the potential in a square well potential). Assuming a factorization of the partition function in several terms, due to the separation of the forces involved, the total Helmholtz free energy results by the addition of all these contributions. Finally, using well-established classic thermodynamics expressions, it is possible to obtain the EoS of such systems and the chemical potential of each of the components.

It is worth mentioning that an associating fluid thermodynamic model alternative to the SAFT approach was also derived by Prausnitz, and it is known as *perturbed hard sphere chain theory* (Song et al. 1994).

When assessing the solubility of a penetrant in a polymer matrix, account for specific polymer–penetrant interactions can be needed. In fact, some penetrants, including water as the main example, can profoundly affect polymer structure by establishing strong interactions (e.g., hydrogen bonding [HB]), which can possibly disrupt the structure of the polymer matrix. However, all the approaches described above are well suited only for systems that do not display specific interactions since; in the case of LF-EoS, only a mean field contribution is considered when constructing the expression of Gibbs free energy for the mixture and, in the case of the original SAFT formulation, interaction among components is accounted for only by a dispersive term in the residual Helmholtz free energy expression. SAFT approach, however, offers the advantage of being built in a way that naturally allows for the inclusion in the model of short-range interactions (association) and long-range electrostatic interactions. In fact, additional parameters can be introduced in order to describe the interactional potential and the distribution of sticky spots for associating contributions, such as HB and ionic interactions. In fact, the theory can be constructed in a way to provide the Helmholtz free energy and the related compressibility factor for both mixtures and pure systems by considering the contribution of several terms, which account, respectively, for the ideal contribution, the hard chain repulsive contribution, a perturbation contribution due to the dispersion attractive forces, and a perturbation contribution due to associating effects. Following this line, several extensions of the original SAFT have been proposed to account for these effects (Gross and Sadowski 2001, 2002a, b; Gross 2005; Kleiner and Gross 2006; Gross and Vrabec 2006; Tsivintzelis et al. 2007a). In fact, starting from the mentioned works due, respectively, to Chapman and coworkers (1988, 1989, 1990) and to Huang and Radosz (1990, 1991), many other theories have been proposed aimed at simplifying or modifying ad hoc the SAFT EoS for specific cases. In particular, Gross and Sadowski (2001) have proposed a modified version of the original SAFT, known as *perturbed chain SAFT (PC-SAFT)*, which enhances the capability to predict the behavior of fluids containing polymeric species. This theory assumes a hard chain reference fluid, since this reference state is more appropriate to describe the behavior of polymer systems. Afterward, it introduces the dispersion forces and the other associating contributions in a way similar to what the ordinary SAFT does.

Also LF theories have been further developed to include the effect of possible self- and cross-interactions in polymer–penetrant systems. In particular, Panayiotou and Sanchez (PS) (1991) have modified the original SL LF-EoS theory to account for the formation of specific interactions, that is, HB, in polymer–penetrant mixtures. PS model assumes that the configurational partition function can be factorized into two separate contributions: one related to mean field interactions and the other accounting for the effects of specific interactions. The first contribution can be expressed, in principle, by using any available mean field LF theory. In particular, the PS model adopts, for the mean field contribution, the simple random mixing SL model, while the effect of HB interactions is accounted for by using a combinatorial approach first proposed by Veytsman (1990, 1998).

It is worth noting that the mean field LF theories described above, both the original ones and those modified to account for specific interactions, are based on a simplified statistical framework, in which the arrangement of r-mers and holes is assumed to be at random. However, in the case of nonathermal contacts between different kind of r-mers and/or holes, such an assumption is likely to be incorrect (Prausnitz et al. 1998). Based on the pioneering work of Guggenheim (1952), several theories have been developed to deal with nonrandomness distribution of contacts in LF systems, first tackling the cases in which occurrence of specific interactions is not accounted for. The basic idea is that the partition function can be factorized into an ideal random contribution and a nonrandom contribution. The latter contribution is obtained treating each kind of contact as a reversible chemical reaction (*quasichemical* approximation). Guggenheim developed the theory for a LF system without holes, and Panayiotou and Vera (PV) further improved it by introducing a compressible LF model accounting for the presence of hole sites (Panayiotou and Vera 1982). In this model, nonrandomness of contacts between mers of the components of the mixture is assumed, but a random distribution of the holes is imposed (free volume random distribution hypothesis).

Later, You et al. (1994) and Taimoori and Panayiotou (2001) have extended this approach allowing for the nonrandomness of all the possible couple of contacts, also including those involving the hole sites, still adopting a nonrandom quasichemical approximation.

Yeom et al. (1999) and Panayiotou et al. (Vlachou et al. 2002; Panayiotou 2003a; Panayiotou et al. 2004; Panayiotou et al. 2007; Panayiotou 2009) extended this nonrandomness approach to include also the contribution of HB interactions, in a way similar to that adopted in the PS model to extend random LF approach. In the following, we will refer to nonrandom model accounting for HB, proposed by the group of Panayiotou in the references (Panayiotou et al. 2004, 2007; Panayiotou 2009) as *nonrandom LF hydrogen bonding* model (NRHB). The reader is referred to Chapter 3 in this book for a detailed description of this model, which, as discussed above, taking into account both the nonrandomness of each kind of contacts in the mean field framework and the possible contribution deriving from specific interactions such as HB (with no limitations of the number of typologies of proton donor and proton acceptor groups in the system), represents an example of compressible LF approach for the description of equilibrium thermodynamics of multicomponent multiphase systems displaying specific self- and cross-interactions.

In the following paragraphs, some literature examples regarding the use of LF models to deal with molten polymer–gas systems are provided. In fact, all the theories mentioned above have been implemented successfully to model gas and/or vapor sorption thermodynamics in molten polymer at high-pressure conditions, which are the working conditions used to perform a foaming process.

Li et al. (2006a) investigated the solubility of carbon dioxide and nitrogen in a polylactide melt at temperatures from 180°C to 200°C and pressures up to 28 MPa by adopting both the SL and SS theories. In particular, the models were used to properly correct the experimental solubilities by predicting the polymer–gas mixture-specific volume in order to determine the buoyancy effect acting on it (more on this later). A similar approach by using the SL model was followed by Pastore Carbone et al. (2012) for the investigation of sorption thermodynamics of carbon dioxide in molten polycaprolactone (PCL).

In a previous publication, Cotugno et al. (2003) have modeled both the sorption thermodynamics of carbon dioxide in molten PCL, by using the SL model, and the mutual diffusivity of the system, by using a combination of the free volume theory—for the prediction of the intra-diffusion coefficients of the two components—and the SL theory—for the prediction of the thermodynamic driving force for the carbon dioxide diffusive flux. Ott et al. have modeled the sorption thermodynamics of supercritical carbon dioxide in polystyrene during the foam formation by using the SAFT (Ott and Caneba 2010).

The reader is referred to Section 4.2.4, focused on the modeling of the interfacial tension of binary two-phase mixtures, for further examples of the use of the SL theory, integrated with the Poser–Sanchez gradient theory approach (Poser and Sanchez 1981) for the simultaneous prediction of the sorption thermodynamics and of the interfacial tension of gas–polymer systems exposed to high-pressure gases (Harrison et al. 1996; Li et al. 2004). The same approach has been also followed by Enders et al. (2005), which compared SL, SAFT, and PC-SAFT models.

As for the application of the NRHB model for the prediction of sorption thermodynamics of gas in polymeric systems, refer to Chapter 3 of this book. We only mention here the contribution by Tsivintzelis et al. (2007b), which provides a proof of the excellent capability of the theory to model the foaming of polystyrene using carbon dioxide as blowing agent. Quite interestingly, the authors show how it is possible to correlate the observed pore population density of the porous polymer matrix to the pressure and temperature saturation conditions.

Both the SAFT-type models and the nonrandomness LF models are characterized by a certain degree of mathematical complexity, which becomes significant, although still tractable, when fully accounting for both nonrandomness of contacts and HB interactions in NRHB theory. A comparison of the capability of NRHB and of a simplified version of SAFT (i.e., simplified perturbed chain SAFT, sPC-SAFT), which includes an association term (Von Solms et al. 2003) in interpreting and correlating fluid phase equilibria in complex systems that exhibit specific interactions, has been reported by Tsivintzelis et al. (Grenner et al. 2008, Tsivintzelis et al. 2008). According to these authors, NRHB and sPC-SAFT approaches, at least for the case of mixtures of low molecular weight compounds, have proven to display similar performances in correlating and predicting phase equilibria in binary mixtures containing several types of associating fluids.

4.2.4 INTERFACIAL TENSION

The capability to predict the interfacial tension of a polymer–gas mixture in equilibrium with an external gas phase is of primary importance for the modeling of gas foaming process. Several approaches have been proposed in the literature to describe the properties of the interface of two-phase pure fluids and two-phase mixtures such as the *parachor* method (MacLeod 1923), the corresponding states principle (Guggenheim 1945; Zuo and Stenby 1997), the thermodynamic correlations (Girifalco and Good 1957; Fowkes 1962; Winterfield et al. 1978), the perturbation theory (Toxvaerd 1972), integral and density functional theories (Bongiorno and Davis 1975; Evans 1979; Nordholm and Haymet 1980; Almeida and Telo da Gama 1989), and the gradient theory (Cahn and Hilliard 1958).

In particular, we briefly summarize here the gradient theory, which currently represents one of the most widely accepted approaches (see Cahn and Hilliard 1958; Bongiorno et al. 1976; Poser and Sanchez 1981; Panayiotou 2002; Panayiotou 2003b; and Miqueu et al. 2004 for a more detailed description of this theory).

The gradient theory approach has been developed, in its general framework, for multicomponent two-phase systems in equilibrium allowing also for an interfacial nonplanar surface. However, in order to operate with workable equations, the assumption of planar interface is usually done in literature and to this case we refer here.

As will be discussed subsequently, this approach consists of a set of differential equations, which must be solved simultaneously to obtain the concentration profiles of the species under consideration in the interfacial region, for a two-phase system at equilibrium. These nonlinear differential equations display coefficients, which are function of that part of the total local density of the Helmholtz energy of the system that is independent on the gradients of local properties. This term can be expressed as a function of the local values of temperature, pressure, and composition of the components under consideration in the framework of classical irreversible thermodynamics. In this respect, the theory must be integrated with a thermodynamic model, which enables the determination of such functions. To this aim, the most used approaches are based on the use of compressible LF models such as SL theory (obtaining the Poser–Sanchez gradient theory model [Poser and Sanchez 1981]), NRHB model (Panayiotou 2002, 2003b; Tsivintzelis et al. 2007b), or different version of SAFT model (Enders et al. 2005; Lafitte et al. 2010; Nino-Amezquita et al. 2010; Fu et al. 2011). All these thermodynamic models have been already briefly presented in Section 4.1.3.

The interface region is inhomogeneous and, based on the principle of local action that allows the definition of a local thermodynamic state, it is assumed that it is possible to define local values of the thermodynamic state variables, that is, temperature, T; pressure, P; concentration of each species i, c_i; density of the Helmholtz energy for the mixture, a; and chemical potential of each species i present in the system, μ_i.

Since the two-phase system is assumed to be at equilibrium and the effects of external force fields are neglected, no gradient of temperature and of the chemical potentials of each species i are present within the bulk of the two phases. Furthermore, due to the planar interface hypothesis, the equilibrium pressure P^{eq} is the same in the bulk of the two phases.

The key assumption of the gradient theory is that, in addition to classical state variables, that is, P, T, density ρ, and the mass concentration of the n species under investigation $\rho_1, \rho_2, \cdots \rho_n$ (among which one can select the variables describing the local equilibrium state), one should also account for local concentration gradients to define the thermodynamic state of the system at equilibrium. Accordingly, a is expressed as the sum of two contributions: the Helmholtz energy of an homogeneous fluid at local values of P, T, and components concentration, $a_0(c)$, and a term that is a function of the local concentration gradients (i.e., $\nabla\rho_1, \nabla\rho_2, \ldots, \nabla\rho_n$). It follows that the total Helmholtz energy of the whole two-phase system is given by (Poser and Sanchez 1981; Miqueu 2004):

$$A = \int_V \left\{ a_0 \left[\underline{\rho}(r), T(r), P(r) \right] + \sum_i \sum_j \frac{1}{2} k_{ij} \underline{\nabla}\rho_i(r) \underline{\nabla}\rho_j(r) \right\} dV \tag{4.76}$$

where:

$\underline{\rho}(r)$ represents the vector whose components are the mass concentration of each component
$\bar{k}_{ij}$ indicates the influence parameters of the model between species i and species j
r is the position vector
V is the total volume of the system

The expression for a provided by the gradient theory (i.e., the integrand in Equation 4.76) holds within the whole volume of the system. This approach intrinsically assumes that the interface does not consist of a sharp discontinuity between the two homogeneous bulk phases, but is a region where all thermodynamic variables (with the important exclusion of chemical potentials and temperature which are uniform) vary significantly attaining their equilibrium values within the two bulk phases as one moves away from the interfacial region.

Based on classical thermodynamic relationship, the interfacial tension γ is given by (Panayiotou 2002, 2003b)

$$\gamma = \frac{\left[P^{eq}V - \left(\sum_{i=1}^t \mu_i^{eq} N_i - A \right) \right]}{S_0} = \frac{A - A^{eq}}{S_0} \tag{4.77}$$

where:

S_0 is the interface surface area
N_i is the number of moles of component i in the system
A and A^{eq} represent, respectively, the effective total Helmholtz energy obtained by performing the integration in Equation 4.76 in the whole volume, V, of the heterogeneous system and the total equilibrium Helmholtz energy one would calculate for the case of an hypothetical system, displaying the same total volume V, made of the two homogeneous phases separated by a sharp, zero volume interface and characterized by the same equilibrium values of pressure, temperature, and chemical potential of each component within the two bulk phases

The concentration profiles, established within the system, are determined by imposing a free energy minimization condition (Cahn and Hilliard 1958) on the integral appearing in Equation 4.76 (Poser and Sanchez 1981; Miqueu et al. 2004). Application of this minimization procedure (a classical variational problem) results in a set of coupled nonlinear differential equations, where the concentration profiles represent the unknown functions.

In the case one assumes the planarity of the interface, the unknown fields defined above are only functions of the coordinate z orthogonal to the interface and are determined by solving the following Eulero–Lagrange equations (see Panayiotou 2002):

$$\sum_j k_{ij} \frac{d^2\rho_j}{dz^2} = \frac{\partial \Delta a}{\partial \rho_i} = \mu_i^0 \left[\underline{\rho}(z), T, P^{eq} \right] - \mu_i^{eq} \left(\rho^{eq}, T, P^{eq} \right) \quad i, j = 1, 2, \ldots n \tag{4.78}$$

where:

$$\Delta a = a_o\left[\underline{\rho}(z), T, P(z)\right] - \frac{A_e}{V} = a_o(z) + P^{eq} - \sum_{i=1}^{n} \mu_i^{eq}\left(\rho^{eq}, T, P^{eq}\right) \cdot \rho_i(z) \text{ and } \mu_i^0\left[\rho(z), T, P^{eq}\right] \text{ repre-}$$

sent the local equilibrium chemical potential of species i at the local value of concentrations $\rho(z)$, at the temperature T and at the bulk equilibrium pressure P^{eq}, while $\mu_i^{eq}\left(\rho^{eq}, T, P^{eq}\right)$ is the uniform equilibrium chemical potential of species i in each bulk phase (it is obviously the same within the two bulk phases coexisting at equilibrium)

Moreover, in the case of a planar interface, A is given by

$$A = S_0 \int_{-\infty}^{+\infty}\left[a_o(z) + \frac{1}{2}\sum_i\sum_j \rho_i(z)k_{ij}\frac{d^2\rho_j}{dz^2}(z)\right]dz \tag{4.79}$$

Integrating by parts the second term in the integral of Equation 4.79 and making the physically sound assumption that the gradients of concentrations in the $+\infty$ and $-\infty$ limits, the following relationship holds

$$A = S_0 \int_{-\infty}^{+\infty}\left[a_o(z) + \frac{1}{2}\sum_i\sum_j k_{ij}\frac{d\rho_i}{dz}(z)\frac{d\rho_j}{dz}(z)\right]dz \tag{4.80}$$

From Equation 4.80, one finally obtains the following expression for the interfacial tension:

$$\gamma = \int_{-\infty}^{+\infty}\left[\Delta a(z) + \frac{1}{2}\sum_i\sum_j k_{ij}\frac{d\rho_i}{dz}(z)\frac{d\rho_j}{dz}(z)\right]dz \tag{4.81}$$

By multiplying Equation 4.78 by $d\rho_i/dz$ and summing over all the species i, it is obtained that (Poser and Sanchez 1981)

$$\Delta a(z) = \frac{1}{2}\sum_i\sum_j k_{ij}\frac{d\rho_i}{dz}(z)\frac{d\rho_j}{dz}(z) \tag{4.82}$$

and, therefore, the following equation holds:

$$\gamma = \int_{-\infty}^{+\infty} 2\Delta a(z)dz \tag{4.83}$$

The *influence* parameters k_{ij} can be correlated with the energy of interaction between the components i and j. However, the determination of such parameters involve quite complex statistical mechanics procedures and a detailed knowledge of the energy of self- (i.e., the case $i = j$) and cross- (i.e., the case $i \neq j$) interactions. On this basis, Poser and Sanchez (1981) proposed a possible procedure to predict the values of the k_{ii} parameters from the characteristic EoS LF parameters of the corresponding component i. Furthermore, for a binary two phase system, a combining rule is proposed by the same authors for the prediction of k_{ij} from the influence parameters k_{ii} and k_{jj}. It is worth noting that this approach is limited to random mixing LF models in which only mean field interactions are taken into account.

More in general, in the gradient theory approach, the *influence* parameters are typically assumed to be independent on the concentration and pressure (and therefore on z). Moreover, since the value of each k_{ii} is correlated with intrinsic energetic properties of the corresponding pure component i, it can be obtained as best fitting parameter of experimental data for interfacial tension, evaluated for pure components at liquid–vapor equilibrium at different temperatures. Furthermore, the cross influence parameters, k_{ij}, are typically treated as adjustable parameters and they can be calculated from best fitting procedures of interfacial tension data for the multicomponent two-phase mixture at different pressure and temperature values.

It is worth noting that, in the present discussion, we have mainly referred the discussion to the operative formulation of the gradient theory developed by Poser and Sanchez (1981). In fact, this formulation is the most widely used in the literature. (Harrison et al. 1996; Li et al. 2004; Enders et al. 2005; Kahl and Winkelmann 2005; Lafitte et al. 2010; Nino-Amezquita et al. 2010, Fu et al. 2011). A slightly different version of the gradient theory has been proposed by Panayiotou (2003b). In this formulation, the author introduces an additional internal parameter, β, which can be obtained by comparison of the two intrinsically equivalent expression derived for the interfacial tension. The first expression was obtained by using an EoS model to calculate the expression of local density of the Helmholtz energy, following the same line of thought adopted by Poser and Sanchez, while the second expression was derived by using the same EoS model to calculate the local expression of the pressure. The parameter β is introduced to take into account, approximately, the contribution to the Helmholtz energy density deriving from gradient of concentration terms beyond the quadratic ones (which are the only terms considered in the Poser–Sanchez approach of the gradient theory). For a detailed discussion of this issue, see Panayiotou (2003a, b) and Panayiotou et al. (2004).

In the following paragraphs, we illustrate how to proceed for the calculation of the concentration profiles, at equilibrium, within the interfacial region and, consequently, of the interfacial tension. For the sake of simplicity, we will refer to the case of a binary two-phase system, which is the most investigated one in the literature.

In this case, to make the theory predictive for the mixtures, the usual geometrical mean assumption for the *influence* parameter k_{12} is done, that is, $k_{12} = (k_{11}k_{22})^{1/2}$. Based on this assumption and by combining the two coupled nonlinear differential Equations 4.78, we will illustrate in the following that a single nonlinear algebraic equation (i.e., Equation 4.84) can be obtained that relates concentration profiles of the two components (Poser and Sanchez 1981):

$$k_{22}^{1/2}\left(\frac{\partial \Delta a}{\partial \rho_1}\right) - k_{11}^{1/2}\left(\frac{\partial \Delta a}{\partial \rho_2}\right) = 0 \tag{4.84}$$

This expression allows to put the concentration profile of one species as a function of the other (this is performed numerically in view of the complexity of this equation). As a consequence, in this case the working equations are Equation 4.84 and one of the two coupled differential equations (Equation 4.78).

Actually, we will also illustrate that still in the case of geometrical mean assumption for the *influence* parameter k_{12}, in the framework of the more general operative approach based upon an ad hoc change of variables described in the following, it is possible to obtain the interfacial tension γ and the mass concentration profiles without solving any differential equation.

We proceed now illustrating the general operative procedure usually performed for the determination the value of γ from Equation 4.83, showing also how Equation 4.84 can be obtained in the case of assumption of geometrical mean rule for k_{12}. In principle, calculation of γ requires the knowledge of the profile of the mass concentration of the two species and these two functions are provided by the solution of the Equation 4.78. These coupled equations, in general, must be solved imposing two boundary conditions for each mass concentration profile, that is, in vectorial terms, for $z = +\infty$: $\underline{\rho}(z) \to \underline{\rho}(z)^{1-eq}$ and for $z = -\infty$: $\underline{\rho}(z) \to \underline{\rho}(z)^{2-eq}$, where $\underline{\rho}(z)^{i-eq}$ is the set of equilibrium mass concentration in the bulk phase i.

It is worth noting that the Equation 4.78 must be solved in an unlimited domain. In fact, the size (thickness) of the interfacial region, identified as that finite domain where one actually needs to solve these differential equations, is unknown since it is, in turn, a function of the unknown concentration profiles. Furthermore, also the integral of Equation 4.83 must be solved in unlimited domain thus involving to perform an analysis of integrability of the integrand function and resulting in a complex numerical implementation.

The analytical and numerical analysis can be made simpler by performing a change of variable. In particular, on reasonable physical grounds, it is assumed that at least one mass concentration profile is monotonic as a function of z, and therefore, inverting this monotonic function, this mass profile variable can be used as the independent variable in place of z. In fact, the assumption of monotonicity of the polymer concentration profile is generally reasonable in a polymer–gas mixture. However, it is, in any case, suggested to perform an *a posteriori* check of this hypothesis (Miqueu 2004; Poser and Sanchez 1981). This change of variable, by using usual chain rules for derivative and integrals, allows to transform the Equations 4.78 and 4.83 defined in the z domain into equations whose domain is in the space of the monotonic mass concentration profile selected (here and subsequently, this selected mass concentration profile is referred to as ρ_1). This results in a great simplification since the transformed equations are now defined in a known finite domain corresponding to $\left[\rho_1^{1-eq}\rho_1^{2-eq}\right]$, where ρ_1^{i-eq} represents the equilibrium mass concentration of the selected variable ρ_1 in the bulk phase i. By convention, we will refer as phase 2 the one in which the equilibrium mass concentration of species 1 is higher.

In this way, the two coupled second-order differential equations with unknown mass concentration profiles, which are function of z (i.e., Equation 4.78), can be combined into one second-order differential equation (i.e., Equation 4.85) with the only unknown function being $\rho_2(\rho_1)$, where ρ_2 represents the mass concentration of the other specie under consideration:

$$\left[k_{11}+2k_{12}\left(\frac{d\rho_2}{d\rho_1}\right)+k_{22}\left(\frac{d\rho_2}{d\rho_1}\right)^2\right]\cdot\left\{\frac{\partial\Delta a}{\partial\rho_2}\cdot\left[k_{11}+k_{12}\left(\frac{d\rho_2}{d\rho_1}\right)\right]-\frac{\partial\Delta a}{\partial\rho_2}\cdot\left[k_{12}+k_{22}\left(\frac{d\rho_2}{d\rho_1}\right)\right]\right\}$$
$$-\left(\frac{d^2\rho_2}{d\rho_1^2}\right)\left[2\Delta a\left(k_{11}k_{22}-k_{12}^2\right)\right]=0 \tag{4.85}$$

Equation 4.85 represents a classical second-order differential boundary problem in which the two conditions are assigned on the unknown function at two different boundary points. Since at each ρ_1^{i-eq}, ρ_2 must take its corresponding value of equilibrium, that is, ρ_2^{i-eq}, that it takes within the bulk phase i.

This equation can be solved by numerical methods. In particular, Miquei et al. (2004) approached the mathematical problem by solving the corresponding set of nonlinear algebraic equations defined on the grid points (i.e., the points on ρ_1 space) associated with a finite difference scheme. The solution of the Equation 4.84 has been taken as first guess values.

The change of independent variable of the problem from z to ρ_1 results in the transformation of Equation 4.83 into the following equation:

$$\gamma=2^{1/2}\int_{\rho_1^{1eq}}^{\rho_1^{2eq}}\left[k_{11}+2k_{12}\left(\frac{d\rho_2}{d\rho_1}\right)+k_{22}\left(\frac{d\rho_2}{d\rho_1}\right)^2\right]^{1/2}\Delta a^{1/2}d\rho_1 \tag{4.86}$$

Therefore, once the function $\rho_2(\rho_1)$ is obtained by numerical solution of Equation 4.85, its derivative $d\rho_2/d\rho_1$ can be then numerically evaluated from it and, in turn, it is possible to calculate γ by

Equation 4.86. It is worth noting that if the mean rule for k_{12} holds, the $\rho_2(\rho_1)$ function is provided by the algebraic Equation 4.84, as already anticipated.

Concerning the determination of the mass concentration profiles, one can proceed by starting from Equation 4.82, which, in the case of a binary system, can be written as

$$\Delta a = \frac{1}{2}\left[k_{11}\left(\frac{d\rho_1}{dz}\right)^2 + 2k_{12}\frac{d\rho_1}{dz}\frac{d\rho_2}{dz} + k_{22}\left(\frac{d\rho_2}{dz}\right)^2 \right] \tag{4.87}$$

Since by applying the chain rule on the derivative, one has

$$\frac{d\rho_2/dz}{d\rho_1/dz} = \frac{d\rho_2}{d\rho_1} \tag{4.88}$$

then, the following equation holds

$$\frac{d\rho_1}{dz} = \pm\left[\frac{2\Delta a}{k_{11} + 2k_{12}\left(d\rho_2/d\rho_1\right) + k_{22}\left(d\rho_2/d\rho_1\right)^2} \right]^{1/2} \tag{4.89}$$

Finally, by integrating Equation 4.89 and taking only the physically meaningful solution, the following equation is obtained:

$$z - z_0 = 2^{1/2}\int_{\rho_1^{1eq}}^{\rho_1(z)}\left[k_{11} + 2k_{12}\left(\frac{d\rho_2}{d\rho_1}\right) + k_{22}\left(\frac{d\rho_2}{d\rho_1}\right)^2 \right]^{1/2}\Delta a^{-1/2}d\rho_1 \tag{4.90}$$

By solving numerically the integral on the right-hand side of Equation 4.90, one obtains the value of $z - z_0$ corresponding to $\rho_1(z)$ (z_0 is the position of the bulk phase 1, that is, the position at which $\rho_1 = \rho_1^{1eq}$). By inverting this relationship, the profile $\rho_1(z)$ is finally obtained, and, in turn, from the relationship $\rho_2(\rho_1)$, the interfacial profile of $\rho_2(z)$ can be calculated. It is noticed how, in the case of assumption of geometrical mean for k_{12}, calculation of mass concentration profiles in the interfacial region does not involve the solution of any differential equation, but requires only the less complex calculation of numerical integrals.

The first attempt of application of gradient theory to binary polymer-low molecular weight compounds two-phase systems, at the best of authors' knowledge, can be found in Poser and Sanchez (1981), with reference to interfacial tension data for the system tetraline-poly(dimethylsiloxane) at 30°C.

Later, Harrison et al. (1996) studied the interfacial tension between supercritical carbon dioxide and polyethylene glycol oligomer (MW600) in the melting state, by using the gradient theory in the framework of SL theory (Poser–Sanchez model) (Poser and Sanchez 1981). In view of the sharp drop of polymer concentration in the extremely thin interface region moving from the polymer-rich bulk phase to the external gas phase, Harrison et al. (1996) assumed the linearity of the relationship $\rho_2(\rho_1)$, in order to overcome the numerical difficulties that could arise when solving the differential equations for the mass concentration profiles. They implemented the Poser–Sanchez model in its predictive version, that is, assuming $k_{12} = (k_{11}k_{22})^{1/2}$. The theoretical predictions resulted to be in a quite good agreement with the experimental data at different carbon dioxide pressures. In particular, they found that in the binary polymer–gas systems investigated, as the density of pure gas phase increases the interfacial tension decreases. This finding is expected since the Helmholtz energy density of the gas phase becomes more similar to the one of the polymer–gas mixture.

Afterward, Li et al. 2004 have used the Poser–Sanchez model, adopting the same simplified assigned linear profile $\rho_2(\rho_1)$, to investigate the interfacial tension of melt polystyrene–carbon dioxide two-phase systems at high temperatures (210°C–230°C) and pressures (form 1 to 140 atm). Also in this case, the Poser–Sanchez model was implemented in its predictive version, evaluating the pure influence parameters k_{ii} as best fitting parameters of the corresponding pure components interfacial tension data and calculating the cross influence parameter k_{ij} according to the geometrical mean rule. They found that the model is able to predict at each temperature the surfactant effect of the gas on the interfacial tension of the system. Moreover, the model, in a qualitative agreement with experimental findings, predicted the decrease of the system interfacial tension as a function of the gas pressure and the decrease of interfacial tension with the temperature at a fixed gas concentration in the polymer-rich phase. More in detail, the model predicted quantitatively the plateau that the experimental data for interfacial tension exhibited in the range of highest gas pressures investigated (80–140 atm), but it strongly underestimated the interfacial tension in comparison to the experimental data in the range of moderate pressure analyzed (20–60 atm). (For instance, the relative error was of approximately 25% at pressures around 20 atm, and it steadily decreased as a function of pressure.)

More recently, Enders et al. (2005), based on the same assumptions of linear behavior for the function $\rho_2(\rho_1)$ and of geometric rule for the cross influence parameter, compared the predictive capability of three different versions of gradient theory based, respectively, on SL, SAFT, and PC-SAFT EoS theories to express the density of the Helmholtz energy of the system and to express local equilibrium thermodynamic relationships. In particular, they investigated the system polystyrene–carbon dioxide from atmospheric pressure up to 25 MPa and from 90°C to 160°C, and they found that all the three models adopted allowed the correct prediction of the experimental trend of interfacial tension with temperature and pressure for the system investigated, but none of the three implemented versions of the models was able to attain a quantitative agreement with the experimental data.

Afterward, Tsivintizelis et al. (2007b) investigated the validity of nucleation theory combined with the NRHB model to predict the morphology of microcellular polystyrene foams using supercritical carbon dioxide as blowing agent. To this aim, they needed to predict the interfacial tension of the polymer–gas two-phase systems in the range of temperature equal to 80°C–120°C and from atmospheric pressure up 40 MPa. Due to the difficulty of correctly calculating the interfacial mass concentration profile of the polymer between the polymer-rich phase and the pure gas phase, they adopted an empirical relationship to predict the mixture interfacial tension as a function of the two pure components interfacial tension (equation proposed by Reid et al. 1988). Finally, they assumed that the pure supercritical gas interfacial tension can be reasonably fixed equal to zero and they calculated the one of the pure polymer by using the gradient theory in the framework of the NRHB model. The influence parameter of the pure polymer and its internal parameter β were obtained as best fitting parameter of the pure polystyrene interfacial tension data (Panayiotou et al. 2004). In particular, the dimensionless parameter β was found, as expected, to be the same of several other pure components investigated (i.e., equal to 2).

It is worth noting that, in the case of pure polymeric species, it is reasonable to transform the problem from the z space to the mass concentration space, as discussed previously (i.e., a monotonic spatial profile is expected for the pure polymer). Once this transformation is made, for a pure component system, it is possible to calculate the interfacial tension, without solving for the concentration profile (Poser and Sanchez 1981). This can be seen considering the Poser–Sanchez formulation of the gradient theory by observing that, if one uses the same procedure followed for binary systems, Equation 4.86 collapses, for a pure component, as follows:

$$\gamma = 2^{1/2} \int_{\rho_1^{1eq}}^{\rho_1^{2eq}} k_{11}^{1/2} \Delta a^{1/2} d\rho_1 \tag{4.91}$$

In a totally similar fashion, the NRHB formulation of the gradient theory leads to the following equation for the pure component interfacial tension (Panayiotou et al. 2004):

$$\gamma = (2+\beta) \int\limits_{\rho_1^{1eq}}^{\rho_1^{2eq}} k_{11}^{1/2} \Delta a^{1/2} d\rho_1 \tag{4.92}$$

Therefore, in both the formulations developed, once the influence parameter is known (and once the β parameter has also been determined, in the NRHB approach), calculation of the interfacial tension needs only the knowledge of the dependence of Δa on the values of local density and pressure at the fixed temperature.

4.3   EXPERIMENTAL METHODS

4.3.1   MEASUREMENT OF GAS SOLUBILITY AND DIFFUSIVITY

4.3.1.1   Gravimetric Methods

The gravimetric sorption method measures the mass uptake of a polymer sample resulting from gas dissolution within the polymer matrix. Many equipments, both commercially available and custom designed, have been proposed in the literature to test gravimetrically the sorption of gas/vapors in polymer systems in molten and/or solid state. These equipments may differ in the principle on which the measurement of mass change of the polymer sample is based.

One of the first method proposed is represented by the so-called McBain balance (McBain and Bakr 1926). In this system, the polymer is suspended to a quartz spring and, as the weight of the sample increases as a consequence of gas dissolution, the spring elongates. It is then possible to correlate the elongation of the spring, monitored optically (for instance, by using a cathetometer or by an automated traveling microscope with digital acquisition of the image of a marker placed on the spring), to the mass of the sample according to Hooke's law, once a calibration of the spring elongation with known weights has been performed. The first versions of the equipments based on the quartz spring operated with gases or vapors at low pressure. An example can be found in some contributions by Duda et al. (Duda and Ni 1978, 1979), which used this technique to measure the solubility of toluene and ethylbenzene in molten polystyrene.

This experimental approach has been extended to analyze gas sorption at higher pressure, as is the case of the analysis of sorption thermodynamics and mutual diffusivity of carbon dioxide in molten PCL at moderately high pressures (slightly less than 100 atm), contributed by Cotugno et al. (2003).

More recently, gas sorption and diffusivity in polymer samples at high temperatures and pressures was investigated using electronic microbalances. For instance, Wong et al. (1998) used an electronic microbalance to investigate the gas solubility and diffusivity of CO_2 and HFC134a in several polymeric systems.

These devices can attain very high accuracy and sensitivity as well as high acquisition rate, making them suitable for sorption measurements involving low solubility gases and for the investigation of fast gas/vapors sorption/desorption processes.

More recently, magnetic suspension microbalances have been introduced to perform these measurements. The main advantage of these systems consists of the separation of the weighing electronics (located in a section of the equipment operating at room conditions) from the test chamber (in which the temperature and pressure test conditions are imposed). The electronics and the stem with hanging sample pan are coupled by a computer-controlled electromagnetic system. The particular arrangement of these devices allows also the measurement of the density of the gas in the measuring chamber. Examples of the use of these systems to measure gas solubilities in polymer samples can be found in several publications (Chaudhary and Johns 1998; Sato et al. 2001; Pastore Carbone et al. 2011).

Independently of the type of device used, gravimetric tests are affected by the intrinsic error induced by the contribution of buoyancy effect, which sums algebraically to the mass sorption contribution leading to an underestimation of the sorbed amount of penetrant. In order to account for this effect to, thus obtaining correct and reliable data on gas solubility in molten polymers, one needs to, thus know the actual volume of the polymer–gas mixture to properly account for buoyancy effect.

To this end, one could use the mixture volume, as estimated by an EoS in a trial-and-error procedure, to correct each experimental solubility data. Li et al. (2006b) showed that for the case of the PP-CO_2 system, the buoyancy effect correction of raw data (apparent solubility) related to volume swelling associated with sorption is rather strong. They also pointed out how the estimated mixture volume and, consequently, the associated correction of the apparent solubility are quite dependent on the specific EoS adopted to perform volume estimation. In particular, they compared the results obtained by correcting experimental data by using SL model (Sanchez and Lacombe 1976a), SS model (Simha and Somcynsky 1969), and PC-SAFT (Gross and Sadowski 2001, 2002a). Later, Park et al. (Li and Park 2009; Mahmood et al. 2014) measured experimentally the volume of molten polymer–gas mixtures showing how SL theory provides a rather large overestimation of mixtures volume, while SS theory supplies values that are in a much better agreement with volumetric experimental data. Based on these results, SS theory can be considered more reliable to correct experimental solubility data whenever an experimental measure of mixture volume should not be available.

Instead of estimating the mixture volume by using EoS, a more reliable procedure consists of measuring both the apparent weight increase of the sample and the volume of polymer–gas solutions, at the same test conditions (Park et al. 2006; Liu and Tomasko 2007; Pastore Carbone et al. 2011, 2012). In addition, the specific volume of the gas phase can be estimated by adopting detailed experimental tables, by using theoretical approaches, like EoS models, or by resorting again to a direct measurement, as is possible in the case of magnetic suspension microbalances. In this respect, the experimental procedure proposed by Pastore Carbone et al. (2011), in contrast to the ones proposed, respectively, by Park et al. (2006) and Liu and Tomasko (2007), has the advantage of allowing the simultaneous measurement of solubility and volumetric data of the polymer–gas mixture within the same chamber. In this way, it is granted that both measurements are conducted in exactly the same conditions.

In principle, the correction procedures described above for the buoyancy effect can be applied also to data collected during transient stages of a sorption/desorption experiments. However, both the procedures based on the use of EoS theories and those based on direct measurement of volume of polymer–gas mixtures are much less reliable if applied in nonequilibrium conditions. As a consequence, the described approaches reported above to account for buoyancy effects are prevalently used to determine data for equilibrium sorption isotherms.

4.3.1.2　Pressure Decay Technique

An alternative way to gather information on the solubility and mass transport properties of a gas/vapor in a polymer is provided by the *pressure decay* method, which has been extensively adopted also in the case of molten gas–polymer mixtures, in all its three different versions, that is, *single-*, *dual-*, and *three-champer* systems.

There are several contributions in the literature where the experimental investigation of gas sorption and diffusion in polymers at high pressures is performed adopting this technique. Among the first examples, we cite here the work of Newitt and Weale (1948), which have investigated the solubility of hydrogen and nitrogen in polystyrene over the pressure range of 8.1–30.4 MPa, and at elevated temperatures (up to 190°C) by using a first rough example of *dual-chamber* system. Lundberg et al., using a *single-chamber* apparatus, obtained the solubility and mutual diffusivities of gases in polymers at pressures between 3 and 71 MPa by using stepwise sorption experiment carried out on molten polymers (Lundberg et al. 1960, 1962; Lundberg 1964). Later, Durrill et al., using a *single-chamber* device (Durrill 1966; Durrill and Griskey 1966, 1969), investigated the solubility and mutual diffusivity coefficients of nitrogen, helium, carbon dioxide, and argon in molten polyethylene, polyisobutylene, and polypropylene up to pressure equal to 2 MPa.

The more simple version of this technique is provided by the *single-chamber* method. In this case, the sample is inserted in a chamber, of known volume and at controlled temperature, which is filled with the gas of interest and in which the pressure is monitored as a function of time. As soon as the system is closed, the mass of gas sorbed in the polymer matrix at each time is assumed to be given by the difference between the initial amount of gas and the gas present at each instant of time in the volume occupied by the gas phase (the assumption is made that the adsorption of gas on the internal surfaces of the chamber is negligible). In fact, once the volume occupied by the gas phase is known, the number of moles of gas present in the gas phase can be calculated using an EoS model as a function of the measured values of pressure and temperature. To this aim, the volume available to the gas phase needs to be evaluated as the difference between the total, known, volume of the chamber and the volume of the polymer–penetrant mixture. Again, as in the case of performing the correction for buoyancy effects in gravimetric methods, one should resort to a trial-and-error procedure based on EoS calculations or to optical measurements of the polymer–gas mixture volume.

It is evident that this technique requires a careful calibrations of the volumes and can only be used for gases whose EoS are accurately known. Normally the sorption test are realized stepwise and a time lap is required for the stabilization of the measured pressure after the introduction of the gas into the sample chamber (Davis et al. 2004). The initial pressure value, which is needed to estimate the initial mass of gas introduced into the system, must be evaluated by extrapolating from the pressure decay curve the value of pressure present at time zero within the measuring chamber. This procedure can result in a significant error in the estimation of the initial mass of gas and of the related determination of the total mass change due to gas sorption in the polymer. In order to circumvent this source of error, a *dual-chamber* system has been proposed. In this case, a reservoir chamber of known volume is previously filled with the gas at a known pressure. This chamber is then connected to the chamber containing the sample by opening a valve. Finally, the valve is closed and the sorption is monitored by following the pressure decay within the sample chamber. The mass sorbed in the polymer sample is then easily calculated as the difference between the mass of gas initially present in the reservoir chamber and the total mass of gas finally present in the measuring and reservoir chambers (see Koros and Paul 1976 for all the details about this experimental approach, with particular reference to the optimization of the ratio of the volumes of the two chambers).

Improvements to the *dual-chamber* technique have then been proposed. For instance, Sato et al. have used a *three-chamber* pressure decay technique to investigate the solubilities of carbon dioxide and nitrogen in molten polystyrene under high-pressure conditions (Sato et al. 1996, 1999), demonstrating that this technique is less affected by cumulative errors in stepwise pressure decay experiments.

4.3.1.3 Inverse Gas Chromatography

There are several examples reported in the literature of sorption data of different gases and vapors in various polymeric materials determined by using inverse gas chromatography (Shiyao et al. 1989; Belov et al. 2006). This technique will not be further discussed here since, at the best of authors' knowledge, these investigations are limited to measurements at pressures much lower than those typical of the operative foaming conditions.

4.3.1.4 Spectroscopic Techniques

Another class of experimental techniques adopted for the quantitative measurement of the solubility and of the diffusivity of gases within polymers are those based on vibrational spectroscopy measurements. Relevant examples are the attenuated total reflectance (ATR), Fourier transformed infrared (FTIR), and Raman spectroscopy. Besides providing information on concentration of gases and vapors absorbed within polymers, vibrational spectroscopy can also be used to evaluate the concentration profiles of penetrants. Moreover, the approaches provide important information on molecular penetrant–polymer interactions (Kazarian 1997; Murphy et al. 2003). There exists an extensive literature dealing with the molecular-level insight that can be gathered by means of IR and Raman spectroscopy, interpreting the modifications occurring in the spectral patterns of both the penetrant and the matrix upon sorption as evidences of molecular interactions (e.g., HB; Lewis acid-base interaction, LA-LB) (Fried and Li 1990; Briscoe and Kelly 1993; Kazarian et al. 1996a, b; Flichy et al. 2002; Kazarian 2002; Nalawade et al. 2006b; Musto et al. 2007, 2012, 2014; Knauer et al. 2013). For instance, investigations based on the use of ATR and FTIR have highlighted that CO_2 interacts with most polymers (Meredith et al. 1996; Nalawade et al. 2008). As reported for the first time by Kazarian et al. for the polymethylmetacrylate (PMMA)–CO_2 system (Kazarian et al. 1996a), CO_2 can participate in LA-LB interactions with carbonyl groups. In fact, in the presence of dissolved CO_2, carbonyl stretching vibrations of several polymers, for example, PMMA, polycarbonate (PC), PCL, polylactic acid (PLLA and PDLLA), have been found to shift in wave number, indicating the interaction of the gas with the polymer (Meredith et al. 1996; Nalawade et al. 2008). In addition, changes in the absorption spectrum of CO_2, in the regions corresponding to the molecule bending mode vibrations, have been observed: the characteristic splitting of the specific band for the dissolved CO_2 is the evidence of the interaction. In fact, due to the interaction of electron lone pairs of the carbonyl oxygen with the carbon atom of the CO_2 molecule, the double degeneracy of the v_2 mode is removed, thus resulting in the peak splitting (Kazarian et al. 1996a; Meredith et al. 1996).

Details on the application of FTIR, ATR, and Raman spectroscopy for the investigation of gas sorption in polymers at high pressures and high temperatures and of the gas–polymer interaction are given in the following sections.

4.3.1.4.1 IR Spectroscopy

In the last two decades, the use of infrared spectroscopy has gained enormous ground in the measurement of gas sorption in polymers. The works by Fieldson and Barbari (1993, 1995; Barbari et al. 1996) and by Kazarian's group (Vincent et al. 1997; 1998; Kazarian et al. 1999; Brantley et al. 2000) represent the pioneering attempts in studying the diffusion of low molecular weight compounds in polymers by using IR spectroscopy. The polymer sample, in the form of a film, is exposed to the gas/vapor atmosphere inside an optical cell, and during the sorption process, specific infrared absorption bands of the penetrant sorbed within the polymer are monitored. The evolution of their intensity with time provides a description of the diffusion mechanism. In fact, the concentration of the penetrant within the polymer is related to the absorbance of the specific bands of the penetrant by the Beer–Lambert law:

$$A = \varepsilon \cdot c \cdot l \tag{4.93}$$

where:
  A is the absorbance
  ε is the molar absorptivity
  c is the concentration of the penetrant
  l is the path length (i.e., the thickness of polymer sample)

In the measurement performed by using the FTIR-ATR, only one side of the film is exposed to vapor, gas, or liquid environment, while the other surface of the sample is in contact with a suitable optical element (crystal) and is probed by using an evanescent wave. The application of FTIR-ATR relies on the spectroscopic observation of penetrant localized within a short distance (0.5–5 μm) from the interface between the sample and the optical element. The efficiency of the contact between the sample and the crystal is the key factor to ensure the quality of the collected spectrum and, consequently, the accuracy of the quantitative analysis. By combining an FTIR apparatus with a high-pressure unit, Kazarian et al. highlighted the interactions, at molecular level, of several polymers with supercritical CO_2 (Kazarian et al. 1996a; Kazarian 2002). In particular, the recent development of ATR using a diamond ATR accessory, which is a miniature pressure cell (Flichy et al. 2002; Nalawade et al. 2008; Novitskiy et al. 2011), has allowed for the direct and simultaneous measurement of the polymer swelling and CO_2 sorption by measuring the intensity of the relevant IR bands (Kazarian 1997). Furthermore, investigations on molten polymers have been performed by using a heating unit, in conjunction with a temperature controller (Pasquali et al. 2008a; Nalawade et al. 2006b). This high-temperature/high-pressure ATR setup allowed for the investigation of interactions between supercritical CO_2 and several polymer, which are commonly used for foaming (e.g., PLA, PCL, and poly(lactide-*co*-glycolide) [PLGA]) (Nalawade et al. 2008; Tai et al. 2010). Furthermore, the combination of the ATR diamond accessory with an FTIR imaging system allowed a qualitative chemical imaging analysis of polymer systems exposed to high-pressure and supercritical environments (Kazarian and Chan 2004; Fleming et al. 2006). An illustration of the high-pressure apparatus and some results obtained from the investigation of polyethyleneglycol (PEG)–CO_2 system are shown in Figures 4.1 and 4.2.

The development of in situ measurements based on FTIR-ATR significantly improved the accuracy of the transport kinetic analysis; however, this technique is characterized by several drawbacks, such as those related to the efficiency of the contact between the sample and the optical element, which is rather difficult to control especially for long-term measurement, and the relationship between absorbance and penetrant concentration, which is more complex than in transmission (Cotugno et al. 2001). By using *in situ* FTIR spectroscopy in the transmission mode, most of the ATR drawbacks can be avoided. For example, the use of transmission spectroscopy does not require information on the refractive index of the sample because the path length remains constant (assuming that elastic deformation of the cell under high pressure is negligible, which is a reasonable assumption) (Guadagno and Kazarian 2004). In the method proposed by Cotugno et al. (2001), a freestanding film is placed in a cell, which is equipped with KBr windows and allows for an accurate control of the environment (e.g., temperature and vapor/gas pressure). During the sorption process, the transmission spectra are collected until the attainment of the equilibrium. The interference of the gas–vapor phase has to be eliminated carefully by using, as background, the spectrum acquired without the polymer sample, at test conditions. This procedure imposes an extremely accurate control of the internal atmosphere of the cell, which can be achieved with a proper design of the sorption cell and of the service equipment (see Cotugno et al. 2001). Studying high-pressure systems may encounter one major obstacle that is the very large—often offscale—absorbances arising from the dense fluid phase, which indicate that essentially all of the IR radiation in these regions is completely absorbed by the dense fluid phase, thus making these regions useless for quantitative analysis. Kazarian et al. (1996b) designed a dedicated cell for measurements at high pressures, which consists of a cell with two parallel optical paths: one for the measurement of the polymer sample immersed in the surrounding fluid and the other for the measurement of only the surrounding fluids. By using this type of cell, the sorption of high-pressure CO_2 and the swelling behavior on PEG and poly(propylene glycol) (PPG) have been investigated (Guadagno and Kazarian 2004).

4.3.1.4.2 Raman Spectroscopy

Raman spectroscopy, an inelastic laser light scattering process, is species sensitive and quantifiable. It has been successfully applied for investigating heat and mass transfer (Knauer et al. 2011), solubility (Rodriguez-Meizoso et al. 2012), as well as diffusion (Yoon et al. 2004) in different fields (Braeuer et al. 2009; Reinhold-López et al. 2010; Rossmann et al. 2012). Knauer and coworkers have recently proposed the adoption of a one-dimensional Raman setup for the analysis of gas sorption in molten polymers (Knauer et al. 2013) and monitored the temporal evolution of spatially

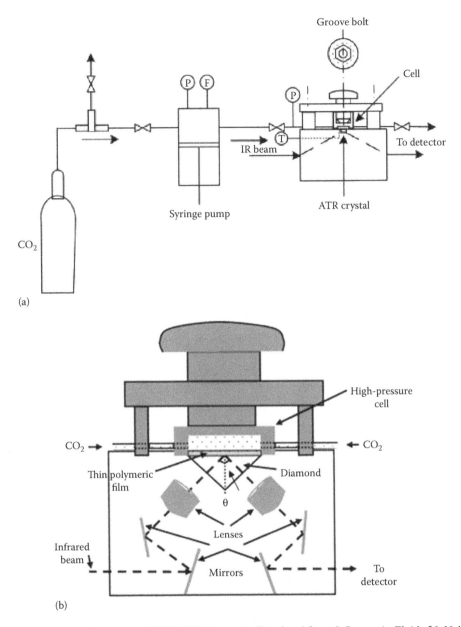

FIGURE 4.1 (a) The high-pressure FTIR-ATR apparatus (Reprinted from *J. Supercrit. Fluid.*, 36, Nalawade, S.P. et al., The FT-IR studies of the interactions of CO_2 and polymers having different chain groups, 236–244, Copyright 2006a, with permission from Elsevier) and (b) a schematic of the diamond ATR accessory. (Reprinted from *J. Supercrit. Fluid.*, 45, Pasquali et al., Measurement of CO_2 sorption and PEG 1500 swelling by ATR-IR spectroscopy, 384–390, Copyright 2008a, with permission from Elsevier.)

resolved CO_2 mass fraction profiles developing into a drop of molten poly(ε-caprolactone) (PCL). In comparison with the Raman depth profiling technique based on confocal Raman microscopy, the main advantage of this one-dimensional simultaneous acquisition of the whole gas mass fraction profile is the elimination of inaccuracies associated with the nonsimultaneous acquisition of data, which is of particular importance when investigating rapidly evolving systems. Furthermore, the Raman depth profiling technique is affected by the uncertainty of the z-position of the laser focus within the sample, due to optical aberrations to the focal volume (Everall 2000), which can

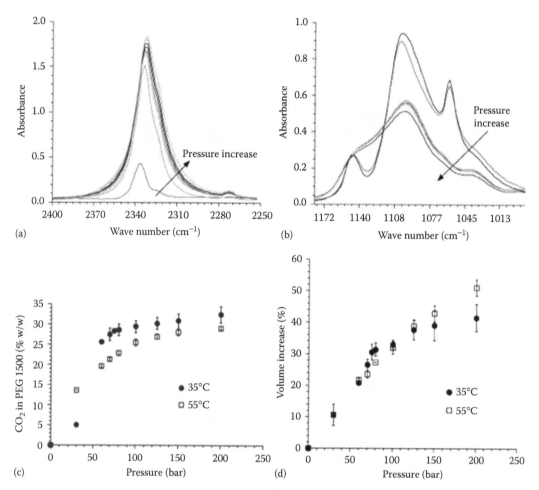

FIGURE 4.2 Simultaneous investigation of CO_2 sorption in PEG and swelling of PEG of the solution: (a) absorbance of the 2338 cm^{-1} band of CO_2 in PEG at 35°C in the range 30–200 bar, (b) absorbance of 1100 cm^{-1} band of PEG at 35°C upon exposure to CO_2 in the range 1–200 bar, (c) sorption isotherm in PEG, and (d) dependence of polymer swelling on CO_2 pressure and temperature. (Reprinted from *J. Supercrit. Fluid.*, 45, Pasquali et al., Measurement of CO_2 sorption and PEG 1500 swelling by ATR-IR spectroscopy, 384–390, Copyright 2008a, with permission from Elsevier.)

be corrected by using complex numerical or experimental devices (Tomba and Pastor 2009). The procedure proposed by Knauer et al. (2013) consists of recording Raman images from the molten polymer before and during high-pressure gas sorption, until the attainment of the equilibrium. Each image consists of spatially resolved Raman spectra, as illustrated in Figure 4.3, and the spatial and temporal evolutions of gas signals are analyzed thus providing a description of mass transport phenomena.

In general, the intensity S_{ij} of the Raman signal, which is scattered from a molecule undergoing a transition from the energy level i (before the scattering process) to j (after the scattering process), is proportional to the molar concentration of scattering molecules C_{ij}:

$$S_{ij} = C_{ij}k_{ij} \qquad (4.94)$$

The proportionality coefficient k_{ij} accounts for several factors, such as the number of excitation photons, the wavelength of the laser excitation radiation, the Raman shift of the transition, the strength

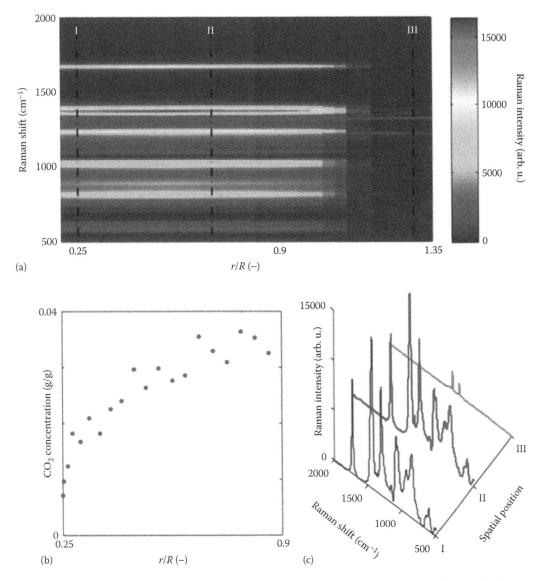

FIGURE 4.3 (a) Raman image of the PCL pendant drop in CO_2 atmosphere, (b) mass fraction profile derived according to Equation 4.97, and (c) Raman spectra at selected radial positions. (Reprinted from *Polymer*, 54, Knauer et al., Investigation of CO_2 sorption in molten polymers at high pressures using Raman line imaging, 812–818, Copyright 2013, with permission from Elsevier.)

of the local electromagnetic field, the differential Raman scattering cross section, the solid angle of the detection optics, the Planck's constant, the speed of light, the Boltzmann constant, the temperature, and an experimental constant taking into account the efficiency of the optical components, the quantum efficiency of the detector, and the efficiency of the spectrometer. The ratio of the signals of two generic Raman transitions ij and kl is proportional to the ratio of the molar concentrations of the two species undergoing these transitions; thus, by substituting the generic transitions ij and kl with the Raman transitions of penetrant (p) and matrix (m), it follows

$$\frac{S_p}{S_m} = \frac{C_p}{C_m} K_{p/m} \tag{4.95}$$

Hence, the ratio of the Raman signal intensities S_p and S_m is directly proportional to the ratio of the molar concentrations of penetrant, C_p, and of matrix, C_m. Thus, by combining Equations 4.94 and 4.95, the mole fraction of the penetrant (x_p) or the mass fraction (ω_p) can be derived:

$$x_p = \frac{C_p}{C_p + C_m} = \frac{S_p/k_p}{S_p/k_p + S_m/k_m} = \frac{S_p}{S_p + S_m \cdot k_p/k_m} = \frac{1}{1 + \left(S_m/S_p\right) \cdot K_{p/m}} \tag{4.96}$$

$$\omega_p = \frac{\rho_p}{\rho_p + \rho_m} = \frac{S_p \cdot M_p/k_p}{\left(S_p \cdot M_p/k_p\right) + \left(S_m \cdot M_m/k_m\right)}$$
$$= \frac{1}{1 + \left(S_m \cdot M_m/S_p \cdot M_p\right)K_{p/m}} \cong \frac{S_p}{S_p + K'_{p/m}} \tag{4.97}$$

where:
    ρ_i represents the mass concentration of component i

By using Equation 4.97, the time evolution of the spectral characteristic of the penetrant can be converted, for each spatial position, into its mass fraction (see Figure 4.4). However, the parameter $K'_{p/m}$ has to be preliminarily evaluated by fitting the equilibrium values of S_p to sorption data obtained for the same system by gravimetric experiments (calibration curve).

4.3.2 Measurement of Rheological Properties of Polymer–Gas Solution

Prevention of gas leakage at elevated pressures, formation of a single-phase solution, and prevention of separation of a single phase into two phases are the major challenges to viscosity measurements in polymer–gas solutions (Nalawade et al. 2006a). Recently, these difficulties were overcome by a number of research groups adopting different experimental approaches. Rheological data of several polymer–blowing agent solutions, in different ranges of temperatures, pressures, and shear rates, were then available to the scientific community (Han et al. 1983; Gerhardt et al. 1997; Elkovitch et al. 1999; Areerat et al. 2002; Royer et al. 2002; Flichy et al. 2003; Qin et al. 2005). The different approaches can be classified as follows: (1) pressure driven, (2) falling bodies, (3) rotational, and (4) vibrating devices. Each type has some significant advantages and disadvantages. Pressure-driven approaches typically use capillary viscometers (Figure 4.5a) or slit dies attached to an extrusion

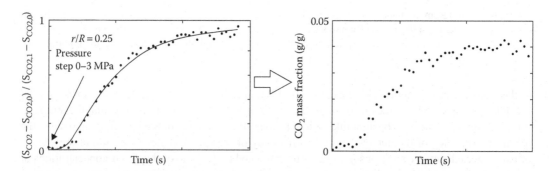

FIGURE 4.4 Normalized integrated Raman signal intensity as a function of time at selected radial positions inside the PCL drop for a pressure increase from 0 to 3.0 MPa at 80°C (left); time evolution of CO_2 mass fraction at exemplary positions of the PCL drop (right). (Reprinted from *Polymer*, 54, Knauer et al., Investigation of CO_2 sorption in molten polymers at high pressures using Raman line imaging, 812–818, Copyright 2013, with permission from Elsevier.)

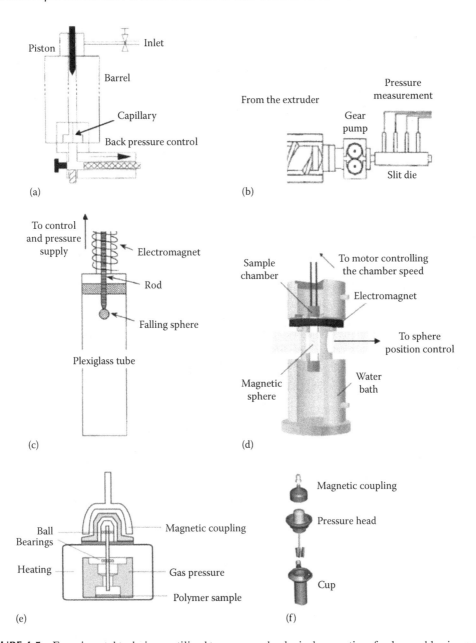

FIGURE 4.5 Experimental techniques utilized to measure rheological properties of polymer–blowing agent solutions. Pressure-driven approach, (a) capillary and (b) in-line extrusion; falling bodies approach, (c) falling sphere and (d) magnetically levitated sphere; rotational approach, (e) rotational and (f) vane-geometry.

foaming line (Figure 4.5b) and are based on the measure of the pressure drop across the capillary and of the volumetric flux of the matter flowing through the capillary. They are accurate and provide simple flow fields, but the pressure control is difficult, and the pumping of high-pressure solutions drop across the capillary can induce flow instabilities, especially in extruders, where fine control metering pump is to be used. Furthermore, the pressure drop across the capillary or slit has to be controlled in order to avoid phase separation, and these approaches are also limited only to relatively high shear conditions. In falling-body (Figure 4.5c) devices, the viscosity is evaluated by measuring the stationary speed of a sinker falling in the polymer–blowing agent solution. They can be operated at constant pressure and at low shear rates, which is, however, fixed by the sinker (e.g., the ball) density and the fluid viscosity, constituting a great limitation. In this case, furthermore, the complex flow field around the sphere makes it difficult to use for non-Newtonian liquids. Royer et al. (2002) designed a magnetically levitated sphere rheometer, partially overcoming some of the difficulties arising in falling body rheometers (mostly, the limited control of the shear rates), having a stationary sphere levitating in a moving cylinder (Figure 4.5d). In this case, the shear rate can be changed by varying the speed of the moving cylinder, while the sphere is kept in position by adjusting the magnetic force. Rotational viscometers are the most versatile viscometers and are well suited for studying non-Newtonian fluids (Figure 4.5e). The main problems associated with their use under high-pressure gases are the magnetic transmission, the achievement of polymer–blowing agent solution and the headspace required to accommodate the swelling of the polymer. Flichy et al. (2003), for instance, utilized a commercial high-pressure, concentric cylinder rotational viscometer, modified with the use of vane geometries, whose scheme is reported in Figure 4.5f, to avoid the problems associated with rotational rheometers.

Data on both nonbiodegradable and biodegradable polymers evidence the expected reduction of the viscosity after solubilization of the blowing agent in the polymer melt, commonly referred to as *plasticization* effect. This plasticization effect results in a reduction of the polymer viscosity, to an extent that depends upon the polymer–gas pair, the gas concentration, and the processing conditions. As an example, in Figure 4.6, the rheological properties in shear of PCL and PCL solutions with nitrogen and carbon dioxide measured with an in-line technique are reported. The rheological curves of the mixtures are only shifted to lower value maintaining a similar dependence on shear rate. The theoretical prediction obtained by using Equation 4.71, where the SS EoS model was used for the prediction of the parameters, is in good agreement with the experimental results (Di Maio et al. 2006).

4.3.3 Measurement of Volumetric (PVT) Properties

Specific volume is another relevant property in the formation of porous and cellular structures. In fact, it affects filling, delivering, and pumping efficiency in the extruder. It also enters in the mass balance describing the growth model, thus contributing in the definition of the final foam morphology. In fact, when the polymer is exposed to a gaseous penetrant, its volume changes as a

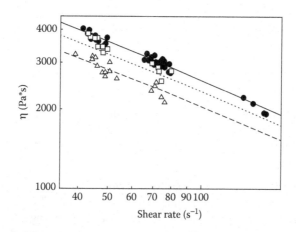

FIGURE 4.6 Effect on shear viscosity of gases dissolved in PCL as measured by in-line rheometry: neat PCL (•), PCL with 2.3 wt% CO_2 (□), and PCL with 1.5 wt% N_2 (△). Power-law fits of neat PCL data, continuous line; model prediction (Equation 4.71) for the PCL–CO_2 solution, dashed line; and model prediction (Equation 4.71) for the PCL/N_2 solution, dotted line. (After Di Maio, E. et al., *J. Polym. Sci. Pol. Phys.*, 44, 1863–1873, 2006.)

consequence of the compressive action of mechanical pressure exerted by the external gas and of the gas solubilization. Typically, at low gas concentration, mass increase is the dominant effect, and a reduction of the specific volume of the solution is observed; at higher gas concentration, volume increase becomes the dominant effect, leading to a bell-shaped curve.

Using a dilatometer is the most common technique to measure the bulk specific volume as a function of temperature and pressure of polymers. The pressure–volume–temperature (PVT) measurements are commonly based on two dilatometer configurations (see Figure 4.7): (1) the piston–die and (2) the confining fluid configuration. In the piston–die configuration, the sample is set in a rigid die and pressurized by means of a piston. During the measurement, both temperature and pressure are varied, and the volume of the material is measured by recording the displacement of the piston. The main advantage of this technique is the simplicity of its design; however, due to the material sticking to the walls of the measuring chamber, the applied pressure is not hydrostatic. The adoption of a confining fluid overcomes this drawback. In fact, in this configuration, the sample is surrounded by a confining fluid inside a variable volume sample cell, which is partially confined by a flexible metal bellows. The internal volume of the cell is filled with a confining fluid under vacuum (mercury or silicon oil), and a hydrostatic pressure is produced in the pressure vessel surrounding the cell by a motorized pump. This pressure is transmitted to the confining fluid sample system by the flexible bellows. The deformation of the bellows is a measure for the cumulative volume change of the confining fluid and of the sample and is measured by a linear variable differential transducer. The absolute specific volume of the polymer can be thus obtained by correcting the relative volume difference with the specific volume of the confining fluid.

The experimental evaluation of specific volume of polymer–gas solutions is indeed much more complex and just few data have been reported so far. For instance, Wulf et al. proposed an approach based on the coupling of the sessile drop and pendant drop methods (which are commonly used for the measurement of interfacial tension) (Wulf et al. 1999); basically, in a high-temperature chamber, both a sessile and a pendant drop of molten polymer are exposed to a gaseous atmosphere (see Figure 4.8a). The gas-saturated molten polymer drop in the sessile configuration was used for the evaluation of interfacial tension and then the analysis of the profile of the pendant drop was used to evaluate the specific volume of the solution, by using an inverted axisymmetric drop shape analysis (ADSA) algorithm. The group of Ohshima proposed an interesting approach based on the use of a modified magnetic suspension balance (see Figure 4.8b) to measure the buoyancy force exerted on a platinum plate submerged in the gas–polymer solution (Funami et al. 2007). In detail, since the buoyancy force exerted on the plate by the polymer–gas solution reduced the apparent weight of the plate, the density of the polymer–gas solution could be straightforwardly calculated by subtracting the true weight of the plate from its measured weight. However, the proposed method has some limitations related to the fact that the accuracy of the readout cannot be guaranteed when solution viscosity is high. Li et al. (2008) proposed an approach for the measurement of PVT properties of polymer melts

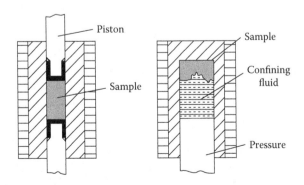

FIGURE 4.7   Schematics of piston-cylinder (left) and confining fluid (right) configurations for dilatometry.

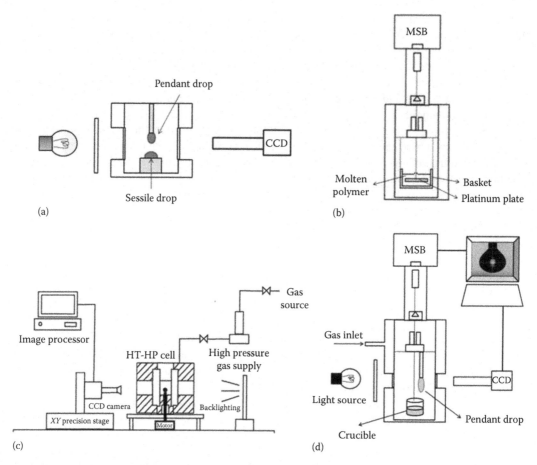

FIGURE 4.8 Schematic illustration of proposed approaches for the measurement of specific volume of polymer–gas solutions: (a) coupling of pendant and sessile drops (Data from Wulf, M. et al., *J. Colloid Interf. Sci.*, 210, 172–181, 1999); (b) measurement of the buoyancy force exerted on a platinum plate submerged in the gas–polymer solution (Data from Funami, E. et al., *J. Appl. Polym. Sci.*, 105, 3060–3068, 2007); (c) observation of the swelling of a pendant drop placed in a view cell (Data from Li, Y.G. et al., *Fluid Phase Equilib.*, 270, 15–22, 2008); and (d) coupling of gravimetric measurement and pendant drop observation. (Data from Pastore Carbone, M.G. et al., *Polym. Test.*, 30, 303–309, 2011).

saturated with high-pressure gas at elevated temperatures, which is based on the observation of the swelling of a pendant drop placed in a view cell (Figures 4.6c and 4.8c) and exposed at gas at several pressures.

Recently, Pastore Carbone et al. have introduced a new methodology to accurately measure the specific volume of molten polymer–gas solution that is based on the coupling of a gravimetric method and ADSA. This approach allows for the simultaneous measurement of solubility, diffusivity, specific volume, and interfacial tension of molten polymer–gas solutions (Pastore Carbone et al. 2011). The proposed methodology is based on the use of a high-pressure and high-temperature magnetic suspension balance, which is equipped with a view cell where both the gravimetric measurement and the observation of a pendant drop are performed at the same time (Figure 4.8d). In this experimental configuration, while the balance is measuring the weight change during sorption, a high resolution digital camera acquires the profile of the pendant drop, at same temperature and pressure, allowing for simultaneous accurate measurement of solubility, diffusivity, specific volume, and interfacial tension of molten polymer–gas solutions. By using this methodology, the specific volume of the PCL–CO_2 solutions has been measured at different pressures and temperatures

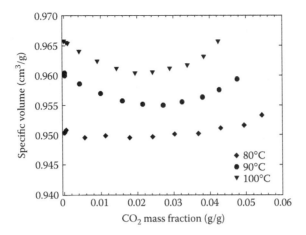

FIGURE 4.9 Specific volume of the solution PCL–CO$_2$ as a function of CO$_2$ mass fraction in the range 80°C–100°C. (Data from Pastore Carbone, M.G. et al., *Polym. Test.*, 30, 303–309, 2011.)

(Pastore Carbone et al. 2012). Figure 4.9 shows these results as a function of CO$_2$ mass fraction, evidencing a nonmonotonic dependence in the investigated experimental pressure (0–4.2 MPa) and temperature range (80°C–100°C).

The proposed method has highlighted an unexpected nonmonotonic trend for the specific volume of the solution PCL–CO$_2$ as shown in Figure 4.9. To justify this result, the authors invoked the spectroscopic findings proving that sorbed CO$_2$ molecules in PCL are present in at least two *populations*: one *polymer-associated*, that is, interacting with the PCL carbonyl groups and the other *nonassociated* (Nalawade et al. 2008). In fact, Raman experiments performed on PCL–CO$_2$ solutions at different pressures have revealed that at low pressures (and hence at low concentration of CO$_2$ inside the polymer) the associated species, characterized by a lower volume of mixing, prevail, which can justify the experimental finding that, at low pressures, the specific volume decreases with pressure (Pastore Carbone et al. 2015).

A collection of data available in the literature (Li et al. 2006a; Liu and Tomasko 2007; Pasquali et al. 2008b; Pastore Carbone et al. 2012) for the swelling of several biodegradable polymers in contact with pressurized blowing agents are shown in Figure 4.10. It is evident the prevailing effect of volume increases at high pressure.

4.3.4 Measurement of Thermal Properties

A relevant effect related to gas dissolution in a polymer phase is the associated depression of the main transition temperatures such as melting, glass transition, and crystallization temperatures. In extrusion foaming, this phenomenon is very important, and it is, in fact, exploited in two ways. First, melting point depression is desirable for lowering processing temperatures, to reduce energy consumption, and, more specifically for biodegradable polymers that are typically less stable than nonbiodegradable, conventional polymers, to limit thermal degradation of the macromolecules. Second, it allows foaming to be conducted at temperatures at which stabilization of the cellular structure, during the formation of cell walls and struts, is more efficient. For instance, it is possible, in principle, to lock-in the newly formed cellular structure isothermally by crystallization (in semicrystalline polymers) or by vitrification (in amorphous polymers) solely by the loss of the plasticizing effect when the blowing agent is released. Data on extrusion foaming die temperatures differ of several tens of Celsius degrees from the ones of the neat polymer in numerous examples, as a result of the dependence on the amount and kind of blowing agent. For example, Di Maio et al. (2005) evidenced that it was possible to extrude CO$_2$-laden PCL at 40°C, N$_2$-laden PCL at 45°C, proving a

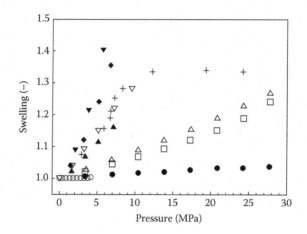

FIGURE 4.10 Swelling of some biodegradable polymers: CO_2 in PCL at 80°C (o) (Data from Pastore Carbone, M.G. et al., *J. Supercrit. Fluids*, 67, 131–138, 2012); N_2 in PLA at 200°C (•); CO_2 in PLA at 200°C (△,◻) (Data from Li, G. et al., *Fluid Phase Equilib.*, 246, 158–166, 2006a); CO_2 in PLA at 30°C, 40°C, and 60°C (▼,♦,▽); CO_2 in PLGA at 60°C (▲) (Data from Liu, D. and Tomasko, J., *J. Supercrit. Fluid.* 39, 416–425, 2007); and CO_2 in PEG at 55 °C (+). (Data from Pasquali, I. et al., *Int. J. Pharm.*, 356, 76–81, 2008b).

TABLE 4.1
Effect of Gas Sorption on Melting Point of Selected Biodegradable Polymers

Polymer	T_m, (K)	Blowing agent, (pressure, MPa)	ΔT_m, (K) (Lian et al. 2006)
PCL	332	CO_2 (9.0)	22
PBS	388	CO_2 (14.5)	14
PEA	328	CO_2 (27.6)	22
PLLA	448	CO_2 (27.6)	55

PCL, polycaprolactone; PBS, poly(butylene succinate); PEA, poly(ethylene adipate); PLLA poly(L-lactide).

more extensive plasticization effect by CO_2 than by N_2, while the neat polymer was only extruded at temperatures as high as 60°C. Specific data on the depression of characteristic temperatures of biodegradable polymers by blowing agents absorption are available in the scientific literature (Lian et al. 2006) and most relevant are reported in Table 4.1. Specific literature on nonbiodegradable polymers and modeling attempts can be found in Quach and Simha (1972), Wissinger and Paulaitis (1991), Chiou et al. (1985), and Condo et al. (1992).

4.3.5 Measurement of Interfacial Tension

The determination of the surface tension of the separation surface between the molten polymer–gas solutions and the surrounding gas itself can be performed by using the well-established ADSA, which is based on the evaluation of the shape of an axisymmetric pendant drop (Wu 1982). This technique consists of fitting the shape of an experimental drop to the theoretical drop profile according to the Laplace equation (Young 1805; Laplace 1805), properly modified to account for the action of the gravitational field (Bashfort and Adams 1892; Cheng 1990). The ADSA procedure provides the surface tension of a polymer–gas solution once the specific volume of the gas saturated polymer drop, the specific volume of the fluid surrounding it, and the coordinates of several points of the drop profile are available. Therefore, in order to evaluate the specific volume of the mixture, both reliable gas solubility data and total volume of the polymer–gas mixture are needed (Liu and Tomasko 2007; Li et al. 2004). As reported in a previous section, very few experimental data on the density of polymer–gas solutions are available in literature (Funami et al. 2007); hence, the application of ADSA at high-pressure/high-temperature conditions is generally based on a trial-and-error analysis of experimental data, relying upon theoretical prediction of the equilibrium mixture density obtained from solution theories grounded on statistical thermodynamics (Park et al. 2006; Liu and Tomasko 2007). Some authors have proposed attempts to circumvent this hybrid approach (Dimitrov at al. 1999; Jaeger et al. 1996; Pastore Carbone et al. 2011). For instance, the approach proposed by Dimitrov et al. (1999) is based on the combination of the observation of the polymer drop placed in the gas atmosphere with a mass balance for the evaluation of the mass of dissolved gas and, in turn, of the density of the gas-saturated polymer. The approach proposed by Pastore Carbone et al. (2011) is based on coupling of simultaneous gravimetric gas sorption measurement and observation of the pendant drop (see Figure 4.8d). The experimental setup consists of a magnetic suspension balance equipped with a high pressure and temperature view cell where the gravimetric measurement and the optical monitoring of the pendant drop are simultaneously performed, at the same temperature and pressure. In this configuration, the magnetic suspension balance measures the weight change of the sample placed in a crucible occurring during sorption and, concurrently, a high-resolution camera records the profile of a pendant drop hanging from a rod fixed to the typical metallic cage protecting the hook-balance coupling, thus minimizing oscillations and associated inertial forces. As already mentioned, this approach allows the concurrent determination of gas solubility and diffusivity as well as of specific volume and surface tension of molten PCL–CO_2, in a single experiment, at several temperatures and CO_2 pressures (Pastore Carbone et al. 2012). In Figure 4.11 is illustrated the typical dependence on pressure of surface tension data of some biodegradable polymers/CO_2 solutions. Surface tension of the solutions was found to decrease with increasing CO_2 pressure and the observed decrease is generally attributed to two concurrent phenomena: (1) as pressure increases, the free energy density of CO_2 becomes closer to that of the polymer phase and the interfacial tension decreases and (2) as gas pressure increases, the concentration of CO_2 in the polymer phase increases, thus, promoting a further decrease of interfacial tension since the two phases in contact become more similar (Harrison et al. 1998; Li et al. 2004).

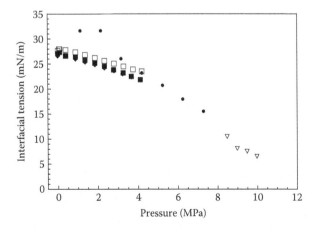

FIGURE 4.11 Surface tension of CO_2 and PCL ($\square$ 80°C, $\blacksquare$, 90°C, $\blacklozenge$ 100°C) (Data from Pastore Carbone, M.G. et al., *J. Supercrit. Fluids*, 67, 131–138, 2012), CO_2 and PS at 200°C ($\bullet$) (Data from Li, G. et al., *Fluid Phase Equilib.*, 246, 158–166, 2004), and CO_2 and PEG600 at 45°C ($\triangledown$). (Data from Harrison, K.L. et al., *Langmuir*, 12, 2637–2644, 1996).

REFERENCES

Almeida, B.S., and M.M. Telo da Gama. 1989. Surface tension of simple mixtures: Comparison between theory and experiment. *J. Phys. Chem.* 93: 4132–4138.

Areerat, S., T. Nagata, M. Ohshima. 2002. Measurement and prediction of LDPE/CO$_2$ solution viscosity. *Polym. Eng. Sci.* 42: 2234–2245.

Barbari, T.A., S.S. Kasargod, G.T. Fieldson. 1996. Effect of unequal transport rates and intersolute solvation on the selective batch extraction of a dilute mixture with a dense polymeric sorbent. *Ind. Eng. Chem. Res.* 35: 1188–1194.

Bashforth, F., and J.C. Adams. 1892. *An Attempt to Test the Theory of Capillary Action.* Cambridge University Press, Cambridge, UK.

Bearman, R.J. 1961. On the molecular basis of some current theories on diffusion. *J. Phys. Chem.* 65 (11): 1961–1968.

Belov, N., Y. Yampolskii, M.C. Coughlin. 2006. Thermodynamics of sorption in an amorphous perfluorinated rubber studied by inverse gas chromatography. *Macromolecules* 39: 1797–1804.

Berry, G.C., and T.G. Fox. 1968. The viscosity of polymers and their concentrated solutions. *Adv. Polym. Sci.* 5: 261–327.

Bird, R.B., W.E. Stewart, E.N. Lightfoot. 1960. *Transport Phenomena.* John Wiley & Sons, New York.

Bondi, A. 1964. Van der Waals volumes and radii. *J. Chem. Phys.* 68: 411–451.

Bongiorno, V., and H.T. Davis. 1975. Modified Van der Waals theory of fluid interfaces. *Phys. Rev. A* 12: 2213–2224.

Bongiorno, V., L.E. Scriven, H.T. Davis. 1976. Molecular theory of fluid interfaces. *J. Colloid Interf. Sci.* 57: 462–475.

Braeuer, A., S.R. Engel, R.F. Hankel, A. Leipertz. 2009. Gas mixing analysis by simultaneous Raman imaging and particle image velocimetry. *Opt. Lett.* 34: 3122–3124.

Brandt, W.W. 1959. Model calculation of the temperature dependence of small molecule diffusion in high polymers. *J. Phys. Chem.* 63: 1080–1085.

Brantley, N.H., S.G. Kazarian, C.A. Eckert. 2000. In situ FTIR measurement of carbon dioxide sorption into poly(ethylene terephthalate) at elevated pressures. *J. Appl. Polym. Sci.* 77: 764–775.

Briscoe, B.J., and C.T. Kelly. 1993. Optical studies of polymers in high pressure gas environments. *Mater. Sci. Eng. A* 168: 111–115.

Cahn, J.W., and J.E. Hilliard. 1958. Free energy of a nonuniform system I. Interfacial free energy. *J. Chem. Phys.* 28: 258–267.

Caruthers, J.M., K.C. Chao, V. Venkatasubrmanian, R. Sy-Siong-Kiao, C.R. Novenario, A. Sundaram. 1998. *Handbook of Diffusion and Thermal Properties of Polymers and Polymer Solutions.* DIPPR, American Institute of Chemical Engineers, New York.

Chapman, W.G., K.E. Gubbins, G. Jackson, M. Radosz. 1989. SAFT: Equation of state solution model for associating fluids. *Fluid Phase Equilib.* 52: 31–38.

Chapman, W.G., K.E. Gubbins, G. Jackson, M. Radosz. 1990. New reference equation of state for associating liquids. *Ind. Eng. Chem. Res.* 29: 1709–1721.

Chapman, W.G., G. Jackson, K.E. Gubbins. 1988. Phase equilibria of associating fluids Chain molecules with multiple bonding sites. *Mol. Phys.* 65: 1057–1079.

Chaudhary, G.I., and A.I. Johns. 1998. Solubilities of nitrogen, isobutane and carbon dioxide in polyethylene. *J. Cell. Plast.* 34: 312–328.

Cheng, P. 1990. *Automation of Axisymmetric Drop Shape Analysis Using Digital Image Processing.* PhD thesis, University of Toronto, Toronto, Canada.

Chiou, J.S., J.W. Barlow, D.R. Paul. 1985. Plasticization of glassy polymers by CO$_2$. *J. Appl. Polym. Sci.* 30: 2633–2642.

Cohen, M.H., and D. Turnbull. 1959. Molecular transport in liquids and glasses. *J. Chem. Phys.* 31: 1164–1169.

Condo, P.D., I.C. Sanchez, C.G. Panayiotou, K.P. Johnston. 1992. Glass transition behavior including retrograde vitrification of polymers with compressed fluid diluents. *Macromolecules* 25: 6119–6127.

Cotugno, S., E. Di Maio, C. Ciardiello, S. Iannace, G. Mensitieri, L. Nicolais. 2003. Sorption thermodynamics and mutual diffusivity of carbon dioxide in molten polycaprolactone. *Ind. Eng. Chem. Res.* 42: 4398–4405.

Cotugno, S., D. Larobina, G. Mensitieri, P. Musto, G. Ragosta. 2001. A novel spectroscopic approach to investigate transport processes in polymers: The case of water-epoxy systems. *Polymer* 42: 6431–6438.

Crank, J. 1975. *The Mathematics of Diffusion*, 2nd Ed. Clarendon Press, Oxford.

Davis, P.K., G.D. Lundy, J.E. Palamara, J.L. Duda, R.P. Danner. 2004. New pressure-decay techniques to study gas sorption and diffusion in polymers at elevated pressures. *Ind. Eng. Chem. Res.* 43: 1537–1542.

DiBenedetto, A.T., D.R. Paul. 1964. An interpretation of gaseous diffusion through polymers using fluctuation theory. *J. Polym. Sci. A.* 2: 1001–1015.

Di Maio, E., S. Iannace, G. Mensitieri, L. Nicolais. 2006. A predictive approach based on the Simha-Somcynsky free volume theory for the effect of dissolved gas on viscosity and glass transition temperature of polymeric mixtures. *J. Polym. Sci. Pol. Phys.* 44: 1863–1873.

Di Maio, E., G. Mensitieri, S. Iannace, L. Nicolais, W. Li, R.W. Flumerfelt. 2005. Structure optimization of PCL foams by using mixtures of CO_2 and N_2 as blowing agents. *Polym. Eng. Sci.* 45: 432–441.

Dimitrov, K., L. Boyadzhiev, R. Tufeu. 1999. Properties of supercritical CO_2 saturated poly(ethylene glycol) nonylphenyl ether. *Macromol. Chem. Phys.* 200: 1626–1629.

Doolittle, A.K., and D.B. Doolittle. 1957. Studies in Newtonian flow. V. further verification of the free-space viscosity equation. *J. Appl. Phys.* 28: 901–905.

Duda, J.L., and Y.C. Ni. 1978. Diffusion of ethylbenzene in molten polystyrene. *J. Appl. Polym. Sci.* 22: 689–699.

Duda, J.L., and Y.C. Ni. 1979. Toluene diffusion in molten polystyrene. *J. Appl. Polym. Sci.* 23: 947–951.

Duda, J.L., and J.M. Zielinski. 1996. Free-volume theory, in *Diffusion in Polymers*. P. Neogi, ed., chap. 3, pp.143–172, Marcel Dekker, New York.

Durrill, P.L. 1966. *The Determination of Solubilities and Diffusion Coefficients of Gases in Polymers at Temperatures above the Softening Point*. PhD thesis in Chemical Engineering, Virginia Polytechnic Institute, Blacksburg, VA.

Durrill, P.L., and R.G. Griskey. 1966. Diffusion and solution of gases in thermally softened or molten polymers. I. Development of technique and determination of data. *AIChE J.* 12: 1147–1151.

Durrill, P.L., and R.G. Griskey. 1969. Diffusion and solution of gases in thermally softened or molten polymers. II. Relation of diffusivities and solubilities with temperature, pressure and structural characteristics. *AIChE J.* 15: 106–110.

Elkovitch, M.D., D.L. Tomasko, L.J. Lee. 1999. Supercritical carbon dioxide assisted blending of polystyrene and poly(methyl methacrylate). *Polym. Eng. Sci.* 39: 2075–2084.

Enders, S., H. Kahl, J. Winkelmann. 2005. Interfacial properties of polystyrene in contact with carbon dioxide. *Fluid Phase Equilib.* 228–229: 511–522.

Evans, R. 1979. The nature of the liquid-vapour interface and other topics in the statistical mechanics of non-uniform, classical fluids. *Adv. Phys.* 28: 143–200.

Everall, N.J. 2000. Confocal Raman microscopy: Why the depth resolution and spatial accuracy can be much worse than you think. *Appl. Spectrosc.* 54: 1515–1520.

Ferry, J.D., and R.A. Stratton. 1960. The free volume interpretation of the dependence of viscosities and viscoelastic relaxation times on concentration, pressure, and tensile strain. *Kolloid-Zeitschrift.* 171: 107–111.

Fieldson, G.T., and T.A. Barbari. 1993. The use of FTIR-ATR spectroscopy to characterize penetrant diffusion in polymers. *Polymer* 34: 1146–1153.

Fieldson, G.T., and T.A. Barbari. 1995. Analysis of diffusion in polymers using evanescent field spectroscopy. *AIChE J.* 41: 795–804.

Fleming, O.S., K.L.A. Chan, S.G. Kazarian. 2006. High-pressure CO_2-enhanced polymer interdiffusion and dissolution studied with in situ ATR-FTIR spectroscopic imaging. *Polymer* 47: 4649–4658.

Flichy, N.M.B., S.G. Kazarian, C.J. Lawrence, B.J. Briscoe. 2002. An ATR–IR study of poly (dimethylsiloxane) under high-pressure carbon dioxide: Simultaneous measurement of sorption and swelling. *J. Phys. Chem. B.* 106: 754–759.

Flichy, N.M.B., C.J. Lawrence, S.G. Kazarian. 2003. Rheology of polypropylene glycol and suspensions of fumed silica in polypropylene glycol under high-pressure CO_2. *Ind. Eng. Chem. Res.* 42: 6310–6319.

Flory, P.J. 1953. *Principles of Polymer Chemistry*. Cornell University Press, Ithaca, NY.

Flory, P.J., A. Orwoll, R.A. Vrij. 1964. Statistical thermodynamics of chain molecule liquids. I. An equation of state for normal paraffin hydrocarbons. *J. Am. Chem. Soc.* 86: 3507–3514.

Fowkes, F.M. 1962. Determination of interfacial tensions, contact angles, and dispersion forces in surfaces by assuming additivity of intermolecular interactions in surfaces. *J. Phys. Chem.* 66: 1863–1867.

Fried, J.R., and W.J. Li. 1990. High-pressure FTIR studies of gas–polymer interactions. *J. Appl. Polym. Sci.* 41: 1123–1131.

Fu, D., X. Hua, Y. Xu. 2011. Cross-association model for the phase equilibria and surface tensions of CO_2-methanol and CO_2-ethanol mixtures. *J. Phys. Chem. C* 115: 3340–3345.

Fujita, H. 1961. Diffusion in polymer-diluent systems. *Fortschr. Hochpolym. Forsch.* 3: 1–47.

Fujita, H., A. Kishimoto, K. Matsumoto. 1960. Concentration and temperature dependence of diffusion coefficients for systems polymethyl acrylate and n-alkyl acetates. *Trans. Faraday Soc.* 56: 424–437.

Funami, E., K. Taki, M. Ohshima. 2007. Density measurement of polymer/CO_2 single-phase solution at high temperature and pressure using a gravimetric method. *J. Appl. Polym. Sci.* 105: 3060–3068.

Gerhardt, L.J., A. Garg, C.W. Manke, E. Gulari. 1998. Concentration-dependent viscoelastic scaling models for polydimethylsiloxane melts with dissolved carbon dioxide. *J. Polym. Sci. B: Polym. Phys.* 36: 1911–1918.

Gerhardt, L.J., C.W. Manke, E. Gulari. 1997. Rheology of polydimethylsiloxane swollen with supercritical carbon dioxide. *J. Polym. Sci. B: Polym. Phys.* 35: 523–534.

Girifalco, L.A., R.J. Good. 1957. A theory for the estimation of surface and interfacial energies. I. Derivation and application to interfacial tension. *J. Phys. Chem.* 61: 904–912.

Grenner, A., I. Tsivintzelis, I.G. Economou, C. Panayiotou, G.M. Kontogeorgis. 2008. Evaluation of the non-random hydrogen bonding (NRHB) theory and the simplified perturbed chain–statistical associating fluid theory (sPC-SAFT). 1. Vapor–liquid equilibria. *Ind. Eng. Chem. Res.* 47: 5636–5650.

Gross, J. 2005. An equation-of-state contribution for polar components: Quadrupolar molecules. *AIChE J.* 51: 2556–2568.

Gross, J., and G. Sadowski. 2001. Perturbed-chain SAFT: An equation of state based on a perturbation theory for chain molecules. *Ind. Eng. Chem. Res.* 40: 1244–1260.

Gross, J., and G. Sadowski. 2002a. Modeling polymer systems using the perturbed-chain statistical associating fluid theory equation of state. *Ind. Eng. Chem. Res.* 41: 1084–1093.

Gross, J., and G. Sadowski. 2002b. Application of the perturbed-chain SAFT equation of state to associating systems. *Ind. Eng. Chem. Res.* 41: 5510–5515.

Gross, J., and J. Vrabec. 2006. An equation-of-state contribution for polar components: Dipolar molecules. *AIChE J.* 52: 1194–1204.

Guadagno, T., and S.G. Kazarian. 2004. High-pressure CO2-expanded solvents: Simultaneous measurement of CO_2 sorption and swelling of liquid polymers with in-situ near-IR spectroscopy. *J. Phys. Chem. B.* 108: 13995–13999.

Guggenheim, E.A. 1945. The principle of corresponding states. *J. Chem. Phys.* 13(7): 253–261.

Guggenheim, E.A. 1952. *Mixtures*. Oxford University Press, Oxford.

Han, C.D., and C.Y. Ma. 1983. Rheological properties of mixtures of molten polymer and fluorocarbon blowing agent. II. Mixtures of polystyrene and fluorocarbon blowing agent. *J. Appl. Polym. Sci.* 28: 851–860.

Harrison, K.L., S.R.P. da Rocha, M.Z. Yates, K.P. Johnston, D. Canelas, J.M. De Simone. 1998. Interfacial activity of polymeric surfactants at the polystyrene–carbon dioxide interface. *Langmuir* 14: 6855–6863.

Harrison, K.L., K.P. Johnston, I.C. Sanchez. 1996. Effects of surfactants on the interfacial tension between supercritical carbon dioxide and polyethylene glycol. *Langmuir* 12: 2637–2644.

Huang, S.H., and M. Radosz. 1990. Equation of state for small, large, polydisperse, and associating molecules. *Ind. Eng. Chem. Res.* 29: 2284–2294.

Huang, S.H., and M. Radosz. 1991. Equation of state for small, large, polydisperse, and associating molecules: Extension to fluid mixtures. *Ind. Eng. Chem. Res.* 30: 1994–2005.

Jaeger, Ph.T., J.V. Schnitzler, R. Eggers. 1996. Interfacial tension of fluid systems considering the nonstationary case with respect to mass transfer. *Chem. Eng. Tech.* 19: 197–202.

Kahl, H., and J. Winkelmann. 2005. Interfacial properties of polystyrene in contact with carbon dioxide. *Fluid Phase Equilib.* 228–229: 511–522.

Kazarian, S.G. 1997. Applications of FTIR spectroscopy to supercritical fluid drying. *Extract. Impreg.* 32: 301–348.

Kazarian, S.G. 2002. Polymers and supercritical fluids: Opportunities for vibrational spectroscopy. *Macromol. Symp.* 184: 215–228.

Kazarian, S.G., N.H. Brantley, C.A. Eckert. 1999. Dyeing to be clean: Use supercritical carbon dioxide. *Chemtech* 29: 36–41.

Kazarian, S.G., and K.L.A. Chan. 2004. FTIR imaging of polymeric materials under high-pressure carbon dioxide. *Macromolecules* 37: 579–584.

Kazarian, S.G., M.F. Vincent, F.V. Bright, C.L. Liotta, C.A. Eckert. 1996a. Specific intermolecular interaction of carbon dioxide with polymers. *J. Am. Chem. Soc.* 118: 1729–1736.

Kazarian, S.G., M.F. Vincent, C.A. Eckert. 1996b. Infrared cell for supercritical fluid–polymer interactions. *Rev. Sci. Instrum.* 67: 1586–1589.

Kelley, F.N., F. Bueche, 1961. Viscosity and glass temperature relations for polymer-diluent systems. *J. Polym. Sci.* 50: 549–556.

Kleiner, M., and J. Gross. 2006. An equation-of-state contribution for polar components: Polarizable dipoles. *AIChE J.* 52: 1951–1961.

Kloczkowski, A., and J.E. Mark. 1989. On the Pace-Datyner theory of diffusion of small molecules through polymers. *J. Polym. Sci. Polym. Phys. Ed.* 27: 1663–1674.

Knauer, O.S., M.C. Lang, A. Braeuer, A. Leipertz. 2011. Simultaneous determination of the composition and temperature gradients in the vicinity of boiling bubbles in liquid binary mixtures using one-dimensional Raman measurements. *J. Raman Spectrosc.*, 42: 195–200.

Knauer, O.S., M.G. Pastore Carbone, A. Braeuer, E. Di Maio, A. Leipertz. 2013. Investigation of CO_2 sorption in molten polymers at high pressures using Raman line imaging. *Polymer* 54: 812–818.

Koros, W.J., and D.R. Paul. 1976. Design considerations for measurement of gas sorption in polymers by pressure decay. *J. Polym. Sci. Pol. Phys.* 14: 1903–1907.

Kwag, C., C.W. Manke, E. Gulari. 1999. Rheology of molten polystyrene with dissolved supercritical and near-critical gases. *J. Polym. Sci. B: Polym. Phys.* 37: 2771–2781.

Kwag, C., C.W. Manke, E. Gulari. 2001. Effects of dissolved gas on viscoelastic scaling and glass transition temperature of polystyrene melts. *Ind. Chem. Res.* 40: 3048–3052.

Lafitte, T., B. Mendiboure, M.M. Pineiro, D. Bessieres, C. Miqueu. 2010. Interfacial properties of water/CO_2: A comprehensive description through a gradient theory-SAFT-VR Mie approach. *J. Phys. Chem. B.* 114: 11110–11116.

Laplace, P.S. 1805. Traité de Mécanique Céleste, Supplement to Book 10, Gauthier–Villars, Paris, France.

Li, G., H. Li, L.S. Turng, S. Gong, C. Zhang. 2006a. Measurement of gas solubility and diffusivity in polylactide. *Fluid Phase Equilib.* 246: 158–166.

Li, G., H. Li, J. Wang, C.B. Park. 2006b. Investigating the solubility of CO_2 in polypropylene using various EoS. *Cell. Polym.* 25: 237–248.

Li, H., L.J. Lee, D. Tomasko. 2004. Effect of carbon dioxide on the interfacial tension of polymer melts. *Ind. Eng. Chem. Res.* 43: 509–514.

Li, Y.G., and C.B. Park. 2009. Effects of branching on the Pressure-Volume-temperature behaviors of PP/CO_2 solutions. *Ind. Eng. Chem. Res.* 48: 6633–6640.

Li, Y.G., C.B. Park, H.B. Li, J. Wang. 2008. Measurement of the PVT property of PP/CO_2 solution. *Fluid Phase Equilib.* 270: 15–22.

Lian, Z., S.A. Epstein, C.W. Blenk, A.D. Shone. 2006. Carbon dioxide-induced melting point depression of biodegradable semicrystalline polymers. *J. Supercrit. Fluid.* 39: 107–117.

Liu, D. and J. Tomasko. 2007. Carbon dioxide sorption and dilation of poly(lactide-*co*-glycolide). *J. Supercrit. Fluid.* 39: 416–425.

Lundberg, J.L. 1964. Diffusivities and solubilities of methane in linear polyethylene melts. *J. Polym. Sci.: Part A* 2: 3925–3931.

Lundberg, J.L., M.B. Wilk., M.J. Huyett. 1960. Solubilities and diffusivities of nitrogen in polyethylene. *J. Appl. Phys.* 31: 1131–1132.

Lundberg, J.L., M.B. Wilk, M.J. Huyett. 1962. Estimation of diffusivities and solubilities from sorption studies. *J. Polym. Sci.* 57: 275–299.

MacLeod, D.B. 1923. On a relation between surface tension and density. *Trans. Faraday Soc.* 19: 38–41.

Mahmood, S.H., M. Keshtkar, C.B. Park. 2014. Determination of carbon dioxide solubility I polylactide acid with accurate pVT properties. *J. Chem. Thermodynamics.* 70: 13–23.

McBain, J.W., and A.M. Bakr. 1926. A new sorption balance. *J. Am. Chem. Soc.* 48: 690–695.

Meares, P. 1954. The diffusion of gases through polyvinyl acetate. *J. Am. Chem. Soc.* 76: 3415–3422.

Meredith, J.C., K.P. Johnston, J.M. Seminario, S.G. Kazarian, C.A. Eckert. 1996. Quantitative equilibrium constants between CO_2 and lewis bases from FTIR spectroscopy. *J. Phys. Chem.* 100: 10837–10848.

Miqueu, C., B. Mendiboure, C. Garcia, J. Lachaise. 2004. Modelling of the surface tension of binary and ternary mixtures with the gradient theory of fluids interfaces. *Fluid Phase Equilib.* 218 (2): 189–203.

Murphy, B., P. Kirwan, P. McLoughlin. 2003. Investigation into polymer-diffusant interactions using ATR-FTIR spectroscopy. *Vibr. Spectrosc.* 33: 75–82.

Musto, P., M. Galizia, M. Pannico, G. Mensitieri. 2014. Time-resolved FTIR spectroscopy, gravimetry and thermodynamic modelling for a molecular level description of water diffusion in poly(ε-caprolactone). *J. Phys. Chem. B.* 118: 7414–7429.

Musto, P., G. Mensitieri, M. Lavorgna, G. Scarinzi, G. Scherillo. 2012. Combining gravimetric and vibrational spectroscopy measurements to quantify first- and second-shell hydration layers in polyimides with different molecular architectures. *J. Phys. Chem. B.* 116: 1209–1220.

Musto, P., G. Ragosta, G. Mensitieri, M. Lavorgna. 2007. On the molecular mechanism of H_2O diffusion into polyimides. A vibrational spectroscopy investigation. *Macromolecules* 40: 9614–9627.

Nalawade, S.P., F. Picchioni, L.P.B.M. Janssen. 2006a. Supercritical carbon dioxide as a green solvent for processing polymer melts: processing aspects and applications. *Prog. Polym. Sci.* 31: 19–43.

Nalawade, S.P., F. Picchioni, L.P.B.M. Janssen, D.W. Grijpma, J. Feijen. 2008. Investigation of the interaction of CO_2 with poly(L-lactide), poly(DL-lactide) and poly(e-caprolactone using FTIR spectroscopy). *J. Appl. Polym. Sci.* 109: 3376–3381.

Nalawade, S.P., F. Picchioni, J.H. Marsman, L.P.B.M. Janssen. 2006b. The FT-IR studies of the interactions of CO_2 and polymers having different chain groups. *J. Supercrit. Fluid.* 36: 236–244.

Newitt, D.M., and K.E. Weale. 1948. Solution and diffusion of gases in polystyrene at high pressures. *J. Chem. Soc.* 1541–1549.

Nino-Amezquita, O.G., S. Enders, P.T. Jaeger, R. Eggers. 2010. Measurements and prediction of interfacial tension of binary mixtures. *Ind. Eng. Chem. Res.* 49: 592–601.

Nordholm, S., and A.D.J. Haymet. 1980. Generalized Van der Waals theory I. Basic formulation and application to uniform fields. *Aust. J. Chem.* 33: 2013–2036.

Novitskiy, A.A., J. Ke, G. Comak, M. Poliakoff, M.W. George. 2011. A modified golden gate attenuated total reflection (ATR) cell for monitoring phase transitions in multicomponent fluids at high temperatures. *Appl. Spectrosc.* 65: 885–891.

Ott, B.A., and G. Caneba. 2010. Solubility of supercritical CO2 in polystyrene during foam formation via statistical associated fluid theory (SAFT) equation of state. *J. Miner. Mater. Char. Eng.* 9: 411–426.

Pace, R.J., and A. Datyner. 1979a. Statistical mechanical model for diffusion of simple penetrants in polymers. I. Theory. *J. Polym. Sci. Polym. Phys. Ed.* 17: 437–451.

Pace, R.J., and A. Datyner. 1979b. Statistical mechanical model for diffusion of simple penetrants in polymers. II. Applications—Nonvinyl polymers. *J. Polym. Sci. Polym. Phys. Ed.* 17: 453–464.

Pace, R.J., and A. Datyner. 1979c. Statistical mechanical model for diffusion of simple penetrants in polymers. III. Applications—Vinyl and related polymers. *J. Polym. Sci. Polym. Phys. Ed.* 17: 465–476.

Pace, R.J., and A. Datyner. 1979d. Statistical mechanical model of diffusion of complex penetrants in polymers. I. Theory. *J. Polym. Sci. Polym. Phys. Ed.* 17: 1675–1692.

Pace, R.J., and A. Datyner. 1980a. Model of sorption of simple molecules in polymers. *J. Polym. Sci. Polym. Phys. Ed.* 18: 1103–1124.

Pace, R.J., and A. Datyner. 1980b. Statistical mechanical model of sorption and diffusion of simple penetrants in polymers. *Polym. Eng. Sci.* 20: 51–58.

Panayiotou, C. 2002. Interfacial tension and interfacial profiles of fluids and their mixtures. *Langmuir* 18: 8841–8853.

Panayiotou, C. 2003a. The QCHB model of fluids and their mixtures. *J. Chem. Thermodyn.* 35: 349–381.

Panayiotou, C. 2003b. Interfacial tension and interfacial profiles: An equation-of-state approach. *J. Colloid Interf. Sci.* 267: 418–428.

Panayiotou, C. 2009. Hydrogen bonding and nonrandomness in solution thermodynamics, in *Handbook of Surface and Colloid Chemistry*, K.S. Birdi, ed., 3rd ed., CRC Press, New York, 45–89.

Panayiotou, C., M. Pantoula, E. Stefanis, I. Tsivintzelis, I.G. Economou. 2004. Nonrandom hydrogen-bonding model of fluids and their mixtures.1. Pure fluids. *Ind. Eng. Chem. Res.* 43: 6592–6606.

Panayiotou, C., and I.C. Sanchez. 1991. Hydrogen bonding in fluids: An equation-of-state approach. *J. Phys. Chem.* 95: 10090–10097.

Panayiotou, C., I. Tsivintzelis, I.G. Economou. 2007. Nonrandom hydrogen-bonding model of fluids and their mixtures. 2. Multicomponent mixtures. *Ind. Eng. Chem. Res.* 46: 2628–2636.

Panayiotou, C., and J.H. Vera. 1982. Statistical thermodynamics of r-mer fluids and their mixtures. *Polym. J.* 14: 681–694.

Park, H., C.B. Park, C. Tzoganakis, K.H. Tan, P. Chen. 2006. Surface tension measurement of polystyrene melts in supercritical carbon dioxide. *Ind. Eng. Chem. Res.* 45: 1650–1658.

Pasquali, I., J.M. Andanson, S.G. Kazarian, R. Bettini. 2008a. Measurement of CO_2 sorption and PEG 1500 swelling by ATR-IR spectroscopy. *J. Supercrit. Fluid.* 45: 384–390.

Pasquali, I., L. Comi, F. Pucciarelli, R. Bettini. 2008b. Swelling, melting point reduction and solubility of PEG 1500 in supercritical CO_2. *Int. J. Pharm.* 356: 76–81.

Pastore Carbone, M.G., E. di Maio, S. Iannace, G. Mensitieri. 2011. Simultaneous experimental evaluation of solubility, diffusivity, interfacial tension and specific volume of polymer/gas solutions. *Polym. Test.* 30: 303–309.

Pastore Carbone, M.G., E. di Maio, P. Musto, A. Braeuer, G. Mensitieri. On the unexpected non-monotonic profile of specific volume observed in PCL/CO_2 solutions. *Polymer* 56: 252–255.

Pastore Carbone, M.G., E. di Maio, G. Scherillo, G. Mensitieri, S. Iannace. 2012. Solubility, mutual diffusivity, specific volume and interfacial tension of molten PCL/CO_2 solutions by a fully experimental procedure: Effect of pressure and temperature. *J. Supercrit. Fluids* 67: 131–138.

Poser, C.I., and I. Sanchez. 1981. Interfacial tension theory of low and high molecular weight liquid mixtures. *Macromolecules* 14: 361–370.

Prausnitz, J.M., R.N. Lichrenthaler, E. Gomes de Azevedo. 1998. *Molecular Thermodynamics of Fluid-Phase Equilibria*, 3rd ed., Prentice Hall, Englewood Cliffs, NJ.

Qin, X., M.R. Thompson, A.N. Hrymak, A. Torres. 2005. Rheology studies of polyethylene/chemical blowing agent solutions within an injection molding machine. *Polym. Eng. Sci.* 45: 1108–1118.

Quach, A., and R. Simha. 1972. Statistical thermodynamics of the glass transition and the glassy state of polymers. *J. Phys. Chem.* 76: 416–421.

Reed, T.M., and K.E. Gubbins. 1973. *Applied Statistical Mechanics*. McGraw-Hill, New York.

Reid, R., J. Prausnitz, B.E. Poling. 1988. *The Properties of Gases and Liquids*, 4th ed., Singapore: McGraw-Hill.

Reinhold-López, K., A. Braeuer, N. Popovska, A. Leipert. 2010. In situ monitoring of the acetylene decomposition and gas temperature at reaction conditions for the deposition of carbon nanotubes using linear Raman scattering. *Opt. Express.* 18: 18223–18228.

Richards, W.D., and R.K. Prud'homme. 1986. The viscosity of concentrated polymer solutions containing low molecular weight solvents. *J. Appl. Polym. Sci.* 31: 763–776.

Rodriguez-Meizoso, I., P. Lazor, C. Turner. 2012. In situ Raman spectroscopy for the evaluation of solubility in supercritical carbon dioxide mixtures. *J. Supercrit. Fluid* 65: 87–92.

Rossmann, M., A. Braeuer, S. Dowy, T.G. Gallinger, A. Leipertz, E. Schluecker. 2012. Solute solubility as criterion for the appearance of amorphous particle precipitation or crystallization in the supercritical antisolvent (SAS) process. *J. Supercrit. Fluid* 66: 350–358.

Royer, J.R., Y.J. Gay, M. Adam, J.M. DeSimone, S.A. Khan. 2002. Polymer melt rheology with high-pressure CO_2 using a novel magnetically levitated sphere rheometer. *Polymer* 43: 2375–2383.

Royer, J.R., Y.J. Gay, J.M. DeSimone, S.A. Khan. 2000. High-pressure rheology of polystyrene melts plasticized with CO_2: Experimental measurement and predictive scaling relationships. *J. Polym. Sci. B: Polym. Phys.* 38: 3168–3180.

Sanchez, I.C., and R.H. Lacombe. 1976a. An elementary molecular theory of classical fluids. Pure fluids. *J. Phys. Chem.* 80: 2352–2362.

Sanchez, I.C., and R.H. Lacombe. 1976b. Statistical thermodynamics of fluid mixtures. *J. Phys. Chem.* 80: 2568–2580.

Sanchez, I.C., and R.H. Lacombe. 1978. Statistical thermodynamics of polymer solutions. *Macromolecules* 2: 1145–1156.

Sato, Y., K. Fujiwara, T. Takikawa Sumarno, S. Takishima, H. Masuoka. 1999. Solubilities and diffusion coefficients of carbon dioxide and nitrogen in polypropylene, high-density polyethylene, and polystyrene under high pressures and temperatures. *Fluid Phase Equilib.* 162: 261–276.

Sato, Y., T. Takikawa, S. Takishima, H. Masuoka. 2001. Solubilities and diffusion coefficients of carbon dioxide in poly(vinyl acetate) and polystyrene. *J. Supercrit. Fluids* 19: 187–198.

Sato, Y., M.I. Yurugi, K. Fujiwara, S. Takishima, H. Masuoka. 1996. Solubilities of carbon dioxide and nitrogen in polystyrene under high temperature and pressure. *Fluid Phase Equilib.* 125: 129–138.

Shiyao, B., S. Sourirajan, F.D.F. Talbot, T. Matsuura. 1989. Gas and vapor adsorption on polymeric materials by inverse gas chromatography. In: *Inverse Gas Chromatography*, D.R. Lloyd, T.C. Ward, H.P. Schreiber, C.C. Pizaña, eds. ACS Symposium Series, Vol. 391. American Chemical Society, Washington, DC. Chapter 6, pp. 59–76.

Simha, R., and T. Somcynsky. 1969. On the statistical thermodynamics of spherical and chain molecule fluids. *Macromolecules* 2: 342–350.

Song, Y.S., M. Lambert, J.M. Prausnitz. 1994. A perturbed hard-sphere-chain equation of state for normal fluids and polymers. *Ind. Eng. Chem. Res.* 33: 1047–1057.

Stern, S.A., S.M. Fang, H.L. Frisch. 1972. Effect of pressure on gas permeability coefficients. A new application of free volume theory. *J. Polym. Sci. Part A2.* 10: 201–219.

Stern, S.A., S.A. Kulkarni, H.L. Frisch. 1983. Tests of a "free-volume" model of gas permeation through polymer membranes. I. Pure CO_2, CH_4, C_2H_4, and C_3H_8 in polyethylene. *J. Polym. Sci. Polym. Phys. Ed.* 21: 467–481.

Tai, H., C.E. Upton, L.J. White, R. Pini, G. Storti, M. Mazzotti, K.M. Shakesheff, S.M. Howdle. 2010. Studies on the interactions of CO2 with biodegradable poly(DL-lactic acid) and poly(lactic acid-co-glycolic acid) copolymers using high pressure ATR-IR and high pressure rheology. *Polymer* 51: 1425–1431.

Taimoori, M., and C. Panayiotou. 2001. The non-random distribution of free volume in fluids: Non-polar systems. *Fluid Phase Equilib.* 192: 155–169.

Tait, P.G. 1888. Report on some of the physical properties of fresh water and sea water. *Phys. Chem.* 2: 1–76.

Tomba, J.P., and J.M. Pastor. 2009. Confocal Raman microspectroscopy: A non-invasive approach for in-depth analyses of polymer substrates. *Macromol. Chem. Phys.* 210: 549–554.

Toxvaerd, S. 1972. Surface structure of a square-well fluid. *J. Chem. Phys.* 57: 4092–4096.

Tsivintzelis, I., A.G. Angelopoulou, C. Panayiotou. 2007b. Foaming of polymers with supercritical CO_2: An experimental and theoretical study. *Polymer* 48: 5928–5939.

Tsivintzelis, I., A. Grenner, I.G. Economou, G.M. Kontogeorgis. 2008. Evaluation of the nonrandom hydrogen bonding (NRHB) theory and the simplified perturbed chain–statistical associating fluid theory (sPC-SAFT). 2. Liquid–liquid equilibria and prediction of monomer fraction in hydrogen bonding systems. *Ind. Eng. Chem. Res.* 47: 5651–5659.

Tsivintzelis, I., T. Spyriouni, I.G. Econoumou. 2007a. Modeling of fluid phase equilibria with two thermodynamic theories: Non-random hydrogen bonding (NRHB) and statistical associating fluid theory (SAFT). *Fluid Phase Equilib.* 253: 19–28.

Veytsman, B.A. 1990. Are lattice models valid for liquids with hydrogen bonds? *J. Phys. Chem.* 94: 8499–8500.

Veytsman, B.A. 1998. Equation of state for hydrogen-bonded systems. *J. Phys. Chem. B.* 102: 7515–7517.

Vincent, M.F., S.G. Kazarian, C.A. Eckert. 1997. Tunable diffusion of D_2O in CO_2-swollen poly(methyl methacrylate) films. *AIChE J.* 43: 1838–1848.

Vincent, M. F., S.G. Kazarian, B.L. West, J.A. Berkner, F.V. Bright, C.L. Liotta, C.A. Eckert. 1998. Cosolvent effects of modified supercritical carbon dioxide on cross-linked poly(dimethylsiloxane). *J. Phys. Chem. B.* 102: 2176–2186.

Vlachou, T., I. Prinos, J.H. Vera, C.G. Panayiotou. 2002. Nonrandom distribution of free volume in fluids and their mixtures: Hydrogen-bonded systems. *Ind. Eng. Chem. Res.* 41: 1057–1063.

Von Solms, N., M.L. Michelsen, G.M. Kontogeorgis. 2003. Computational and physical performance of a modified PC-SAFT equation of state for highly asymmetric and associating mixtures. *Ind. Eng. Chem. Res.* 42: 1098–1105.

Vrentas, J.S., J.L. Duda. 1977a. Diffusion in polymer-solvent systems: I. re-examination of the free-volume theory. *J. Polym. Sci. Polym. Phys. Ed.* 15: 403–416.

Vrentas, J.S., J.L. Duda. 1977b. Diffusion in polymer-solvent systems: II. A predictive theory for the dependence of the diffusion coefficient on temperature, concentration and molecular weight. *J. Polym. Sci. Polym. Phys. Ed.* 15: 417–439.

Vrentas, J. S., and J.L. Duda. 1986. Diffusion. In *Encyclopedia of Polymer Science,* J.I. Kroschwitz, ed. 2nd ed., Vol. 5, John Wiley & Sons, New York, pp. 36–68.

Vrentas, J.S., J.L. Duda, S.T. Ju, H.T. Liu. 1982. Prediction of diffusion coefficients for polymer-solvent systems. *AIChE J.* 28: 279–285.

Vrentas, J.S., and C.M. Vrentas. 1993. A new equation relating self-diffusion and mutual diffusion coefficients in polymer-solvent systems. *Macromolecules* 26: 6129–6131.

Vrentas, J.S., and C.M. Vrentas. 1996. Solvent self-diffusion in rubbery polymer-solvent systems. *Macromolecules* 27: 4684–4690.

Wertheim, M.S. 1984a. Fluids with highly directional attractive forces. I. Statistical thermodynamics. *J. Stat. Phys.* 35: 19–34.

Wertheim, M.S. 1984b. Fluids with highly directional attractive forces. II. Thermodynamic perturbation theory and integral equations. *J. Stat. Phys.* 35: 35–47.

Wertheim, M.S. 1986a. Fluids with highly directional attractive forces III. Multiple attraction sites. *J. Stat. Phys.* 42: 459–476.

Wertheim, M.S. 1986b. Fluids with highly directional attractive forces IV. Equilibrium polymerization. *J. Stat. Phys.* 42: 477–492.

Wertheim, M.S. 1986c. Fluids of dimerizing hard spheres, and fluid mixtures of hard spheres and dispheres. *J. Chem. Phys.* 85: 2929–2936.

Wertheim, M.S. 1987. Thermodynamic perturbation theory of polymerization. *J. Chem. Phys.* 87: 7323–7331.

Williams, M.L., R.F. Landel, L.D. Ferry. 1955. The temperature dependence of relaxation mechanisms in amorphous polymers and other glass-forming liquids. *J. Am. Chem. Soc.* 77: 3701–3707.

Winterfield, P., L.E. Scriven, H.T. Davis. 1978. An approximate theory of interfacial tensions of multicomponent systems: Applications to binary liquid–vapor tensions. *AIChE J.* 24: 1010–1014.

Wissinger, R.G, M.E. Paulaitis. 1991. Molecular thermodynamic model for sorption and swelling in glassy polymer-carbon dioxide systems at elevated pressures. *Ind. Eng. Chem. Res.* 30: 842–851.

Wong, B., Z. Zhang, Y.P. Handa. 1998. High-precision gravimetric technique for determining the solubility and diffusivity of gases in polymers. *J. Polym. Sci. Polym. Phys.* 36: 2025–2032.

Wu, S. 1982. *Polymer Interface and Adhesion*, Marcel Dekker, New York and Basel, Chapter 11, pp. 360–370.

Wulf, M., S. Michel, K. Grundke, O.I. Del Rio, D.Y. Kwok, A.W. Neumann. 1999. Simultaneous determination of surface tension and density of polymer melts using axisymmetric drop shape analysis. *J. Colloid Interf. Sci.* 210: 172–181.

Yeom, M.S., K. Yoo, B.H. Park, C.S. Lee.1999. A nonrandom lattice fluid theory for phase equilibria of associating systems. *Fluid Phase Equilib.* 158–160: 143–149.

Yoon, J.H., T. Kawamura, S. Takeya, S. Jin, Y. Yamamoto, T. Komai, M. Takahashi, A. V. Nawaby, Y. P. Handa. 2004. Probing Fickian and non-Fickian diffusion of CO_2 in poly(methyl methacrylate) using in situ Raman spectroscopy and microfocus x-ray computed tomography. *Macromolecules* 37: 9302–9304.

You, S.S., K.P. Yoo, C.S. Lee. 1994. An approximate nonrandom lattice theory of fluids-general derivation and application to pure fluids. *Fluid Phase Equilib.* 93: 193–213.

Young, T. 1805. An essay on the cohesion of fluids. *Philos Trans. R. Soc. Lond.* 95: 65–87.

Zuo, X.Y., and E.H. Stenby. 1997. Corresponding-states and parachor models for the calculation of interfacial tensions. *Can. J. Chem. Eng.* 75: 1130–1137.

5 Heterogeneous Cell Nucleation Mechanisms in Polylactide Foaming

Mohammadreza Nofar and Chul B. Park

CONTENTS

5.1 INTRODUCTION AND BACKGROUND

At present, most polymeric products are derived from fossil fuels, and they become nondegradable waste materials in the environment. Consequently, global efforts are being made to create green polymers from renewable resources, which are also biodegradable and compostable. Poly(lactic acid) or polylactide (PLA) is a biodegradable and biocompatible polymer produced from renewable resources such as cornstarch and sugarcane (Grijpma and Pennings 1994; Perego et al. 1996; Sinclair 1996; Tsuji and Ikada 1998; Martin and Averous 2001). It is a thermoplastic aliphatic polyester that is synthesized through the ring-opening polymerization of lactide and lactic acid monomers (Lunt 1998; Drumright et al. 2000; Gupta et al. 2007), as shown in Figure 5.1. Over the last decade, PLA has attracted extensive industrial and academic interests as a potential substitute for petroleum-based polymers, which are used in both commodity and biomedical applications. This is not only due to its green and biodegradable features but also because it releases no toxic components during the manufacture.

Due to its competitive material and processing costs, and comparable mechanical properties, this environmentally friendly biopolymer is considered as a promising replacement for polystyrene (PS), especially for PS foam products in such commodity applications as packaging, cushioning, construction, thermal and sound insulation, and plastic utensils (Ajioka et al. 1995; Fang and Hanna 2001; Garlotta 2001; Auras et al. 2004; Bandyopadhyay and Basak 2007; Lim et al. 2008; Nofar and Park 2014b). It would be environmentally very attractive to replace the commodity PS foam products with PLA foams because the amount of landfill for the large volume PS foam waste has been a serious global concern.

PLA foaming has mostly been conducted by dissolving a physical blowing agent in the PLA matrix and blowing it. Cell nucleation and cell growth can occur through thermodynamic instability

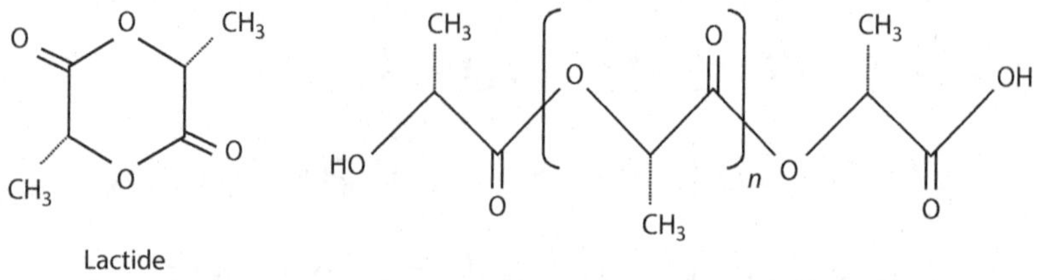

Lactide

FIGURE 5.1 Synthesis of polylactide: ring-opening polymerization. (Adapted from Gupta, B. et al., *Prog. Polym. Sci.*, 32, 455–482, 2007; Lunt, J., *Polym. Deg. Stab.*, 59, 145–152, 1998.)

generated from the supersaturation of the blowing agent (i.e., a pressure drop or a temperature increase). The foam structure is then produced by expelling the dissolved gas from the PLA/gas mixture. Foam products are made with cell stabilization as the temperature reaches below the PLA's T_g, which is around 60°C (Mikos et al. 1994; Di et al. 2005a; Lim et al. 2008).

Currently, the mass production of low-density PLA foams with a uniform cell morphology using supercritical carbon dioxide and nitrogen as the physical blowing agent is still quite challenging. This is mainly due to PLA's low melt strength (Dorgan and Williams1999; Dorgan et al. 2000; Palade et al. 2001; Dorgan et al. 2005; Lim et al. 2008; Dorgan 2010). Introducing a chain extender to create a branched structure (Carlson and Dubois 1998; Dorgan and Williams 1999; Di et al. 2005b; Sungsanit et al. 2010); modifying the L-lactide/D-lactide ratio configuration of the PLA molecules (Dorgan et al. 2005; Dorgan 2010); varying the PLA's molecular weight (Dorgan and Williams 1999; Dorgan et al. 2000, 2005; Dorgan 2010); and compounding the PLA with different types of additives (Krishnamoorti and Yurekli 2001; Ray et al. 2002; Ray and Okamoto 2003; Di et al. 2005a; Pluta 2006; Ray 2006; Wu et al. 2006; Gu et al. 2007, 2009; Wang et al. 2011) have been recognized as efficient methods to improve PLA's poor melt strength, and consequently, its foamability. Also, enhancing the PLA's slow crystallization kinetics has proven to be a significant parameter for improving its inherently low melt strength and for extending its applications (Rasal et al. 2010; Saeidlou et al. 2012). The enhanced crystallization compensates for the PLA's low melt strength during processing and thereby its foamability (Mihai et al. 2009, 2010; Wang et al. 2012; Keshtkar et al. 2014). The enhanced crystallization can further improve the final product's mechanical properties and the PLA's low heat deflection temperature (i.e., service temperature) (Ameli et al. 2013; Srithep et al. 2013).

This chapter discusses cell nucleation mechanisms in PLA foaming. First, we address the fundamental studies on heterogeneous cell nucleation. Then, we discuss the mechanisms of cell nucleation in PLA foams through micro/nano-sized additives and crystals. The influence of crystallization on the cell nucleation behavior of PLA foams is further discussed through various foam-processing technologies (i.e., bead foaming, extrusion foaming, and foam injection molding).

5.2 HETEROGENEOUS CELL NUCLEATION

According to the classical nucleation theory (Gibbs 1961; Leung et al. 2009), bubbles that are larger than the critical radius (R_{cr}) grow spontaneously, whereas those that are smaller than R_{cr} collapse. The expression of R_{cr} is derived as follows (Equation 5.1):

$$R_{cr} = \frac{2\gamma_{lg}}{P_{bub,cr} - P_{sys}} \tag{5.1}$$

where:

γ_{lg} represents the surface tension at the liquid-gas interface

$P_{bub,cr}$ and P_{sys} represent the pressure inside a critically sized bubble and the system pressure, respectively

The term ($P_{bub,cr} - P_{sys}$) represents the supersaturation level

Furthermore, by assuming that the polymer-gas solution is a weak solution, $P_{\text{bub,cr}}$ can be estimated by Henry's law. Consequently, Equation 5.1 can be rewritten as follows (Equation 5.2):

$$R_{\text{cr}} = \frac{2\gamma_{\text{lg}}}{HC - P_{\text{sys}}} \tag{5.2}$$

where:

H is Henry's constant
C is the gas concentration dissolved in the polymer

Based on molecular kinetics and the classical nucleation theory, the nucleation rate (J) can be expressed as follows (Equation 5.3) (Blander and Katz 1975):

$$J = A\exp\left(-\frac{W}{k_{\text{b}}T_{\text{sys}}}\right) \tag{5.3}$$

where:

A is the preexponential term
W is the free energy barrier that needs to be overcome for cell nucleation to occur
k_{b} is Boltzmann's constant
T_{sys} is the system temperature

The free energy barrier, W, is often lower if a bubble is nucleated on the surface of a second phase (heterogeneous nucleation), such as solid additives and crystals and the equipment wall surface, when compared to the case where the bubble is nucleated in the bulk phase of the polymer-gas mixture (homogeneous nucleation). Therefore, additives such as talc are often added to the polymer-gas mixture as a bubble nucleating agent. Recently, it has been demonstrated that the crystals nucleated in the polymer-gas mixture also acts as bubble nucleating agent (Wong et al. 2013a). The effectiveness of a nucleating agent depends on its surface geometry and the surface properties. In order to account for the surface toughness and irregularity of various additives in the micro- and nano-scale, Leung et al. (2006) modeled the nucleating sites on these additives as conical cavities with random semiconical angles, β. Consequently, the energy barrier, W, was modeled as follows (W_{hom} for homogeneous nucleation and W_{het} for heterogeneous nucleation) (Equations 5.4 and 5.5):

$$W_{\text{hom}} = \frac{16\pi\gamma_{\text{lg}}^3}{3\left(P_{\text{bub,cr}} - P_{\text{sys}}\right)^2} = \frac{16\pi\gamma_{\text{lg}}^3}{3\left(HC - P_{\text{sys}}\right)^2} \tag{5.4}$$

$$W_{\text{het}} = W_{\text{hom}}F\left(\theta_{\text{c}},\beta\right) = \frac{W_{\text{hom}}}{4}\left[2 - 2\sin\left(\theta_{\text{c}} - \beta\right) + \frac{\cos\theta_{\text{c}}\cos^2\left(\theta_{\text{c}} - \beta\right)}{\sin\beta}\right] \tag{5.5}$$

where:

$F(\theta_{\text{c}},\beta)$ is the ratio of the volume of the nucleated bubble on a nucleating agent to the volume of a spherical bubble with the same radius of curvature
θ_{c} is the contact angle between the bubble surface and the solid surface measured in the liquid phase

Alternatively, other researchers have suggested that potential nucleating sites can be found in free volumes within the saturated polymer-gas mixture, as well as preexisting cavities due to incomplete wetting between different phases in the mixtures (i.e., polymer, additives, crystals, blowing agents, and equipment surfaces) (Harvey et al. 1944; Levy 1981; Ward and Levart 1984; Taki et al. 2011).

After a rapid decrease in P_{sys}, R_{cr} would decrease due to the increased supersaturation level. As the R_{cr} decreases beyond the sizes of these microvoids, the latter would grow spontaneously to become nucleated cells.

Furthermore, in a plastic devolatilization study (Albalak et al. 1990) with a falling strand apparatus, some micro-sized bubbles were observed along the surface of bigger bubbles in a series of scanning electron micrographs of foamed plastic strands. They proposed that bubble expansion could generate tensile stresses in the surrounding melted plastic that results in decreased local P_{sys}. This would increase the supersaturation level and cause secondary micro-bubbles to nucleate around the bubble. Using the batch foaming view-cell system (Guo et al. 2006), a similar bubble growth-induced cell nucleation phenomenon was observed *in situ* in the foaming of PS-talc composites with CO_2 (Leung et al. 2012). It is believed that in the presence of growing bubbles, pressure fluctuations were generated around the nearby talc particles due to the biaxial stretching of the polymer-gas solution near the bubble surface. Some local regions, (i.e., around the side of the talc particles along the direction of the biaxial stretch) would be subjected to a pressure lower than the overall P_{sys}. Consequently, the supersaturation level increased, which ultimately increased the cell nucleation rate. To account for this pressure fluctuation, the expression for R_{cr} (Equations 5.1 and 5.2) and W (Equation 5.4) were modified in Equations 5.6, 5.7, and 5.8 (Okamoto et al. 2001; Wong et al. 2011):

$$R_{cr} = \frac{2\gamma_{lg}}{P_{bub,cr} - \left(P_{sys} + \Delta P_{local}\right)} = \frac{2\gamma_{lg}}{HC - \left(P_{sys} + \Delta P_{local}\right)} \tag{5.6}$$

$$W_{hom} = \frac{16\pi\gamma_{lg}^3}{3\left[P_{bub,cr} - \left(P_{sys} + \Delta P_{local}\right)\right]^2} = \frac{16\pi\gamma_{lg}^3}{3\left[HC - \left(P_{sys} + \Delta P_{local}\right)\right]^2} \tag{5.7}$$

$$W_{het} = W_{hom}F\left(\theta_c, \beta\right) \tag{5.8}$$

where ΔP_{local} is the difference between the overall and local P_{sys}. ΔP_{local} is positive if the local region experiences a compressive stress and negative if the local region is under an extensional stress. In regions with extensional stresses, the cell nucleation rate would increase due to an increase of the supersaturation level and the growth of existing microvoids. As said, crystals in semi-crystalline polymers might have a similar effect as the solid fillers to induce local pressure variations for cell nucleation. Wong et al. (2013a) observed that crystals have similar effects as solid fillers to cause bubble growth-induced cell nucleation under static conditions.

It is well known that adding nanoparticles to neat polymers can significantly increase the cell density in plastic foaming (Okamoto et al. 2001; Ray and Okamoto 2003; Nofar et al. 2012a; Srithep et al. 2012; Keshtkar et al. 2014). As a result, nanoparticles have been regarded collectively as an effective nucleating agent. As such, they improve the cell density of polymeric foams, and their effectiveness as a nucleating agent depends on the degree of exfoliation in the polymeric matrix (Fujimoto et al. 2003; Nofar et al. 2012a; Srithep et al. 2012; Keshtkar et al. 2014). There may be two explanations for this. First, the nanoparticles seem to act as a cell-nucleating agent to promote more cell nuclei. Second, the dispersed nanoparticles seem to stabilize the nucleated bubbles more effectively by decreasing cell coalescence with an increase of melt strength (Zheng et al. 2010; Keshtkar et al. 2014). It seems that the nanoparticles also decrease the cell-ripening phenomenon (Zhu et al. 2008) by decreasing the gas diffusion rate and by increasing the melt stiffness, and thereby contribute to the cell stability. As a result, a larger number of nucleated cells will last longer during cell growth with nanoparticles present. Improvements in foaming have been observed with the addition of nanoparticles to neat polymers, but most of these studies were conducted using the batch foaming process. Further studies are required, therefore, to identify the fundamental mechanisms of cell nucleation and growth in the presence of shear and elongational flows of polymer, which are severe when processing equipment such as extruders and injection-molding machines are used. Also, as previously noted, crystal formation is critical during foam processing. The presence of a physical foaming agent, such as carbon dioxide, has been shown to favorably modify PLA crystallization kinetics (Nofar et al. 2012b, 2013b, c, 2014a; Nofar and Park 2014b). In addition, to prevent premature phase separation (i.e., foaming) prior to exiting the die, the extrusion process must be conducted under controlled pressure. The high-pressure conditions needed to maintain the appropriate gas concentration in a solution are expected to influence the crystallization kinetics. Another parameter to consider is that the added nanoparticles will also affect the crystallization rate. Thus, careful attention should be paid to the impact of these variables on the crystallization kinetics, and adequate control of these parameters during extrusion foaming should be undertaken.

5.2.1 EFFECTS OF NANO- OR MICRO-SIZED ADDITIVES ON CELL NUCLEATION

5.2.1.1 PLA Composites with Micro-Sized Additives

Some researchers have studied foaming of PLA composites using micro-sized natural and synthetic additives (Kramschuster et al. 2007; Pilla et al. 2009; Kang et al. 2009; Matuana et al. 2010a, b; Boissard et al. 2012; Ameli et al. 2013). The presence of micro-sized additives can indeed increase the cell nucleation rate during foaming because they act as heterogeneous cell nucleation agents, as shown in Equations 5.6, 5.7, and 5.8. The subsequent two paragraphs introduce the studies in which micro-sized additives were used.

In order to manufacture fully degradable biocomposite foams with enhanced final mechanical properties, a few studies have investigated the foaming behavior of PLA using several natural fibers/additives such as flax fiber (Pilla et al. 2009), silk fibroin powder (Kang et al. 2009), wood flour (Matuana et al. 2010a, b), and microfibrillated cellulose (Boissard et al. 2012). Pilla et al. (2009) reported on the foam injection molding of PLA biocomposites using flax fiber contents of 1, 10, and 20 wt%. Silane was also used as a coupling agent to create a strong interface bonding between the PLA and the fibers. The foaming results showed that the cell density and the crystallinity of the foams were both improved with the fiber content. The average cell size as low as 3 μm was achieved when 20 wt% fibers were used. Kang et al. (2009) also investigated the effects of 1, 3, 5 and 7 wt% of silk fibroin powder on the PLA's foaming behavior in batch foaming, with CO_2 as the blowing agent. The biocomposites were prepared using a solution casting technique. They showed that the addition of silk fibroin powder decreased the average cell size from 52 μm in neat PLA to around 15 μm in PLA with 7 wt% silk. Further, Matuana et al. (2010b) explored PLA's batch foaming behavior with 10, 20, 30, and 40 wt% of wood fiber. Silane was also used as the coupling agent. They reported that an increase in the wood flour content reduced the expansion ratio from around 10-fold in neat PLA to twofold in PLA-40 wt% wood flour due to the matrix's high stiffness. However, the cell size was reduced to an average cell size of around 35 μm. Boissard et al. (2012) described the production of microfibrillated cellulose-reinforced PLA biocomposite foams using supercritical CO_2. Similar to the previous study, the expansion ratio of the neat PLA was reduced from 6.8-fold to 3.8-fold in PLA with 5 wt% microfibrillated cellulose and the cell density promoted with the addition of microfibrillated cellulose.

In another study, Kramschuster et al. (2007) investigated the PLA's foam injection molding behavior with recycled paper shopping bag fibers. The fiber contents of 10 and 30 wt% were used, with silane as the coupling agent. They showed that the addition of fibers up to 30 wt% enhanced the cell density of the foamed samples. Ameli et al. (2013) also studied PLA's foam injection molding behavior with talc particles up to 5 wt%. They confirmed that the presence of talc, together with the improved crystallinity, created more uniform foam morphology with larger cell density and smaller cell sizes. They also attributed the enhanced cell density to the promoted crystallization kinetics though addition of talc.

5.2.1.2 PLA Nanocomposites

The presence of a large number of dispersed nanoparticles in polymers can also significantly promote cell density in foam samples. Similar to the micro-sized additives, the nanoparticles act as heterogeneous cell nucleating sites (Okamoto et al. 2001; Srithep et al. 2012; Keshtkar et al. 2014) through the local pressure variations that are created around the larger number of small-sized (i.e., nano-sized) additives (Wong et al. 2011, 2013b, Leung et al. 2012). Therefore, the effectiveness of nanoparticles as cell nucleating agents strongly depends on how well they disperse and exfoliate in the polymer matrix (Fujimoto et al. 2003; Wong et al. 2013b; Keshtkar et al. 2014). While nanoparticles can promote cell density, the dispersed particles can simultaneously enhance the melt strength of the polymer matrix, thereby stabilizing the nucleated cells by minimizing cell coalescence (Zheng et al. 2010). At the same time, the reduced gas diffusion rate can minimize the cell-ripening phenomenon (Zhu et al. 2008). Consequently, a larger number of nucleated cells will last longer during cell growth, and the final expansion ratio of the foam products will also be promoted.

In the past decade, several studies have investigated the effects of nanoparticles on PLA's foaming behaviors (Ray and Okamoto 2003; Di et al. 2005a; Keshtkar et al. 2014; Nofar and Park 2014b). Some of these studies used batch foaming apparatus to explore the influence of nanoparticles on PLA's foaming behavior. These studies used nanoclay due to its long aspect ratio and platelet-shaped structure, which provides superior viscoelastic behavior in the polymer matrix. They found that the PLA–clay nanocomposite foams resulted in producing foams with homogeneous cell morphology and average cell sizes ranging from 165 nm to 25 µm. The majority of the nucleated cells appeared to be of a closed-cell structure. In fact, due to the biaxial stretching during cell growth, the long aspect ratio platelet-shaped nanoclay particles were oriented along the cell walls, and the increased cell wall strength inhibited cell rupture (Okamoto et al. 2001). All of these studies found that the nanoparticles behaved as heterogeneous cell nucleating agents, and that a varying range of cell sizes were obtained by using different nanoclay loadings. The cell density of the foamed samples increased as the amount of nanoclay particles increased. Thereby, the cell size of PLA foamed samples could be controlled by the amount of nanoparticles loaded. Further, Tsimpliaraki et al. (2011) reported that PLA cell nucleation behavior depended not only on the nanoclay loading, but could also be significantly affected by chemically modifying the nanoparticles. They observed that adding organically modified nanoclay particles resulted in a more heterogeneous cell nucleation, with smaller cell sizes. This was due to the nanoclay's different degree of dispersion in the PLA nanocomposite (Fujimoto et al. 2003; Keshtkar et al. 2014).

5.2.1.2.1 Effects of Nanoclay Particles

A few studies also explored the continuous extrusion foaming behavior of PLA nanocomposites using supercritical CO_2 (Matuana et al. 2010a; Keshtkar et al. 2014). As extensively studied by Keshtkar et al. (2014), this section describes more in detail the development of PLA's foaming behavior with the use of nanoclay using a tandem line extrusion system (Figure 5.2) and supercritical CO_2 as the blowing agent.

A commercial linear PLA (Ingeo 2002D) with a D-lactide content of 4.5% supplied by NatureWorks LLC, Minnetonka, Minnesota, was used. The nanoclay (Cloisite 30B: alkyl quaternary ammonium bentonite) was provided by Southern Clay Products, Texas.

The PLA and nanoclay were oven-dried at 65°C for 24 hours. They were dry-blended and then melt-extruded using a counter-rotating twin-screw extruder with 10 distributive and 10 dispersive zones at 180°C and at a rotator speed of 150 rpm. A master batch of 5 wt% of PLA/clay nanocomposite (PLACN5) was prepared. A series of PLA/clay nanocomposites with a clay content of 0.5, 1, and 2 wt% was then prepared by diluting the master batch with PLA (PLACN0.5, PLACN1, and PLACN2). Figure 5.3 shows the transmission electron micrographs and the X-ray diffraction graphs of PLACN samples. The results showed that the clay nanoparticles were well intercalated and exfoliated to a high degree within PLA. The d-spacing between the nanosilicates increased from 1.84 to 3.48 nm and 3.57 nm for PLACN5 and PLACN1, respectively. Also, in PLACN0.5 almost no peak was detected, which generally indicates a high degree of exfoliation.

As shown in Figure 5.2, a tandem extrusion system was used for extrusion foaming experiments. The first extruder plasticizes the polymer resin and disperses the blowing agent into the polymer melt. The second extruder provides further mixing and initial cooling of the melt, and the heat

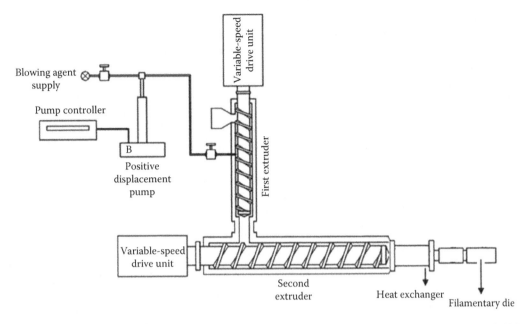

FIGURE 5.2 Schematic of the tandem extrusion line. (Adapted from Wang, J. et al., *Chem. Eng. Sci.*, 75, 390–399, 2012.)

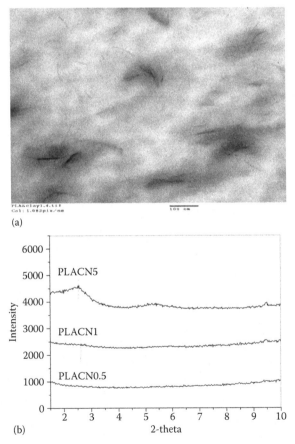

FIGURE 5.3 Transmission electron micrographs of PLACN1 (a) and X-ray diffraction results of PLACN5, PLACN1, and PLACN0.5. (Image and data adapted from Keshtkar, M. et al., *Polymer*, 55, 4077–4090, 2014.)

exchanger removes the remaining heat for testing at a specific temperature. CO_2 with a purity of 99.5% (Linde gas) was used as a physical blowing agent. The CO_2 content injected into the barrel was fixed to 9 wt% throughout the foaming process. Foamed samples at various die temperatures were collected for characterization.

The experimental results in this study revealed that the foam properties (i.e., the cell density and expansion ratio) were not only affected by the nanoclay particles, but were also influenced by the expedited crystallization along the second extruder. It was demonstrated that the nanoclay (Nofar et al. 2013c; Keshtkar et al. 2014), the dissolved CO_2 (Nofar et al. 2012b, 2013b, c, 2014a; Nofar and Park 2014b), and the shearing effect (Keshtkar et al. 2014) during extrusion expedited the PLA's slow crystallization kinetics. It was reported that the nanoparticles, in conjunction with the induced crystallization during the foam processing, behaved as cell-nucleating sites and thereby created foams with a larger cell density, confirming Mihai et al.'s earlier results (Mihai et al. 2009, 2010). These two factors further enhanced the PLA's melt strength, and low-density microcellular foams were subsequently achieved.

Figure 5.4 shows the cell density and the expansion ratio for PLA and PLACNs at various die temperatures. Figure 5.5 also shows the cell morphologies of PLA and PLACNs at die temperatures of 115°C, 125°C, and 135°C. By increasing the clay content, both the expansion ratio and the cell density of the foam samples increased and, in the case of PLACN5, higher expansion ratios with finer cells were obtained during a wider processing window. As seen, nanoclay is regarded as an effective nucleating agent that improves the cell density of polymeric foams. As will also be shown more in detail in Section 5.2.2, both the cell nucleation and growth behaviors of PLA in foaming, were also significantly affected by the nucleated crystals along the second extruder.

5.2.1.2.2 Effects of Nanoclay Dispersion on the Cell Nucleation Behavior

The effects of nanoclay dispersion on the PLA's foaming behavior by using nanoclay Cloisite 20A particles with a poor dispersion-ability in the PLA have further been investigated. Transmission electron micrographs of PLACN1 prepared by using Cloisite 20A (Figure 5.6a,b) and Cloisite 30B (Figure 5.6c,d) show the intercalation of Cloisite 20A particles, while Cloisite 30B particles are partially exfoliated. Foams obtained at various die temperatures using these two materials were characterized, and the results showed that both the expansion ratio and the cell density were higher with the Cloisite 30B, as shown in Figure 5.7.

Fujimoto el al. (2003) also looked at the effects of nanoparticle dispersion on the foaming properties of PLA nanocomposites using a batch foaming system. They showed that the quality of the dispersion played a vital role in the final foam's cell density and cell size. They observed that cell density increased dramatically with a better dispersion of the intercalated silicate layers. Keshtkar's results were also consistent with their findings. The transmission electron micrographs seen in

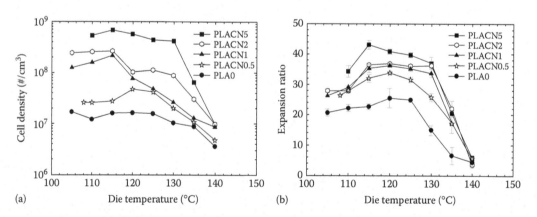

(a)   (b)

FIGURE 5.4 (a) Cell density and (b) expansion ratio of PLA and PLACN foams obtained at various die temperatures. (Adapted from Keshtkar, M. et al., *Polymer*, 55, 4077–4090, 2014.)

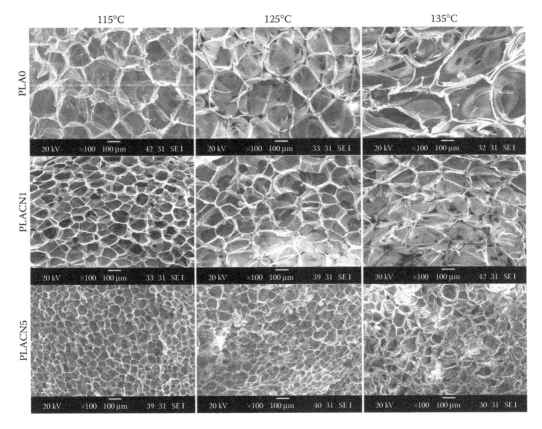

FIGURE 5.5 Cell morphologies of PLA and PLACN foams obtained at die temperatures of 115°C, 125°C, and 135°C. (Adapted from Keshtkar, M. et al., *Polymer*, 55, 4077–4090, 2014.)

Figure 5.6 show that in the PLACN1 samples prepared using Cloisite 30B a semi-exfoliated structure was obtained. A higher degree of nanoparticle dispersed at a given weight fraction resulted in greater nanoparticle density. Thus, larger heterogeneous cell nucleation sites were created.

On the other hand, unlike the findings of Fujimoto et al., who observed a higher foam density for well-dispersed intercalated silica layers (expansion ratio ≈ 2.5), Wong et al. (2013b) also found that a better dispersion resulted in a higher expansion, that is, a lower foam density. Moreover, Wong et al. observed that a wider processing window was provided for PLA foam production using Cloisite 30B, which has a greater degree of dispersion.

Figure 5.8 compares the complex viscosity of PLACN1 using Cloisite 20A and Cloisite 30B, with each sample having undergone small amplitude oscillatory shear at 100°C after having being annealed for 10 min at 180°C. No pre-shearing was applied. The results showed that even the crystallization rate was more rapid for the PLACN1, which contained Cloisite 30B. Consequently, in their early formation stages, more crystal domains can further enhance cell nucleation in PLA nanocomposites with Cloisite 30B than with Cloisite 20A.

5.2.2 Effects of Crystallization on Cell Nucleation

During PLA foam processes, the PLA's inherently low melt strength (Lim et al. 2008) leads to cell coalescence and cell rupture during cell growth. Moreover, its low melt strength facilitates gas loss during foam expansion, which causes severe shrinkage (Lee et al. 2008b). Enhancing PLA's crystallization kinetics during foam processing has been recognized as an effective way to overcome

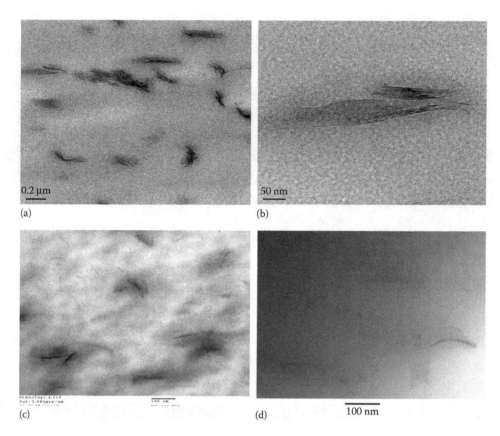

FIGURE 5.6 Transmission electron micrographs of PLACN1 (a and b) prepared using Cloisite 20A and (c and d) prepared using Cloisite 30B. (Adapted from Keshtkar, M. et al., *Polymer*, 55, 4077–4090, 2014.)

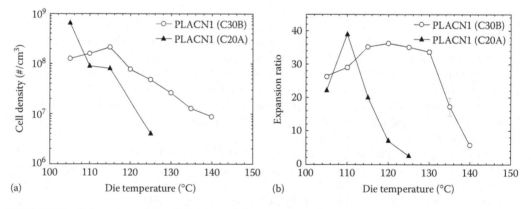

FIGURE 5.7 Comparison of (a) cell density and (b) expansion ratio of PLACN1 prepared using Cloisite 20A (C20A) and Cloisite 30B (C30B) at various die temperatures. (Adapted from Keshtkar, M. et al., *Polymer*, 55, 4077–4090, 2014.)

its weak viscoelastic properties and to improve its foaming behaviors (i.e., cell nucleation and expansion) (Mihai et al. 2009, 2010; Wang et al. 2012; Keshtkar et al. 2014). Crystallization during the foam processes can promote PLA's low melt strength through the network of nucleated crystals. This will consequently increase the PLA's ability to expand by minimizing the gas loss and cell coalescence. It should be also noted that too high a crystallinity will also suppress foam expansion.

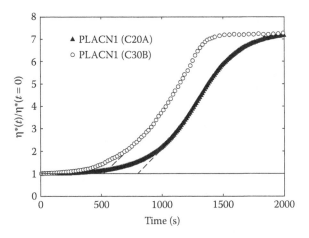

FIGURE 5.8 Normalized complex viscosity of PLACN1 measured at a frequency of 1 Hz and at a strain amplitude of 0.01 at 100°C after being annealed at 180°C for 10 min using Cloisite 20A (C20A) and Cloisite 30B (C30B). (Adapted from Keshtkar, M. et al., *Polymer*, 55, 4077–4090, 2014.)

This would be due to the resultant excessive stiffness and to less gas dissolution (Lee et al. 2008b; Marrazzo et al. 2008). On the other hand, according to the heterogeneous cell nucleation theory (Wong et al. 2013a, 2013; Leung et al. 2012), cell nucleation can be promoted around the nucleated crystal (Marrazzo et al. 2008; Liao et al. 2010; Taki et al. 2011; Wong et al. 2013a) through local pressure variations (Wong et al. 2013a) and thereby the final cell density of the foamed samples will also be significantly improved.

Taki et al. (2011) visualized the cell nucleation behavior of a poly-L-lactide acid (PLLA)/CO_2 mixture in a batch process. They showed that the presence of spherulites in the PLLA matrix promoted the number of cell nuclei. Figure 5.9 shows that as depressurization occurred, the cells were nucleated around the growing PLLA spherulites. Taki et al. (2011) also showed that the number of nucleated cells was significantly promoted as the spherulites density increased. They explained that the growing spherulites expelled the dissolved CO_2 from the interface phase between the amorphous and spherulites sections of the PLLA. Therefore, an increased CO_2 concentration in the interface caused cell nucleation around these growing spherulites. Moreover, they demonstrated that the cell density increase was a function of the spherulites area.

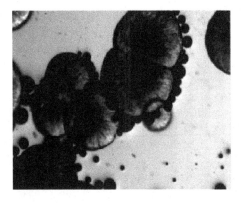

FIGURE 5.9 Foam visualization of PLLA/CO_2 system under an optical microscope. The saturation pressure and temperature were 11 MPa and 180°C, respectively. The crystallization and bubble nucleation occurred at 110°C after 6.1 s. (Adapted from Taki, K. et al., *Ind. Eng. Chem. Res.*, 50, 3247–3252, 2011.)

Liao et al. (2010) also explored the effects of crystallization on the foaming behavior of PLLA samples with various crystallinities and crystallite sizes. The PLLA samples were first saturated with various CO_2 pressures at 25°C to induce different degrees of crystallinity in the PLLA samples. Then, all of the saturated samples were quenched and re-saturated with 2.8 MPa CO_2 at 0°C. Before the re-saturation, five PLLA samples were achieved with the crystallinity and crystallite size, ranging 12%–30% and 2.2–12.7 nm, respectively. The samples were subsequently foamed at temperatures between 50°C and 100°C. The trends of the obtained foam densities revealed that a higher amount of preexisting crystals with larger sizes created a higher foam density (i.e., close to the neat PLLA's density), although finer cells were achieved in the foamed regions. This means that the increased stiffness throughout the large crystalline domain restricted cell growth and further expansion. On the other hand, with a lower amount of preexisting crystallinity with smaller crystallite sizes, both cell nucleation and cell growth were promoted, although the cell growth rate dominated the cell nucleation rate. In other words, a lower degree of crystallinity facilitated both cell growth and the eventual foam expansion ratio.

5.2.2.1 Bead Foaming Process

In recent studies done by Nofar et al. (Park and Nofar 2014; Nofar and Park 2014b; Nofar et al. 2015a, 2015b), for the first time, expanded PLA (EPLA) bead foaming was introduced with a double-crystal melting peak structure. In this technique, the high-melting-temperature-peak crystals that form during the isothermal saturation step in a batch-based bead foaming process maintain the bead geometry (Braun 2004; Sasaki et al. 2005). The formation of this high-melting-temperature crystal is due to the crystal perfection during which the saturation occurs. The gas saturation occurs around the melting temperature of the PLA at which the unmelted crystals become more perfect with a higher melting temperature (Choi et al. 2013; Nofar et al. 2013a). These induced high-melting-temperature crystals can significantly affect cell nucleation and the expansion behaviors of PLA bead foams. As explained earlier, cell nucleation can be promoted around the formed crystals (Taki et al. 2011; Wong et al. 2013a). On the other hand, the low melt strength PLA molecules become high-melt strength material. This increases the PLA's ability to expand by minimizing both gas loss and cell coalescence. But too high a crystallinity will depress the foam's expansion ability due to the increased stiffness of the matrix (Marrazzo et al. 2008; Liao et al. 2010).

A small autoclave foaming chamber and a lab-scale bead foaming chamber (Nofar et al. 2015a) were used to produce PLA bead foams with double crystal melting peak structure. The ratio of the induced crystal melting peaks could be controlled by varying the saturation temperature, time, and pressure (Nofar et al. 2015a, 2015b).

As the saturation temperature (T_s) increased, a less amount of crystals was encountered to crystal perfection during the annealing and thereby a smaller area was generated as the second peak at high temperatures. However, the generated second peak appeared at higher temperatures due to the increased molecular mobility and easier molecular retraction (i.e., better perfection) (Nofar et al. 2013a). In contrast, the samples annealed at higher temperatures revealed a larger area as the low-melting-peak crystals due to the increased amount of melt available for crystallization during cooling. Moreover, the samples, which were annealed for a longer time, showed a larger amount of perfect crystals due to the longer diffusion time for molecular rearrangement. Compared to the lower saturation pressures, the double peak structure was generated at a lower range of T_s when the CO_2 pressure was increased. This was due to the plasticization effect of the dissolved CO_2, which suppressed the required T_s range (Nofar et al. 2013a).

Figure 5.10 shows the differential scanning colorimetry (DSC) heating thermograms of the EPLA bead foams produced under various saturation pressures and temperatures at a given saturation time of 60 min. The generated high-melting-temperature-peak crystals (i.e., second peak) in the PLA beads, the total crystallinity, and the second peak crystallinity are also presented in Figure 5.11. The bead foam results showed that the PLA samples saturated at 435 psi CO_2 pressure were notable to foam.

Figures 5.12 and 5.13 show the expansion ratio and the cell morphology of the EPLA bead foams obtained using various saturation pressures and temperatures. As seen, after saturation at 870 and

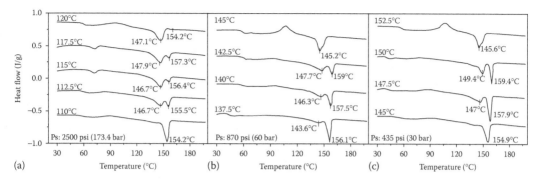

FIGURE 5.10 DSC heating thermograms of the PLA bead samples achieved at various saturation conditions (2500, 870, and 435 psi). (Adapted from Nofar, M., A. Ameli, C.B. Park, 2015a. Development of polylactide bead foams with double crystal melting peaks. *Polymer* 69: 83–94.)

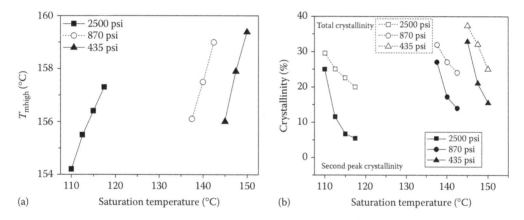

FIGURE 5.11 The second high-melting temperature peak (T_{mhigh}) (a) and total crystallinity and second peak crystallinity (b) of the PLA bead samples when different saturation pressures and temperatures were applied. (Adapted from Nofar, M., A. Ameli, C.B. Park, 2015a. Development of polylactide bead foams with double crystal melting peaks. *Polymer* 69: 83–94.)

2,500 psi, with the increased saturation temperature, bead foams with expansion ratios of 3 to 25 times and 3 to 30 times, respectively, were obtained. The corresponding average cell sizes were also observed between 500 nm–500 μm and 700 nm–15 μm, respectively. As seen, nanocellular bead foams with expansion ratios around three times were also able to reach at the lowest applied saturation temperatures. This was most likely due to the enhanced heterogeneous cell nucleation rate that occurred around the larger amount of perfected crystals formed during the saturation. As Figure 5.11b shows, after saturation at 870 and 2,500 psi at the lowest corresponding saturation temperatures, 25%–30% of crystallinity was induced during the saturation. These large amount of crystals were involved in promoting the cell nucleating toward nano-sized cells; in fact, the cell nucleation dominated the cell growth. Therefore, regardless of the bead foaming purposes with double crystal melting peak, this manufacturing method can be suggested as a novel way to produce nanocellular foams mainly for superinsulation applications (Thiagarajan et al. 2010).

After saturation at CO_2 pressures of 870 and 2,500 psi, as the saturation temperature increased, the expansion ratio of the PLA bead foams was enhanced up to 25 times and 30 times, respectively. This was due to the reduced stiffness of the PLA matrix, which facilitated the cell growth. In other words, with the increased saturation temperatures, the amount of induced perfect crystals (i.e., the second peak crystals) started to decrease from around 25%–30% to less than 15%. Therefore, due to

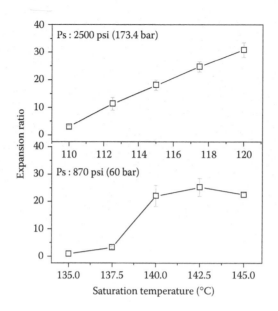

FIGURE 5.12 Expansion ratio of the foamed PLA beads after saturation at various CO_2 pressures and temperatures. (Adapted from Nofar, M., A. Ameli, C.B. Park, 2015a. Development of polylactide bead foams with double crystal melting peaks. *Polymer* 69: 83–94.)

the reduced crystallinity of the matrix, the heterogeneous cell nucleation rate was also suppressed and cell growth governed the foaming mechanism.

As Figure 5.13 shows, after saturation at 870 and 2,500 psi, the obtained cell density was decreased with the increased saturation temperature. According to the DSC analyses, after saturation at 137.5°C (870 psi) and 110°C (2,500 psi), 27% and 25% of crystallinity was induced during the saturation, respectively. The large amount of induced crystallinity significantly promoted the heterogeneous cell nucleation rate through the local pressure variations (Wong et al. 2011; Leung et al. 2012) around the crystals (Wong et al. 2013a). However, the cell growth was noticeably hindered due to the increased stiffness of the PLA matrix. On the other hand, after saturation at higher temperatures, although the reduced amount of perfect crystals facilitated the cell growth, it also reduced the cell nucleation density because of the reduced heterogeneous cell nucleation sites (i.e., perfect crystals induced during the saturation).

Overall, the scanning electron micrographs illustrate that after saturation at 2,500 psi, very uniform cell morphologies with more closed-cell contents were achieved with the average cell sizes of 700 nm to 15 μm when the saturation temperature was increased from 110°C to 117.5°C. However, after saturation at 870 psi, the cell uniformity was suppressed with more open-cell structure and the cell size was varied from 500 nm at 137.5°C to around 500 μm at 145°C.

5.2.2.2 Extrusion Foaming

Several researchers have investigated the effect of crystallization on the foaming behavior of PLA during the foam extrusion process (Mihai et al. 2009, 2010; Wang et al. 2012; Keshtkar et al. 2014). It is well known that isothermal melt crystallization occurs at temperature ranges between the PLA's T_m and T_g (Nofar et al. 2012b). As explained earlier, during extrusion foaming, where the PLA/gas mixture can encounter isothermal melt crystallization, a certain amount of nucleated crystals can affect the PLA's foaming behavior by enhancing the cell density and the expansion ratio. Using a twin-screw extruder, Mihai et al. (2009) showed that the presence of nucleated crystals in the flowing PLA/CO_2 mixture inside the extruder barrel acted as cell nucleation sites during the foaming

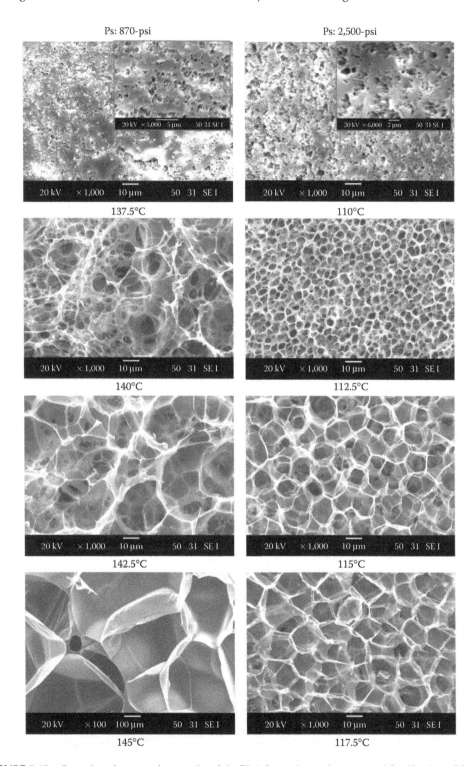

FIGURE 5.13 Scanning electron micrographs of the PLA foamed samples saturated for 60 min at CO_2 pressures of 870 psi and 2,500 psi and at various saturation temperatures. (Adapted from Nofar, M., A. Ameli, C.B. Park, 2015a. Development of polylactide bead foams with double crystal melting peaks. *Polymer* 69: 83–94.)

step and enhanced the PLA's melt strength and thereby its expansion ratio. Specifically, with a higher gas content, the CO_2-induced crystallization created larger cell nucleation sites. This eventually resulted in higher expansion and increased cell density in the PLA foams. As Figure 5.14 shows, with a lower CO_2 content, the cell coalescence was severe due to the PLA's low melt strength, and the final foam expansion was not considerable. However, as the CO_2 content increased, the foaming behavior improved, and the expedited crystal nucleation rate caused by the gas also increased, thereby promoting cell density and melt strength (i.e., the expansion ratio). Similar findings have also been confirmed in work done by Mihai et al. (2010), where they foamed various PLA materials with different branching degrees using a twin-screw extruder. They demonstrated that the crystallizable PLA had a higher expansion ratio and finer cell density due to the presence of nucleated crystals during extrusion foaming. The nucleated crystals, specifically at a higher CO_2 content, must have enhanced the PLA's melt strength and provided more cell nucleation sites before the foaming action (Figure 5.15).

Wang et al. (2012) also conducted extrusion foaming of a linear and two branched PLA materials with a D-lactide content of 4.5%. The experiments were done in a tandem extrusion system consisting of two single extrusion lines (Figure 5.2). A resistance die was used after the tandem line and prior to the foaming step to induce further isothermal melt crystallization. The crystallization was controlled by varying the length of the die reservoir. They showed that when a longer die residence time was used, the isothermally induced crystallization improved, and the high expanded foams were obtained in a wider processing window. This was due to the increased melt strength of the PLA/CO_2 mixture with the molecules connected through crystals, which reduced gas loss during foam expansion. According to the reported scanning electron micrographs, the induced crystallization also provided more nucleated cells and more closed-cell content (i.e., minimized cell coalescence) due to the increased melt strength.

| 5% CO_2/expansion = 2.8 | 7% CO_2/expansion = 35.6 | 9% CO_2/expansion = 32.6 |

FIGURE 5.14 Scanning electron micrographs and expansion ratios of the linear PLA (D-content = 2) samples with foamed at various CO_2 contents. (Adapted from Mihai, M. et al., *J. Appl. Polym. Sci.*, 113, 2920–2932, 2009.)

| Expansion = 31.7 | Expansion = 35.6 | Expansion = 35.6 |

FIGURE 5.15 Scanning electron micrographs and expansion ratios of the branched PLA (D-content = 2) samples with 2% chain extender foamed at various CO_2 contents. (Adapted from Mihai, M. et al., *Polym. Eng. Sci.*, 50, 629–642, 2010.)

TABLE 5.1

Various Temperature Profiles during Extrusion Foaming in the Second Extruder of the Tandem-Screw Extruder (Figure 5.2)

	Zone 1	Zone 2	Zone 3	Heat Exchanger	Die
Profile 1	180°C	140°C	130°C	130°C	120°C
Profile 2	180°C	135°C	125°C	125°C	120°C
Profile 3	180°C	130°C	120°C	120°C	120°C

As discussed in Section 5.2.1.2.1, Keshtkar et al. (2014) also used the same tandem extrusion system to investigate the induction of isothermal melt crystallization along the second extruder in a tandem line. They varied its temperature profile, and therefore, they controlled the extruded PLA foam properties by varying the amount of induced crystallization along the tandem line. Table 5.1 shows the temperature profile variations of the second extruder and Figures 5.16 and 5.17 shows the cell morphology, as well as the cell density and expansion ratio variations of the PLA–clay foamed samples, which are controlled by the induced crystallinity from varying the second extruder's temperature profile. They showed that reducing the temperature profile along the second extruder, with a constant residence time, significantly enhanced the cell density and expansion ratio of the foamed samples. This was most likely due to the accelerated isothermal crystallization of PLA and PLACN along the extruder. Therefore, by choosing a lower temperature profile, we could expect a greater

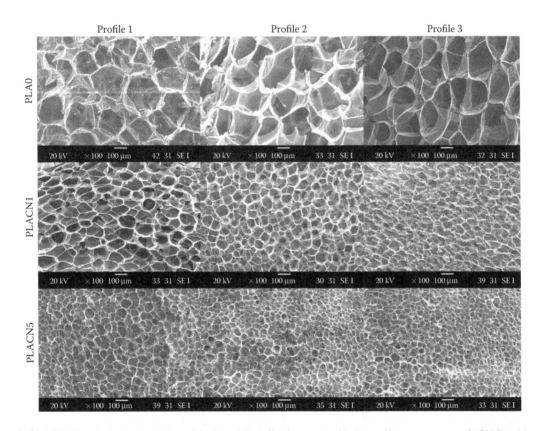

FIGURE 5.16 Cell morphologies of PLA and PLACN foams obtained at a die temperature of 120°C, with varying temperature profiles in second extruder, as shown in Table 5.1. (Adapted from Keshtkar, M. et al., *Polymer*, 55, 4077–4090, 2014.)

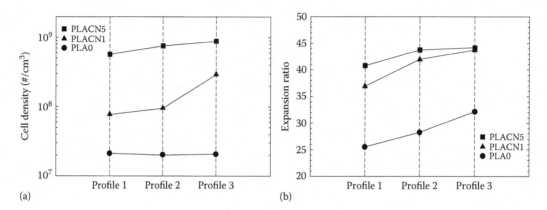

FIGURE 5.17   (a) Cell density and (b) expansion ratio of PLA and PLACN foams obtained by varying temperature profiles in second extruder, as shown in Table 5.1. (Adapted from Keshtkar, M. et al., *Polymer*, 55, 4077–4090, 2014.)

degree of crystals and thereby enhanced cell nucleation rate through promoted heterogeneous cell nucleation around the crystals and increased expansion ratio via improved melt strength.

5.2.2.3   Foam Injection Molding

Ameli et al. (2013, 2014, 2015) have investigated the PLA foam injection molding by incorporating talc and nanoclay to enhance the cell density and to improve the cell uniformity along the foam injected molded samples. It has been shown that not only the talc and more significantly nanoclay particles but also the expedited crystallization through the additives increased the cell density and enhanced the cell uniformity of the injected molded PLA foams.

They used a 50-ton Arburg Allrounder 270/320C injection molding machine (Lossburg, Germany) with a 30 mm diameter screw and equipped with MuCell® technology (Trexel, Inc., Woburn, Massachusetts). The mold contained a rectangular cavity of 3.1 mm thickness with a fan gate connecting the cavity to the sprue. For samples with a void fraction of 30%, regular foam injection molding (RFIM) was used, while the samples with 55% void fraction were fabricated using foam injection molding with mold opening (FIM + MO). In the case of FIM + MO, a cavity gas counter pressure of about 6 MPa was also applied using a control module of Caropreso Associates, Massachusetts. The mold cavity was then completely filled by injection (i.e., full shot) and expansion was induced later by releasing the gas from the mold cavity and opening the mold in the thickness direction after 9 s delay.

5.2.2.3.1   Regular Foam Injection Molding

Figure 5.18 shows the cellular morphology in the transition and core regions of PLA, PLA/talc, and PLA/clay foams with 30% void fraction made using RFIM. In general, relatively poor cell morphology was observed in the samples fabricated using RFIM. Since there was no resisting pressure in the cavity with 70 vol.% partial filling as in the structural foam molding (Xu et al. 2008), the pressure drop of the polymer/gas mixture obtained at the gate decides the cell nucleation rates (Lee et al. 2008b). This sudden pressure drop, to a pressure lower than the solubility pressure of 9.5 MPa for 0.6 wt% N_2 in the PLA (Kwag et al. 2001), resulted in cell nucleation at the gate location (Lee et al. 2008b; Ameli et al. 2013). Since the melt temperature was still high at the gate location, the nucleated cells were prone to grow. Meanwhile, the injected melt had to travel in the machine direction to fully fill the cavity. The combination of these two actions of cavity filling and cell growth resulted in shear stresses on the growing cells in the machine direction of the transition region. Therefore, the cells were elongated in that direction. On the other hand, the core region moved with less shear, and therefore, the cells were more spherical.

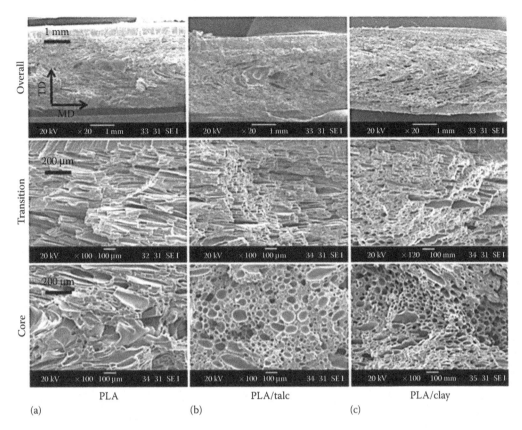

FIGURE 5.18 Representative scanning electron micrographs of (a) PLA, (b) PLA/talc, and (c) PLA/clay foams with 30% void fraction fabricated using regular foam injection molding. The final thickness of the samples was 3.1 mm and the entire thickness is shown in the upper micrographs. Lower micrographs show the cell morphologies of the transition and core regions. Magnification is the same for each row. MD and TD represent machine direction and thickness direction, respectively. (Adapted from Ameli, A. et al., *Comp. Sci. Tech.*, 90, 88–95, 2014; Ameli, A. et al., *Chem. Eng. J.*, 262, 78-87, 2015.)

Compared to the PLA composites, the more severe cell elongation was observed in neat PLA (Figure 5.18a). This was due to the low melt strength and poor cell nucleating power of neat PLA. In the PLA/talc samples, the addition of 5 wt% talc improved the PLA's foam structure, specifically in the core region (Figure 5.18b). In the RFIM samples, the presence of talc improved the PLA's cell nucleating power and melt strength. Consequently, more cells could nucleate and retain the spherical shape during growth (Figure 5.18b).

In the PLA/clay samples, the addition of 5 wt% nanoclay further improved the cell morphology (Figure 5.18c) and yielded the highest cell density and the lowest cell size compared to the other cases. The larger number of the uniformly dispersed nanoclay particles together with the larger number of the crystals nucleated through the nanoparticles (Nofar et al. 2013c) further improved the cell nucleation rate as well as the melt strength of the PLA/gas mixture.

5.2.2.3.2   Foam Injection Molding with Mold Opening

Figure 5.19 shows the cellular morphology in the transition and core regions of the PLA and PLA composite foams with 55% void fractions fabricated using FIM + MO method. Very different cellular morphology was obtained in FIM + MO samples, compared to RFIM because of the differences of two processes.

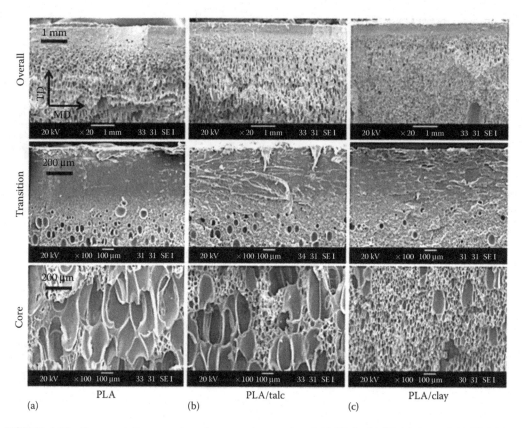

FIGURE 5.19 Representative scanning electron micrographs of (a) PLA, (b) PLA/talc, and (c) PLA/clay foams with 55% void fraction fabricated using FIM + MO. The final thickness of samples was 7.0 mm and half of the thickness is shown in the upper micrographs with centerline of the samples being at the bottom of the image. Lower micrographs show the cell morphologies of the transition and core regions. Magnification is the same for each row. (Adapted from Ameli, A. et al., *Comp. Sci. Tech.*, 90, 88–95, 2014; Ameli, A. et al., *Chem. Eng. J.*, 262, 78–87, 2015.)

One major difference between RFIM and FIM+MO was their foaming mechanisms. For RFIM, the shot size is short and the rapid pressure drop occurring at the gate decides the cell nucleation rates. In contrast, for FIM + MO, the shot size is full and the cavity was charged with high pressure gas (i.e., gas counter pressure) before filling, and this limited cell nucleation and cell growth during the injection and delay time. In general, the pressure drop occurring during mold opening decides the cell nucleation rates, depending on the magnitude of gas counter pressure. The solubility pressure corresponding to 0.6 wt% N_2 is 9.5 MPa at 170°C (Li et al. 2006), so the 6 MPa gas counter pressure charged in the mold suppressed cell nucleation significantly, although some cells were nucleated during filling. So most gas molecules were remaining in the melt when filling is done. During the next step, when the mold was opened and gas counter pressure was released, the pressure of the polymer/gas mixture dropped and the previously nucleated cells further grew while secondary cell nucleation was also occurred. Consequently, a bimodal structure was developed. The large cells were nucleated during filling and the small cells during mold opening.

Another major difference between the RFIM and FIM + MO methods was the melt temperature at which cell nucleation and growth occurred. In RFIM, cell nucleation occurred at the gate where the melt temperature was relatively high and thus very little crystallization could take place before cell nucleation. However, in FIM + MO, foaming was introduced after a noticeable time delay (9 s), which should have decreased the melt temperature significantly. This cooling of the melt provided a chance for PLA to further crystallize, especially with dissolved gas (Nofar et al. 2012b, Nofar and Park 2014b), before foaming, and thus to enhance the melt strength and cell nucleating power (Wong et al. 2013a).

Another difference between RFIM and FIM + MO was the direction of expansion, and thereby the direction of cell elongation. In RFIM, the melt traveled in the machine direction during cell growth. However, in FIM + MO, cell growth occurred by mold opening in the thickness direction, resulting in cell elongation in that direction. This affects the mechanical and insulation properties of the foam parts (Ameli et al. 2014).

On the other hand, the fillers also played an important role for the cell morphology of the foams made using FIM + MO. PLA, PLA/talc, and PLA/clay samples of FIM + MO showed different cellular morphologies in the core region. In PLA (Figure 5.19a), the large cells in the bimodal structure were in the range of 100 microns and the small cells in the range of tens of micron, an order of magnitude smaller. The addition of 5 wt% talc slightly improved the bimodal structure and increased the number of smaller cells (Figure 5.19b). On the other hand, the addition of 5 wt% nanoclay yielded a uniformly distributed cellular structure with the highest cell density and the lowest cell size compared to the other cases (Figure 5.19c) in its bimodal structure. PLA/clay foams possessed cells with relatively uniform spherical shape, with average cell sizes of less than 50 μm and cell densities of more than 2×10^7 cells/cm^3. The average cell density and cell size of PLA, PLA/talc and PLA/clay samples made using RFIM and FIM + MO are given in Figure 5.20. The quality of foams based on the smaller cell size and higher cell density can thus be ordered as PLA/clay > PLA/talc > PLA. In PLA, the average cell size increased from 94 to 224 μm when the void fraction was increased from 30% to 55%. However, when the nanoclay was used, the small cell size could be maintained with an increase in the void fraction.

In FIM + MO process, the secondary nucleation occurring at mold opening might have been facilitated by the crystallization of PLA during the delay time in addition to the unused gas concentration with gas counter pressure. With crystallization, the volume shrinks, and therefore, the tensile stress is induced around the crystals, which is favorable for cell nucleation (Wong et al. 2011, 2013a). Also, during the delay time before mold opening, as the crystals are formed, the dissolved gas is expelled out of the crystallized region and this creates high concentration gas around the crystals, which was again favorable for cell nucleation (Leung et al. 2009; Taki et al. 2011). Therefore, as Equations 5.6 through 5.8 show, the tensile stress created surrounding the crystals (i.e., the negative P_{local} term), as well as the high concentration (the increased H term) decreased the critical bubble radius (R_{cr}), and therefore, further cell nucleation would occur more favorably when the mold was opened.

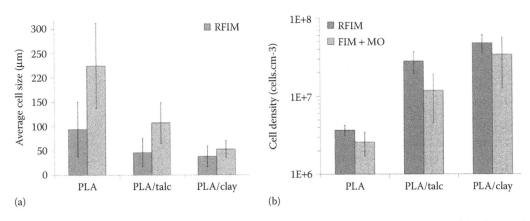

(a) (b)

FIGURE 5.20 (a) Average cell size and (b) average cell density of PLA, PLA/talc, and PLA/clay foams of RFIM and FIM + MO, measured in the core region. Error bars show ± one standard deviation. (Adapted from Ameli, A. et al., *Comp. Sci. Tech.*, 90, 88–95, 2014; Ameli, A. et al., *Chem. Eng. J.*, 262, 78–87, 2015.)

It should also be noted that the plasticization effect of N_2 on enhancing the crystallization kinetics of the PLA and PLA composites should have been pronounced. Consequently, the plasticization effect of N_2 should be more pronounced during the injection molding process and thereby the PLA's crystallization kinetics can be further improved.

Compared to neat PLA, the secondary cell nucleation was greater in PLA/talc. This may have stemmed not only from the talc with high nucleating ability (Nofar et al. 2011, 2012b, 2013c) but also from the enhanced crystallinity of PLA in the presence of talc with faster crystal growth rate compared to that in neat PLA. In the case of PLA/clay, a much larger number of small-sized crystals (Nofar et al. 2011, 2013c) created during the delay time not only might have resulted in an increased melt strength that limited the growth of the primary cells but also yielded a uniform cell nucleation with a large number in the secondary stage. Consequently, uniform cell morphology with a higher cell density and a smaller cell size was obtained in PLA/clay samples.

5.3 CONCLUSION AND PERSPECTIVES

Investigations of PLA foaming behaviors showed that adding micro-sized additives and more effectively nanoparticles improved the PLA foams cell nucleation rate and the final cell density by enhancing the heterogeneous cell nucleation power through the local pressure variations around these additives. Moreover, the introduced additives can also expedite the crystal nucleation rate, which could further improve the cell density through the similar path that the additives affect. In general, the enhanced crystallization kinetics can significantly increase the expansion ratio and the cell density of the PLA foamed samples. The crystal nuclei that are induced during foam processing can increase the PLA's inherently low melt strength through the crystal-to-crystal network. Simultaneously, the crystal nuclei act as heterogeneous cell nucleating agents and can promote cell density.

It should be noted that although the cell density of the PLA foams can be promoted by introducing micro-/nano-sized additives and crystallinity, too high an additive content and/or crystallinity can hinder the expansion ratio of the PLA foams due to the PLA's increased stiffness.

REFERENCES

Ajioka, M., K. Enomoto, A. Yamaguchi, K. Suzuki, T. Watanabe, Y. Kitahara. 1995. Degradable foam and use of same. US Patent, US5447962.

Albalak, R.J., Z. Tadmor, Y. Talmon. 1990. Polymer melt devolatilization mechanisms. *AIChE J.* 36: 1313–1320.

Ameli, A., D. Jahani, M. Nofar, P. Jung, C.B. Park. 2013. Processing and characterization of solid and foamed injection-molded polylactide with talc. *J. Cell. Plast.* 49: 351–374.

Ameli, A., D. Jahani, M. Nofar, C.B. Park. 2014. Development of high void fraction polylactide composite foams using injection molding: Mechanical and thermal insulation properties. *Comp. Sci. Tech.* 90: 88–95.

Ameli, A., M. Nofar, D. Jahani, C.B. Park. 2015. Development of high void fraction polylactide composite foams using injection molding: Crystallization and foaming Properties. *Chem. Eng. J.* 262: 78-87.

Auras, R., B. Harte, S. Selke. 2004. An overview of polylactides as packaging materials. *Macromol. Biosci.* 4: 835–864.

Bandyopadhyay, A., G.C. Basak. 2007. Studies on photocatalytic degradation of polystyrene. *Mater. Sci. Technol.* 23: 307–314.

Blander, M., J.L. Katz. 1975. Bubble nucleation in liquids. *AIChE J.* 21: 833–848.

Boissard, C.I., P.E. Bourban, C.J.G. Plummer, R.C. Neagu, J.A.E. Manson. 2012. Cellular biocomposites from polylactide and microfibrillated cellulose. *J. Cell. Plast.* 48: 445–458.

Braun, F. 2004. Method for producing expanded or expandable polyolefin particles. US Patent 2004, US6723760 B2.

Carlson, D., P. Dubois. 1998. Free radical branching of polylactide by reactive extrusion. *Polym. Eng. Sci.* 38: 311–321.

Choi, J.B., M.J. Chung, J.S. Yoon. 2013. Formation of double melting peak of poly(propylene-co-ethylene-co-1-butene) during the preexpansion process for production of expanded polypropylene. *Ind. Eng. Chem. Res.* 44: 2776–2780.

Di, Y., S. Iannace, E. Di Maio, L. Nicolais. 2005a. Poly(lactic acid)/organoclay nanocomposites: Thermal, rheological properties and foam processing. *J. Polym. Sci. Part B Polym. Phys.* 43: 689–698.

Di, Y., S. Iannace, E. Di Maio, L. Nicolais. 2005b. Reactively modified poly(lactic acid): Properties and foam processing. *Macromol. Mater. Eng.* 290: 1083–1090.

Dorgan, J. 2010. Rheology of poly(lactic acid). In *Poly(lactic acid): Synthesis, Structures, Properties, Processing, and Applications*, ed. R. Auras, L.T. Lim, S.E.M. Selke, and H. Tsuji. New York: John Wiley & Sons.

Dorgan, J., J. Janzen, M. Clayton, S. Hait, D. Knauss. 2005. Melt rheology of variable L-content poly(lactic acid). *J. Rheol.* 45: 607–619.

Dorgan, J., R.H. Lehermeier, M. Mang. 2000. Thermal and rheological properties of commercial-grade poly(lactic acid). *J. Polym. Environ.* 8: 1–9.

Dorgan, J., J. Williams. 1999. Melt rheology of poly(lactic acid): Entanglement and chain architecture effects. *J. Rheol.* 43: 1141–1155.

Drumright, R.E., P.R. Gruber, D.E. Henton. 2000. Polylactic acid technology. *Adv. Mater.* 12: 1841–1846.

Fang, Q., M.A. Hanna. 2001. Characteristics of biodegradable Mater-Bi®-starch based foams as affected by ingredient formulations. *Ind. Crops Pdts.* 13: 219–227.

Fujimoto, Y., S.S. Ray, M. Okamoto, A. Ogami, K. Yamada, K. Ueda. 2003. Well-Controlled biodegradable nanocomposite foams: From microcellular to nanocellular. *Macromol. Rapid Commun.* 24: 457–461.

Garlotta, D.J. 2001. A literature review of poly(lactic acid). *J. Polym. Environ.* 9: 63–84.

Gibbs, J.W. 1961. *The Scientific Papers of J. Willard Gibbs Volume 1.* New York: Dover Publications Inc.

Grijpma, D.W., A.J. Pennings. 1994. (Co)polymers of L-lactide, 2. Mechanical properties. *Macromol. Chem. Phys.* 195: 1649–1663.

Gu, S.Y., J. Ren, B. Dong. 2007. Melt rheology of polylactide/montmorillonite nanocomposites. *J. Polym. Sci. Part B Polym. Phys.* 45: 3189–3196.

Gu, S.Y., C.Y. Zou, K. Zhou, J. Ren. 2009. Structure-rheology responses of polylactide/calciumcarbonate composites. *J. Appl. Polym. Sci.* 114: 1648–1655.

Guo, Q., J. Wang, C.B. Park, M. Ohshima. 2006. A microcellular foaming simulation system with a high pressure-drop rate. *Ind. Eng. Chem. Res.* 45: 6153–6161.

Gupta, B., N. Revagade, J. Hilborn. 2007. Poly(lactic acid) fiber: An overview. *Prog. Polym. Sci.* 32: 455–482.

Harvey, E.N., D.K. Barnes, W.D. McElroy, A.H. Whiteley, D.C. Pease, K.W. Cooper. 1944. Bubble formation in animals. I. physical factors. *J. Cell Comp. Physiol.* 24: 1–22.

Kang, D.J., D. Xu, Z.X. Zhang, K. Pal, D.S. Bang, J.K. Kim. 2009. Well-controlled microcellular biodegradable PLA/silk composite foams using supercritical CO_2. *Macromol. Mater. Eng.* 294: 620–624.

Keshtkar, M., M. Nofar, C.B. Park, P. Carreau. 2014. Extruded PLA/clay nanocomposite foams blown with supercritical CO_2. *Polymer* 55: 4077–4090.

Kramschuster, A., S. Pilla, S. Gong, A. Chandra, L.S. Turng. 2007. Injection molded solid and microcellular polylactide compounded with recycled paper shopping bag fibers. *Int. Polym. Process.* 22: 436–445.

Krishnamoorti, R., K. Yurekli. 2001. Rheology of polymer layered silicate nanocomposites. *Curr. Opin. Colloid Interf. Sci.* 6: 464–470.

Kwag, C., C.W. Manke, E. Gulari. 2001. Effects of dissolved gas on viscoelastic scaling and glass transition temperature of polystyrene melts. *Ind. Eng. Chem. Res.* 40: 3048–3052.

Lee, J.W.S., J. Wang, J.D. Yoon, C.B. Park. 2008a. Strategies to achieve a uniform cell structure with a high void fraction in advanced structural foam molding. *Ind. Eng. Chem. Res.* 47: 9457–9464.

Lee, S.T., L. Leonard, J. Jun. 2008b. Study of thermoplastic PLA foam extrusion. *J. Cell. Plast.* 44: 293–305.

Leung, S.N., C.B. Park, H. Li. 2006. Numerical simulation of polymeric foaming processes using modified nucleation theory. *Plast. Rubber Comp. Macromol. Eng.* 35: 93–100.

Leung, S.N., A. Wong, Q. Guo, C.B. Park, J.H. Zong. 2009. Change in the critical nucleation radius and its impact on cell stability during polymeric foaming processes. *Chem. Eng. Sci.* 64: 4899–4907.

Leung, S.N., A. Wong, C. Wang, C.B. Park. 2012. Mechanism of extensional stress-induced cell formation in polymeric foaming processes with the presence of nucleating agents. *J. Supercrit. Fluids* 63: 187–198.

Levy, S. 1981. *Advances in Plastics Technology.* New York: Van Nostrand Reinhold.

Li, G., H. Lia, L.S. Turng, S. Gongc, C. Zhang. 2006. Measurement of gas solubility and diffusivity in polylactide. *Fluid Phase Equilib.* 246: 158–166.

Liao, X., A.V. Nawaby, P.S. Whitfield. 2010. Carbon dioxide-induced crystallization in poly(L-lactic acid) and its effect on foam morphologies. *Polym. Int.* 59: 1709–1718.

Lim, L.T., R. Auras, M. Rubino. 2008. Processing technologies for poly (lactic acid). *Prog. Polym. Sci.* 33: 820–852.

Lunt, J. 1998. Large-scale production, properties and commercial applications of polylactic acid polymers. *Polym. Deg. Stab.* 59: 145–152.

Marrazzo, C., E. Di Maio, S. Iannace. 2008. Conventional and nanometric nucleating agents in poly(e-caprolactone) foaming: Crystals vs. bubbles nucleation. *Polym. Eng. Sci.* 48: 336–344.

Martin, O., L. Averous. 2001. Poly(lactic acid): Plasticization and properties of biodegradable multiphase systems. *Polymer* 42: 6209–6219.

Matuana, L.M., C.A. Diaz. 2010a. Study of cell nucleation in microcellular poly(lactic acid) foamed with supercritical CO_2 through a continuous-extrusion process. *Ind. Eng. Chem. Res.* 49: 2186–2193.

Matuana, L.M., O. Faruk. 2010b. Effect of gas saturation conditions on the expansion ratio of microcellular poly(lactic acid)/wood-flour composites. *Express Polym. Lett.* 4: 621–631.

Mihai, M., M.A. Huneault, B.D. Favis. 2009. Crystallinity development in cellular poly(lactic acid) in the presence of supercritical carbon dioxide. *J. Appl. Polym. Sci.* 113: 2920–2932.

Mihai, M., M.A. Huneault, B.D. Favis. 2010. Rheology and extrusion foaming of chain-branched poly(lactic acid). *Polym. Eng. Sci.* 50: 629–642.

Mikos, A.G., A.J. Thorsen, L.A. Czerwonka, Y. Bao, R. Langer, D.G. Winslow, J.P. Vacanti. 1994. Preparation and characterization of Poly (L-lactide acid) foams. *Polymer* 35: 1068–1077.

Nofar, M., A. Ameli, C.B. Park. 2014a. Thermal behavior of polylactide with different D-contents: Effect of CO_2 dissolved gas. *Macromol. Mater. Eng.* 299 (10): 1232–1239.

Nofar, M., A. Ameli, C.B. Park. 2015a. Development of polylactide bead foams with double crystal melting peaks. *Polymer* 69: 83–94.

Nofar, M., A. Ameli, C.B. Park, 2015b. A novel technology to manufacture biodegradable polylactide bead foam products. *Materials and Design* 83: 413–421.

Nofar, M., Y. Guo, C.B. Park. 2013a. Double crystal melting peak generation for expanded polypropylene bead foam manufacturing. *Ind. Eng. Chem. Res.* 52: 2297–2303.

Nofar, M., K. Majithiya, T. Kuboki, C.B. Park. 2012a. The foamability of low-melt-strength linear polypropylene with nanoclay and coupling agent. *J. Cell. Plast.* 48: 271–287.

Nofar, M., C.B. Park. 2014b. Poly(lactic acid) foaming. *Prog. Polym. Sci.* 39: 1721–1741.

Nofar, M., C.B. Park. A method for the preparation of PLA bead foams, Int. Appl. No.: PCT/NL2013/050231, WO2014158014 A1, Pub Date: October 2, 2014.

Nofar, M., A. Tabatabaei, A. Ameli, C.B. Park. 2013b. Comparison of melting and crystallization behaviors of polylactide under high-pressure CO_2, N_2, and He. *Polymer* 54: 6471–6478.

Nofar, M., A. Tabatabaei, C.B. Park. 2013c. Effects of nano-/micro-sized additives on the crystallization behaviors of PLA and PLA/CO_2 mixtures. *Polymer* 54: 2382–2391.

Nofar, M., W. Zhu, C.B. Park. 2012b. Effect of dissolved CO_2 on the crystallization behavior of linear and branched PLA. *Polymer* 53: 3341–3353.

Nofar, M., W. Zhu, C.B. Park, J. Randall. 2011. Crystallization kinetics of linear and long-chain-branched polylactide. *Ind. Eng. Chem. Res.* 50: 13789–13798.

Okamoto, M., P.H. Nam, P. Maiti, T. Kotaka, T. Nakayama, M. Takada, M. Ohshima, A. Usuki, N. Hasegawa, H. Okamoto. 2001. Biaxial flow-induced alignment of silicate layers in polypropylene/clay nanocomposite foam. *Nano Lett.* 1: 503–505.

Palade, L.I., H. Lehermeier, J. Dorgan. 2001. Melt rheology of high L-content poly(lactic acid). *Macromolecules* 34: 1384–1390.

Perego, G., G.D. Gella, C. Bastioli. 1996. Effect of molecular weight and crystallinity on poly(lactic acid) mechanical properties. *J. Appl. Polym. Sci.* 59: 37–43.

Pilla, S., A. Kramschuster, J. Lee, G.K. Auer, S. Gong, L.S. Turng. 2009. Microcellular and solid polylactide-flax fiber composites. *Compos. Interfaces* 16: 869–890.

Pluta, M. 2006. Melt compounding of polylactide/organoclay: Structure and properties of nanocomposites. *J. Polym. Sci. Part B Polym. Phys.* 44: 3392–3405.

Rasal, R.M., A.V. Janorkar, D.E. Hirt. 2010. Poly(lactic acid) modifications. *Prog. Polym. Sci.* 35: 338–356.

Ray, S.S. 2006. Rheology of polymer/layered silicate nanocomposites. *J. Ind. Eng. Chem.* 12: 811–842.

Ray, S.S., P. Maiti, M. Okamoto, K. Yamada, K. Ueda. 2002. New polylactide/layered silicate nanocomposites. 1. preparation, characterization, and properties. *Macromolecules* 35: 3104–3110.

Ray, S.S., M. Okamoto. 2003. New polylactide/layered silicate nanocomposites, 6 melt rheology and foam processing. *Macromol. Mater. Eng.* 288: 936–944.

Saeidlou, S., M. Huneault, H. Li, C.B. Park. 2012. Poly(lactic acid) crystallization. *Prog. Polym. Sci.* 37: 1657–1677.

Sasaki, H., K. Ogiyama, A. Hira, K. Hashimoto, H. Tokoro. 2005. Production method of foamed polypropylene resin beads. US Patent 2005, US6838488 B2.

Sinclair, R.G. 1996. The case for polylactic acid as a commodity packaging plastic. *J. Macromol. Sci. Pure Appl. Chem.* 33: 585–597.

Srithep, Y., P. Nealey, L.S. Turng. 2013. Effects of annealing time and temperature on the crystallinity, heat resistance behavior, and mechanical properties of injection molded poly(lactic acid) (PLA). *Polym. Eng. Sci.* 53: 580–588.

Srithep, Y., L.S. Turng, R. Sabo, C. Clemons. 2012. Nanofibrillated cellulose (NFC) reinforced polyvinyl alcohol (PVOH) nanocomposites: Properties, solubility of carbon dioxide, and foaming. *Cellulose* 19: 1209–1223.

Sungsanit, K., N. Kao, S.N. Bhattacharya, S. Pivsaart. 2010. Physical and rheological properties of plasticized linear and branched PLA. *Korea-Aust. Rheol. J.* 22: 187–195.

Taki, K., D. Kitano, M. Ohshima. 2011. Effect of growing crystalline phase on bubble nucleation in poly(L-Lactide)/CO_2 batch foaming. *Ind. Eng. Chem. Res.* 50: 3247–3252.

Thiagarajan, C., R. Sriraman, D. Chaudhari, M. Kumar, A. Pattanayak. 2010. Nano-cellular polymer foam and methods for making them. US Patent 2010, US7838108 B2.

Tsimpliaraki, A., I. Tsivintzelis, S.I. Marras, I. Zuburtikudis, C. Panayiotou. 2011. The effect of surface chemistry and nanoclay loading on the microcellular structure of porous poly(d,l lactic acid) nanocomposites. *J. Supercrit. Fluids* 57: 278–287.

Tsuji, H., Y. Ikada. 1998. Blends of aliphatic polyesters. II. Hydrolysis of solution-cast blends from poly(L-lactide) and poly(E-caprolactone) in phosphate-buffered solution. *J. Appl. Polym. Sci.* 67: 405–415.

Wang, B., T. Wan, T. Zeng. 2011. Dynamic rheology and morphology of polylactide/organic montmorillonite nanocomposites. *J. Appl. Polym. Sci.* 121: 1032–1039.

Wang, J., W. Zhu, H. Zhang, C.B. Park. 2012. Continuous processing of low-density, microcellular poly(lactic acid) foams with controlled cell morphology and crystallinity. *Chem. Eng. Sci.* 75: 390–399.

Ward, C.A., E. Levart. 1984. Conditions for stability of bubble nuclei in solid surfaces contacting a liquid-gas solution. *J. Appl. Phys.* 56: 491–500.

Wong, A., R.K.M. Chu, S.N. Leung, C.B. Park, J.H. Zong. 2011. A batch foaming visualization system with extensional stress-inducing ability. *Chem. Eng. Sci.* 66: 55–63.

Wong, A., Y. Guo, C.B. Park. 2013a. Fundamental mechanisms of cell nucleation in polypropylene foaming with supercritical carbon dioxide—Effects of extensional stresses and crystals. *J. Supercrit. Fluids* 79: 142–151.

Wong, A., S.F.L. Wijnands, T. Kuboki, C.B. Park. 2013b. Mechanisms of nanoclay-enhanced plastic foaming processes—Effects of nanoclay intercalation and exfoliation. *J. Nanopart. Res.* 15: 1815–1829.

Wu, D., L. Wu, L. Wu, M. Zhang. 2006. Rheology and thermal stability of polylactide/clay nanocomposites. *Polym. Degrad. Stab.* 91: 3149–3155.

Xu, X., C.B. Park, J.W.S. Lee, X. Zhu. 2008. Advanced structural foam molding using a continuous polymer/gas melt flow stream. *J. Appl. Polym. Sci.* 109: 2855–2861.

Zheng, W., Y.H. Lee, C.B. Park. 2010. Use of nanoparticles for improving the foaming behaviors of linear PP. *J. Appl. Polym. Sci.* 117: 2972–2979.

Zhu, Z, C.B. Park, J. Zong. 2008. Challenges to the formation of nano cells in foaming processes. *Int. Polym. Process.* 23: 270–276.

6 Solid-State Microcellular Poly (Lactic Acid) Foams

Vipin Kumar and Krishna V. Nadella

CONTENTS

6.1 INTRODUCTION

Poly(lactic acid) (PLA) is a plant-based, biodegradable plastic made from the fermented sugars of crops such as corn, beet, and sugarcane. The L- and D-isomers of lactic acid are used, either in combination or separately, to synthesize high-molecular weight PLA plastic, which is approved by the Food and Drug Administration for use in food packaging and biomedical applications. PLA has a glass transition temperature and melt temperature of about 55°C and 175°C, respectively. It is colorless, glossy, and rigid and has properties that are similar to polystyrene. It is an aliphatic polyester, which means that it lacks the highly stable, six-carbon rings typically associated with hard and extremely durable polyesters (such as polyethylene terephthalate), which are generally nonbiodegradable. In the presence of moisture and heat, PLA breaks down into smaller hydrocarbon compounds that can be absorbed and metabolized by microorganisms in a process called *chemical hydrolysis*. This process can take about six months to two years (Garlotta 2001). This is the primary reason that PLA is considered as an attractive substitute for petroleum-based plastics that are used in many single-use disposable packagings, which end up in landfills and pollute waterways for centuries. In the last several years, vast amounts of biodegradable solid PLA has been extruded for thermoforming and molding into food and beverage packaging applications. Solid PLA, however, requires more material and is heavier than plastic foams. As the food and film packaging industries are looking for material that reduces the amount of polymer used, microcellular foams are considered a viable alternative to solid plastics.

Microcellular foams refer to thermoplastic foams with cells of the order of 10 μm in size. Typically these foams are rigid, closed-cell structures. The idea to incorporate such small bubbles in

179

thermoplastics goes back to early work at Massachusetts Institute of Technology (Martini-Vvdensky et al. 1982, 1984), where the advent of microcellular polystyrene was described using nitrogen as the blowing agent. The invention was in response to a challenge by food and photographic film companies to reduce the amount of polymer used for packaging their products. Thus was born the idea to create microcellular foam, where we could have, for example, 100 bubbles across a 1 mm thickness, and expect to have a reasonable strength for the intended applications. The basic solid-state batch process developed for polystyrene (Martini-Vvdensky et al. 1984) has been used to create microcellular foams from a number of amorphous and semicrystalline polymers, such as polyvinyl chloride (Kumar and Weller 1993), polycarbonate (Kumar and Weller 1994; Collias et al. 1994), acrylonitrile-butadiene-styrene (Murray et al. 2000; Nawaby and Handa 2004), polyethylene terephthalate (Shimbo et al. 2007; Handa et al. 1999), and crystallizable polyethylene terephthalate (Kumar et al. 2000), to name a few. The cell sizes achieved in these studies typically range from 1 to 50 µm.

There are two basic steps in the solid-state foaming of thermoplastic polymers. The first step consists of saturation of the polymer with gas under high pressure. This step is normally carried out at room temperature. Given sufficient time for diffusion of gas into the polymer, the gas attains an equilibrium concentration that is consistent with the solubility of gas in the polymer and the gas pressure. In the second step, bubbles are nucleated in the gas-polymer system by creating a thermodynamic instability. This is achieved by either a sudden drop in pressure (Goel and Beckman 1993) or a sudden increase in temperature (Kumar and Weller 1993, 1994). Both strategies suddenly reduce the solubility of the gas, driving the gas out of the polymer matrix and into nucleated bubbles. One consequence of dissolving gas in the polymer is plasticization, reducing the polymer's glass transition temperature (Chow 1980; Zhang and Handa 1998). After saturation, the temperature of the gas-saturated polymer only needs to be raised to the glass transition temperature of the gas-polymer system to nucleate bubbles. Hence the phrase *solid-state foam* is used to describe such foams, as opposed to the conventional foams produced from a polymer melt.

The solid-state foaming process has been employed in a number of studies on PLA and its blends, composites, biocomposites, and nanocomposites. The PLA-CO_2 system has been investigated by several researchers (Wang et al. 2007, 2012; Matuana 2008). Some examples of studies where PLA is blended with fillers include PLA biocomposites reinforced with cellulose fiber (Ding et al. 2011); PLA-silica nanocomposites (Ji et al. 2013); foaming of PLA thin films (Lu et al. 2008); and injection molding of microcellular PLA and its blends (Peng et al. 2011; Zhao et al. 2013). Several researchers have used ultrasound cavitation to break the closed-cell structure of PLA foams to create an interconnected porous structure for tissue-engineering applications (Wang et al. 2006, 2009; Guo et al. 2013). In an innovative study, Wang et al. (2011) used ultrasonic irradiation in combination with solid-state process to enhance cell nucleation and increase PLA foam expansion.

This chapter presents a review of microcellular foaming of PLA with CO_2 gas by the solid-state process. It is based on a series of experimental studies conducted on solid-state microcellular PLA foams at the University of Washington's Microcellular Plastic Laboratory in collaboration with MicroGREEN Polymers, Inc, Arlington, Washington,. The studies include all aspects of the solid-state process: gas sorption/desorption, diffusion, foaming behavior, microstructure, crystallization, and mechanical properties. An objective of our studies is to determine whether commercially viable microcellular PLA foams could be produced using the solid-state process. As a result, we are able to give specific guidance on the optimal processing conditions for manufacturing microcellular PLA foams of a consistent density and quality suitable for thermoforming applications such as in food packaging.

6.2   MATERIALS AND EQUIPMENT

Solid PLA material from Ex-Tech Plastics, Richmond, Illinois, which was extruded from a resin by NatureWorks™ (PLA2002D), Minnetonka, Minnesota, was used in the as-received condition to prepare the different samples for the studies. The PLA material had a thickness of 0.60 mm (0.024 in), density of 1.24 g/cm³, and a glass transition temperature of 55°C. A pressure vessel regulated by a process controller was used to saturate the specimens. Pressurized CO_2 gas served as the foaming agent. A digital scale with 10 µg (2.2E-9 lb) readability was used to measure gas solubility and relative density. Samples were foamed in either a circulating hot liquid bath or an infrared (IR) oven comprising top and bottom panels with inner/outer heating zones. IR samples were monitored with a pyrometer attached to a computer-based data acquisition system that recorded the surface temperature of the sample every second. A temperature-controlled freezer was used in some experiments to store gas-saturated samples prior to foaming. Microstructural analysis of PLA foams was conducted with a scanning electron microscope, crystallinity measurements by differential scanning calorimetry, tensile test using an Instron machine and data collection software, flexural test using the Instron with a three-point bending fixture according to ASTM Standard D790-03, and impact testing on a Gardner impact tester according to ASTM Standard D5420-93.

6.3   SORPTION AND DESORPTION OF CO_2 GAS IN PLA

6.3.1   SORPTION

To establish accurate processing parameters, both sorption and desorption behaviors need to be characterized for the PLA-CO_2 system. Both saturation pressure and time influence the migration of CO_2 gas through the PLA polymer in terms of the rate at which gas diffuses into the polymer and the equilibrium gas concentration that is absorbed by the polymer. A sorption study was conducted to determine which pressure level(s) yields the desired gas concentration, reduces saturation time, and slows down or speeds up gas desorption. Five levels of saturation pressure (1, 2, 3, 4, and 5 MPa) were investigated. PLA samples were dimensioned, weighed, and interleaved with a gas-channeling cellulose material. A batch of samples at a time was saturated with CO_2 gas in a pressure chamber at a given pressure and in room temperature (20°C). Periodically, the chamber was depressurized and a single sample permanently removed and weighed for gas concentration based on the difference between the current and initial weight. It was recognized that de-pressurization for a fast-diffusing polymer such as PLA could lead to reading error; consequently, the chamber was immediately re-pressurized every time a sample was removed to minimize gas desorption. Saturation halted when sample weight leveled off, indicating that equilibrium concentration for a given pressure was reached. For lower pressures (1 and 2 MPa), which took longer to attain equilibrium, more than one batch of samples were required. Key parameters that govern sorption behavior are summarized in Table 6.1.

Increased saturation pressure is attended by a dramatic rise in equilibrium concentration and significant drop in saturation time. At 1 MPa, it took 32 h to achieve an equilibrium concentration of about 4% by weight compared to 20% by weight in an hour for 5 MPa. When equilibrium concentration is plotted against saturation pressure, a linear trend clearly emerges as shown in Figure 6.1.

TABLE 6.1
Summary of PLA Sorption Parameters

Saturation Pressure (MPa)	Equilibrium Saturation Time (h)	Equilibrium Gas Concentration (mg g⁻¹ of PLA)	Average Diffusion Coefficient (cm² s⁻¹)
1	32.00	35.41	5.42E-09
2	16.00	83.53	1.52E-08
3	3.25	114.88	7.23E-08
4	2.50	165.71	1.52E-07
5	1.00	204.76	2.76E-07

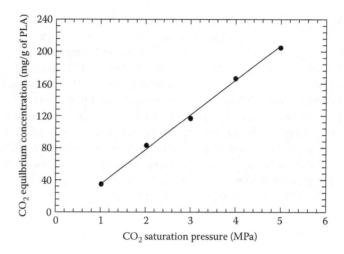

FIGURE 6.1 Equilibrium concentration in solid PLA as a function of saturation pressure at 20°C.

That equilibrium concentration in PLA increases linearly with increasing saturation pressure is consistent with Henry's Law of sorption for pressures between 1 and 5 MPa for a given saturation temperature, which is expressed as:

$$C = sP \tag{6.1}$$

where:
C is the concentration of CO_2 gas in (mg/g of PLA)
s is the gas solubility coefficient in (mg/g of PLA * MPa)
P is the saturation pressure (MPa)

The solubility coefficient of the PLA-CO_2 system is 42.93 mg/g for 1–5 MPa. Figure 6.2 shows gas concentration plots of PLA samples saturated at five pressure levels in room temperature until equilibrium.

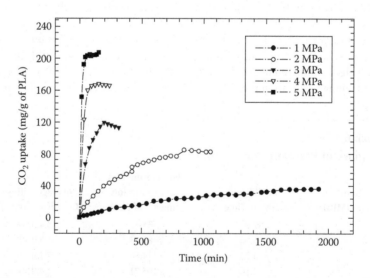

FIGURE 6.2 Plots of CO_2 uptake of 0.60 mm thick PLA samples for each pressure as a function of saturation time.

Experiments at 3, 4, and 5 MPa show rapid CO_2 uptake within a short period of time, leveling off after peaking and dipping slightly thereafter. In contrast, 1 and 2 MPa trend more gradually arriving at equilibrium much later. The dip in gas concentration for all pressures, except 1 MPa, may be attributed to CO_2 gas-induced crystallization. As more gas is introduced into the polymer, the gas transition temperature (T_g) of the gas-polymer system drops causing plasticization and formation of crystallites that encroach on the amorphous regions of the polymer where most of the gas is believed to reside.

Average CO_2 diffusivity in PLA was obtained using the equation:

$$D = \frac{0.049}{\left(t/l^2\right)_{1/2}} \tag{6.2}$$

where:

D is the average diffusion coefficient (in cm^2 s^{-1})

t is the sorption time (in seconds)

l is the thickness of the sheet (in centimeter), and the term $(t/l^2)_{1/2}$ denotes the value of (t/l^2) when half of the equilibrium gas concentration gets absorbed into the polymer matrix (Crank 1975).

When the value of D for each pressure taken from Table 6.1 is plotted against equilibrium concentration levels, as shown in Figure 6.3, it is clear that the diffusion rate increases exponentially the higher the gas concentration.

6.3.2 Desorption

Once PLA is saturated with CO_2 at a given pressure, the rate at which CO_2 exits the polymer is an important process parameter. A desorption study was conducted using the same pressure levels (1, 2, 3, 4, and 5 MPa) to determine the gas diffusion rate at desorption in order to establish the process window for desorption prior to foaming the polymer and thereby control the quality of the foam. PLA samples were saturated at a given pressure in room temperature until equilibrium and then set aside to desorb in atmospheric condition at room temperature. At periodic intervals samples were weighed to determine the amount of gas desorbed over time. Figure 6.4 shows the desorption plots of PLA for different saturation pressures.

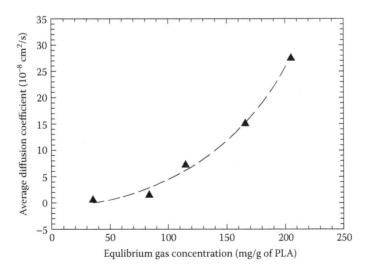

FIGURE 6.3 Average diffusion coefficient as a function of CO_2 concentration in PLA at 20°C.

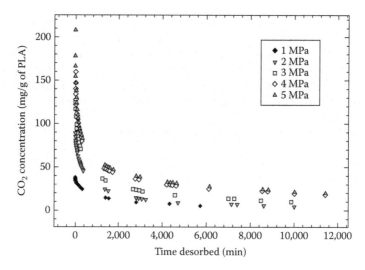

FIGURE 6.4 Plots of CO_2 diffusion rate of 0.60 mm thick PLA as a function of desorption time.

As expected, the higher the pressure, the faster the desorption rate, but the longer it took for the higher pressures to attain equilibrium. At 200 desorption hours (12,000 min) 1 MPa samples reached near-zero gas concentration, whereas 5 MPa samples still had 20 mg of gas/g of PLA remaining. To estimate average diffusivity from the desorption curve, the fraction of the gas leftover ($M_t/M\infty$) in the PLA sample is plotted as a function of the $\sqrt{t/l^2}$, where M_t is the amount of gas remaining at time t, $M\infty$ is the equilibrium concentration for a given saturation pressure, t is the desorption time, and l is the thickness of the sample. A linear curve fit of the plots for the SQRT(t/l^2) values of up to 1000 was performed in order to capture the early phase of desorption as shown in Figure 6.5.

The slopes were calculated from the linear fit of the plot curves and were plugged into the following equation to estimate the diffusion coefficient (Crank 1975):

$$D = \frac{\pi}{16} R^2 \qquad (6.3)$$

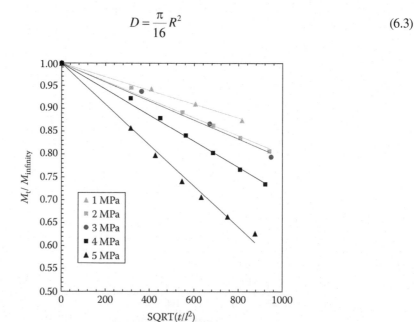

FIGURE 6.5 Plots of the fraction of the remaining gas in PLA for different saturation pressures.

where:

   D is the average diffusion coefficient (in cm^2 s^{-1})

   R is the slope of the curve

The rate at which CO_2 diffuses out of PLA increases exponentially the higher the gas concentration as shown in Figure 6.6.

Table 6.2 summarizes the results of the PLA desorption study. While the diffusion coefficient values estimated from both sorption and desorption data clearly indicate gas-concentration dependency, the D values for desorption are consistently lower than those for sorption, suggesting that CO_2 diffusion is dependent on both saturation pressure and gas concentration.

6.3.3 Qualitative Observations of PLA Sorption and Desorption Samples

All 1 MPa samples irrespective of saturation time and gas concentration remain optically clear. 2 MPa samples retain their optical clarity and stiffness, but a noticeable curve at the top corners of the samples, whether due to depressurization and/or propping of samples against the chamber wall, suggests that some plasticization occurred during saturation. Samples saturated less than nine hours look essentially unchanged, whereas samples saturated for more than nine hours show a curve that

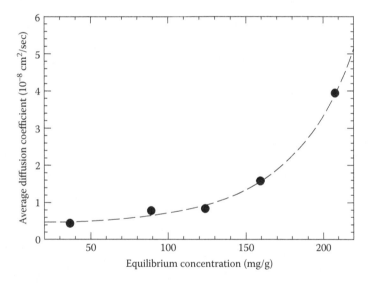

FIGURE 6.6   Plot of the average diffusion coefficient as a function of CO_2 concentration in PLA.

TABLE 6.2
Summary of PLA Desorption Parameters

Saturation Pressure (MPa)	Equilibrium Saturation Time (h)	Equilibrium Gas Concentration (mg g^{-1} of PLA)	Average Diffusion Coefficient (cm^2 s^{-1})
1	32.00	36.53	4.41E-09
2	16.00	88.90	7.85E-09
3	3.25	124.19	8.41E-09
4	2.50	159.12	1.58E-08
5	1.00	207.76	3.94E-08

FIGURE 6.7 The 3 MPa sample continued to nucleate bubbles as it desorbed under atmospheric pressure and room temperature, evidence of the drop in the T_g of the gas–polymer system to below room temperature during saturation.

becomes more pronounced with prolonged saturation. The 3 MPa samples developed a hazy appearance and felt slightly cool to the touch. Sorption samples progressed from translucent to opaque as saturation lengthened. It is believed that about three hours into saturation, the amount of CO_2 gas dissolved into the polymer was enough to drop the T_g of the gas-polymer system below room temperature as to induce cell nucleation. A scattering of large bubbles were found in all samples that reached equilibrium solubility, including desorption samples that continued to form bubbles even under ambient condition and temperature (Figure 6.7).

Only the first 4 MPa sorption sample showed any nucleation after 30 min of saturation. Longer than 30 min, 4 MPa samples appear uniformly translucent, show no bubbles, are initially soft and cool to the touch and, unlike 2 MPa, do not develop a curve probably due to the advent of gas-induced crystallization that stiffens the polymer matrix. Only the first 5 MPa sorption sample shows bubble nucleation after 15 min of saturation. Beyond 15 min, 5 MPa samples are uniformly translucent like the 4 MPa. Compared to the lower pressure samples, 5 MPa samples are at time of removal the coldest tactilely, most pliable, and deform easily even at room temperature.

6.4 SOLID-STATE FOAMING OF PLA

To determine the effects of saturation pressure, foaming temperature, and foaming treatment on PLA foaming behavior and morphology, five saturation pressures (1, 2, 3, 4, and 5 MPa), five foaming temperatures (20°C, 40°C, 60°C, 80°C, and 100°C), and two foaming methods (hot liquid bath versus IR oven) were investigated. For bath foaming, PLA samples were saturated at a given pressure in room temperature until equilibrium, foamed in a liquid bath at a given temperature for two min, quenched, and weighed at regular intervals to measure for desorption over time. Initial 20°C and 40°C bath treatment had no effect on samples and were eliminated from later bath foaming experiments. Consequently, 1, 4, and 5 MPa samples were heated at all five foaming temperatures; 2 MPa samples at 40°C, 60°C, 80°C, and 100°C; and 3 MPa samples at 60°C, 80°C, and 100°C. For IR foaming, PLA samples were saturated at a given pressure in room temperature until equilibrium, stored in a freezer at 0°C until time of foaming, and IR foamed on a spring frame until their surface reached a target temperature of 40°C, 60°C, 80°C, or 100°C. No desorption measurement for IR-foamed samples were taken.

6.4.1 BATH FOAMING

Plots of the average relative densities as a function of foaming temperature (Figure 6.8) indicate that as foaming temperature increases, relative density decreases until it reaches a minimum threshold

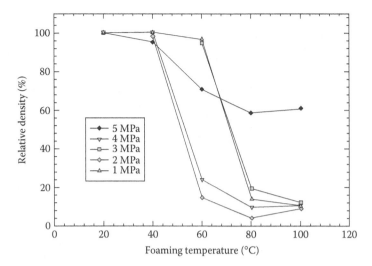

FIGURE 6.8 Plots of relative density as a function of foaming temperature for 1–5 MPa samples foamed in a hot water bath. 1, 4, and 5 MPa have data points for all foaming temperatures, 2 MPa for 40°C–100°C, and 3 MPa for 60°C–100°C.

beyond which a further rise in temperature would result in cell collapse and higher densities. Since 4 and 5 MPa samples did not appear to foam at 20°C and 40°C, we assumed that no density reduction takes place at these lower temperatures for all pressure conditions. For all pressure levels excepting 5 MPa, samples achieved lower relative densities to as much as 10% the higher the foaming temperature. There are no clear trends among 1–4 MPa samples that were foamed at 60°C. At 80°C and 100°C, all pressure levels except 5 MPa achieved relative densities below 20%. The higher relative densities of 5 MPa at 60°C, 80°C, and 100°C indicate cell collapse and densification following volume expansion. Because bath foaming was conducted immediately after gas saturation, it is believed that high saturation pressure and short desorption times caused rapid and violent outgassing when samples were heated to foaming temperature. In addition, CO_2 gas exited the polymer so quickly that there was no opportunity for bubbles to nucleate and expand properly. We observed that rapid foaming led to uneven and uncontrolled growth, surface blisters, and breaching of cell walls.

At 1 MPa, the 60°C sample discloses an unfoamed microstructure with a few scattered bubbles near the surface region. Both 80 and 100°C samples contain large bubbles between 100 and 300 μm in diameter with bubbles protruding from beneath an insubstantially thin skin. 100°C sample reveals bubbles having corrugated walls that are consistent with what was observed to be a weak and spongy macrostructure (Figure 6.9).

At 2 MPa, 60°C, 80°C, and 100°C samples disclose bubbles between 100 and 300 μm in diameter. 60°C and 80°C exhibit an uneven surface marked by large bubbles jutting from beneath the

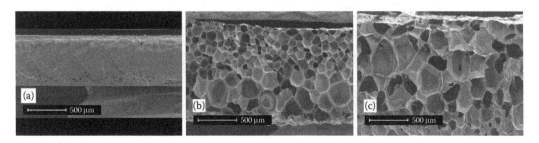

FIGURE 6.9 Scanning electron micrographs of 1 MPa samples bath foamed at 60°C (a), 80°C (b), and 100°C (c).

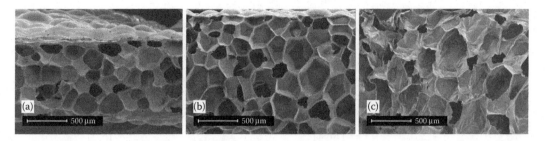

FIGURE 6.10 Scanning electron micrographs of 2 MPa samples bath foamed at 60°C (a), 80°C (b), and 100°C (c).

thin surface layer and rough textured cell walls likely formed by coalescent smaller bubbles. 100°C reveals corrugated cell walls with a weak spongy structure similar to the 1 MPa sample (Figure 6.10).

At 3 MPa, the 60°C sample shows negligible cell density with unusually shaped cells that are only visible at 1000× magnification. 80°C and 100°C samples disclose very similar foam structures with bubbles in the order of 20 μm, relatively thick cell walls in some regions, and a smooth skin layer (Figure 6.11).

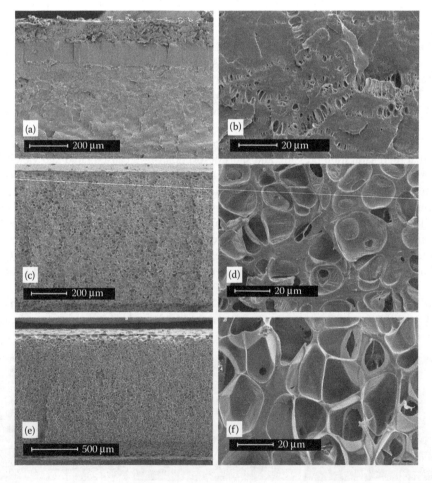

FIGURE 6.11 Scanning electron micrographs of 3 MPa samples bath foamed at 60°C (a, b), 80°C (c, d), and 100°C (e, f).

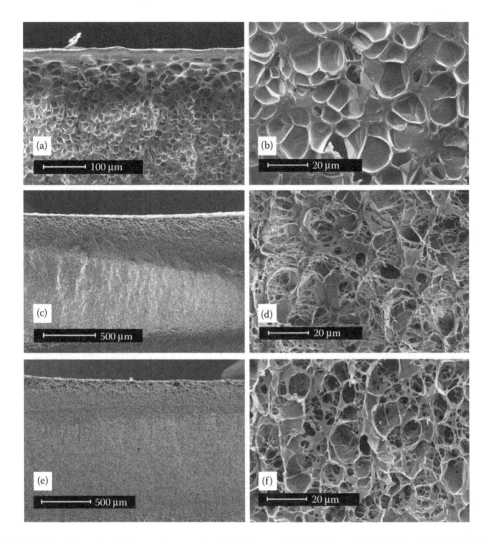

FIGURE 6.12   Scanning electron micrographs of 4 MPa samples bath foamed at 60°C (a, b), 80°C (c, d), and 100°C (e, f).

At 4 MPa, the 60°C sample discloses microcellular bubbles in the order of 10 µm in diameter and appreciably thick skin layers. 80°C and 100°C samples show web-like microstructures with bubbles so indistinguishable that they can only be seen at 1000× magnification. Samples for all three foaming temperatures contain larger bubbles near the surface region that are three times the diameter of the bubbles nearer the interior region (Figure 6.12).

At 5 MPa, 60°C, 80°C, and 100°C samples reveal limited foaming around the surface regions. 80°C and 100°C have blister-like structures in the form of nodules comprising a number of small bubbles (Figure 6.13).

The pore size of bath foamed samples as a function of foaming temperature and saturation pressure is shown in Figures 6.14 and 6.15, respectively. Both plots indicate a sharp transition in cell size as saturation pressure increases from 2 to 3 MPa. At 2 MPa or lower large size cells in the 150–300 µm range are produced; at 3 MPa or higher, micron cells in the 10–20 µm range are produced. This abrupt transition is attributed to CO_2-induced crystallization, which occurs at saturation pressures of 3 MPa and higher. The properties of the foams produced at 2 MPa versus 3 MPa also prove to be quite different.

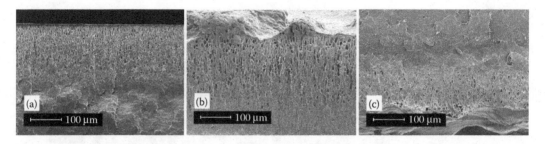

FIGURE 6.13 Scanning electron micrographs of 5 MPa samples bath foamed at 60°C (a), 80°C (b), and 100°C (c).

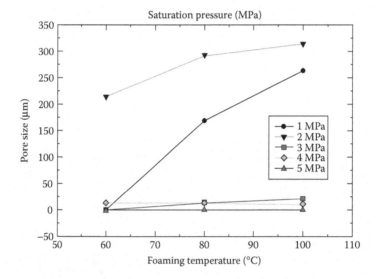

FIGURE 6.14 Pore size as a function of foaming temperature for PLA bath foamed samples for different saturation pressures.

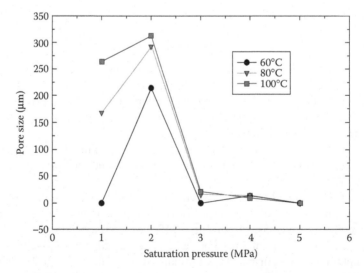

FIGURE 6.15 Pore size as a function of saturation pressure for PLA bath foamed samples for different foaming temperatures.

6.4.2 IR FOAMING

The relative density of IR-foamed samples is given in Figure 6.16. Similar to bath foamed samples, the 5 MPa IR-foamed samples do not show significant density reduction, evidence that high saturation pressure and short desorption time cause violent foaming and outgassing that inhibit cell nucleation and induce collapse of newly formed cell structures. For all other pressures, samples foamed at 100°C achieved relative densities of less than 20%. The 4 MPa samples show a linear reduction in density with increased foaming temperature. The 3 MPa samples outperformed the other samples with about 40% and 20% relative densities at 60°C and 80°C, respectively. Finally, it was observed that IR samples at different saturation pressures appear very different even when their relative densities are about the same.

At 1 MPa, the sample shows no cell nucleation at 60°C, contains sporadic bubbles near the surface region at 80°C, and is foamed throughout at 100°C, comprising large bubbles in the order of 200 µm in diameter (Figure 6.17).

At 2 MPa, the sample contains a few sporadic bubbles at 60°C. The sample reveals thick cell walls surrounded by a considerable amount of unnucleated mass with only a few large bubbles of about 200 µm in diameter in the center region and smaller bubbles of about 50 µm near the surface region at 80°C. Similar to 80°C, at 100°C, the sample has thick cell walls but more foamed content dominated by large bubbles in the order of 300 µm and an uneven thin surface layer due to the obtruding bubbles (Figure 6.18).

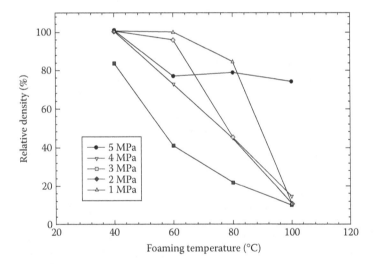

FIGURE 6.16 Plots of relative density as a function of foaming temperature for 1–5 MPa samples foamed in an IR oven.

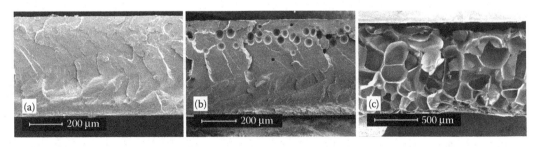

FIGURE 6.17 Scanning electron micrographs of 1 MPa samples IR foamed at 60°C (a), 80°C (b), and 100°C (c).

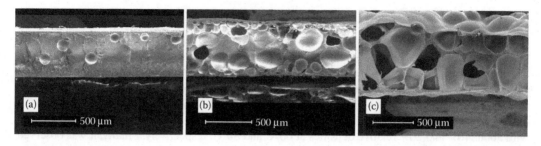

FIGURE 6.18 Scanning electron micrographs of 2 MPa samples IR foamed at 60°C (a), 80°C (b), and 100°C (c).

FIGURE 6.19 Scanning electron micrographs of 3 MPa samples IR foamed at 60°C (a), 80°C (b), and 100°C (c).

At 3 MPa, at 60°C, 80°C, and 100°C, the samples disclose much smaller bubbles than 1 and 2 MPa samples, ranging from 20 to 80 μm in diameter across the microstructure. At 80°C and 100°C, the samples have a layer of very small bubbles near the surface region; at 60°C and 80°C, the samples have cells walls that appear much thicker than at 100°C (Figure 6.19).

At 4 MPa, at 60°C, 80°C, and 100°C, the samples show interconnected microstructures indicative of active bubble nucleation and coalescence. For all three temperatures, cell size range from 15 to 30 μm in diameter. At 60°C and 80°C, the samples show considerable unfoamed mass interposed between the cells. At 100°C, the sample shows an uneven surface due to a skin layer that is even more insubstantial than the lower temperatures. Higher foaming temperature correlates with incidence of higher void fraction (Figure 6.20).

At 5 MPa, the sample discloses no presence of bubbles at 60°C. At 80°C and 100°C, the samples have low cell density with very small bubbles ranging from 1 to 5 μm in diameter (Figure 6.21).

The pore size of IR-foamed samples as a function of foaming temperature and saturation pressure is shown in Figures 6.22 and 6.23, respectively. IR-foaming trends prove to be very similar to those observed in bath foaming (Figures 6.14 and 6.15). Saturation pressures at 2 MPa or lower yield macrocellular foams, and yield microcellular foams at 3 MPa or higher.

6.5 EFFECTS OF CRYSTALLINITY

Polymer crystallinity is an important factor in solid-state processing since it affects the expansion and morphology of the foamed polymer including cell size, cell density, surface texture, and skin thickness, which impart to the polymer various physical and mechanical properties (e.g., permeability, density, impact resistance, flexural strength, and melting temperature), that can be exploited and controlled to meet product requirements. Differential scanning calorimetry was performed to measure percentage crystallinity of bath and IR-foamed samples for all saturation pressures (1, 2, 3, 4, and 5 MPa) and three foaming temperatures (40°C, 60°C, and 80°C). We investigated the effects of saturation pressure and foaming temperature on crystallinity in PLA samples since pressure and

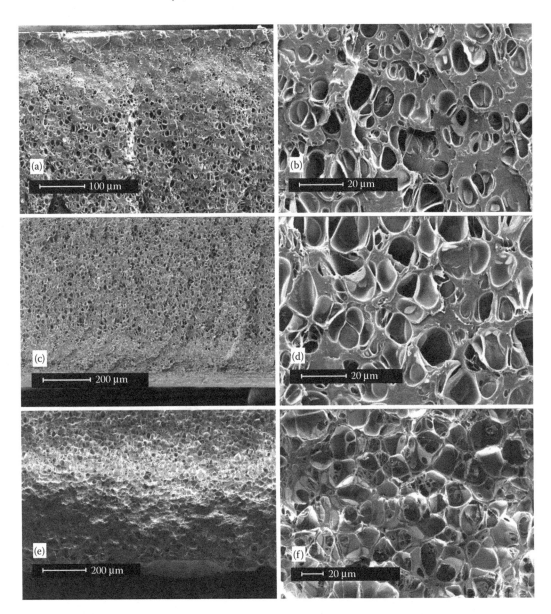

FIGURE 6.20 Scanning electron micrographs of 4 MPa samples IR foamed at 60°C (a, b), 80°C (c, d), and 100°C (e, f).

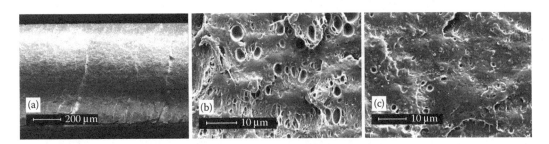

FIGURE 6.21 Scanning electron micrographs of 5 MPa samples IR foamed at 60°C (a), 80°C (b), and 100°C (c).

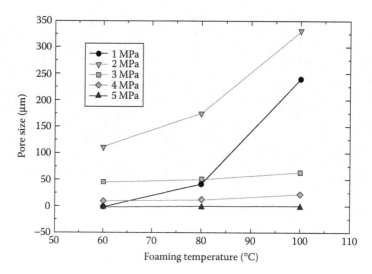

FIGURE 6.22   Pore size as a function of foaming temperature for IR-foamed samples for different saturation pressures.

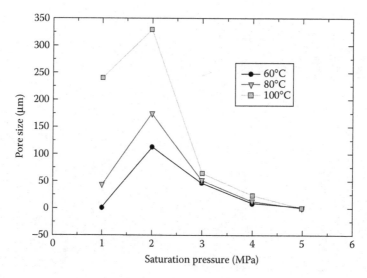

FIGURE 6.23   Pore size as a function of saturation pressure for IR-foamed samples for different foaming temperatures.

temperature are known causes of plasticization that imbues the polymer's molecular chains with greater mobility to arrange themselves into more orderly crystalline structures. Enthalpies of fusion and crystallinity of PLA samples were determined according to ASTM D 3417-97 standard. Samples were given time to desorb gas to less than 1% by weight in order to get a true reading of crystallinity. An enthalpy of fusion of 93 J g^{-1} for 100% crystalline PLA was used to calculate crystallinity values.

Graphs of bath and IR-foamed samples are presented in Figures 6.24 through 6.27 showing the percentage crystallinity in the polymer as a function of foaming temperature and saturation pressure. Although not quite comparable, a comparison of the percentage crystallinity between bath and IR-foamed samples bears looking at. For both bath and IR foaming, crystallinity levels spike when saturation pressure increases from 2 to 3 MPa. The 3, 4, and 5 MPa samples for bath and IR-foaming register about 20% crystallinity and up for all foaming temperatures. The maximum crystallinity

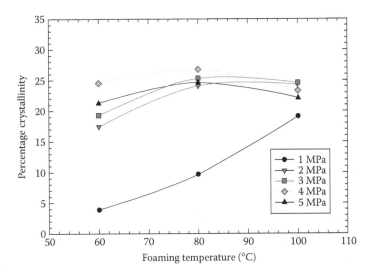

FIGURE 6.24 Graph of percentage crystallinity as a function of foaming temperature for bath foamed samples saturated at different pressures.

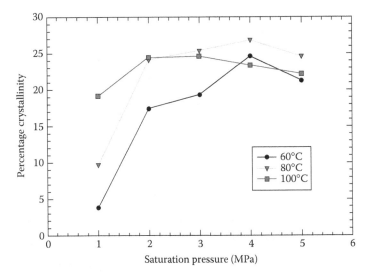

FIGURE 6.25 Graph of percentage crystallinity as a function of saturation pressure for bath foamed samples at different foaming temperatures.

obtained is about 27% for a 4 MPa sample bath foamed at 80°C. Saturation pressure clearly has a significant effect on microcellular PLA crystallization. However, at 3 MPa and higher for all foaming temperatures, there is no clear evidence that higher pressures or temperatures contribute any further to the polymer's crystallization. In fact, samples appear to have reached their upper bound limits for these processing conditions. Overall, lower saturation pressures have lower crystallinity values than higher pressures. This finding is particularly true for 1 and 2 MPa IR-foamed samples, which register no greater than 10% crystallinity, the only exception being a 2 MPa sample that foamed at 100°C to yield 20% crystallinity. For bath foaming, 1 and 2 MPa samples that were foamed at 60°C and 80°C yield much higher crystallinity levels than IR-foamed samples for the same pressures and temperatures. The huge difference in crystallinity between bath and IR-foamed samples suggests that other factors than saturation pressure and foaming temperature are at play. For instance, bath

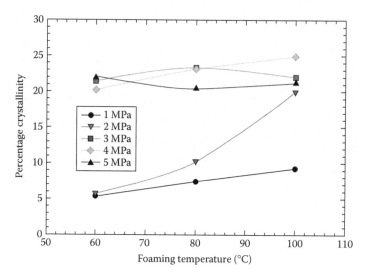

FIGURE 6.26 Graph of percentage crystallinity as a function of foaming temperature for IR-foamed samples at different saturation pressures.

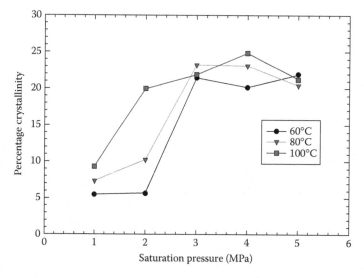

FIGURE 6.27 Graph of percentage crystallinity as a function of saturation pressure for IR-foamed samples at different foaming temperatures.

foamed samples were exposed to a hot liquid bath for two min whereas IR-foamed samples were heated for only a few seconds as that was enough time for the surface to reach target temperature. Lengthier heat exposure attended by greater strain (expansion) may explain the higher crystallinity in the bath samples.

6.6 MECHANICAL PROPERTIES OF PLA MICROCELLULAR FOAMS

Microcellular PLA foams were evaluated for their tensile, flexural, and impact properties. Six solid PLA sheets, which were used to prepare samples for testing, were saturated at 3 ± 1.38 MPa until equilibrium and foamed in an IR oven until they reached their target temperature. We narrowed the target temperature range between 55°C and 112°C, which based on our earlier findings is associated with producing uniform foams of a certain relative density. However, there were issues with foaming over 100°C as that caused samples to either overheat or over expand before reaching their target relative density for a given temperature. The six microcellular PLA sheets achieved relative densities of 0.05, 0.07, 0.14, 0.16, 0.24, and 0.39, respectively.

Tensile testing was conducted on PLA samples for each of the relative densities according to ASTM standard D638-03, Type IV on the Instron to measure for tensile strength, Young's modulus, and elongation at break. Stress-strain curve was used to characterize sample behavior under controlled tension expressed as the ratio of the tensile property over the sample's relative density. Overall trend lines of tensile properties for microcellular PLA are similar to that of solid PLA. For tensile strength and modulus, a positive linear relationship exists between strength and relative density as shown in Figures 6.28 and 6.29. If the trend line is extrapolated out to 1.0 relative density, it would approximate the values given in NatureWorks technical sheet, meaning that microcellular PLA foams would have the same tensile resistance as solid PLA of the same density.

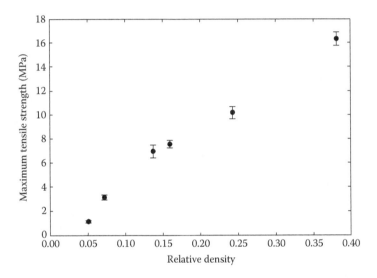

FIGURE 6.28   Tensile strength versus relative density of PLA-foamed samples.

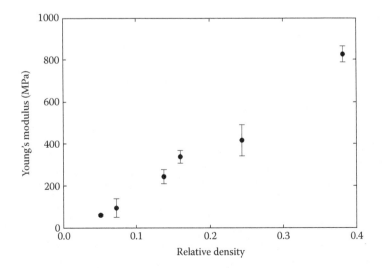

FIGURE 6.29   Young's modulus versus relative density of PLA-foamed samples.

Results of the elongation at break test (Figure 6.30) show that percentage elongation increases up to around 0.25 relative density and then begins to drop after that point. This concave trend is expected given that the elongation for solid PLA is known to have a nominal value of 6%, also reported by NatureWorks. The trend obtained for microcellular PLA foam may be related to its composite profile of a foam core structure interposed between solid skin layers.

Flexural testing was conducted on PLA samples for each of the relative densities according to ASTM Standard D 790-03 of Procedure A on the Instron. Samples were subjected to a three-point bend test at a strain rate of 0.1 mm/min. Cross-head displacement was used to calculate strain. Results of the flexural test show a positive correlation between flexural modulus and relative density (Figure 6.31). Like the Young's modulus, the flexural modulus exhibits a linear trend reflecting the basic principle that any percentage of voids introduced into the polymer diminishes its flexural strength by the same percentage.

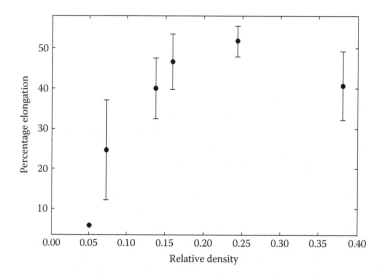

FIGURE 6.30   Percentage elongation versus relative density of PLA-foamed samples.

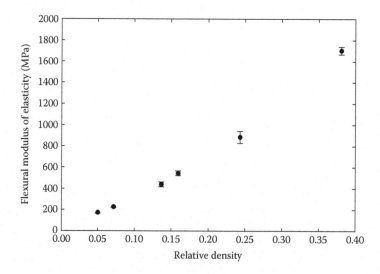

FIGURE 6.31   Flexural modulus as a function of relative density of PLA-foamed samples.

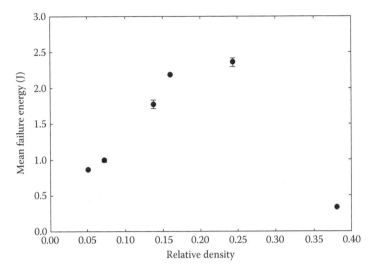

FIGURE 6.32   Impact resistance as a function of relative density of PLA-foamed samples.

Impact testing of PLA samples for each of the relative densities was carried out on the Garner impact tester according to ASTM Standard D5420-93. From the dropping test, drop height was determined based on the even distribution of failed and nonfailed samples. Mean failure energy (MFE) was then calculated based on drop height and drop weight. As the graph in Figure 6.32 shows, MFE increases with higher relative densities until a maximum limit is reached, in this case 2.36 J at a relative density of 0.24, at which point the MFE curve trends downward. Samples with a relative density of 0.39 failed at the lowest drop height, registering an MFE of about 0.04 J, indicating that densities greater than 0.39 have a mean failure height of less than 0.75 for a 4 lb weight. The concave plot suggests an optimal range for relative density that yields polymer foams with high impact strength. At the same time, it was recognized that other competing factors could alter the material's impact resistance including the energy absorbing property of the foam structure and the multilayered components (solid versus foam layers) of the material.

6.7   PRODUCTION OF LOW-DENSITY PLA FOAMS FOR THERMOFORMING

To determine the processing window for producing PLA foams of a consistent and uniform quality for thermoforming, we investigated the effects of saturation pressure, desorption time, desorption temperature, and foaming temperature on the relative density of the foam. These parameters are key to relative density since they determine the amount of gas concentration in the polymer at the time of foaming. A total of 230 samples were tested at three saturation pressures (3, 4, and 5 MPa), two desorption temperatures (0°C and −20°C), and two foaming temperatures (80°C and 100°C), for a total of 12 test conditions. Samples were prepared and saturated with CO_2 at a given pressure in room temperature up to equilibrium. Afterward, samples were weighed and placed inside a freezer to slow desorption. Periodically, a sample was removed from the freezer, weighed, and foamed for two min at a set foaming temperature in a hot liquid bath. The foamed sample was given time to desorb any residual gas until equilibrium or near equilibrium and then measured for density following ASTM standard D792-91.

Observations on the quality of the foams were reported. One, for all test groups, except groups listed in Table 6.3, a number of samples were observed to have internal blisters that measured between 2 and 20 mm (Figure 6.33). It appears that relatively short desorption times and high gas concentration account for the blistering.

TABLE 6.3

**Test Condition Groups That Experienced No
Internal Blistering**

Saturation Pressure (MPa)	Desorption Temperature (°C)	Foaming Temperature (°C)
3, 4, and 5	0	80
3 and 4	−20	80

FIGURE 6.33 Example of internal blistering experienced by a number of samples during foaming.

Second, certain foamed samples display small blisters on the surface to give it a rough texture and appearance. These blisters, which are especially noticeable on 5 MPa samples that desorb for a short time (Figure 6.34), are attributed to violent outgassing from the high internal pressure released during foaming. Lower pressure samples with short desorption times also show a somewhat rough texture surface due to bubbles protruding from beneath the thin surface layer (Figure 6.35). Third, most 5 MPa samples experienced nonuniform foaming.

Plots of relative density as a function of desorption time for samples from the 12 test conditions are shown in Figure 6.36.

At very short desorption times, 5 MPa samples for all desorption and foaming temperatures show higher relative densities of about 0.3–0.4 than samples for all other testing conditions. However, as desorption time lengthens, saturation pressure takes a back seat. The 3, 4, and 5 MPa plots are clustered together by foaming temperature and desorption temperature, indicating that for samples with longer desorption times, desorption and foaming temperature have greater effect on relative density than saturation pressure. All plots display a similar trend where the curves are shifted to the right with increasing foaming temperature and decreasing desorption temperature, indicating that samples are behaving as expected: An increase in foaming temperature reduces relative density, while a decrease in desorption temperature slows gas desorption to yield foams with a lower relative density. Hence, irrespective of saturation pressure, samples that desorbed at 0°C and foamed at 80°C reached higher relative densities that were closer to solid plastic and at a faster rate than the other test conditions. In contrast, samples that desorbed at −20°C and foamed at 100°C had the lowest relative density even beyond 100 h of desorption. The graph further indicates that between

FIGURE 6.34 5 MPa sample with small surface blisters.

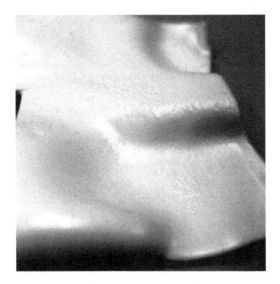

FIGURE 6.35 4 MPa sample with a rough surface due to bubbles protruding from beneath the surface layer.

desorption and foaming temperature, desorption temperature is the predominant factor. One explanation is that a lower desorption temperature slows down the rate of gas diffusion from the polymer, maintaining gas concentration levels longer and thereby lengthening the processing window for producing PLA foams with low or lower relative densities. Hence, samples that desorbed at −20°C performed very well, achieving densities to as low as 10% with a desorption time window of three, six, and eight h for 3, 4 and 5 MPa samples, respectively. Figures 6.37 and 6.38 illustrate desorption behavior of samples that desorbed at 0°C and −20°C.

The −20°C desorption temperature was able to maintain concentration levels to about 80 mg of CO_2/g of PLA beyond 100 h compared to 0°C desorption temperature, which dropped below 40 mg of CO_2/g of PLA beyond 100 h. Storing gas-saturated samples in lower temperatures can lengthen the desorption time window for achieving low-density foams. These findings all confirm

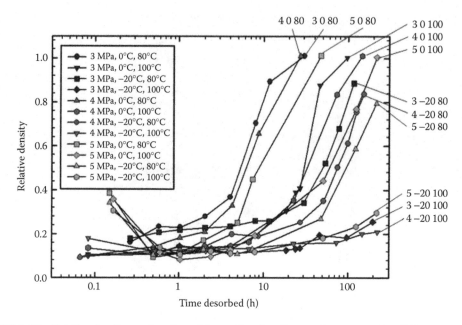

FIGURE 6.36 Relative density as a function of desorption time for samples processed at different saturation pressures, desorption temperatures, and foaming temperatures. Desorption time is given on a log scale to better depict the effects at short desorption times. Each data point is the average relative density of two samples for the same test conditions.

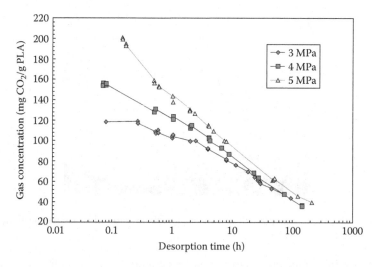

FIGURE 6.37 Gas concentration of samples desorbed at 0°C prior to foaming as a function of desorption time.

that density is gas-concentration dependent. Figures 6.39 and 6.40 show plots of relative density as a function of gas concentration for samples foamed at 80°C and 100°C.

The graphs illustrate an inverse relationship between density and gas concentration: Lower relative density corresponds with higher gas concentration and vice versa. In short, gas concentration is a good predictor of PLA foams regardless of the other processing conditions. This finding gives the solid-state process a great deal of flexibility in achieving the desired foam density by manipulating any number of processing variables that may affect concentration levels. It should be noted that

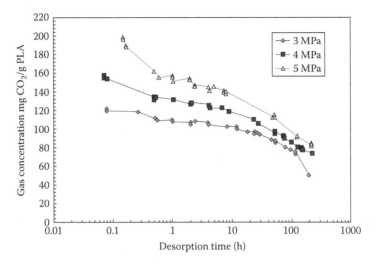

FIGURE 6.38 Gas concentration of samples desorbed at −20°C prior to foaming as a function of desorption time.

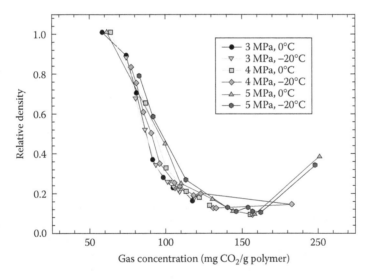

FIGURE 6.39 Relative density as a function of gas concentration for samples foamed at 80°C for different saturation pressures and desorption temperatures.

this result may not apply to lower saturation pressures. As was earlier reported in the crystallinity experiments, 3–5 MPa samples have a relatively uniform crystalline profile. Below these pressures, crystallinity levels appear to be much lower. Crystallinity has an effect on relative density insofar as it limits the diffusion of gas into the polymer and influences the polymer's expansion. As a further note, there is probably a gas concentration threshold beyond which an expanding polymer is unable to support. For instance, the morphology of the 5 MPa foamed samples that desorbed for only a short time before foaming is attributed to the samples' high gas concentration. At about 16% by weight of CO_2 and higher, 5 MPa samples experience a collapse of their cell structure that result in higher relative densities. The optimal gas concentration for producing thermoformable PLA articles appears to fall in the range of 10%–15% CO_2 by weight.

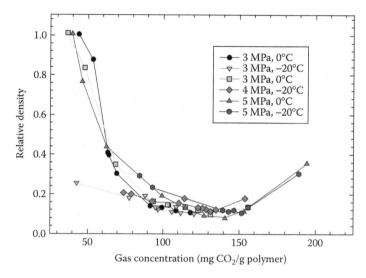

FIGURE 6.40 Relative density as a function of gas concentration for samples foamed at 100°C for different saturation pressures and desorption temperatures.

6.8 CONCLUDING REMARKS

In this chapter, we presented a comprehensive study of the PLA-CO_2 system. The study was designed and conducted with the view of exploring the feasibility of manufacturing PLA foams of various densities via the solid-state foaming method, which has been successfully applied to a large number of polymers. A particular focus of the study was to develop a process for producing low-density microcellular PLA foams that could be used in food packaging applications. Some key findings were derived from this study:

Sorption and desorption behavior of CO_2 in PLA was studied over CO_2 pressures in the 1–5 MPa range. CO_2 is shown to be highly soluble in PLA. At 5 MPa 20% of CO_2 by weight can be absorbed by the polymer. It was found that the time needed for a PLA specimen to absorb the equilibrium amount of CO_2 is highly pressure dependent. Thus, for example, a 0.6 mm thick PLA sheet reached equilibrium solubility in just one h at 5 MPa pressure compared to 32 h at 1 MPa. Analysis of the data showed that the diffusion coefficient for CO_2 in PLA increases exponentially with CO_2 pressure.

Studies on foaming behavior yielded may useful insights. In addition to foaming temperature and foaming time, it was found that desorption time is an important process parameter. For short desorption times, more volume expansion was seen at 4 MPa saturation than at 5 MPa. Foams with relative densities of around 10% can be made at different saturation pressures in the 2–4 MPa range. With respect to heating methods, both the hot liquid bath and IR oven had a relatively small effect on PLA expansion. The lowest foam densities were achieved with saturation at 3 and 4 MPa.

Microstructure studies showed that PLA foams made at 1 and 2 MPa saturation pressures had large bubbles in the 200–400 range and a thin solid skin. At 3 MPa and higher, cell size decreased by an order of magnitude with bubbles ranging from 10 to 20 µm. This is attributed to a steep increase in PLA crystallinity induced by CO_2 as pressure is increased from 2 to 3 MPa. The crystallites provide a large number of nucleation sites that lead to higher cell density and smaller cell size. Typical crystallinity for 3–5 MPa foam specimens was in the 20%–25% range. The highest percentage crystallinity achieved in foamed PLA was 27%.

Mechanical properties of PLA foams show trends that are similar to microcellular foams from other polymers. Tensile strength, flexural modulus, and Young's modulus all scale linearly with relative density to a good first approximation. Thus, for example, a 25% relative density foam has about 25% of the strength of solid PLA. No attempt was made to relate the mechanical properties to cell size in this study.

Toward developing a viable manufacturing process, a set of experiments that included desorption time and desorption temperature as additional process variables was conducted. One of the issues investigated related to the high diffusion rate of PLA. A significant amount of CO_2 can be lost from the time the pressure vessel is depressurized to the time the gas-impregnated PLA is heated and foamed. In addition, manufacturing operations may require storing the gas-saturated PLA for a length of time before it can be heated. We tested CO_2 loss at storage (or desorption) temperatures of $0°C$ and $-20°C$. It was found that the storage at $-20°C$ significantly increases the processing window for producing microcellular foams. The experiments determined that the best conditions for producing low-density microcellular foams of a consistent quality that could be thermoformed are as follows: (a) saturation in the 3–5 MPa range, (b) desorption or storage temperatures from room temperature to $-20°C$, (c) gas concentration below 16% by weight at the time of foaming, and (d) foaming temperatures in the $40°C–100°C$ range. These processing conditions are a part of a patent (Kumar et al. 2008).

Future studies need to focus on increasing the service temperature of PLA articles in order to address the hot food and microwave applications. This will require development of novel PLA formulations that remain stable at higher temperatures.

ACKNOWLEDGMENTS

The authors thank the graduate students Stephen Probert and Xiaoxi Wang, and the undergraduate students John Lu, Matt Olson, and Jordan Lee of the University of Washington for their participation in the experiments reported herein. The tireless work by Shirley Lee of MicroGREEN Polymers, Inc. in integrating individual technical reports into one cohesive document is gratefully acknowledged.

REFERENCES

Chow, T. S. Molecular interpretation of the glass transition temperature of polymer-diluent systems, *Macromolecules* 13, no. 2 (1980): 362–364.

Collias, D.I., D.G. Baird, and R.J.M. Borggreve. Impact toughening of polycarbonate by microcellular foaming, *Polymer* 35, no. 18 (1994): 3978–3983.

Crank, J. *The Mathematics of Diffusion* (Oxford: Oxford University Press, 1975).

Ding, W.-D., T. Kuboki, R. Koyama et al. Solid-state foaming of cellulose nanofiber reinforced polylactic acid biocomposites. *ANTEC, Conference Proceedings*, Boston, MA, vol. 1, pp. 756–761. Society of Plastics Engineers, 2011.

Garlotta, D. A literature review of poly(lactic acid). *Journal of Polymers and the Environment* 9, no. 2 (2001): 63–84.

Goel, S.K., and E.J. Beckman. Generation of microcellular polymers using supercritical CO_2. *Cellular Polymers* 12, no. 4 (1993): 251–274.

Guo, G., Q. Ma, B. Zhao et al. Ultrasound-assisted permeability improvement and acoustic characterization for solid-state fabricated PLA foams. *Ultrasonics Sonochemistry* 20, no. 1 (2013): 137–143.

Handa, Y.P., B. Wong, Z. Zhang et al. Some thermodynamic and kinetic properties of the system PETG-CO_2, and morphological characteristics of the CO_2-blown PETG foams, *Polymer Engineering and Science* 39, no.1 (1999): 55–61.

Ji, G., W. Zhai, D. Lin et al. Microcellular foaming of poly(lactic acid)/silica nanocomposites in compressed CO_2: Critical influence of crystallite size on cell morphology and foam expansion. *Industrial and Engineering Chemistry Research* 52, no. 19 (2013): 6390–6398.

Kumar, V., R.P. Juntunen, and C. Barlow. Impact strength of high relative density solid state carbon dioxide blown crystallizable poly(ethylene terephthalate) microcellular foams, *Cellular Polymers* 19, no. 1 (2000): 25–37.

Kumar, V., K.V. Nadella, and S.M. Probert. Method for making shapeable microcellular poly lactic acid articles. U.S. Patent no. 8926876, filed August 15, 2008.

Kumar, V., and J.E. Weller. A process to produce microcellular PVC. *International Polymer Processing* 8, no. 1 (1993): 73–80.

Kumar, V., and J.E. Weller. Production of microcellular polycarbonate using carbon dioxide for bubble nucleation, *Journal of Engineering in Industry* 116 (1994): 413–420.

Lu, J.C., V. Kumar, and H.G. Schirmer. Exploratory experiments on solid-state foaming of PLA films and COC/LDPE multi-layered films. *ANTEC, Conference Proceedings*, Milwaukee, WI, vol. 3, pp. 1890–1894. Society of Plastics Engineer, 2008.

Martini-Vvdensky, J.E., N.P. Suh, and F.A. Waldman. Microcellular closed cell foams and their method of manufacture. U.S. Patent 4,473,665, issued September 25, 1984.

Martini-Vvdensky, J.E., F.A. Waldman, and N.P. Suh. The production and analysis of microcellular thermoplastic foam. *SPE ANTEC Technical Papers* 28 (1982): 674.

Matuana, L.M. Solid state microcellular foamed poly(lactic acid): Morphology and property characterization. *Bioresource Technology* 99, no. 9 (2008): 3643–3650.

Murray, R.E., J.E. Weller, and V. Kumar. Solid-state microcellular acrylonitrile-butadiene-styrene foams. *Cellular Polymers* 19, no. 6 (2000): 413–425.

Nawaby, A., and Y. Paul Handa. Fundamental understanding of the ABS-CO_2 interactions. *ANTEC, Conference Proceedings,* Chicago, IL, vol. 2, pp. 2532–2536. Society of Plastics Engineers, 2004.

Peng, J., J. Wang, K. Li et al. Effect of expandable thermoplastic microspheres on microcellular injection molded polylactic acid (PLA): Microstructure, surface roughness, and tensile properties. *ANTEC, Conference Proceedings*, Boston, MA, vol. 3, pp. 2643–2648. Society of Plastics Engineers, 2011.

Shimbo, M., I. Higashitani, and Y. Miyano. Mechanism of strength improvement of foamed plastics having fine cell. *Journal of Cellular Plastics* 43, no. 2 (2007): 157–167.

Wang, J., W. Zhai, J. Ling et al. Ultrasonic irradiation enhanced cell nucleation in microcellular poly(lactic acid): A novel approach to reduce cell size distribution and increase foam expansion. *Industrial and Engineering Chemistry Research* 50, no. 24 (2011): 13840–13847.

Wang, X., V. Kumar, and W. Li. Low density sub-critical CO_2-blown solid-state PLA foams. *Cellular Polymers* 26, no. 1 (2007): 11–35.

Wang, X., V. Kumar, and W. Li. Development of crystallization in PLA during solid-state foaming process using sub-critical CO_2. *Cellular Polymers* 31, no. 1 (2012): 1–18.

Wang, X., W. Li, and V. Kumar. A method for solvent-free fabrication of porous polymer using solid-state foaming and ultrasound for tissue engineering applications. *Biomaterials* 27, no. 9 (2006): 1924–1929.

Wang, X., W. Li, and V. Kumar. Creating open-celled solid-state PLA foams using ultrasound. *Journal of Cellular Plastics* 45, no. 4 (2009): 353–369.

Zhang, Z., and Y.P. Handa. An in situ study of plasticization of polymers by high-pressure gases *Journal of Polymer Science Part B: Polymer Physics* 36, no. 6 (1998): 997–982.

Zhao, H., Z. Cui, X. Sun et al. Morphology and properties of injection molded solid and microcellular polylactic acid/polyhydroxybutyrate-valerate (PLA/PHBV) blends. *Industrial and Engineering Chemistry Research* 52, no. 7 (2013): 2569–2581.

7 Starch Foams

Jim Song

CONTENTS

7.1 INTRODUCTION TO STARCH

Starch, as natural polysaccharide, is one of the most abundant renewable resources widely used for industrial applications. In addition to food uses, the nonfood applications includes chemical and/or biological conversion to fine chemicals; pharmaceuticals or chemical building blocks; manufacturing of paper and boards; and production of adhesives, bioplastics, and biofuels. In the bioplastic sector, starch has been the key raw material for production of polylactic acid (e.g., www.natureworksllc.com), polyhydroxyalkanoates (e.g., www.meredianpha.com) via biological routes, and a wide range of starch-based biopolymers such as Mater-Bi (www.materbi.com) via compounding/grafting routes. Starch materials used in the latter process has been shown to retain its inherent high biodegradability and home compostability, a distinctive advantage to facilitate biological waste management, as opposed to conventional polymers (Song et al. 2009).

In this section, some background knowledge about starch is described briefly. It covers molecular and crystalline structure in native starch and behavior in gelatinization and retrogradation so as for the readers to appreciate the key characteristics of starch materials and the necessary considerations in structural and property control during formulation, processing, manufacturing, and application.

7.1.1 MOLECULAR AND CRYSTALLINE STRUCTURE

Starch is a natural polymer of D-glucose for storage of energy from photosynthesis in many plants. Agro-plants cultivated for starch uses include wheat, barley, oats, rice, corn, potatoes, and cassava. In its native state, starch exists as granules. Both the size of starch granules (typically ranging from 2 to 100 μm) and their shape are dependent on the botanic origin (Galliard and Bowler 1987; Buleon et al. 1998).

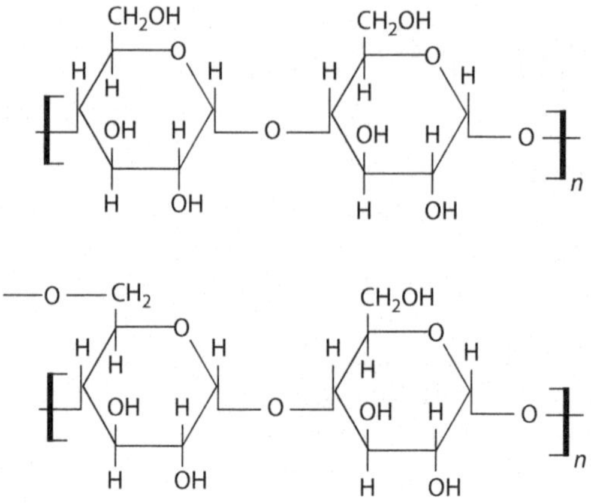

FIGURE 7.1   Molecular structures of amylose and amylopectin. (Data from Blanshard 1987.)

Starch consists of two major components: amylose and amylopectin. Amylose is a nearly linear polymer of α-(1,4) linked D-glucose with a molar mass in the order of 10^5 g mol^{-1}, while amylopectin is composed of short α-(1,4) linked chains connected to each other by α-(1,6) glucosidic linkages to form branched molecular structure with molecular mass in the order of 10^8–10^{10} g mol^{-1} (Blanshard 1987; Galliard and Bowler 1987). The molecular structures of both amylose and amylopectin are shown in Figure 7.1.

The relative amounts, structures, and molecular masses of amylose and amylopectin are determined by plant genetic, as well as environment factors, during the biosynthesis process, and therefore wide variation occurs among plants and origins. Most wheat, maize, and potato starches contain 20–30 wt% amylose. Some waxy starches contain very little (<1%) amylose, and amylomaize starches contain around 65% amylose (Ahmad et al. 1999; Parker and Ring 2001). The relative amylose to amylopectin concentration has been show to influence rheological behavior of starch melts and change of mechanical behavior with time (aging) of starch materials.

Starch granules consist of alternating 120–400 nm thick amorphous and semicrystalline layers (Buleon et al. 1998). The crystalline regions give rise to the property of birefringence, which is evidenced in the form of a Maltese cross when viewed under a crossed polar microscope. The crystalline structure can be classified into three main types (known as A-, B-, and C-type) by their distinct X-ray diffraction patterns shown in Table 7.1. Cereal starches give a characteristic A-type diffraction pattern. Tuber starches usually show a B-type diffraction pattern. Certain root and seed starches have C-type pattern, which is considered as an intermediate structure of A and B (Van Soest et al. 1996). The overall crystallinity of the native starch varies from 15% to 45% depending not only on the origin and hydration of starch but also on the technique used to determine it (Zobel 1988; Buleon et al. 1998).

Industrial uses of starch are normally in relatively purified state as opposed to flours for food or animal feed uses, which contain significant amount of nonstarch constitutes with nutritional values. Table 7.2 shows comparison of compositions of materials prepared from wheat grains: whole grain, milled flour, and purified starch. It is worth noting that small amounts of noncarbohydrate constituents such as protein and lipids, up to 1.2% and 0.6%, respectively, are usually still present in a commercial starch, depending on sources of the starch (Galliard and Bowler 1987; Buleon et al. 1998).

7.1.2 GELATINIZATION AND DESTRUCTORIZATION OF STARCH

Native starch granules are insoluble in cold water. However, when it is heated in the presence of water, starch *gelatinization*, an irreversible order-disorder transition, takes place and the process is of significant importance to cooking of starchy foods and processing of starch-based materials include foaming. Starch gelatinization is the collapse (disruption) of molecular orders within the starch

---

TABLE 7.1

X-Ray Diffraction Parameters of Three Types of Crystalline in Native Starch

Type	2θ (°)	Intensity
A	15.3	Strong
	17.1	Strong
	18.2	Less than strong
	23.5	Strong
B	5.59	Medium
	17.2	Strong
	22.2	Medium
	24.0	Less than medium
C	5.73	Weak
	15.3	Strong
	17.3	Strong
	18.3	Medium
	23.5	More than medium

TABLE 7.2

Composition of Materials Prepared from Wheat Grain as wt% of Dry Solids

Component	Whole Grain	Milled Flour	Purified Starch
Starch	75	80	90
Protein	12	11	0.2
Fiber	8.0	1.5	0.5
Lipid	2.5	1.5	1.0
Ash	1.5	1.0	0.5

Source: Guy, R., Raw materials for extrusion cooking. In *Extrusion Cooking Technologies and Application*, R. Guy, ed., Woodhead Publishing, Cambridge, pp. 5–28, 2001b.

---

granule accompanied by water absorption, granular swelling, crystallinity loss, amylose leaching, and viscosity development. The resulting paste becomes a macromolecular aqueous solution composed of water, amylose leached from starch granules, and swollen starch granules. Viscosity is strongly influenced by the swelling of starch granules (Tattiyakul and Rao 2000; Parker and Ring 2001) and peak viscosity occurs when the swollen granule reaches its largest size before rupture.

Temperature and water content are the primary factors influencing the gelatinization in addition to the botanical origin of starch. Starch gelatinization has been studied by several methods based on the observation of changes in micro- and/or macroscopic changes in structure and properties of starch granules when they are heated in aqueous solution. These include differential scanning calorimetry (Burt and Russell 1983; Noel and Ring 1992), nuclear magnetic resonance (Cooke and Gidley 1992), microscopy (Burt and Russell 1983; Liu et al. 1991), small-angle neutron scattering (Jenkins and Donald 1998) and small and wide angle X-ray scattering (Cooke and Gidley 1992; Jenkins and Donald 1998). The differential scanning calorimetry method allows the investigation of gelatinization over a wide range of water content and hence has been widely used to characterize gelatinization of starches (Jacobs et al. 1998; Jenkins and Donald 1998; Takaya et al. 2000). In the presence of excessive water, the differential scanning calorimetry endothermic peak temperature of a starch due to gelatinization is located in the range of 55°C–70°C depending on the botanical source (Jenkins and Donald 1998). As water content decreases, the second endotherm at higher temperature is developed (Jenkins and Donald 1998; Barron et al. 2000; Maaruf et al. 2001). One explanation for the two endotherms is that the swelling of the amorphous regions of the granule at higher water level pulls molecular chains from the surface of crystallites. As water becomes limited, some part of the crystallites cannot be disrupted in this manner; they undergo melting at higher temperature and give rise to the second endotherm (Donovan 1979; Jenkins and Donald 1998; Maaruf et al. 2001).

Starch gelatinization gives rise to a starch gel. For common starches containing both amylose and amylopectin, when the starch concentration is low and in the absence of mechanical shear, the gel consists of fully swollen amylopectin-rich granules dispersed in an amylose-water matrix. At higher starch concentration, the granules may be partly swollen and tightly packed.

If shear stress is applied to the system, the granule structure may be partially or even totally destroyed and becomes as *destructurized* starch. The behavior of the gel will depend on the amount of amylose and amylopectin (Svegmark and Hermansson 1991; Parker and Ring 2001). The *destructurization* is desirable to achieve homogeneous compounding of starch with other additives and/or polymers for preparation of subsequent processes, for example, extrusion, molding, and foaming.

The rheological behavior of starch gels is rather complex. In general, most starch gels are non-Newtonian fluids (Nguyen et al. 1998); their viscosity not only depends on starch concentration and temperature but also on shear rate, time, and degree of destructurization (Nguyen et al. 1998; Nurul et al. 1999; Lagarrigue and Alvarez 2001). The development of viscosity in starch gels is attributed to the gelation and crystallization of amylose and takes only a few hours (Fredriksson et al. 1998, 2000). Therefore, the higher the amylose content the starch has, the greater the gelling capability. Gelling of amylose also gives rise to opacity and syneresis in starch paste.

Being of hydroscopic nature arising from the hydroxyl groups in its molecular structure, starch gels is inherently sensitive to moisture content. Water is a natural plasticizer for starch, which affects the glass transition, T_g of starch. Dried starch can give rise to T_g in an estimated range of 230°C–250°C (Biliaderis et al. 1986; Orford et al. 1989) and the loss of plasticization renders the material difficult to process. For starch equilibrated at 50% relative humidity and 23°C, it contains typically 13%–14% moisture and T_g drops considerably to temperatures below 100°C, while higher moisture levels, for example, at ~22% can lead to T_g below room temperature (Zeleznak and Hoseney 1987), making starch a rather flexible and rubbery material.

7.1.3 RETROGRADATION AND AGING OF STARCH

Retrogradation is a term used for the reassociation of starch chains (amylose and amylopectin) in gelatinized starch gel that occurs on cooling or storage and leads to the structural changes from an initially amorphous or disordered state to a more ordered or crystalline state (Perera and Hoover 1999), leading to thickening of the starch gels. It is a slower process than gelatinization and takes place over a period of several days or weeks (Miles et al. 1985; Karim et al. 2000; Fredriksson et al. 2000). Generally, the longer the length and abundance of the short chain fraction of amylopectin, the greater the tendency to retrograde and crystallize from aqueous solution (Parker and Ring 2001).

The kinetics of starch retrogradation exhibit a strong temperature dependence because the nucleation rate increases exponentially with decreasing temperature down to the glass transition temperature, T_g, while the propagation rate increases exponentially with increasing temperature up to the melting temperature, T_m (Silverio et al. 2000). The rate of retrogradation is expected to be at a maximum approximately mid way between the T_g and T_m.

Depending on the processing conditions, the gelatinized starch could be composed of residual swollen granular starch (partially broken and partially melted granules showing residual native A-, B-, or C-type crystallinity), completely molten starch, and process-induced crystalline structures, which arise from recrystallization of single-helical structures of amylose during cooling

TABLE 7.3
Diffraction Parameters of V-Type and
E-Type Crystals in Extruded Starches

Type	2θ (°)	Intensity
V	7.1/7.4	Strong
	12.6/13.0	Strong
	19.4/20.0	Very strong
	22.1	Medium
E	6.6/6.9	Medium
	11.6/12.0	Medium
	18.0/18.5	Strong

(Van Soest et al. 1996). The most frequently observed structure is denoted as V-type, which is resulted when starches are extruded in the presence of lipids. Another structure termed *E-type* has been also observed, which is characterized by three diffraction peaks slightly displaced from those of the V-type, as shown in Table 7.3 (Van Soest et al. 1996; Cairns et al. 1997). The differences between the two structures have been attributed to different interaxial distances between helices (Shogren et al. 1993). The E variant is formed under conditions of reduced water availability and higher temperatures and can be converted to the stable V-type by increasing the moisture content of the extrudate (Cairns et al. 1997).

It has been found that the amount of V-type crystallinity is proportional to the amount of amylose (Van Soest et al. 1996). Starch source and thereby lipid contents are important factors affecting the formation of these processing-induced structures. Since the V-type and E-type structures come from the amylose-lipid complexes, none of them have been observed in potato starch (which is lipid free) or in waxy maize starch (which contains no amylose).

Physical aging in thermoplastic starch (see Section 7.2.1) also leads to the material embrittlement (Lee et al. 1998). *Aging* refers to the change in mechanical properties of starch thermoplastics with time. Generally, modulus and strength of the materials increase and impact strength or flexibility decreases. The exact mechanisms of aging are still unclear, but it has been attributed to phase separation of plasticizers from starch, starch molecule re-orientation, or crystallization of amylopectin (Van Soest and Knooren 1997; Parker and Ring 2001; Myllarinen et al. 2002).

7.2 PREPARATION OF STARCH FOR FOAMING

Starch in its native state can hardly be directly used as a material. It is not a thermoplastic polymer because when heated, it will thermally degrade without forming a melt (Villar et al. 1995; Simmons and Thomas 1995) and thus cannot be directly processed by conventional melt processing technology for thermoplastics. Therefore, it is necessary, in a majority of process relying on foaming of molten starch, to modify the structure of the native starch. Such processes are to achieve homogeneous constituency with thermoplastic characteristics for subsequent foaming process and to incorporate processing agents for improvement of certain properties of starch such as ductility, strength, and stiffness, and sometimes water and aging resistance.

Almost all starch foaming processes involve a mixing process where starch is incorporated with processing additives and/or property modifiers such as plasticizers, surfactant, blowing agents, fillers, pigments, and possibly other polymers and compatibilizers. In making such preparations of starch materials, considerations may also need to be given to the potential loss of biodegradability of the material. This section describes the general approaches in preparation of thermoplastic starch focusing on the use of plasticizers and polymeric processing agents.

7.2.1 THERMOPLASTIC STARCH

Preparation of thermoplastic starch is commonly carried out by extrusion under controlled pressure and temperature with water or other plasticizers to destructurize the starch granule structure with a combination of mechanical shearing and heating to obtain a homogeneous material that behaves as a thermoplastic (Funke et al. 1998; Lorcks 1998). The presence of enough water and/or other plasticizers is necessary, so that the starch will melt below its thermal decomposition temperature and yield a continuous polymeric phase (Shogren et al. 1993). The thermoplastic material can then be transformed into useful articles in different forms (sheet/film, moldings, or foams) using appropriate technologies. The type of native starch, moisture content in the starch, type and amount of plasticizer and additives, as well as the processing conditions, are all influencing factors to the conversion process (Bastioli 1998; Matzinos et al. 2001). The following three structural changes may take place (Poutanen and Forssell 1996):

1. Fragmentation of starch granules
2. Hydrogen-bond cleavage between starch molecules, leading to loss of crystallinity
3. Partial depolymerization of the starch polymer chains

Depending on the processing conditions, the extrudates could be composed of residual swollen granular starch (partially broken and partially melted granules), completely molten starch, and recrystallized starch. Therefore, residual native A-, B-, or C-type crystallinity and processing-induced crystallinity (V- and E-types) could be present in the extrudates.

7.2.2 PLASTICIZATION OF STARCH

The role of a plasticizer added to a polymer is to assist the flow of the polymer melt and to increase the ductility of the final material. In the context of thermoplastic starch, plasticizers are expected influence the starch in two ways: (1) by controlling the degree of gelatinization and destructurization during processing and (2) by acting as softeners in the processed material (Poutanen and Forssell 1996).

Water is the most commonly used plasticizer in starch processing, and the physical properties of starch materials are greatly influenced by the amount of water present. However, its volatility results in changes of moisture content in the materials and hence changes in properties of materials under environment humidity during storage and usage. Equilibrated at mean ambient environment (20°C and 50% relative humidity), the starch material plasticized by water alone typically possess characteristic of a glassy material with glass transition temperature above the ambient temperature and are thus often too brittle to be of any use.

For more permanent plasticization, a wide range of alternative plasticizers has been studied. These include glycerol (Lourdin et al. 1997b; Van Soest and Knooren 1997; Averous et al. 2000; Myllarinen et al. 2002), glucose (Ollett et al. 1991), sorbitol (Lourdin et al. 1997b; Funke et al. 1998; Gaudin et al. 1999), lactic acid sodium (Lourdin et al. 1997b), amino acids (Stein et al. 1999), and urea (Lourdin et al. 1997b). It had been demonstrated that some of these can effectively reduce the glass transition temperature of the starch material and improve its toughness. Ideally, the plasticizer should be nonvolatile; reduce the sensitivity of the material properties of starch to fluctuations in moisture content; and reduce the sensitivity of the materials to aging through crystallization (Moates et al. 2001).

The use of glycerol as a plasticizer has also received much attention in the production of thermoplastic starches. Many studies have been focused on the selection and stability of plasticizers during preparation of starch-based materials by extrusion or other methods, such as casting, and characterization of their effects on material properties.

Lourdin et al. (1997b) studied the use of glycerol as plasticizer for potato starch at different water content. The glass transition temperature of starch casting films decreased with increasing content of glycerol at water content ranging from about 10 to 30 wt%. It has been observed that there is a *antiplasticization* effect in the starch–glycerol system (Lourdin et al. 1997a): when the glycerol content was below 12%, there was an unexpected decrease in elongation (from 7% to 3%) as plasticizer level is increased, while elongation increased markedly from approximately 3% to about 40% when glycerol content increased above 12%.

Van Soest and Knooren (1997) reported the influence of glycerol and water content on the mechanical properties of extruded potato starch during storage. For the compositions based on the mass ratio of starch:glycerol:water = 100:25:22, the equilibrium moisture content was about 14 wt% at a controlled environment humidity of approximately 60% relative humidity. The elongation at break decreased and the elastic modulus and the tensile strength increased rapidly during the first week. For the samples stored over one week, only slight changes were observed in elongation and tensile strength. But the elastic modulus continued to increase during the aging process from 30 to 70 MPa. These changes of mechanical properties were attributed to the formation of helical structures and crystals, which resulted in a reinforcement of the starch network by physical cross-linking.

Averous et al. (2000) reported the mechanical behavior of a wheat thermoplastic starch plasticized by glycerol. Elastic modulus and tensile strength were seen to decrease with increasing of glycerol content. Depending on the plasticizer content, the elongation at break first increased with the glycerol concentrations and then decreased at high glycerol concentrations. The maximum elongation at break, about 126%, was obtained at a glycerol:starch ratio about 0.2.

Myllarinen et al. (2002) studied the effect of glycerol on amylose and amylopectin cast films. It was found that the behavior of amylose and amylopectin plasticized by glycerol was quite different. The amylopectin films showed more brittle character. Generally, under low glycerol contents, elongation at break of the films decreased with increasing of glycerol content. Above 20 wt% glycerol, however, amylose films were much ductile than the low glycerol content films and was still strong, while the amylopectin still produced very low-strength and nonflexible films. The highly plasticized amylose film (30 wt% glycerol) was rather strong showing a tensile stress of 10 MPa (similar to polyethylene film), but the corresponding amylopectin film had lost its strength entirely.

Water plays an important role in the plasticization of a starch-glycerol system. By investigating the mechanical properties of compression-molded thermoplastic starch from native corn, potato, waxy corn, and wheat starch with constant weight ratio of glycerol/dry starch at 0.30, the influence of the amount of water present (from about 11.2% to 42.2%) during compression molding was evaluated (Hulleman et al. 1998). It was found that varying the water content in the premixes led to large changes in strain and stress at break. Although there were obvious differences in strain and stress at break among the materials made from these different starch sources, the maximum of elongation at break for all the materials was observed at water content between 20 and 25 wt%, except the potato starch at 30–35 wt%.

Lourdin et al. (1997b) showed that the T_g of potato starch films was reduced when sorbitol or lactic acid sodium were added as plasticizers. However, the work of Gaudin et al. (1999) showed that there was antiplasticization effect below 27 wt% sorbitol. Beyond this value, sorbitol acted as a plasticizer and increased the elongation at break with increasing sorbitol concentration. This seems to indicate the poor plasticizing performance of sorbitol as compared with glycerol much higher concentration is needed to avoid the so-called antiplasticization phenomena.

Stein et al. (1999) blended 20 natural and synthetic amino acids (five cyclic and 15 acyclic) with a starch-glycerol mixture and extruded as ribbons. It was found that the amino acids were exceptionally good plasticizers of the starch-based materials, especially cyclic amino acids with nitrogen in the rings. For instance, pipecolic acid increased the elongation at break of the starch-glycerol sample by tenfold, and piperidinepropionic acid increased the elongation at break nearly by eight-fold. Nipecotic acid and L-proline were also good plasticizers, increasing the elongation at break by sixfold and fourfold, respectively.

7.2.3 Modifications of Starch with Polymeric Additives

Starch is hydrophilic and the moisture content within the material varies with the relative humidity of the service environment. Because of this, mechanical properties of the material vary and it is often necessary to modify starch for many practical applications where waster resistance is important. Though some chemical modification of starch, for example, by acetylation, is feasible (Fringant et al. 1996) to reduce the hydrophilic nature of the chains, it often resulted in inferior mechanical properties and greater product cost (Averous et al. 2000). Alternatively, thermoplastic starch may be blended with or without crafting to other polymer chains. Thermoplastic starch has been blended with synthetic polymers for two major purposes: (1) to improve the properties of thermoplastic starch, such as mechanical properties (e.g., melt strength and brittleness) and water resistance, which hinder its more widespread applications and (2) to reduce the cost of bioplastics since starch is inexpensive.

Blending starch with conventional polymers is known for several decades. These include low-density polyethylene (St-Pierre et al. 1997; Prinos et al. 1998; Matzinos et al. 2001) and polypropylene (Gonsalves et al. 1991; Zuchowska et al. 1998) and many others. It is aimed primarily at assisting fragmentation and certain biodegradation of the materials for short-term applications. The starch content was limited primarily by acceptable level of reduction in mechanical properties. In some of work, starch was only used in granular form as fillers (Griffin 1977, 1994; Willett 1994) to give some biodegradable characteristics to the thermoplastic polymers. However, inclusion of starch in granular form is limited to typical ~10 wt% (Shogren et al. 1993) without mechanical property deterioration. In comparison, starch content can be increased to ~50 wt% by using the destructurized (or gelatinized) starches in polyvinyl chloride (PVC), acrylonitrile butadiene syrene (ABS), and polyethylene (PE) (Griffin 1978; Otey and Westhoff 1982). However, this class of materials is not strictly biodegradable and fragments of synthetic polymers will still exist and may cause problems associated with the residues, such as deterioration of soil quality when such materials are used, for example, as mulching films.

Bioplastics such as poly(ethylene-*co*-vinyl alcohol) (George et al. 1994; Simmons and Thomas 1995; Villar et al. 1995; Stenhouse et al. 1997), poly(ethylene-*co*-acylic acid) (Fanta et al. 1992), poly(vinyl alcohol) (Lenk 1980; Lawton and Fanta 1994; Lawton 1996; Stenhouse et al. 1997; Shogren et al. 1998a; Liu et al. 1999; Willett and Shogren 2002), polycaprolactone (PCL) (Koenig and Huang 1995; Averous et al. 2000), poly(lactic acid) (Park and Im 2000; Martin and Averous 2001; Willett and Shogren 2002), and poly(hydroxyester ether) (Zhou et al. 2001; Walia et al. 2002; Willett and Doane 2002; Willett and Shogren 2002) were also used in blends with starches. Some starch-based blends have been commercialized, such as the Mater-Bi® by Novamont, Italy (Bastioli 1998) and the Bioplast® by Biotec, Germany (Lorcks 1998).

Due to the hydrophilic nature of starch, blends with hydrophobic polymers have poor mechanical properties due to lack of compatibility and poor interfacial adhesion. Direct grafting of polymeric chains onto the starch backbone provides an important method for preparing starch-polymer composites. By grafting two materials together (usually using coupling agents), a primary chemical bonding between two or more phase can be established and so a singular continuous phase containing material may be achieved (Uyama et al. 1998). Since starch and polymeric additives are held together by chemical bonding, as opposed to merely existing as a physical mixture in blends, the two dissimilar polymers could be intimately associated, and phase separation is less likely to occur.

7.3 STARCH FOAMS AND FOAMING TECHNOLOGIES

Starch foams are not new and have played an essential role in our everyday foods. To obtain necessary sensory needs and assist digestion, starch is used to produce food products in cellular solid forms in many food types, ranging from bread to snacks, confectionaries, to bread and cakes, and thus starch foaming has remained an important subject in food processing technologies (Guy 2001a, b). Industrial use of starch forms, however, is relatively new. In conjunction with the development of bioplastics from renewable resources in the last few decades or so, research and development of bioplastic foams or biofoams, including starch foams, has received increasing attention and established considerable foundation for applications in, for example, protective or thermal packaging, acoustic damping, and biomedical applications such as biodegradable scaffolding or wound dressing. With engineered formulations and cell structures, starch foams have been demonstrated to possess distinctive advantages in sustainability and unique characteristics and functionalities unattainable from conventional plastic foams.

In the following subsections, technologies developed for the production of starch foams for various nonfood applications are described and discussed in some details. It is worth noting that due to natural compatibility between starch and water, water plays multiple roles in many of the foaming techniques: as a solvent, a processing aids/plasticizer, and a blowing or co-blowing agent.

7.3.1 Bake Foaming

Bake foaming or backing technique has originated from the backing technology in food processing to produce, for example, wafer or ice-cream cones. A starch-water batter is backed inside a closed mold with temperature of 175°C–235°C. Water acts as a gelatinization agent, plasticizer, and a blowing agent to produce thin-walled objects (Shogren et al. 1998b). The foamed objects have a typical dense outer skin, due to rapid loss of water from the foam surface in contact with the hot mold, and less dense interior with large, mostly open cells.

Glenn et al. (2001b, c) also studied effects of inclusion of fiber and $CaCO_3$ fillers on the properties of bake foamed sheet. The foam sheet was made by baking the dough samples for 3–5 min in a mold preheated to 160°C–180°C. The properties of the starch/fiber foams, such as density, flexural strength, and flexural strain at maximum force, were within the range of commercial food containers made of extruded polystyrene or coated paperboard, but the addition of $CaCO_3$ to the starch/fiber foam composites did not improve the foam properties.

Due to the high water content with the material (primarily for assisting material flow and mold filling), the baking time, which varies with the size and thickness of the products, are rather long and typically around 60 s, resulting in low production rate and limiting foam thickness in this process. Multiple cavity molds are often used to mitigate the disadvantages. Baked foam trays/punnets have been adapted in food service and packaging of a range of food products as exampled in Figure 7.2 (http://www.potatoplates.com).

A variation of this process has also been developed as PaperFoam™ (http://www.paperfoam.nl). The starch-water batter with pulp fiber reinforcement is injection molded and baked to produce foamed thin-shell objects as shown in Figure 7.3. The injection molding pressure improves initial mold filling and can thus reduce water content in the batter and the bake foaming cycle time. The product combines a fiber-reinforced foam core with a smooth outer surface, offering good impact

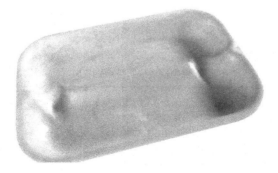

FIGURE 7.2 Example of starch foam tray made by bake foaming process. (Data from http://www.potatoplates.com.)

FIGURE 7.3 Example of PaperForm for packaging of electronic products. (Data from http://www.paper-foam.nl.)

protection, thermal insulation, surface patterning or printing, and anti-static properties. Currently, PaperFoam is used to pack electronics, cosmetics, medical, and dry foods.

7.3.2 PUFF FOAMING

Puff foaming or puffing technique can be traced to a similar technique for making rice cakes. Thermoplastic starch pellets typically prepared by extrusion and pelletizing are heated and pressurized in a mold and expanded by the residual water in the pellets on sudden release of the mold pressure. In Glenn and Orts's work (2001a), starch pellet (1–3 mm) are prepared by extrusion and conditioned to moisture levels 8–20 wt%. The conditioned aggregates were loaded in an aluminum compression mold heated to 230°C and compressed for 10 s under a pressure of 3.5 MPa. The pressure was then released rapidly resulting in an explosive release of steam as the starch feedstock expanded and filled the mold. The starch foam contains mostly closed cells less than 1 mm in diameter. Compared with the baking method, the water content in starch is much lower and thus the puffing process is much quicker, typically a few seconds. However, due to poor flow of the foaming material and lack of control of expansion, this process is limited to molding of simple shapes with a rough surface finish.

7.3.3 EXTRUSION FOAMING

Extrusion processing technology has been used to develop starch-based foams since the late 1980s (Lacourse and Altieri 1989). This is a continuous process typically carried out using a twin-screw extruder to generate thermoplastic starch melt compounded with necessary modification additives and to create desired foam structure in the extrudate (Kokini et al. 1992; Guy 2001c). Starch is commonly produced directly from raw starch in an extrusion foaming process, where the starch granules are converted into a thermoplastic melt *in situ* with addition of plasticizer and other processing additives. Water is used for gelatinization and plasticization, as well as a blow agent. As the melt exits the die nozzles, the sudden release of pressure causes the melt to become superheated and water rapidly vaporize. This process enables bubbles to be created and expanded in the extruded melt. The increase in the viscosity of the extrudate due to loss of water and cooling stabilizes the bubbles resulting in solid starch foams. The ability to create the sudden pressure drop is one of the key requirements to prevent premature foaming, to generate uniform and well-expanded cell structure. This imposes constraints on die design and limitations to the diameter of extruded foam rod or thickness of sheet foam.

Considerable effort has been made to study the influence of extrusion conditions, moisture content, and composition on the cell structure and physical properties of starch-based foams (Bhatnagar and Hanna 1995a, b; Cha et al. 2001; Fang and Hanna 2001a, b; Willett and Shogren 2002). Many patents on extruded starch blends foams have been filed (Lacourse and Altieri 1989, 1991; Bastioli et al. 1994, 1998a, b; Xu and Doane 1997, 1998). The polymers used to blend with starch include poly(vinyl alcohol), poly(ethylene vinyl alcohol), PCL, poly(ethylene-*co*-acrylic acid), cellulose acetate, and hydroxy-functional polyesters.

Bhatnagar and Hanna (1995a) made starch-based foams using corn starch with 25% amylose and polystyrene or poly(methyl methacrylate). All formulations were based on a starch to plastic ratio of 70:30. The foams were produced with a twin-screw extruder at 120°C barrel temperature, 140 rpm screw speed and 6% moisture content. These produced foams performed well compared to the commercially available expanded polystyrene (EPS) loose-fills and better functional properties than 100% starch loose-fills.

Fang and Hanna (2000a, 2001a, b) blended PLA, Mater-Bi ZF03U and Easter™ Bio Copolyester® 14766 to regular (25% amylose) and waxy corn starch at different levels (10, 25, and 40 wt%). The blends were then extruded into loose-fill packaging foams using a co-rotating twin-screw extruder with peak temperature of 150°C and screw speed of 150 rpm. It was found that foams produced with waxy starch had greater radial expansions and lower foam and loose-fill bulk densities than the regular starch. Addition of higher levels of both Mater-Bi and Easter Bio Copolyester reduced radial expansion and increase foam and bulk densities.

Cha et al. (2001) produced starch-based foams by blending 49 wt% wheat and corn starch, 33% synthetic polymers [poly(ethylene-*co*-vinyl alcohol), polystyrene, and poly(styrene/maleic anhydride)], 10.5% water, 7% blowing agents (methanol), and 0.5% nucleating agent (silicon dioxide) and extruded at 100°C–160°C with 100 rpm screw speed using a single-screw extruder. Bulk density of the starch-based loose-fill foams was found to decrease with higher extrusion temperature. The highest expansion occurred when the blends containing normal wheat starch were extruded at 140°C.

Willett and Shogren (2002) reported their work on the extrusion foaming process and properties of various starch (corn, wheat, high amylose (70%), and potato) blended with several biodegradable polymers. The addition of polylactic acid and poly(hydroxybutyrate-*co*-valerate) significantly lowered densities and increased radial expansion ratios of extruded foams. Other polymers studied, such as PCL, poly(ester amide), and cellulose acetate, also decreased foam density but to a lesser extent. Compressive strength of the loose-fill foams depended primarily on the foam density, not on starch type or polymer additives.

The most common product is chopped short cylinders (of typical 20–40 mm diameter) and used as an alternative to EPS counterpart for loose-fill packaging as shown in Figure 7.4 (e.g., http://www.greenlightproducts.co.uk/). Compared with EPS loose-fills, the starch loose-fills have in general higher bulk density. They are dominated by an open cell structure, which is responsible for the lower resiliency than EPS foams, though there is no significant difference on compressive strength (Tatarka and Cunningham 1998).

FIGURE 7.4   Example of starch loose-fill packaging. (Data from www.greenlightproducts.co.uk.)

Only relatively thin sheet foams (typically 10–20 mm thick) can be extruded. To mitigate such limitation, starch foams in the form of corrugated sheet known as GreenCell™ has also been manufactured (e.g., http://www.bioviron.com) and found applications in transit packaging and postal shipping of a wide range of products from electronic goods to auto parts (Figure 7.5). Shipping coolers have also been developed for temperature sensitive goods.

Starch loose-fills form natural adhesion when the surface is moistened and are brought into contact with each other under low pressure. This leads to studies of a number of postextrusion processes to produce bulk starch foams. In a *compression bonded loose-fills* (Bonin 2010), starch loose-fills were wetted by passing through a water mist chamber and compressed to blocks with densities ranging from 30 to 100 kg m^{-3} (Figure 7.6). The compressive and tensile strength, dynamic impact properties, and creep behavior of the block foams under different humidity levels were studied systematically and compared with a range of polymer foams. Despite of its low tensile strength and sensitivity to high humidity, the starch block foams were shown to perform well in general for cushioning and thermal insulation applications.

Extruded starch foam as continuous cylindrical rods have been processed further using a similar method known as *regular packing and stacking (RPS)* process (Kang and Song 2009). A pultrusion

FIGURE 7.5 Example of a cool box with GreenCell foam as thermal insulation. (Data from http://www .bioviron.com.)

FIGURE 7.6 An example of block foam made from compression bonded loose-fills process. (Data from Bonin, M., *An Investigation into the Properties of Starch-Based Foams*. PhD thesis, Brunel University, London, 2010.)

FIGURE 7.7 An example of block foams made from RPS process. (Data from Kang and Song, 2009.)

treatment was used to transform the circular section to square so as to assist wetting of the flat surfaces by a contact water absorption method. A purpose-built RPS equipment was used to bring multiple rods into contact and form self-adhesion between them to form foam planks. This enables fabrication of laminated block foams with a regular network of interfaces acting as reinforcement (Figure 7.7). The RPS bulk foams consist of foams enclosed in an ordered network of interfaces and can be regarded as *macro-composite* foam and the mechanical properties of which can be manipulated by designing the interface network and the thickness of the interfaces via variation of the amount of water absorbed (Kang and Song 2009). By aligning the extrusion direction of the foams with that of the compressive load, the RPS block foams exhibit optimum compressive performance for energy absorption in cushion packaging applications at high stress levels, while alignment of extrusion direction perpendicular to the load gives rise to more soft RPS bulk foams suitable for packaging of low-fragility products. Combination of the foam orientations can be utilized to produce bulk foam with intermediate properties.

In addition, the RPS starch foam has also been shown to be an excellent thermal insulator with low thermal conductivity comparable to expanded polystyrene counterpart, which makes it desirable alternative to polystyrene or polyethylene foams in thermal as thermal insulation packaging (Wang et al. 2010).

7.3.4 Microwave Foaming

Microwave has been used in the food industry to expand starch-based half-products (e.g., pellet snacks) prepared by extrusion cooking (Boehmer et al. 1992; Messager and Despre 1998; Lee et al. 2000). In a series of studies, microwave foaming of expandable starch pellets has been investigated as a method for molding of starch foams.

Preparation of expandable starch pellets is a crucial stage to enable effective heat generation under microwave radiation and generation of internal water vapor pressure as a blowing agent. The foaming mechanisms and effects of some additives on the dielectric and thermo-mechanical properties of starch pellets were investigated using a microwave calorimeter (Peng et al. 2013a) and thermo-mechanical analysis (Peng et al. 2013b). Extrusion destructurization of starch and its composition have significant impact on their microwave foamability. For extruded wheat flour/starch without additives, a rapid increase in dielectric loss factor has been observed at an onset temperature of 75°C–95°C, approximate to the glass transition temperature, and the foaming temperature is approximately 10°C–20°C above the glass transition temperature. This was not observed in the raw starch, indicating the importance of thermoplastic transition of starch. In comparison with purified wheat starch, wheat flour compositions gave rise to relatively lower expansion, which may be attributed to impurities such as proteins. The incorporation of organic additives (glycerol and polyvinyl alcohol) in the starch pellets has generally negative effects on their microwave foamability. Concentration of the additives over 10 wt% led to the deterioration of their microwave expansion. This was attributed to the restriction to the movement of starch molecules caused by interaction among starch, additives, and water—the inability to release the water as the blow agent from the hydrophilic additives and higher constraint to foam cell expansion by these additives, or a combination of both.

Experimental studies have also been conducted in a process known as *microwave assisted molding* by Zhou et al. (2006, 2007). Wheat flour and starch pellets containing salts as microwave absorber and talc as nucleation agent were prepared with twin-screw extrusion and conditioned to various levels of moisture content. Microwave expansion and molding were conducted within a polytetra-fluoroethylene mold using a microwave and hot air combinational oven. The physical properties of microwave-foamed starch pellets, including density, porosity, cell structure, water absorption characteristics, and mechanical properties, were characterized. It was found that the physical properties of these starch foams produced by microwave heating are highly dependent on the raw materials and additives. An optimum moisture content around 10 wt% exists for maximum expansion. Foam density decreased significantly with addition of 5.5%–10.5% w/w salts, while foams containing nucleation agent (talc) resulted in higher density but refined cell structure. The addition of salts also increased the water sorption of foams and plasticized cell walls. Mechanical properties of the foamed pellets in the elastic region, as well as under large deformation (at 40% strain), all follow a power–law relationship with foam density. At room temperature and 50% relative humidity, some mechanical properties, such as compressive strength, compressive modulus of elasticity, and deformation energy at 40% strain, are comparable to some commercial EPS foams. When the microwave expansion was conducted in mold, pretreatment of pellets and control of the initial loading of pellets in the mold cavity were found to be important to achieve a uniformly foamed block with good fusion strength and integrity. The bonding between foamed pellets in a block can be significantly enhanced by soaking the pellets in a NaCl solution prior to microwave foaming. There exists an optimum initial loading of pellets in the mold for a given pellet formulation, which allows sufficient expansion to achieve an acceptable extent of interfacial bonding and mold filling. The work demonstrated the feasibility and potential applications. Free-flowing foamed balls may be produced for loose-fill packaging. When the pallets are foamed in a mold, lightweight moldings can be produced in forms of containers, end caps, and edge or corner cushion pads for protective packaging, which are difficult to produce with extrusion foaming technology.

7.3.5 NANOSTRUCTURE STARCH FOAMS

The cell sizes of starch foams produced by technologies described so far are typically in tens of micron meters to a few millimeters. They are predominantly developed for production of light-weight and biodegradable materials as alternative to polymer foams. The applications are predominantly for packaging (e.g., containers, trays for foods, and cushion planks for energy dissipation against impact damages or thermal insulation). Using starch as a renewable resource, a method has been developed as an alternative to conventional hard or soft template routes for nanostructure starch foams as a precursor for production of mesoporous carbon (pore size between 2 and 50 nm). The activated carbonaceous materials aim at advanced applications, including separation science, heterogeneous catalyst supports, water purification filters, and stationary phase materials, as well as energy generation and storage applications (White et al. 2008, 2009).

The technology involves three stages as shown in Figure 7.8: (a) gelatinization of a starch using microwave heating to prepare an aqueous gel, (b) production of solid mesoporous starch foam via solvent exchange/drying, and (c) thermal carbonization/dehydration to produce mesoporous carbons known as *starbon*.

The properties of the material are dependent on the preparation temperature of the aquagel, allowing regulation of the textural properties, particle morphology, and polysaccharide ordering of

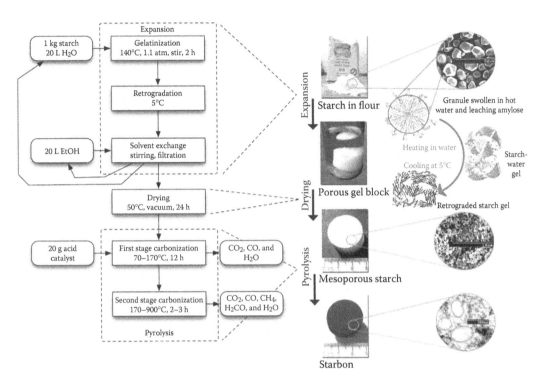

FIGURE 7.8 Schematic representation of the main processing steps in the production of starch-derived carbonaceous materials. (Data from White et al., 2009.)

the precursor. The resulting carbonaceous materials are highly porous, with high specific surface area around 200 m^2 g^{-1}, and a bimodal pore size distribution with the majority of porosity lying in the mesopore region (2–50 nm). At temperature above 700°C, the carbonization process leads to the synthesis of robust mesoporous carbons with tuneable nanostructure. By selection of gelation conditions, polysaccharide type and carbonization temperature, a wide range of carbonaceous materials may be synthesized using inexpensive and readily available renewable sugar-based precursors.

7.4   SUMMARY

Starch as a natural polysaccharide polymer is a desirable choice for production of foams for a range of applications. It is featured by its renewable origins, low cost, and availability for industrial-scale exploitations. Starch foams also have excellent biodegradability and thus enabling biological waste management via home/municipal compositing or anaerobic digestion. As a result, most starch foams are designed for short-service life applications. Packaging is currently the dominating application of such foams with products commercially available for compostable packaging, cushion protection, and thermal insulation purposes. However, it may well extend to other applications as demonstrated by the application of nanostructured starch foam as a precursor for producing mesoporous carbon.

Starch is inherently hydroscopic and can hardly be used directly as a material without modification of the native molecular structure. In most foaming processes, preparation of starch is a necessary step during which native starch structure is destructurized and transformed to a thermoplastic starch, enabling foaming of starch in a molten state. In addition, additives such as plasticizers, nucleation agents, and blowing agents, as well as property enhancers, including those for water resistance can be incorporated at this stage.

Each starch foaming technology is capable to manufacture foams in a particular form. The bake foaming and related processes are suitable for molding thin shell foam products such as clamshell and trays. The extrusion foaming has been successful in the production of loose-fills and sheets but extendable to block foams with application of some postextrusion process. The microwave foaming is capable of producing free flowing spherical loose-fills and molded foam components. The aquagel route allows creation of nanostructured starch foam precursor for mesoporous carbon production.

From view point of commercialization, the baked foam products, extruded foam loose-fill, and sheets are currently available on the market, while the other processes are in the stage of further development or scaling-up and given the attractions of starch foams, such as low cost from renewable resource, good functionalities, and biodegradability, one would expect increasing effort in the near future in research and development and emerging of new processing technologies and products.

REFERENCES

Ahmad, F. B., Williams, P. A., Doublier, J., Durand, S., and Buleon, A. (1999) Physico-chemical characterisation of sago starch. *Carbohydr. Polym.* 38, 361–370.

Averous, L., Moro, L., Dole, P., and Fringant, C. (2000) Properties of thermoplastic blends: Starch-polycaprolactone. *Polymer* 41, 4157–4167.

Barron, C., Buleon, A., Colonna, P., and Della Valle, G. (2000) Structural modifications of low hydrated pea starch subjected to high thermomechanical processing. *Carbohydr. Polym.* 43, 171–181.

Bastioli, C. (1998) Properties and applications of mater-bi starch-based materials. *Polym. Degrad. Stab.* 59, 263–272.

Bastioli, C., Bellotti, V., Del Giudice, L., Lombi, R., and Rallis, A. (1994) Expanded articles of biodegradable plastic materials. U.S. Patent 5,360,830.

Bastioli, C., Bellotti, V., Del Tredici, G., Montino, A., Ponti, R. (1998b) Biodegradable foamed plastic materials. U.S Patent 5736586.

Bastioli, C., Bellotti, V., Del Tredici, G., and Rallis, A. (1998a) Biodegradable foamed articles and process for preparation thereof. U.S Patent 5,801,207.

Bhatnagar, S. and Hanna, M. A. (1995a) Properties of extruded starch-based plastic foam. *Ind. Crops Prod.* 4, 71–77.

Bhatnagar, S. and Hanna, M. A. (1995b) Physical, mechanical, and thermal properties of starch-based plastic foams. *Trans. ASAE* 38, 567–571.

Biliaderis, C. G., Page, C. M., Maurice, T. T., and Juliano, B. O. (1986) Thermal characterisation of rice starches—A polymeric approach to phase-transitions of granular starch. *J. Agric. Food Chem.* 34, 6–14.

Blanshard, J. M. V. (1987) Starch granule structure and function: A physicochemical approach. In *Starch Properties and Potential* (Galliard, T., ed.), John Wiley & Sons, London, pp. 16–54.

Boehmer, E. W., Bennet, W. L., Guanella, T. J., and Levine, L. (1992) Microwave expandable half product and process for its manufacture. U.S. Patent 5,165,950.

Bonin, M. 2010. *An Investigation into the Properties of Starch-Based Foams.* PhD thesis, Brunel University, London.

Buleon, A., Colonna, P., Planchot, V., and Ball, S. (1998) Starch granules: Structure and biosynthesis. *Int. J. Biol. Macromol.* 23, 85–112.

Burt, D. J. and Russell, P. L. (1983) Gelatinisation of low water content wheat starch mixture—A combined study by differential scanning calorimetry and light microscopy. *Starch/Starke* 35, 354–360.

Cairns, P., Morris, V. J., Singh, N., and Smith, A. C. (1997) X-ray diffraction studies on extruded maize grits. *J. Cereal Sci.* 26, 223–227.

Cha, J. Y., Chung, D. S., Seib, P. A., Flores, R. A., and Hanna, M. A. (2001) Physical properties of starch-based foams as affected by extrusion temperature and moisture content. *Ind. Crops Prod.* 14, 23–30.

Cooke, D. and Gidley, M. J. (1992) Loss of crystalline and molecular order during starch gelatinisation—Origin of the enthalpic transition. *Carbohydr. Res.* 227, 103–112.

Donovan, J. W. (1979) Phase transition of starch-water system. *Biopolymer* 18, 263–275.

Fang, Q. and Hanna, M. A. (2000) Functional properties of polylactic acid starch-based loose-fill packaging foams. *Cereal Chem.* 77, 779–783.

Fang, Q. and Hanna, M. A. (2001a) Characteristics of biodegradable Mater-Bi®-starch based foams as affected by ingredient formulations. *Ind. Crops Prod.* 13, 219–227.

Fang, Q and Hanna, M. A. (2001b) Preparation and characterisation of biodegradable copolyester-starch based foams. *Bioresour. Technol.* 78, 115–122.

Fanta, G. F., Swanson, C. L., and Shogren, R. L. (1992) Starch-poly (ethylene co-acrylic acid) composite films. Effect of processing conditions on morphology and properties. *J. Appl. Polym. Sci.* 44, 2037–2042.

Fredriksson, H., Bjorck, I., Andersson, R., Liljeberg, H., Silverio, J., Eliasson, A.-C., and Aman, P. (2000) Studies on α-amylase degradation of retrogradaded starch gels from waxy maize and high-amylopectin potato. *Carbohydr. Polym.* 43, 81–87.

Fredriksson, H., Silverio, J., Andersson, R., Eliasson, A.-C. and Aman, P. (1998) The influence of amylose and amylopectin characteristics on gelatinisation and retrogradation properties of different starches. *Carbohydr. Polym.* 35, 119–134.

Fringant, C., Desbrieres, J., and Rinaudo, M. (1996) Physical properties of acetylated starch-based materials: Relation with their molecular characteristics. *Polymer* 37, 2663–2673.

Funke, U., Bergthaller, W., and Lindhauer, M. G. (1998) Processing and characterisation of biodegradable products based on starch. *Polym. Degrad. Stab.* 59, 293–296.

Galliard, T. and Bowler, P. (1987) Morphology and composition of starch. In *Starch Properties and Potential* (Galliard, T., ed.), John Wiley & Sons, London, pp. 55–78.

Gaudin, S., Lourdin, D., Le Botlan, D., Ilari, J. L., and Colonna, P. (1999) Plasticisation and mobility in starch-sorbitol films. *J. Cereal Sci.* 29, 273–284.

George, E. R., Sullivan, T. M., and Paek, E. H. (1994) Thermoplastic starch blends with a poly(ethylene-co-vinyl alcohol): processability and physical properties. *Polym. Eng. Sci.* 34, 17–23.

Glenn, G. M. and Orts, W. J. (2001a) Properties of starch-based foam formed by compression/explosion processing. *Ind. Crops Prod.* 13, 135–143.

Glenn, G. M., Orts, W. J., and Nobes, G. A. R. (2001b). Starch, fibre and CaCO$_3$ effects on the physical properties of foams made by a baking process. *Ind. Crops Prod.* 14, 201–212.

Glenn, G. M., Orts, W. J., Nobes, G. A. R., and Gray, G. M. (2001c) In situ laminating process for baked starch-based foams. *Ind. Crops Prod.* 14, 125–134.

Gonsalves, K. E., Patel, S. H., and Chen, X. (1991) Development of potentially degradable materials for marine applications. II. Polypropylene-starch blends. *J. Appl. Ploym. Sci.* 43, 405–415.

Griffin, G. J. L. (1977) Synthetic resin sheet material. US Patent 4,021,388.

Griffin, G. J. L. (1978) Synthetic/resin based compositions. US Patent 4,125,495.

Griffin, G. J. L. (1994) Starch polymer blends. *Polym. Degrad. Stab.* 45, 241–247.

Guy, R. (2001a) Snack foods. In *Extrusion Cooking Technologies and Application* (Guy, R. ed.), Woodhead Publishing, Cambridge, pp. 161–181.

Guy, R. (2001b) Raw materials for extrusion cooking. In *Extrusion Cooking Technologies and Application* (Guy, R. ed.), Woodhead Publishing, Cambridge, pp. 5–28.

Guy, R. (2001c) *Extrusion Cooking Technologies and Applications* (Guy, R. ed.), Woodhead Publishing, Cambridge.

Hulleman, S. H. D., Janssen, F. H. P., and Feil, H. (1998) The role of water during plasticization of native starches. *Polymer* 39, 2043–2048.

Jacobs, H., Mischenko, N., Koch, M. H. J., Eerlingen, R. C., Delcour, J. A., and Reynaers, H. (1998) Evaluation of the impact of annealing on gelatinisation at intermediate water content of wheat and potato starches: A differential scanning calorimetry and small angle x-ray scattering study. *Carbohydr. Res.* 306, 1–10.

Jenkins, P. J. and Donald, A. M. (1998) Gelatinisation of starch: A combined SAXS/WAXS/DSC and SANS study. *Carbohydr. Res.* 308, 133–147.

Kang, Y.-G. and Song, J. (2009) Structure and compression behavior of macro-composite starch foams. *J. Cell. Plast.* 45, 33.

Karim, A. A., Norziah, M. H., and Seow, C. C. (2000) Methods for the study of starch retrogradation. *Food Chem.* 71, 9–36.

Koenig, M. F. and Huang, S. J. (1995) Biodegradable blends and composites of polycaprolactone and starch derivatives. *Polymer* 36, 1877–1882.

Kokini, J. L., Ho, C.-T., and Karwe, M. V. (1992) *Food Extrusion Science and Technology* (Kokini, J. L., Ho, C.-T., and Karwe, M. V. eds.), Marcel Dekker, New York.

Lacourse, N. L. and Altieri, P. A. (1989) Biodegradable packaging material and the method of preparation thereof. US Patent 4,863,655.

Lacourse, N. L. and Altieri, P. A. (1991) Biodegradable shaped products and the method of preparation thereof. US Patent 5,043,196.

Lagarrigue, S. and Alvarez, G. (2001) The rheology of starch dispersions at high temperatures and high shear rates: A review. *J. Food Eng.* 50, 189–202.

Lawton, J. W. (1996) Effect of starch type on the properties of starch containing films. *Carbohydr. Polym.* 29, 203–208.

Lawton, J. W. and Fanta, G. F. (1994) Glycerol-plasticised films prepared from starch poly (vinyl alcohol) mixtures—Effect of poly(ethylene-co-acrylic acid). *Carbohydr. Polym.* 23, 275–280.

Lee, E. Y., Lim, K. I., Lim, J.-K., and Lim, S.-T. (2000) Effects of gelatinization and moisture content of extruded starch pellets on morphology and physical properties of microwave-expanded products. *Cereal Chem.* 77, 769–773.

Lee, H. A., Kim, N. H., and Nishinari, K. (1998) DSC and rheological studies of the effects of sucrose on the gelatinisation and retrogradation of acorn starch. *Thermochim. Acta* 322, 39–46.

Lenk, R. S. (1980) An unsupported water soluble and heat sealable film from predominantly non-fossil raw materials. *Polymer* 21, 371–373.

Liu, H., Lelievre, J., and Ayoung-Chee, W. (1991) A study of starch gelatinization using differential scanning calorimetry, x-ray, and birefringence measurements. *Carbohydr. Res.* 210, 79–87.

Liu, Z, Feng, Y., and Yi, X. (1999) Thermoplastic starch/PVAl compounds: Preparation, processing, and properties. *J. Appl. Polym. Sci.* 74, 2667–2673.

Lorcks, J. (1998) Properties and applications of compostable starch-based plastic material. *Polym. Degrad. Stab.* 59, 245–249.

Lourdin, D., Bizot, H., and Colonna, P. (1997a) "Antiplasticization" in starch-glycerol films? *J. Appl. Polym. Sci.* 64, 1047–1053.

Lourdin, D., Coignard, L., Bizot, H., and Colonna, P. (1997b) Influence of equilibrium relative humidity and plasticizer concentration on the water content and glass transition of starch materials. *Polymer* 38, 5401–5406.

Maaruf, A. G., Che Man, Y. B., Asbi, B. A., Junainah, A. H., and Kennedy, J. F. (2001) Effect of water content on the gelatinisation temperature of sago starch. *Carbohydr. Polym.* 46, 331–337.

Martin, O. and Averous, L. (2001) Poly(lactic acide): Plasticization and properties of biodegradable multiphase systems. *Polymer* 42, 6209–6219.

Matzinos, P., Bikiaris, D., Kokkou, S., and Panayiotou, C. (2001) Processing and characterisation of LDPE/starch products. *J. Appl. Polym. Sci.* 79, 2548–2557.

Messager, A. and Despre, D. J. C. (1998) Microwave-puffable pellet-shaped food product in a susceptorpackage, and process for making it. Europe Patent EP0836807.

Miles, M. J., Morris, V. J., Orford, P. D., and Ring, S. G. (1985) The roles of amylose and amylopectin in the gelation and retrogradation of starch. *Carbohydr. Res.* 135, 271–281.

Moates, G. K., Noel, T. R., Parker, R., and Ring, S. G. (2001) Dynamic mechanical and dielectric characterisation of amylose-glycerol films. *Carbohydr. Polym.* 44, 247–253.

Myllarinen, P., Partanen, R., Seppala, J., and Forssell, P. (2002) Effect of glycerol on behaviour of amylose and amylopectin films. *Carbohydr. Polym.* 50, 355–361.

Nguyen, Q. D., Jensen, C. T. B., and Kristensen, P. G. (1998) Experimental and modelling studies of the flow properties of maize and waxy maize starch pastes. *Chem. Eng. J.* 70, 165–171.

Noel, T. R. and Ring, S. G. (1992) A study of the heat capacity of starch/water mixtures. *Carbohydr. Res.* 227, 203–213.

Nurul I. M., Azemi, B. M. N. M., and Manan, D. M. A. (1999) Rheological behaviour of sago (*Metroxylon sagu*) starch paste. *Food Chem.* 64, 501–505.

Ollett, A.-L., Parker, R., and Smith, A. C. (1991) Deformation and fracture behaviour of wheat starch plasticized with glucose and water. *J. Mater. Sci.* 26, 1351–1356.

Orford, P. D., Parker, R., Ring, S. G., and Smith, A. C. (1989) Effect of water as a diluent on the glass transition behaviour of malto-oligosaccharides, amylose and amylopectin. *Int. J. Biol. Macromol.* 11, 91–96.

Otey, F. H. and Westhoff, R. P. (1982) Biodegradable starch-based blown films. US Patent 4,337,181.

Park, J. M. and Im, S. S. (2000) Biodegradable polymer blends of poly(L-lactic acid) and gelatinised starch. *Polym. Eng. Sci.* 40, 2539–2550.

Parker, R. and Ring, S. G. (2001) Aspects of the physical chemistry of starch. *J. Cereal Sci.* 34, 1–17.

Peng, X., Song, J., Nesbitt, A., and Day, R. (2013a) Microwave foaming of starch-based materials (I) dielectric performance. *J. Cell. Plast.* 49(3), 245–258.

Peng, X., Song, J., Nesbitt, A., and Day, R. (2013b) Microwave foaming of starch-based materials (II) thermo-mechanical performance. *J. Cell. Plast.* 49(2), 147–160.

Perera, C. and Hoover, R. (1999) Influence of hydroxypropylation on retrogradation properties of native, defatted and heat-moisture treated potato starches. *Food Chem.* 64, 361–375.

Poutanen, K. and Forssell, P. (1996) Modification of starch properties with plasticizers. *Trends Polym. Sci.* 4, 128–132.

Prinos, J., Bikiaris, D., Theologidis, S., and Panayiotou, C. (1998) Preparation and characterisation of LDPE/starch blends containing ethylene/vinyl acetate copolymer as compatibilizer. *Polym. Eng. Sci.* 38, 954–964.

Shogren, R. L., Fanta, G. F., and Doane, W. M. (1993) Development of starch based plastics—A reexamination of selected polymer systems in historical perspective. *Starch/Starke* 45, 276–280.

Shogren, R. L., Lawton, J. W., Doane, W. M., and Tiefenbacher, K. F. (1998b) Structure and morphology of baked starch foams. *Polymer* 39, 6649–6655.

Shogren, R. L., Lawton, J. W., Tiefenbacher, K. F., and Chen, L. (1998a) Starch-poly(vinyl alcohol) foamed articles prepared by a baking process. *J. Appl. Polym. Sci.* 68, 2129–2140.

Silverio, J., Fredriksson, H., Andersson, R., Eliasson, A.-C., and Aman, P. (2000) The effect of temperature cycling on the amylopectin retrogradation of starches with different amylopectin uni-chain length distribution. *Carbohydr. Polym.* 42, 175–184.

Simmons, S. and Thomas, E. L. (1995) Structural characteristics of biodegradable thermoplastic starch/poly(ethylene-vinyl alcohol) blends. *J. Appl. Polym. Sci.* 58, 2259–2285.

Song, J. H., Murphy, R. J., Narayan, R., and Davies, G. B. H. 2009. Biodegradable and compostable alternatives to conventional plastics. *Philos. Trans. R. Soc. B* 364, 2127–2139.

Stein, T. M., Gordon, S. H., and Greene, R. V. (1999) Amino acids as plasticizers II. Use of quantitative structure-property relationships to predict the behaviour of mono ammonium mono carboxylate plasticizers in starch-glycerol blends. *Carbohydr. Polym.* 39, 7–16.

Stenhouse, P. J., Ratto, J. A., and Schneider, N. S. (1997) Structure and properties of starch/poly(ethylene-co-vinyl alcohol) blown films. *J. Appl. Polym. Sci.* 64, 2613–2622.

St-Pierre, N., Favis, B. D., Ramsay, B. A., Ramsay, J. A., and Verhoogt, H. (1997) Processing and characterization of thermoplastic starch/polyethylene blends. *Polymer* 38, 647–655.

Svegmark, K. and Hermansson, A. M. (1991) Distribution of amylose and amylopectin in potato starch paste—Effects of heating and shearing. *Food Struct.* 10, 117–129.

Takaya, T., Sano, C., and Nishinari, K. (2000) Thermal studies on the gelatinisation and retrogradation of heat-moisture treated starch. *Carbohydr. Polym.* 41, 97–100.

Tatarka, P. D. and Cunningham, R. L. (1998) Properties of protective loose-fill foams. *J. Appl. Polym. Sci.* 67, 1157–1176.

Tattiyakul, J. and Rao, M. A. (2000) Rheological behaviour of cross-linked waxy maize starch dispersions during and after heating. *Carbohydr. Polym.* 43, 215–222.

Uyama, Y., Kato, K., and Ikada, Y. (1998) Surface modification of polymers by grafting. *Adv. Polym. Sci.* 137, 1–39.

Van Soest, J. J. G., Hulleman, S. H. D., De Wit, D., and Vliegethart, J. F. G. (1996) Crystallinity in starch bioplastics. *Ind. Crops Prod.* 5, 11–22.

Van Soest, J. J. G. and Knooren, N. (1997) Influence of glycerol and water content on the structure and properties of extruded starch plastic sheets during aging. *J. Appl. Polym. Sci.* 64, 1411–1422.

Villar, M. A., Thomas, E. L., and Armstrong, R. C. (1995) Rheological properties of thermoplastic starch and starch/poly(ethylene-co-vinyl alcohol) blends. *Polymer* 36, 1869–1876.

Walia, P. S., Lawton, J. W., and Shogren, R. L. (2002) Mechanical properties of thermoplastic starch/poly(hydroxy ester ether) blends: Effect of moisture during and after processing. *J. Appl. Ploym. Sci.* 84, 121–131.

Wang, Y., Gao, Y.-X., Song, J., Bonin, M., Guo, M., and Murphy, R. (2010) Assessment of technical and environmental performances of wheat based foams in thermal packaging applications. *J. Pack. Technol. Sci.* 23, 363–382.

White, R. J., Budarin, V. L., and Clark, J. H. 2008. Tuneable Mesoporous Materials from α-D-Polysaccharides. *ChemSusChem* 1, 408–411.

White, R. J., Budarin, V., Luque, R., Clark, J. H., and Macquarriea, D. J. 2009. Tuneable porous carbonaceous materials from renewable resources. *Chem. Soc. Rev.* 38, 3401–3418.

Willett, J. L. (1994) Mechanical properties of LDPE/granular starch composites. *J. Appl. Polym. Sci.* 54, 1685–1695.

Willett, J. L. and Doane, W. M. (2002) Effect of moisture content on tensile properties of starch/poly(hydroxyester ether) composite materials. *Polymer* 43, 4413–4420.

Willett, J. L. and Shogren, R. L. (2002) Processing and properties of extruded starch/polymer foams. *Polymer* 43, 5935–5947.

Xu, W. and Doane, W. M. (1997) Biodegradable polyester and natural polymer compositions and expanded articles therefrom. US Patent 5,665,786.

Xu, W. and Doane, W. M. (1998) Biodegradable polyester and natural polymer compositions and expanded articles therefrom. US Patent 5,854,345.

Zeleznak, K. J. and Hoseney, R. C. (1987) The glass transition in starch. *Cereal Chem.* 64, 121–124.

Zhou, G., Willett, J. L., and Carriere, C. J. (2001) Effect of starch content on viscosity of starch-filled poly(hydroxy ester ether) composites. *Polym. Eng. Sci.* 41, 1365–1372.

Zhou, J., Song, J., and Parker, R. (2006) Structure and properties of starch-based foams prepared by microwave heating from extruded pellets. *Carbohydr. Polym.* 63, 466–475.

Zhou, J., Song, J., and Parker, R. (2007) Microwave-assisted moulding using expandable extruded pellets from wheat fours and starch. *Carbohydr. Polym.* 69, 445–454.

Zobel, H. F. (1988) Molecules to granules—A comprehensive starch review. *Starch/Starke* 40, 44–50.

Zuchowska, D., Steller, R., and Meissner, W. (1998) Structure and properties of degradable polyolefin-starch blends. *Polym. Degrad. Stab.* 60, 471–480.

8 Bio-Based Aerogels by Supercritical CO$_2$

Ciro Siviello and Domenico Larobina

CONTENTS

8.1 INTRODUCTION: BACKGROUND

Aerogels are a class of solid nano-porous materials characterized by open pore structure and a very high specific surface area. The term *aerogel* was originally proposed by Kistler (1931, 1932) to describe gels subjected to supercritical drying condition, where the liquid within the skeleton of the gel was, by a suitable process, replaced by air. In his original works, Kistler used supercritical water condition in autoclave to obtain aerogels from oxide hydrogels. However, at the high pressure and temperature used to reach the supercritical water condition, the stability of the gel was impaired by the mineralizing effect of water. To overcome this problem, Kistler was then forced to exchange water with solvents having milder critical value, such as alcohol or ether. Successive studies on the topic, by Teichner and coworkers (Nicolaon and Teichner 1968; Teichner et al. 1976), have enlarged the technique of synthesis allowing the oxide gel to be formed directly in an alcoholic medium; consequently, exchange of solvent was no longer necessary. It is worth noticing that the earlier techniques where all ending to a process of supercritical drying involving an alcohol, whose critical condition are in the range $T_c = 512–514$ K and $P_c = 6–8$ MPa.

The extension of supercritical drying process to more gentile organic gels required then an additional step in the preparation, in order to further reduce the supercritical condition. Thanks to the mild critical temperature and pressure ($T_c = 304.1$ K $P_c = 7.38$ MPa) along with a sufficient solvent power, CO$_2$ finally replaced alcohol in the preparation of an aerogel. Supercritical CO$_2$ is also attractive for its inert, nontoxic, and inexpensive properties, which make the organic aerogel suitable for biomedical application.

To make an aerogel by supercritical carbon dioxide, a series of steps must be accomplished. Focusing on bio-based aerogels, such steps include (1) gelation of a polymer solution, where the solvent could be either CO$_2$ miscible (i.e., ethanol and acetone) or CO$_2$ immiscible (i.e., water); (2) eventual replacement of the CO$_2$ immiscible solvent with a miscible one (i.e., replacing the water with alcohol); (3) displacement of the solvent with supercritical CO$_2$ (drying step); and (4) at the last, depressurization to atmospheric pressure. To fix the idea, we will assume in the following that the CO_2 miscible solvent to be displaced in step (3) is an alcohol; that is, the gel to be dried is actually an alcogel.

This chapter is arranged as follows: in Section 8.2, we introduce the basic principles of supercritical drying. Due to their importance in the selection of proper process condition, we will dwell on both thermodynamics and kinetics of aerogel preparation. In Section 8.3, we will describe the technologies in use to prepare aerogel by supercritical carbon dioxide (sc-CO_2). We then report in Section 8.4 a brief summary of most of the bio-based aerogels produced by sc-CO_2 extraction during the last decay. Finally, in Section 8.5, we will review the main properties and applications of bio-based aerogels.

8.2 BASIC PRINCIPLES OF SUPERCRITICAL DRYING

Traditional drying procedures, such as air drying, are not suitable to form an aerogel since they are unable to avoid the solid structure collapsing. Collapsing is a consequence of formation of capillary forces inside the material during the solvent removal. In fact, during air drying, a liquid–vapor meniscus sets in determining the instauration of a capillary pressure gradient in the pore walls, which, in turn, produces the collapse of the gel structure. Monolithic and crack-free aerogels can only be obtained by removing the solvent in the absence of even small capillary forces.

To give an idea of the capillary pressure (P_c) that develops inside a pore, it must be considered that:

$$P_c = \frac{-\gamma_{lv}}{\left(r_p - \delta\right)} \qquad (8.1)$$

where:

γ_{lv} is the surface tension of the liquid in the pore
δ is the thickness of a surface adsorbed layer (see Brinker and Scherer 1990)
r_p is the pore radius, expressed by the equation

$$r_p = \frac{2V_p}{S_p} \qquad (8.2)$$

where:

V_p and S_p are the pore volume and the surface area, respectively

From Equation 8.1, we can deduce in case of nonoptimizing CO_2 drying condition, for instance, when the surface tension takes the value of $\gamma_{lv} \approx 73 \, mN/m$, that a small pore size (~4 μm) can contribute to a capillary pressure of about 20 kPa.

Actually, in the air-drying process, capillary tension may reach the exceptional value of 100–200 MPa (Smith et al. 1995), causing consequent shrinkage and cracking phenomena. Approximately, we can state that when the pore size is smaller than 20 nm, there exists a considerable liquid tension in the pore, which can cause material fracture; instead, when the pore size is larger than 20 nm, the shrinkage will be less, and we have a lower probability that cracking will occur (Venkateswara Rao and Haranath 1999). On the other hand, in some cases, small pore size gels (of about 4 nm) can be easier to dry than larger pore as explained by the theory of cavitation, which consists in a homogeneous nucleation of bubbles of vapor.

It has been demonstrated (Scherer and Smith 1995) that cavitation could have an important influence on the stresses developed during the drying step. This is due to vapor bubbles formation that tends to reduce the capillary tension inside the liquid. This translates into the possibility of drying monolithic gels also at unusually high rates if the drying is conducted at conditions that permit the internal nucleation of bubbles (Sarkar et al. 1994). Moreover, it has been shown that cavitation can take place only if pores have a characteristic size, which is small enough to give rise to a capillary tension greater than a certain value, as reported by the classical nucleation theory, and bubbles can attain growth, by reaching macroscopic dimensions, only after the overcoming of the critical point of drying (Scherer and Smith 1995).

To overcome the above difficulties, related mainly to capillary force, supercritical drying, and in particular supercritical CO$_2$, has been developed. This procedure is able to prevent the problems encountered in traditional drying methods, previously mentioned, because the use of a supercritical condition permits to annihilate the liquid surface tension responsible of pores collapsing. In supercritical conditions, there is no liquid–vapor interface and, thus, we have a null capillary pressure, as can be seen by Equation 8.1 (Scherer 1992).

Supercritical drying consists of essentially three stages: (1) pressurization/heating, (2) extraction, and (3) depressurization/cooling. During these stages, both thermodynamics and kinetics are at play. Thus, several parameters must be controlled (e.g., pressure, temperature, drying time, and ramp rate of pressurization/depressurization, to name a few) to have the desired final product. In fact, if the drying process is not conducted in a correct way, cracking can also occur during supercritical CO$_2$ extraction. For example, if the desired temperature is increased too quickly, a thermal expansion of the liquid, yet entrapped into the gel sample, induces a stretching and a consequent breaking of the nanostructure.

To fix the idea we will consider the sc-CO$_2$ drying of an alcogel. For this system, we report on Figure 8.1 the thermodynamic maps of CO$_2$ (a) and CO$_2$-ethanol (b) system for some consideration.

To obtain the supercritical conditions, the liquefied CO$_2$ is pressurized above its critical pressure (73 bar) and, before entering the extractor, the temperature is heated above its critical temperature (31°C).

At pressure and temperature below the critical point, we find the presence of a two-phase region of vapor (CO$_2$) and liquid (ethanol or methanol). The presence of this vapor–liquid phase results in capillary forces that cause significant internal stresses and a consequent sample cracking. The depressurization phase then represents a critical step. In fact, entering the two-phase region during the depressurization of the system can cause cracking phenomena. If the diffusion time is not long enough, the residual ethanol can form a liquid phase during the depressurization; as a consequence, we observe the evaporation of the liquid into the gas phase and this results in considerable capillary forces that are able to crack or damage the aerogel structure.

In Figure 8.1b is reported the vapor–liquid equilibrium diagram for the systems ethanol-CO$_2$. The three-dimensional surfaces in the temperature-pressure-composition graphic are obtained as interpolation of the literature data at six different temperatures (i.e., 30°C, 35°C, 40°C, 50°C, 55°C, and 60°C) (Chiehming et al. 1997; Yeo et al. 2000; Joung et al. 2001; Stievano and Elvassore 2005; Secuianu et al. 2008; Knez et al. 2008). Considering the evolution of the mixture composition inside the hydrogel during the supercritical drying, we can observe that at the beginning of the process, thus on the left-hand side of the diagram, the CO$_2$-solvent mixture contains a large ethanol content. As the CO$_2$ diffuses into the gel sample, its molar fraction progressively increases; thus, the mixture composition moves to the right-hand side of the equilibrium diagram where the mixture critical point is located to higher value of pressure (at a fixed temperature). Only when drying is conducted in these conditions we have a supercritical CO$_2$-solvent mixture throughout the process.

In most cases, the gelation of the polymer is obtained in a water solution. This requires an intermediate step before drying, in which a CO$_2$-soluble solvent, usually an organic polar solvent (generally ethanol, methanol, or acetone), replaces the CO$_2$-immiscible water. To prevent (or minimize) the sample shrinkage, naturally arising as a consequence of the different solvent power between alcohol and water, gel samples are gradually dehydrated by immersion in a set of hydro-alcoholic solutions of increasing alcohol concentration (e.g., 10%, 30%, 50%, 70%, 90%, and 100%).

Among the different operating parameters, the time of fluid exchange and the time of drying play a fundamental role on aerogel formation. This consideration implies the fundamental importance of an adequate knowledge of the diffusion processes that take place during the drying step.

In the supercritical drying, we can identify two main steps involving a mass transfer of sc-CO$_2$ and gel solvent to and from the pores of the wet gel. In a first step, the process is predominantly

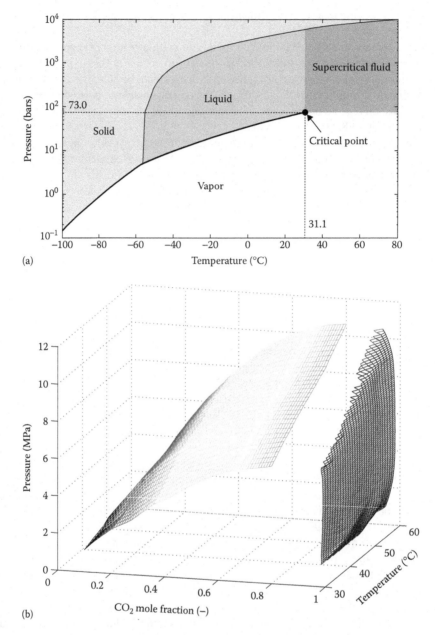

FIGURE 8.1 Phase diagram of (a) pure CO_2 and (b) ethanol-CO_2 binary system.

characterized by a high dissolution of sc-CO_2 in the liquid gel solvent; then, the formation of an expanded liquid causes the emission of the excess liquid volume from the gel network. In a second step, instead, we observe a progressive increasing of CO_2 content in the pores gel until the attainment of the supercritical conditions for the fluid mixture present in the network pores avoiding any intermediate vapor–liquid transition. In this way, we have an absence of surface tensions in the system overcoming the problem of pores collapsing during the solvent extraction. Another advantage of this procedure is that in the supercritical state, the diffusion coefficient D, which value is typically of the order of 10^{-7} m^2 s^{-1}, is about two orders of magnitude greater than that of normal liquids, offering a significant reduction in drying time.

TABLE 8.1

CO$_2$ Diffusion Coefficients in Ethanol at Different Temperatures

T(K)	$D \times 10^9 \left(\text{m}^2\text{s}^{-1}\right)$
298	4.11
308	4.70
313	5.05
323	5.71
333	6.51

In a study conducted by Snijder and coworkers (1995), an empirical relationship, which describes the temperature dependence for the CO$_2$ diffusion coefficient in ethanol, is reported:

$$D_{CO_2} / \left(\text{m}^2 \cdot \text{s}^{-1}\right) = 336.5 \times 10^{-9} \exp\left(\frac{-1314.7}{T/K}\right) \quad (8.3)$$

It has been observed that this equation models faithfully the experimental evidence, provided it is applied in the temperature range 298–333 K for ethanol. Table 8.1 reports a prospectus of the CO$_2$ diffusion coefficients values in ethanol at different temperatures.

It is important that sc-CO$_2$ is not released too quickly from the extraction apparatus; otherwise, the fluid inside the sample will not be able to diffuse out of the gel network at a comparable rate causing possible gel swelling or breaking. Or more, cracking of the aerogel could appear also if the diffusion duration is not long enough because of a possible incomplete diffusion of gas from the core of the aerogel sample.

8.3 SUPERCRITICAL DRYING TECHNIQUES

There are two different techniques of supercritical drying: *dynamic supercritical drying*, in which the solvent extraction is done by a continuous flow of the supercritical fluid, and *static supercritical drying*. In the first case (Rangarajan and Lira 1991; García-González et al. 2011), the wet gel is loaded in a thermo-stated chamber (an autoclave or extractor) and subjected to a continuous flow of CO$_2$ at a temperature and a pressure above its critical point. The supercritical fluid escapes from the reactor enriched in gel solvent, thus allowing the extraction of it from the gel pores. After a certain time under sc-CO$_2$ continuous flow, the system is depressurized to atmospheric pressure to obtain the aerogel. The duration of the diffusion also affects the quality of the final aerogel samples which, depending on it, can vary from cracked (in the case of short diffusion times) to crack-free aerogels (in the case of long diffusion times). Moreover, it is generally difficult to predict the exact duration of the drying step, because in most cases, the diffusion coefficients of the liquid in the sample are not known.

In the second case (static supercritical drying), instead, the wet gel is first exposed to sc-CO$_2$ for a certain period of time (typically 6–12 h) (Placin et al. 2000; Tan et al. 2001) and, successively, sc-CO$_2$ flow is applied in order to replace the environment surrounding the gel (a CO$_2$ enriched in gel solvent) with pure sc-CO$_2$. Finally, the system is depressurized until atmospheric pressure and the temperature is decreased until room temperature. Batch type process seems to lead to good results in some cases, provided a sufficient time is allowed for diffusion. However, it has been demonstrated (Czakkel et al. 2013) that no considerable differences between the two techniques are observed on the final sample morphology if they are both run in optimized conditions: the final aerogel structures are very similar in the two cases and both shrinkage and macroscopic density are not influenced by the process type.

About the depressurization step, the importance of maintaining the depressurization rate at a fixed and constant value to obtain restrained shrinkages and reproducible aerogel samples should be remarked (Amaral-Labat et al. 2012).

To obtain a monolithic aerogel through the supercritical drying, the alcogel is put in the extractor, which is heated and pressurized until it reaches the CO_2 supercritical conditions. It is of fundamental importance that this step is conducted by never overcoming the liquid–vapor equilibrium curve (Figure 8.2). The sample is left under these conditions for an appropriate time in both batch and continuous flow conditions. Then the system is depressurized by venting the gas from the extractor; at the same time, the gas inside the gel expands and flows out of the gel network. For this reason, there is a limit to the depressurization rate to prevent sample cracking or damaging. In fact, if the depressurization is too fast, the gas inside the gel sample has no sufficient time for escaping out of the network; accordingly, it expands causing network swelling.

A lot of studies have been conducted in order to analyze the effects of the process parameters, which characterize the different process steps, on the characteristics of the final gels and, in particular, on its final nanostructure. Nevertheless, the structural evolution of the gels during the drying step is poorly known. In this respect, an interesting work, recently conducted by Czakkel and coworkers (2013), presents an *in situ* small-angle X-ray scattering (SAXS) study of the changes that occur in the structure of a gel networks (in particular, in the case of a soft resorcinol-formaldehyde polymer gel) during supercritical drying under various experimental conditions. By observing *in situ*, the effect of process parameters on the evolution of the gel structure, the study investigated the influence of pressure, temperature, ramp rate on both pressurization washing and depressurization phases, as well as the influence of the choice of the process type: batch-type or continuous flow.

The results reveal clearly that the pressurization step is independent of the ramp rate of CO_2 and very similar network structures are obtained. By contrast, the critical step is the depressurization one. The work remarks the importance of imposing low depressurization rate: after slow depressurization, samples undergo a very small shrinkage; instead, rates higher than about 4 bar/min induce higher and irreversible network shrinkage. This densification preferentially regards the larger pores, whereas the smaller pores are less affected. Faster depressurization (20 bar/min) causes bubbles to develop in the final network and is able to increase the macroscopic density of the gel by about

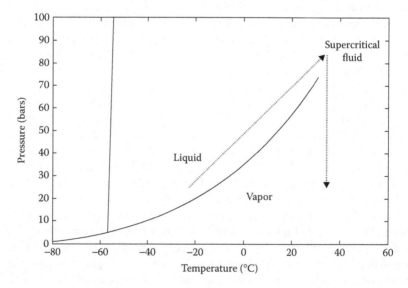

FIGURE 8.2 Example of supercritical drying path.

2 orders of magnitude. If the pressurization process is too fast, the fluid inside the gel does not have a sufficient time to flow out of the network; as a consequence, it expands within the gel causing cracking and network collapsing.

At large **k** values, that is, small characteristic length (where **k** is the scattering wave vector, see Equation 8.15) the effect of depressurization rate appears similar, with essentially a sigmoidal shape in the range 100–1 atm. However, at high rates of depressurization, the rapid upturn delay, appearing shifted toward lower pressure values. The situation changes for the small **k** values (i.e., large characteristic length). The slopes here are constant in the case of lower depressurization rate, while displaying a broad maximum for higher rates. This effect is a consequence of the nucleation and growth process occurring inside the gel and appears also when depressurization is initiated from the liquid state. We remind here that for the liquid CO$_2$ state, the fluid system enters the two-phase regime independently of the depressurization rate, and bubbles nucleation and growth occurs unavoidably.

Liquid CO$_2$ is also reported as an effective drying method, which is able to preserve the gel network structure, provided that the depressurization stage always starts from the CO$_2$ supercritical state. However, the faster solvent exchange of sc-CO$_2$ method, due to its higher rate of molecular diffusion, makes this method the generally preferred one (Czakkel et al. 2013).

A typical experimental apparatus for drying process can be schematically represented as in Figure 8.3.

As previously anticipated, from the Table 8.2, in which we reported some of the several examples of bio-based aerogel production, we can observe a discrepancy in operating conditions. In fact, we have seen that there are a lot of parameters to be considered in order to optimize the extraction process (temperature, pressure, process time, solvent exchange, etc.). Moreover, there are two aspects that must be taken into account to find the best supercritical drying conditions: sample shrinkage and solvent residues. In literature, shrinkage percentages between 5% and 20% are generally reported, but choosing the most appropriate process parameters lower shrinkages (0.3%–3%) and negligible solvent residues are possible.

An estimate of the volume shrinkage can be calculated as follows:

$$\text{Volume shrinkage}(\%) = \frac{V_0 - V}{V_0} \times 100\% \tag{8.4}$$

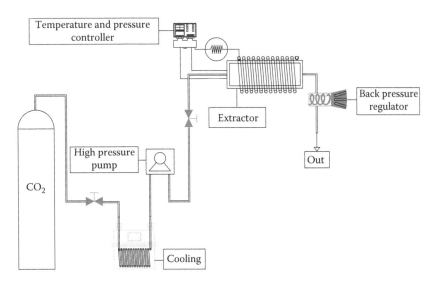

FIGURE 8.3 Schematic apparatus for CO$_2$ supercritical drying.

TABLE 8.2

Brief Summary of Most of the Bio-Based Aerogels Produced by Supercritical CO_2 Drying during the Last Decade

Materials	Operating Conditions	Morphology	Applications	Reference
Alginate	Flow rate of 0.6 kg/h, 150 bar, and 37°C	Small particle shrinkage, preservation of the hydrogel internal porous texture, and nano-pores diameters around 200 nm.	Nonsteroidal antiinflammatory drugs controlled delivery systems	Del Gaudio et al. 2013
Alginate	Flow rate of 100–200 g/min, 60 bar, and 40°C, for eight h in a 4L autoclave	Surface area of 680 m² g⁻¹, pore volume 4 cm³ g⁻¹, and pore radius of 15 nm.	Potential biodegradable, lightweight, and oil-absorptive materials	Alnaief et al. 2011
Acidic and ionotropic alginate	Polaron 3100 apparatus, 74 bar, and 31.5°C	Slight density and a porosity higher than 98%.	—	Valentin et al. 2006
Ca-alginate	Polaron 3100 apparatus, 74 bar, and 31°C	Preservation of the fibrillar structure of the hydrogels, with an open network of fibrils 200–300 nm long with 5–8 nm diameter.	—	Robitzer et al. 2008
Chitosan	Polaron 3100 apparatus, 74 bar, and 31.5°C	The material has a porous and fibrous character.	—	Valentin et al. 2003
Chitosan-L-glutamic	100 (200) bar and 40°C	The struts are formed at the junction of the micro-porous cells and oriented randomly in all direction. The pore dimensions are nonuniform and randomly dispersed in the polymer matrix.	Biomedical applications	Singh et al. 2009
Genipin-cross-linked chitosan	200 bar and 40°C	A poly-phasic micro-porous structure having nonuniform pore dimensions with thin walls.	Scaffolds for tissue engineering applications	Rinki and Dutta 2010
Chitosan–montmorillonite bio-based aerogel	Polaron 3100 apparatus, 73.8 bar, and 31.5°C	An open framework of fibrillar chitosan–montmorillonite, in which chitosan is petrified by the clay platelets.	Adsorption, sensing, catalysis, or drug delivery systems	Ennajih et al. 2012
Beta-glucan	One h at 15 MPa and 40°C, and four h in sc-CO_2 flow at 1 L/min	Aerogels of 5% β-glucan had a significantly lower pore volume ($p \leq .05$) compared to 7% β-glucan aerogels, while 6% aerogels were similar to both 5% and 7% aerogels. Pore diameters similar between the different concentrations, and at a relative pressure of $p_0/p = .35$; the pore diameters ranged from 2.7 to 2.8 nm.	Application as delivery vehicles for nutraceuticals or pharmaceuticals	Comin et al. 2012

(Continued)

TABLE 8.2 (Continued)
Brief Summary of Most of the Bio-Based Aerogels Produced by Supercritical CO$_2$ Drying during the Last Decade

Materials	Operating Conditions	Morphology	Applications	Reference
Peptide aerogels comprising self-assembling nanofibrils	• Two h at 90 bar and 40°C, and two to three min in sc-CO$_2$ flow, degassing at 20°C for over five min • 12 h (opposed to two h) at 90 bar and 40°C, and degassing for one h (opposed to five min)	Collapse of peptide organo-gel networks, presumably because of peptide insolubility in carbon dioxide. Freeze drying of peptide hydrogels proved to be a more efficient method of removing the solvent without destroying the self-assembled fibrillar network, leading to a microscopic aligned lamellar structure consisting of thousands of stacked peptide nano-fibrils.	Catalysis, sensing, separation and filtration applications, tissue engineering, and hybrid biopolymer	Scanlon et al. 2007
Cellulose	Autoclave, 100 bar, and 40°C, for two h	Macro-pore system having diameters of up to 100 mm and a meso-porous substructure with pore diameters in the nanometer range.	—	Liebner et al. 2010
Chitin	Different conditions of pressure and temperature: 80–300 bar and 40–80°C; the system initially was kept at static conditions for 90 min, and then in continuous flow of CO$_2$ for two h with a mass flow of CO$_2$ of 0.42 ± 0.09 g/min.	Porosity higher than 85%, surface area greater than 220 m^2 g^{-1}, and low density (<0.23 g cm^{-3}). The surface area and the porosity increase (smaller pore size) by increasing pressure or decreasing temperature.	Carbon aerogel precursors and thermal and sound insulation	Tsioptsias et al. 2009
Protein-based aerogels	Continuous flow of sc-CO$_2$ at 40°C, 115–120 bar and 150–250 L/h during four to six h	Meso-porous structure with pore diameter ≤27.4 nm.	Drug carriers for life science applications	Betz et al. 2012
Cellulose	1 L autoclave, at 85 bars, 40°C and 5 kg CO$_2$/h flow for four h. Slowly (0.1 bar/min) and isothermally (40°C) depressurization	Low-density materials (from 0.25 to 0.85 cm^3 g^{-1}), specific surface areas between 140 and 250 m^2 g^{-1} and characteristic pore sizes between 13 and 25 nm, with specific porous volumes of about 3.30 cm^3 g^{-1}.	*Green* materials for thermal insulation application	Fischer et al. 2006
Cellulose	Comparison of ethanol and methanol alcogel, at 200 bar and 40°C. Depressurization to atmospheric pressure at a controlled pressure-time gradient	Densities in a range from 0.014 to 0.5 g cm^{-3} and internal surface areas ranging from 50 to 420 m^2 g^{-1}. Pores in the 1–100 nm regime.	Insulating biodegradable material	Innerlohinger et al. 2006

(Continued)

TABLE 8.2 (Continued)
Brief Summary of Most of the Bio-Based Aerogels Produced by Supercritical CO$_2$ Drying during the Last Decade

Materials	Operating Conditions	Morphology	Applications	Reference
Starch	50°C and 15 MPa in a 50 ml autoclave	Porous and homogeneous structure composed of fibers with a diameter range of 30–50 nm entangled with each other to form an interconnected network. Broad pore size distribution with a maximum around 35.5 nm.	Template for hierarchically ordered TiO$_2$ networks preparation	Miao et al. 2008
Polylactic acid	200 bar and 33°C for four h, with depressurization time of 10 min up to atmospheric pressure	Nano-fibrillar internal structure (mean diameter lower than 200 nm) and highly porous structure. Porosity of about 95%. Grade of interconnection of about 100% and large surface areas (about 45 m^2 g^{-1}).	Scaffolds for tissue replacement	Reverchon et al. 2008
Chitosan	200 bar and 35°C	Uniform nano-metric network with a porosity higher than 91% and a mean diameter of the fiber of about 50 nm.	Nano-porous structures for tissue engineering applications	Cardea et al. 2010
Chitosan	200 bar and 40°C for two h, initially in static conditions and then in continuous CO$_2$ flow	Poly-phasic micro-porous structure with a nonuniform distribution of pore dimensions. Surface area of 19 m^2 g^{-1}, total pore volume of 0.076096 cm^3 g^{-1}, average pore diameter of 13.3347 nm, and median pore diameter of 35.0967 nm.	Engineered tissue systems, scaffolds, and biomedical applications	Rinki et al. 2009
Alginate	Comparison of ethanol and acetone: • Ethanol: equilibration batch for five h and sc-CO$_2$ flow of 0.6 kg/h for 90 min at 38°C and 150 bar • Acetone: equilibration batch for five h and sc-CO$_2$ flow of 0.6 kg/h for 90 min at 38°C and 100 bar	Uniform and homogeneous internal nanostructure.	Controlled delivery systems for pharmaceutical application or as scaffolds in tissue engineering	Della Porta et al. 2013

where:

V_0 is the volume of wet hydrogel

V is the volume of the aerogel after drying

The operating conditions must be chosen so as to be sufficiently above the supercritical condition of the CO_2-solvent mixture throughout the duration of the extraction (Della Porta et al. 2013). In this respect, it is necessary to consider the evolution of the CO_2-solvent mixtures, which are forming during the process.

8.4 MATERIALS

In the literature, it is possible to find several examples of aerogel production from bio-based materials. The principle of aerogel production is the same in each case, but sometimes we can observe a discrepancy in operating conditions. Some examples, their eventual applications, and the authors who conducted the studies are resumed in Table 8.2.

8.5 PROPERTIES AND APPLICATIONS

Since their production, aerogels have been used for a large number of practical applications, such as adsorbent material, porous electrode, heat and acoustic insulator, catalyst support, controlled drug delivery, and ultrafiltration membrane. Their success is due to the ability of combining properties of lightweight, active sorption/release with a good mechanical behavior. Earlier studies were focused mainly to inorganic aerogels (SiO_2, TiO_2, ZrO_2, etc.). However, in the last few decades, an increasing interest has been reserved to bio-based aerogels and, in particular, to chemically or physically cross-linked polysaccharides (including cellulose, hemicellulose, marine polysaccharides, and starch). Compared to inorganic aerogels, these natural materials have certainly the drawbacks of lower mechanical properties, limited processability, a greater heterogeneity, and a certain variability of properties from batch to batch; nevertheless, their additional benefits and new possible applications surpass widely their disadvantages.

Subsequently, we report a summary of the principal characteristics along with the main applications of this interesting class of material.

8.5.1 TEXTURE AND STRUCTURE

To have an adequate control of aerogel properties and fully exploit the potentiality of these materials, it is necessary to have a thorough knowledge of the structure–property relationships. On this respect, the key structural parameters that characterize an aerogel are the total fraction of solid and porous phases, the pores interconnectivity, and the characteristics of the pore–solid interface. Actually, majority of structural considerations reported below refers to inorganic aerogels; however, their conclusion can be applied to aerogels in general and, thus, also to bio-based aerogels.

Undoubtedly, the main characteristic of an aerogel is its volume fraction, or equivalently its apparent density (ρ_{app}). Obviously, the apparent density can only give a mean value of the internal porosity; actually, internal texture can range from micro to macro-porosity, presenting different specific area. Porosity distribution would be a more appropriate parameter to identify aerogel with equal ρ_{app}. Apparent density can be easily determined by mercury porosimetry (Pirard et al. 1995). Hg porosimetry is based on the physical principle according to which a nonreactive and nonwetting liquid will not penetrate pores until a sufficient pressure will be applied against the opposing force of the liquid's surface tension. Hence, the nonwetting property of mercury, combined with its high surface tension, makes it the most qualified liquid for this use.

From the external pressure needed to force the liquid into a pore, it is possible to determine the pore size. In fact, by considering a cylindrical pore of radius R_{pore}, the external pressure p_{ext}

required to penetrate the pore can be expressed by the Washburn equation (Washburn 1921) as follows:

$$p_{ext} = -\frac{2\gamma \times \cos(\theta)}{R_{pore}} \qquad (8.5)$$

where:

γ is the mercury surface tension
θ the contact angle at the mercury–pore interface

Typically, the contact angle between mercury and most solids is in the range between 135° and 142°; thus, an average value of 140° can be safely assumed. By estimating the value of the surface tension of mercury at 25°C as equal to 0.485 N m^{-1}, the pore size of the aerogels can directly related to the p_{ext}. Operatively, by measuring the applied external pressure p_{ext} and the volume of mercury that penetrates into the material pores (by a mercury penetrometer), the volume of pores in the corresponding size class is determined. It is worth noticing that the pressurization rate must be carefully chosen in such a way that the aerogel is at equilibrium at each pressure imposed, in order to prevent pores size underestimation errors.

The main drawback in the use of Hg porosimetry is related to the densification of the material (aerogel collapse) (Pirard et al. 1998). The pressures involved, which are usually not critical for stiff porous materials, could be critical in the case of aerogels producing a collapse of the pores. In fact, the compression moduli of aerogels, with porosity higher than 80%, are typically included in the range 0.1–50 MPa. This effect could lead to mistakes in attributing the consequent change in volume (due to compression) to a wide contribution of macro-pores (Broecker et al. 1986). Compression can be partially or totally irreversible, depending on the aerogel type; for instance, unmodified silica-based aerogels are irreversibly deformed. Two ways are available when compression effect occurs: (1) exploit this effect and determine the mechanical bulk modulus of the aerogel and (2) try to mathematically correct for the related contribution. In the latter case, we refer to the specialized literature (Pirard et al. 2002; Job et al. 2006).

Gas sorption porosimetry (Gregg and Sing 1982; Condon 2006) is another effective technique for aerogels structure and texture characterization. It provides reliable information on porosity (also for pores sizes in the range of 0.3–100 nm), pore volume, and surface area (to below 0.01 m^2 g^{-1}). The technique is based on the interaction between a gas (generally N$_2$ at 77.3 K for aerogels) and the adsorbent, the aerogel in the present case. Measurement consists on the evaluation of the volume V of adsorbed gas at various pressures lower than the saturation pressure p_s. The volume V_m of a monolayer adsorbed is then theoretically deduced.

To obtain a sorption isotherm, the amount of absorbed (or released) gas is quantified by a volumetric or a gravimetric method. In the first case, the amount absorbed (or released) is derived from the measure of the pressure increase (or decrease) induced in the gas, while in the gravimetric method, the weight variation of the adsorbent is directly measured. When an isotherm is recorded, the assumption of the equilibrium between the adsorbent (the gas phase) and the adsorbate (the solid phase) at each point of the curve is made. However, for the aerogels, which have a very high specific pore volume, it has been shown (Reichenauer and Scherer 2000, 2001a, b) that this assumption may not always be considered rigorously true. Hence the isotherm could be mistaken in the micro-pores and meso-pores range.

Several theoretical analysis have been developed in order to extract quantitative information from isotherm curves; the most common among these are the density functional theory for micro-pores and meso-pores, the Dubinin–Radushkevich equation for the extraction of characteristic parameters on micro-pores, the t-plot and the αs-plot for the separation of surface area located in micro- and nonmicro-pores, the Brunauer-Emmet-Teller (BET) method to calculate the so-called BET surface area, and the Barret-Joyner-Halenda relationship that provides access to the meso-pore size distribution.

For the specific surface area calculation, the most popular evaluation method is the so-called BET method (developed by Brunauer, Emmett, and Teller; see Condon 2006). The BET model relates the volume adsorbed V_{ads} to the relative pressure p/p_0 as follows:

$$\frac{V_{ads}\left(p/p_0\right)}{V_{ml,BET}} = \left[\frac{p/p_0}{\left(1-p/p_0\right)}\right] \Big/ \left[\frac{1}{C}+\left(\frac{C-1}{C}\right)\left(\frac{p}{p_0}\right)\right] \tag{8.6}$$

$V_{ml,BET}$ and C are two parameters representing the specific monolayer capacity and the interaction of the adsorbent surface and the adsorbate, respectively. C can be written in terms of adsorption energy ΔE_{ads} as follows:

$$C \approx \exp\left(\frac{\Delta E_{ads}}{RT}\right) \tag{8.7}$$

where:
   ΔE_{ads} is the difference in adsorption energy in the first and the higher layers
   R is the gas constant
   T is the absolute temperature

The BET equation can be rewritten in a more convenient form as

$$\frac{\left(p/p_0\right)}{V_{ads}\left(1-p/p_0\right)} = \frac{1}{V_{ml,BET}C} + \frac{C-1}{V_{ml,BET}C}\cdot\left(p/p_0\right) \tag{8.8}$$

By plotting the left-hand side versus the reduced pressure p/p_0, the two parameters of the model can be obtained from the intercept $1/V_{ml,BET}C$ and the slope $C-1/V_{ml,BET}C$ of the linear fitting.

From the knowledge of $V_{ml,BET}$, it is possible to obtain the BET-specific surface area S_{BET} as:

$$S_{BET} = \frac{V_{ml,BET}S_{Gas}N_A}{V_{mole,STP}} \tag{8.9}$$

where:
   S_{Gas} is the area taken by a single adsorbate molecule
   N_A is the Avogadro constant
   $V_{mole,STP}$ is the molar volume at standard temperature and pressure ($=22,414$ cm^3[STP]mol^{-1})

It is worth noticing that the BET parameters obtained by the fitting of the isotherm strongly depend on the range of relative pressure used. Therefore, as a general recommendation, for aerogels that do not contain a significant amount of micro-pores, 0.05–0.25 relative pressure is preferred, while for aerogels with high micro-porosity, the range of relative pressures should be shifted to lower values, until the C value becomes positive. As a simple rule, the range of relative pressure should be chosen in the interval between the minimum accessible value and the maximum in the plot of $V_{ads}\left(1-p/p_0\right)$ versus $\left(p/p_0\right)$.

Once S_{BET} is known, the simplest method to estimate the average pores size, d, consists of applying the relationship

$$d = \frac{4V_{pore}}{S_{BET}} \tag{8.10}$$

where:
   V_{pore} is the specific total pores volume

In the case of meso- or macro-pore size in micro-porous samples, it is needed to replace V_{pore} with $\left(V_{pore} - V_{micro}\right)$ and S_{BET} with the external surface area.

To identify the total pores volume in micro- and meso-pores, Gurvich rule (Gregg and Sing 1982) assumes that all the gas adsorbed up to the saturation value (e.g., at a relative pressure of about 0.99) is present in a liquid-like state. By this assumption, the specific pore volume $V_{pore,\,micro+meso}$ is obtained from the amount of gas adsorbed V_{ads}, reported in $cm^3(STP)\ g^{-1}$, by

$$V_{pore,\,micro+meso}\left(cm^3 / g\right) = V_{ads}\left[cm^3\left(STP\right)/g\right]c \tag{8.11}$$

where:

c is a gas-dependent constant that converts gas phase into liquid phase volumes

From the specific volume, it is possible to derive the bulk density ρ of an aerogel (for nonmacro-porous samples with well-accessible pores) as

$$\frac{1}{\rho} = V_{tot} = V_{pore,\,micro+meso} + \frac{1}{\rho_s} \tag{8.12}$$

where:

ρ_s is the density of the solid phase
V_{tot} is the total mass specific volume

It may happen that the apparent density determined macroscopically (from the macroscopic dimensions and the total mass of the aerogel) results in a value that is higher than the density value determined by Equation 8.12; this discrepancy could arise from a residual of gases in the sample. Conversely, if the apparent density is lower than value derived by Equation 8.12, it could be that (1) the sample is not sufficiently equilibrated (Reichenauer et al. 2001a and 2001b), (2) part of porosity is not well accessible, (3) there may be unwanted macro-pores, and (4) a substantial amount of the adsorbate shows a density lower than the free liquid.

Other techniques to measure aerogel apparent density include helium pycnometry; thermoporometry can be found in the specialized literature (Aegerter et al. 2011).

Microscopy is an effective technique that allows observing the structure of an aerogel sample down to few nanometers. In the context of this technique, there are different types of measures that can be performed, such as light microscopy, atomic force microscopy, scanning electron microscopy, and transmission electron microscopy. However, the first two techniques, even if they can in principle be applied to aerogels, turn out to be critical and inadequate: light microscopy has a limited resolution (about 0.5 μm), and is not able to resolve the features of most aerogels; atomic force microscopy, on the contrary, is limited by the surface topology, which generally appears to be highly irregular at the aerogel fracture surface. The last two techniques (scanning electron microscopy and transmission electron microscopy) prove to be the most useful for structure aerogels investigation (Figure 8.4).

The scanning electron microscopy is characterized by a maximum resolution in the range of few nanometers, and it is used to observe the three-dimensional structure of the aerogel network. Instead, the transmission electron microscopy, which has a higher resolution, is used to better analyze the substructure of the particles, along with the presence of eventual additional phases incorporated in it.

Connected with aerogel porous structure is its characteristic fractal geometrical architecture (Mandelbrot 1977). Literature distinguishes between mass fractals, M, and surface fractals, A, dimension. If we consider a sphere of radius R with its center coincident with a random point in the aerogel network, M and A are connected to the radius and the related fractal dimension by the following laws:

$$M \approx R^f \quad \text{and} \quad A \approx R^{fs} \tag{8.13}$$

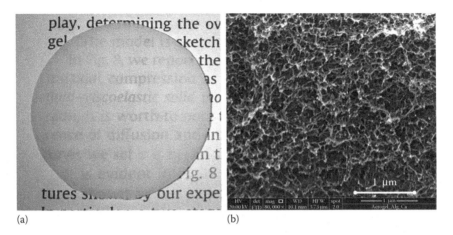

(a)   (b)

FIGURE 8.4 Ca-alginate aerogel obtained by CO$_2$ supercritical drying: (a) photograph showing the aerogel semi-transparency and (b) scanning electron micrograph showing its characteristic nano-porous structure.

The fractal dimension f can be experimentally determined by SAXS or through adsorption experiments with various molecules of different cross-sectional area. SAXS is a quantitative, nondestructive technique for structural analysis. Due to its noninvasive characteristic, SAXS can also be applied for *in situ* study of aerogel structural changes, for example, during aging or drying (Czakkel et al. 2013).

In the case of scattering experiments, the scattered intensity $I(\mathbf{k})$ can be expressed by a Porod type law

$$I(\mathbf{k}) \approx \mathbf{k}^{-2f+fs} \tag{8.14}$$

where the wave vector $\mathbf{k}$ is defined, at a scattering angle θ and an X-ray wavelength λ as

$$\mathbf{k} = \frac{4\pi}{\lambda \sin(\theta/2)} \tag{8.15}$$

Instead, in the case of adsorption experiments method, the surface fractal dimension f_s can be determined referring to a law of the type

$$w \approx \sigma^{-\left(f_{s,a}/2\right)} \tag{8.16}$$

where:
  w is the adsorbed mass of an adsorbate
  σ is its cross-sectional area

Effectively, an aerogel can be regarded as a fractal only in a limited range of R and approximately in the range 1–50 nm. It should be remarked that the supercritical drying method affects mainly the larger pores, which have dimensions larger than the fractal scale. Hence, from this point of view, aerogels not differ much from xerogels. In Table 8.3, the principal microstructural properties of some polysaccharides of interest are resumed.

8.5.2 Mechanical Properties

As stated in Section 8.5, practical applications of bio-based aerogels are often limited due to their poor mechanical properties (fragility and hydrophilicity), even if, under certain conditions, they can reach mechanical properties that are comparable to the inorganic ones.

TABLE 8.3

Typical Microstructural Data for Some Polysaccharide Aerogels of Interest

Aerogel	Meso-Pore Volume (cm³g⁻¹)	Surface Area (m² g⁻¹)	Reference
Ca-alginate	1.16	570	Robitzer et al. 2011b
Cu-alginate	2.2	680	Robitzer et al. 2011a
Ba-alginate	2.6	376	Mallepally et al. 2013
β-chitin	1.0	560	Robitzer et al. 2011a
Chitosan from α-chitin	0.5	330	Robitzer et al. 2011a
Nanofibrillated cellulose	–	250	Hoepfner et al. 2008
Starch	0.37	90	Mehling et al. 2009
Barley β-glucan	0.8	166	Comin et al. 2012
Cellulose monolith	–	289	Tan et al. 2001
k-carrageenan	0.76	200	Quignard et al. 2008

Typical mechanical characterization methods on aerogels are quasi-static, dynamic, and fatigue tests. All tests are aimed to suitably simulate the actual service conditions. Usual configurations include compression, bending, tension, torsion, and multiaxial loading conditions. However, owing to the usual brittleness of natural aerogels, which makes difficult to grip the specimens, tension and torsion tests are seldom conducted. In these cases, compression and flexural tests are generally preferred. According to the current regulations, cylindrical specimens are generally adopted for compression tests (ASTM D695). It is important, in the preparation of specimens, to obtain samples with smooth surfaces and free of defects, thus avoiding to impair the accuracy of the measurements. Moreover, the cylindrical end surfaces have to be perfectly parallel to one another, since a small misalignment could also result in significant measurement errors.

When bio-based aerogel are used at service temperature (T_{serv}) lower than T_g, physical aging needs to be properly characterized in order to avoid device failure (Knauss et al. 2008). On this point, creep- and stress-relaxation tests are fundamentals characterization tools. Creep or relaxation test is conducted on samples having experienced different physical aging times, or have been exposed to different environmental condition, for example, water, moisture, and ultraviolet light. Combined environmental exposure, such as periodic moisture/ultraviolet exposure along with temperature histories, is also used to simulate service conditions.

Particular attention must be paid when nano-indentation tests are performed. Potentially, these types of tests allow detecting in-homogeneities on a nanoscale by probing the mechanical properties at many different locations (e.g., 100 points on a sample). However, preparation of the specimen surface and selection of correct indenter tips play a crucial role on the final result. For instance, measurements of effective Young's modulus of an aerogel must be performed with a spherical indenter with a tip radius considerably larger than the aerogel pore size (Wingfield et al. 2009). Nano-indentation test on aerogels are often made to determine the suitability for aerospace, defense, and energy applications.

An uncommon, although quite effective, technique used to get simultaneously the elongation modulus E and Poisson ratio v is the ultrasonic echo tests. In these tests, the time of flight of an ultrasonic pulse is measured when traveling throughout an aerogel specimen. Actually, the pulse, generated by an ultrasonic transducer, travels back and forth between the opposite faces of the specimen and its echo is monitored by the transducer itself. Wave amplitude attenuates in time at each bouncing and its delay time is related to the speed of sound in the aerogel and twice the specimen thickness. For samples with lateral dimensions larger then thickness, longitudinal C_L and shear C_s wave speed are given (Krautkramer 1969; Ensminger 1988) as

$$C_{\mathrm{L}} = \sqrt{\frac{E(1-\nu)}{\rho_{\mathrm{b}}(1+\nu)(1-2\nu)}} \quad \text{and} \quad C_{\mathrm{s}} = \sqrt{\frac{E}{2\rho_{\mathrm{b}}(1+\nu)}} = \sqrt{\frac{G}{\rho_{\mathrm{b}}}}$$

(8.17)

where:

ν is the Poisson ratio
E is the Young's modulus
G is the shear modulus
ρ_{b} is the bulk density of the aerogel

Measuring both C_{L} and C_{s}, and knowing ρ_{b}, it is possible to solve for E and ν:

$$E = 2\rho_{\mathrm{b}}C_{\mathrm{s}}^2(1+\nu) \quad \text{and} \quad \nu = \frac{1 - 2(C_{\mathrm{s}}/C_{\mathrm{L}})^2}{2 - 2(C_{\mathrm{s}}/C_{\mathrm{L}})^2}$$

(8.18)

It is worth to note that a careful analysis of the data is required to properly evaluate whether the mechanical behavior of the aerogel has frequency or rate dependency.

Other mechanical test include deflection–compression set tests, to determine the dimensional stability; loading/unloading/reloading test at different strain levels, to measure the dissipated energy density in a loading cycle; dynamical mechanical analysis test, which in the temperature scanning investigation allows the evaluation of the glass transition temperature (T_{g}). All tests are usually conducted over a range of temperatures representative of the range of service temperatures of the aerogel.

Although bio-based aerogels have inferior mechanical properties compared to inorganic aerogels, they show mechanical performances, depending on the network connectedness and aerogel density, which allow their practical uses. It is evident that mechanical strength and stiffness of aerogels are very relevant for most of planned applications. As a consequence, a lot of studies are known in the literature that have dealt with the determination of the mechanical properties of bio-based aerogels of interest (including the Young's modulus, compressive strength, compressive modulus, bending strength, and tensile strength) or with the possibility of improving the performance reinforcing the aerogel matrix (e.g., with cellulosic fibers and cellulose nano-crystals). Table 8.4 resumes the principal properties of some of them.

TABLE 8.4
Typical Mechanical Properties of Some Bio-Based Aerogels of Interest

Aerogel	Mechanical Properties	Reference
Ca-alginate monolith	Compressive strength 20–40 kPa	Cheng et al. 2012
Ca-alginate monolith reinforced with N,N'-methylenebisacrylamide	Compressive strength 40–70 kPa	Cheng et al. 2012
Ca-alginate monolith reinforced with carboxymethylcellulose	Compressive strength 20–60 kPa	Cheng et al. 2012
Starch-chitosan	Tensile strength 1.0 N mm^{-2}	Salam et al. 2011
Starch reinforced with nanofibrillated cellulose	Young's modulus 1.7–7.0 MPa	Svagan et al. 2008
Starch citrate-chitosan	Tensile strength 1.8 N mm^{-2}	Salam et al. 2011
Nanofibrillated cellulose in hydrated calciumthiocyanate	Bending strength 2 MPa	Hoepfner et al. 2008
Commercial cellulose foam	Tensile strength 23.8 N mm^{-2}	Salam et al. 2011
Glucuronoxylan-chitosan	Tensile strength 0.54 N mm^{-2}	Salam et al. 2011

8.5.3 Sorption Capacity

The characteristic structure of aerogels, and in particular, the very high surface area, confers a remarkable sorption capacity to all aerogels. In particular, for bio-based aerogel, the occurrence of functional groups easily available for interactions with external compounds further enhances the dissolution of polar molecules, water, saline solutions, and surfactants. For instance, water can be rapidly absorbed into bio-based aerogel by capillary force up to several tens of grams per gram of aerogel (Mallepally et al. 2013) (see Table 8.5). Table 8.5 resumes the principle sorption capacity obtained for polysaccharide aerogels.

Bio-based aerogels have also the remarkable properties to rapidly recover the initial shape if subjected to repeated compression/immersion cycles in water. For example, aerogel of polyamide-epichlorohydrin resin, cross-linked with nano-fibrillated cellulose (NFC), has shown such ability (Zhang et al. 2012). Analogous sorption properties have been observed for alginate aerogels exposed to water vapor (Trens et al. 2007). For alginate, a dependence of sorption capacity from the divalent cation used to form the gel has also been reported, appearing to be the highest in the case of calcium ion.

The main application of bio-based aerogels is to remove even traces of polar substances from nonpolar gas or liquid stream. For instance, alginate aerogels have been exploited to remove 1-hexanol from liquid hydrocarbons (dodecane) (Rodriguez Escudero et al. 2009); in the same way, chitosan aerogels were used to adsorb anionic surfactant, sodium dodecylbenzenesulfonate, from waste water (Chang et al. 2008).

8.5.4 Insulation Properties

Aerogels, because of their nano-porous structure, are excellent thermal insulator and good convective inhibitors. The gas trapped into the nano-porous structure is not allowed to circulate inside the materials, opposing then an extraordinary resistance to the passage of heat flow. Because of its characteristic structure, the effective total thermal conductivity of an aerogel, λ_{eff}, can be expressed as the sum of three contributions:

$$\lambda_{eff}(T, p_g) = \lambda_s(T) + \lambda_g(T, p_g) + \lambda_r(T) \tag{8.19}$$

where:

λ_s is the contribution due to the solid thermal conductivity of the solid network

λ_g is the thermal conductivity of the gaseous phase

λ_r is the radiative conductivity

The solid thermal conductivity, λ_s, is found to scale with the aerogel density, ρ, according to the relationship

---

TABLE 8.5
Typical Sorption Capacities of Some Bio-Based Aerogels of Interest

Aerogel Material	Saline Solution (Grams Absorbed per Gram of Aerogel)	Water (Grams Absorbed per Gram of Aerogel)	Reference
Citric acid-cross-linked with xylanechitosan	100	80	Salam et al. 2011
Alginate	120	20	Mallepally et al. 2013
Polyamide-epichlorohydrin resin-cross-linked NFC	–	98	Zhang et al. 2012

$$\lambda_s \approx \rho^\gamma, \quad \gamma = 1.2 - 1.8 \tag{8.20}$$

To fix the ideas, a monolithic silica aerogel, with a typical density ρ of 120 kg m^{-3}, is characterized by a value of λ_s of about 5 mW m^{-1} K^{-1}. Furthermore, because of the occurrence of the Knudsen effect, the nano-porous structure, and hence the very small pore size (lower than the mean free path of air), determines a very low gaseous thermal conductivity, λ_g, which at ambient pressure is typically lower than 10 mW m^{-1} K^{-1}.

It is noteworthy that this equation is independent of direction and, hence, it is only true in the case of aerogels that show isotropic thermal properties.

Several research studies have been conducted in order to determine the optimal conditions to which an aerogel has the lowest value of the total effective thermal conductivity (Hrubesh and Pekala 1994). The studies show the importance of using a starting material (organic or inorganic), which already possesses a small thermal conductivity; hence, in this respect, organic materials having smaller thermal conductivity values are preferred. Moreover, the authors showed that reducing the pores size allows to inhibit the heat transfer through the gaseous phase, thereby producing a further decrease in the thermal conductivity, the term λ_g in the Equation 8.19 being negligible.

It should be remarked that in their works, Hrubesh and Pekala always neglected any interplay between the heat transfer within the solid and the gaseous phases of the aerogel. This coupling effect, in fact, could significantly increase the total thermal conductivity; hence, it is important to obtain aerogel structures that minimize this effect, also at ambient conditions (Swimm et al. 2009).

Aerogels are also known for their excellent acoustic insulation properties. Again, this property derives principally from the characteristic structure and density and also depends on pressure and the interstitial gas type. In fact, due to the particular structure that characterizes these materials, the amplitude and velocity propagation of the acoustic waves is reduced by a partial loss of the acoustic propagation during the process of energy transfer from the gas phase to the solid one of the aerogel. The typical longitudinal acoustic velocity of an aerogel is in the order of about 100 m s^{-1} (Pierre and Pajonk 2002).

8.5.5 Function as Reagent Carrier

Because of their very high porosity and surface area, combined with their biodegradability, bio-based aerogels find an extensive use as reagent carrier in pharmaceutical applications (Broecker et al. 1986), biotechnological processes (Peralta-Perez et al. 2011), and in food industry. In fact, an active agent can be incorporated in the initial hydrogel, through an appropriate procedure that depends on both the structural characteristics of the gel matrix and the specific agent (Kadib et al. 2011). The incorporated agent can then be retained into the final aerogel by an appropriate drying process, leading to a device that can be used as delivery system. Indeed, exploiting the water solubility of certain polysaccharides aerogels, the agent can be effectively delivered by immersing the loaded aerogel in an aqueous media (Guari et al. 2006).

8.5.6 Tissue Engineering Applications

One of the most important applications of bio-based aerogels is in tissue engineering; in fact, they can be successfully used as a bio-material for the production of scaffolds for tissue replacement. In the literature, a great variety of proposals of natural porous solid support for scaffolding applications are available, such as aerogels based on linear aliphatic polyester (poly-lactic acid and poly-glycolic acid and their copolymers), proteins (collagen, for example) or polysaccharides (alginate, chitosan, and hyaluronate) (Ma 2004). In fact, the use of these bio-materials as support for cells proliferation, and hence for tissue and organs regeneration, allows in avoiding a whole series of drawbacks related to problems of donor scarcity, transfer of pathogens, and immune rejection.

To be used in the realization of temporary scaffolds, the materials must possess properties that are able to confer to them some essential characteristics:

- Biocompatibility
- Biodegradability and a degradation rate that is comparable to that of the neo-formed tissue
- Adequate mechanical properties, which allow to maintain the designed scaffold structure
- A reproducible three-dimensional structure
- A controlled and high porosity (>90%) and an open-pore structure that permit cells growth and reorganization
- High internal surface areas, which permit adequate cell adhesion and proliferation

It has been widely observed that the supercritical CO_2 drying method confers to bio-based aerogels properties (previously listed) that are suitable for their application in this field (Reverchon et al. 2008).

REFERENCES

Aegerter, M.A., N. Leventis, and M.M. Koebel. 2011. *Aerogels Handbook*. Springer, New York.

Alnaief, M., M.A. Alzaitoun, C.A. García-González, and I. Smirnova. 2011. Preparation of biodegradable nanoporous microspherical aerogel based on alginate. *Carbohydrate Polymers*. 84: 1011–1018.

Amaral-Labat, G., A. Szczurek, V. Fierro, E. Masson, A. Pizzi, and A. Celzard. 2012. Impact of depressurizing rate on the porosity of aerogels. *Microporous and Mesoporous Materials*. 152: 240–245.

Betz, M., C.A. García-Gonzálezb, R.P. Subrahmanyam, I. Smirnova, and U. Kulozik. 2012. Preparation of novel whey protein-based aerogels as drug carriers for life science applications. *Journal of Supercritical Fluids*. 72: 111–119.

Brinker, C.J., and G.W. Scherer (Eds.). 1990. *Sol-Gel Science: The Physics and Chemistry of Sol-Gel Processing*. Academic Press Inc., San Diego, CA.

Broecker, F.J., W. Heckmann, F. Fischer, M. Mielke, J. Schroeder, and A. Stange. 1986. Structural analysis of granular silica aerogels. In *Aerogels* (pp. 160–166). Springer, Berlin, Germany.

Cardea, S., P. Pisanti, and E. Reverchon. 2010. Generation of chitosan nanoporous structures for tissue engineering applications using a supercritical fluid assisted process. *Journal of Supercritical Fluids*. 54: 290–295.

Chang, X., D. Chen, and X. Jiao. 2008. Chitosan-based aerogels with high adsorption performance. *Journal of Physical Chemistry B*. 112: 7721–7725.

Cheng, Y., L. Lu, W. Zhang, J. Shi, and Y. Cao. 2012. Reinforced low density alginate-based aerogels: Preparation, hydrophobic modification and characterization. *Carbohydrate Polymers*. 88: 1093–1099.

Chiehming, J.C., C.Y. Day, C.M. Ko, and K.L. Chiu. 1997. Densities and P-x-y diagrams for carbon dioxide dissolution in methanol, ethanol, and acetone mixtures. *Fluid Phase Equilibria*. 131: 243–258.

Comin, L.M., F. Temelli, and M.D. Saldaña. 2012. Barley beta-glucan aerogels via supercritical CO_2 drying. *Food Research International*. 48: 442–448.

Condon, J.B. 2006. *Surface Area and Porosity Determinations by Physisorption: Measurements and Theory*. Elsevier, Amsterdam, the Netherlands.

Czakkel, O., B. Nagy, E. Geissler, and K. László. 2013. In situ SAXS investigation of structural changes in soft resorcinol–formaldehyde polymer gels during CO_2-drying. *The Journal of Supercritical Fluids* 75: 112–119.

Del Gaudio, P., G. Auriemma, T. Mencherini, G. Della Porta, E. Reverchon, and R.P. Aquino. 2013. Design of alginate-based aerogel for nonsteroidal anti-inflammatory drugs controlled delivery systems using prilling and supercritical-assisted drying. *Journal of Pharmaceutical Sciences*. 102: 185–194.

Della Porta, G., P. Del Gaudio, F. De Cicco, R.P. Aquino, and E. Reverchon. 2013. Supercritical drying of alginate beads for the development of aerogel biomaterials: Optimization of process parameters and exchange solvents. *Industrial & Engineering Chemistry Research*. 52(34): 12003–12009.

Ennajih, H., R. Bouhfid, E.M. Essassi, M. Bousmina, and A. El Kadib. 2012. Chitosan–montmorillonite bio-based aerogel hybrid microspheres. *Microporous and Mesoporous Materials*. 152: 208–213.

Ensminger, D. 1988. *Ultrasonics: Fundamentals, Technology, Applications, Revised and Expanded*. Marcel Dekker Inc., New York.

Fischer, F., A. Rigacci, R. Pirard, S. Berthon-Fabry, and P. Achard. 2006. Cellulose-based aerogels. *Polymer*. 47: 7636–7645.

García-González, C.A., M. Alnaief, and I. Smirnova. 2011. Polysaccharide-based aerogels—Promising biodegradable carriers for drug delivery systems. *Carbohydrate Polymers*. 86(4): 1425–1438.

Gregg, S.J., and K.S.W. Sing. 1982. Adsorption, Surface Area and Porosity, 2nd ed. Academic Press, London.

Guari, Y., J. Larionova, K. Molvinger, B. Folch, and C. Guérin. 2006. Magnetic water-soluble cyano-bridged metal coordination nano-polymers. *Chemical Communications*. 24: 2613–2615.

Hoepfner, S., L. Ratke, and B. Milow. 2008. Synthesis and characterisation of nanofibrillar cellulose aerogels. *Cellulose*. 15: 121–129.

Hrubesh, L.W., and R.W. Pekala. 1994. Thermal properties of organic and inorganic aerogels. *Journal of Materials Research*. 9(03): 731–738.

Innerlohinger, J., H.K. Weber, and G. Kraft. 2006. Aerocellulose: Aerogels and aerogel-like materials made from Cellulose. *Macromolecular Symposia*. 244: 126–135.

Job, N., R. Pirard, J.P. Pirard, and C. Alié. 2006. Non intrusive mercury porosimetry: Pyrolysis of resorcinol-formaldehyde xerogels. *Particle & Particle Systems Characterization*. 23(1): 72–81.

Joung, S.N., C.W. Yoo, H.Y. Shin, S.Y. Kim, K.-P. Yoo, C.S. Lee, and W.S. Huh. 2001. Measurements and correlation of high-pressure VLE of binary CO$_2$–alcohol systems (methanol, ethanol, 2-methoxyethanol and 2-ethoxyethanol). *Fluid Phase Equilibria*. 185: 219–230.

Kadib, A.E., K. Molvinger, T. Cacciaguerra, M. Bousmina, and D. Brunel. 2011. Chitosan templated synthesis of porous metal oxide microspheres with filamentary nanostructures. *Microporous and Mesoporous Materials*. 142(1): 301–307.

Kistler, S.S. 1931. Coherent expanded aerogels and jellies. *Nature* 127: 741.

Kistler, S.S. 1932. Coherent expanded-aerogels. *J. Phys. Chem.* 36(1): 52–64.

Knauss, W.G., I. Emri, and H. Lu. 2008. Mechanics of polymers: Viscoelasticity. In *Handbook of Experimental Solid Mechanics*. Sharpe, Jr. and N. William (Eds.), Springer, Boston, MA.

Knez, Z., M. Skerget, L. Ilic, and C. Lutge. 2008. Vapor–liquid equilibrium of binary CO$_2$–organic solvent systems (ethanol, tetrahydrofuran, ortho-xylene, meta-xylene, para-xylene). *Journal of Supercritical Fluids*. 43: 383–389.

Krautkramer, K. 1969. *Ultrasonic Testing of Materials*. Springer-Verlag, New York.

Liebner, F., E. Haimer, M. Wendland, M.A. Neouze, K. Schlufter, P. Miethe, T. Heinze, A. Potthast, and T. Rosenau. 2010. Aerogels from unaltered bacterial cellulose: application of scCO$_2$ drying for the preparation of shaped, ultra-lightweight cellulosic aerogels. *Macromolecular Bioscience*. 10: 349–352.

Ma, P.X. 2004. Scaffolds for tissue fabrication. *Materials Today*. 7(5): 30–40.

Mallepally, R.R., I. Bernard, M.A. Marin, K.R. Ward, and M.A. McHugh. 2013. Super-absorbent alginate aerogels. *The Journal of Supercritical Fluids*. 79: 202–208.

Mandelbrot, B.B. 1977. Fractals: Form, chance and dimension. Freeman, San Francisco, CA.

Mehling, T., I. Smirnova, U. Guenther, and R.H.H. Neubert. 2009. Polysaccharide-based aerogels as drug carriers. *Journal of Non-Crystalline Solids*. 355: 2472–2479.

Miao, Z., K. Ding, T. Wu, Z. Liu, B. Han, G. An, S. Miao, and G. Yang. 2008. Fabrication of 3D-networks of native starch and their application to produce porous inorganic oxide networks through a supercritical route. *Microporous and Mesoporous Materials*. 111: 104–109.

Nicolaon, G.A., and S.J. Teichner. 1968. On a new process of preparation of silica xerogels and aerogels and their textural properties. *Bulletin de la Société Chimique de France* 5: 1900.

Peralta-Pérez, M.R., M.A. Martínez-Trujillo, G.V. Nevárez-Moorillón, R. Pérez-Bedolla, and M. García-Rivero. 2011. Immobilization of *Aspergillus niger* sp. in sol gel and its potential for production of xylanases. *Journal of Sol-Gel Science and Technology*. 57(1): 6–11.

Pierre, A.C., and G.M. Pajonk. 2002. Chemistry of aerogels and their applications. *Chemical Reviews*. 102(11): 4243–4266.

Pirard, R., C. Alié, and J.P. Pirard. 2002. Characterization of porous texture of hyperporous materials by mercury porosimetry using densification equation. *Powder Technology*. 128(2): 242–247.

Pirard, R., S. Blacher, F. Brouers, and J.P. Pirard. 1995. Interpretation of mercury porosimetry applied to aerogels. *Journal of Materials Research*. 10: 2114–2119.

Pirard, R., B. Heinrichs, O. Van Cantfort, and J.P. Pirard. 1998. Mercury porosimetry applied to low density xerogels; relation between structure and mechanical properties. *Journal of Sol-Gel Science and Technology*. 13(1–3): 335–339.

Placin, F., J.P. Desvergne, and F. Cansell. 2000. Organic low molecular weight aerogel formed in supercritical fluids. *Journal of Materials Chemistry*. 10(9): 2147–2149.

Quignard, F., R. Valentin, and F. DiRenzo. 2008. Aerogel materials from marine polysaccharides. *New Journal of Chemistry*. 32: 1300–1310.

Rangarajan, B., and C.T. Lira. 1991. Production of aerogels. *The Journal of Supercritical Fluids*. 4(1): 1–6.

Reichenauer, G., and G.W. Scherer. 2000. Nitrogen adsorption in compliant materials. *Journal of Non-Crystalline Solids*. 277(2): 162–172.

Reichenauer, G., and G.W. Scherer. 2001a. Nitrogen sorption in aerogels. *Journal of Non-Crystalline Solids*. 285(1): 167–174.

Reichenauer, G., and G.W. Scherer. 2001b. Effects upon nitrogen sorption analysis in aerogels. *Journal of Colloid and Interface Science*. 236(2): 385–386.

Reverchon, E., S. Cardea, and C. Rapuano. 2008. A new supercritical fluid-based process to produce scaffolds for tissue replacement. *The Journal of Supercritical Fluids*. 45(3): 365–373.

Rinki, K., and P.K. Dutta. 2010. Physicochemical and biological activity study of genipin-crosslinked chitosan scaffolds prepared by using supercritical carbon dioxide for tissue engineering applications. *International Journal of Biological Macromolecules*. 46: 261–266.

Rinki, K., P.K. Dutta, A.J. Hunt, J.H. Clark, and D.J. Macquarrie. 2009. Preparation of chitosan based scaffolds using supercritical carbon dioxide. *Macromolecular Symposia*. 277: 36–42.

Robitzer, M., L. David, C. Rochas, F. Di Renzo, and F. Quignard. 2008. Supercritically-dried alginate aerogels retain the fibrillar structure of the hydrogels. *Macromolecular Symposia* 273: 80–84.

Robitzer, M., F. Di Renzo, and F. Quignard. 2011a. Natural materials with high surface area. Physisorption methods for the characterization of the texture and surface of polysaccharide aerogels. *Microporous and Mesoporous Materials*. 140: 9–16.

Robitzer, M., A. Tourrette, R. Horga, R. Valentin, M. Boissière, J.M. Devoisselle, F. Di Renzo, and F. Quignard. 2011b. Nitrogen sorption as a tool for the characterisation of polysaccharide aerogels. *Carbohydrate Polymers*. 85: 44–53.

Rodriguez Escudero, R., M. Robitzer, F. Di Renzo, and F. Quignard. 2009. Alginate aerogels as adsorbents of polar molecules from liquid hydrocarbons: Hexanol as probe molecule. *Carbohydrate Polymers*. 75: 52–57.

Salam, A., R.A. Venditti, J.J. Pawlak, and K. El-Tahlawy. 2011. Crosslinked hemicellulose citrate-chitosan aerogel foams. *Carbohydrate Polymers*. 84: 1221–1229.

Sarkar, A., S.R. Chaudhuri, S. Wang, F. Kirkbir, and H. Murata. 1994. Drying of alkoxide gels—Observation of an alternate phenomenology. *Journal of Sol-Gel Science and Technology*. 2(1–3): 865–870.

Scanlon, S., A. Aggeli, N. Boden, R.J. Koopmans, R. Brydson, and C.M. Rayner. 2007. Peptide aerogels comprising self-assembling nanofibrils. *Micro & Nano Letters*. 2(2): 24–29.

Scherer, G.W. 1992. Stress development during supercritical drying. *Journal of Non-Crystalline Solids*. 145: 33–40.

Scherer, G.W., and D.M. Smith, 1995. Cavitation during drying of a gel. *Journal of Non-Crystalline Solids* 189(3): 197–211.

Secuianu, C., V. Feroiu, and D. Geana. 2008. Phase behavior for carbon dioxide + ethanol system: Experimental measurements and modeling with a cubic equation of state. *Journal of Supercritical Fluids*. 47: 109–116.

Singh, J., P.K. Dutta, J. Dutta, A.J. Hunt, D.J. Macquarrie, and J.H. Clark. 2009. Preparation and properties of highly soluble chitosan–L-glutamic acid aerogel derivative. *Carbohydrate Polymers*. 76: 188–195.

Smith, D.M., G.W. Scherer, and J.M. Anderson. 1995. Shrinkage during drying of silica gel. *Journal of Non-Crystalline Solids*. 188(3): 191–206.

Snijder, E.D., M.J. te Riele, G.F. Versteeg, and W.P. van Swaaij. 1995. Diffusion coefficients of CO, CO_2, N_2O, and N_2 in ethanol and toluene. *Journal of Chemical and Engineering Data*. 40(1): 37–39.

Stievano, M., and N. Elvassore. 2005. High-pressure density and vapor–liquid equilibrium for the binary systems carbon dioxide–ethanol, carbon dioxide–acetone and carbon dioxide–dichloromethane. *Journal of Supercritical Fluids*. 33: 7–14.

Svagan, A., M.A.S. Azizi Samir, and L.A. Berglund. 2008. Biomimetic foams of high mechanical performance based on nanostructured cell walls reinforced by native cellulose nanofibrils. *Advanced Materials*. 20: 1263–1269.

Swimm, K., G. Reichenauer, S. Vidi, and H.P. Ebert. 2009. Gas pressure dependence of the heat transport in porous solids with pores smaller than 10 μm. *International Journal of Thermophysics*. 30(4): 1329–1342.

Tan, C., B.M. Fung, J.K. Newman, and C. Vu. 2001. Organic aerogels with very high impact strength. *Advanced Materials*. 13(9): 644–646.

Teichner, S.J., G.A. Nicolaon, M.A. Vicarini, and G.E.E. Gardes. 1976. Inorganic oxide aerogels. *Advances in Colloid and Interface Science* 5(3): 245–273.

Trens, P., R. Valentin, and F. Quignard. 2007. Cation enhanced hydrophilic character of textured alginate gel beads. *Colloids and Surfaces A: Physicochemical and Engineering Aspects.* 296: 230–237.

Tsioptsias, C., C. Michailof, G. Stauropoulos, and C. Panayiotou. 2009. Chitin and carbon aerogels from chitin alcogels. *Carbohydrate Polymers.* 76: 535–540.

Valentin, R., R. Horga, B. Bonelli, E. Garrone, F. Di Renzo, and F. Quignard. 2006. FTIR spectroscopy of NH$_3$ on acidic and ionotropic alginate aerogels. *Biomacromolecules.* 7: 877–882.

Valentin, R., K. Molvinger, F. Quignard, and D. Brunel. 2003. Supercritical CO$_2$ dried chitosan: An efficient intrinsic heterogeneous catalyst in fine chemistry. *New Journal of Chemistry.* 27(12): 1690–1692.

Venkateswara Rao, A., and D. Haranath. 1999. Effect of methyltrimethoxysilane as a synthesis component on the hydrophobicity and some physical properties of silica aerogels. *Microporous and Mesoporous Materials.* 30(2): 267–273.

Washburn, E.W. 1921. The dynamics of capillary flow. *Physical Review* 17(3): 273.

Wingfield, C., A. Baski, M.F. Bertino, N. Leventis, D.P. Mohite, and H. Lu. 2009. Fabrication of sol-gel materials with anisotropic physical properties by photo-cross-linking. *Chemistry of Materials.* 21(10): 2108–2114.

Yeo, S.D., S.J. Park, J.W. Kim, and J.C. Kim. 2000. Critical properties of carbon dioxide+methanol,+ethanol,+ 1-propanol, and+1-butanol. *Journal of Chemical and Engineering Data.* 45: 932–935.

Zhang, W., Y. Zhang, C. Lu, and Y. Deng. 2012. Aerogels from cross-linked cellulose nano/micro-fibrils and their fast shape recovery property in water. *Journal of Materials Chemistry.* 22: 11642.

9 Clay-Based Bionanocomposite Foams

Eduardo Ruiz-Hitzky, Francisco M. Fernandes,
Bernd Wicklein, and Pilar Aranda

CONTENTS

9.1 INTRODUCTION

Foams based on polymer nanocomposites are becoming an ever-growing field of research in view of the development of novel approaches and new materials for multifunctional applications. Such foam composite materials have been gaining terrain in a myriad of applications, from biomaterials for tissue engineering to core materials for insulating sandwich panels in automobile, railway, and aerospace vehicles.

Typical procedures for the preparation of composite foams are based on the combination of inorganic fillers with diverse types of polymers, using as processing additives surfactant species and solvents in some cases. The use of naturally occurring polymers (biopolymers) with nontoxic and renewable characteristics is of great importance from the environmental point of view. Hence, polysaccharides and proteins appear as widely available biopolymers, displaying biocompatible and biodegradable properties. The resulting materials are thus considered as green composite materials and are generally referred to as *biocomposites*. When the interaction between biopolymers and inorganic fillers takes place at the nanometric scale, the resulting materials are named as *bionanocomposites*, reflecting the dimensionality of the inorganic filler and their degree of interaction with the biopolymer matrix (Darder et al. 2007; Ruiz-Hitzky et al. 2005, 2008, 2009, 2010). Research on clay-based biohybrids is a topic dealing with the assembly of diverse smectites, sepiolite, halloysite, and other silicates to organic species of biological origin (Darder et al. 2007; Ruiz-Hitzky et al. 2007).

As hybrid materials are organic–inorganic assemblies resulting from the combination of organic molecules and inorganic solids, biohybrids and biocomposites structured at the nanometric range are an emerging group of materials showing synergetic properties between both components, the

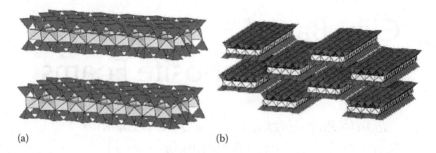

FIGURE 9.1 The scheme showing the structural arrangement of lamellar smectites (e.g., montmorillonite) (a) and the fibrous clay sepiolite (b). (Reprinted with permission from Wicklein, B. et al., *Langmuir*, 26, 5217–5225, 2010. Copyright 2010 American Chemical Society.)

inorganic substrate and the biological molecules. These bioinorganic hybrids are bionanomaterials studied within the scope of the *bionanotechnology of materials*, which could be considered as a new interdisciplinary discipline resulting from the interphase between biotechnology, materials science, and nanotechnology. Most of bionanocomposite materials are prepared as powders and films; their processing as foams has recently gained momentum in view of diverse applications.

Although the majority of biocomposite materials processed as foams uses phosphates (e.g., hydroxyapatite) and carbonates as inorganic phase, silicates including clay minerals have been started to be used alone or in combination with other fillers, especially in tissue engineering applications (Aranda et al. 2011). Actually, most of polymer nanocomposites incorporate clays as disperse phase, which act as nanofillers of diverse morphology with different composition and crystal structure. Typically, lamellar (smectite clays such as montmorillonite [MMT], hectorite, and saponite) and fibrous (sepiolite and palygorskite) clays (Figure 9.1), and to a minor extent tubular (halloysite), have been assembled to a large variety of thermosetting or thermoplastic polymers, as well as biopolymers of different origin (Alexandre and Dubois 2000; Avérous and Pollet 2012; Detellier and Letaief 2013; Lambert and Bergaya 2013; Lvov and Abdullayev 2013; Ray and Okamoto 2003; Ruiz-Hitzky and Fernandes 2013; Ruiz-Hitzky and Van Meerbeek 2006; Ruiz-Hitzky et al. 2013a, b). In general, incorporation of clay nanoparticles to biopolymers may improve properties such as toughness, strength, barrier properties, and thermal stability (Bordes et al. 2009; Frydrych et al. 2011a; Ruiz-Hitzky and Van Meerbeek 2006), being of special relevance when processed as foams. Interestingly, the presence of clays in bionanocomposite foams introduces critical characteristics such as active centers for adsorption and reaction, contributing to the porous hierarchical organization in these systems associated with the microporosity of those silicates.

The aim of this chapter is to summarize recent progresses in the development of biocomposite foams based on clays of diverse nature, emphasizing aspects related to the role of the clay filler during foam formation, as well as the active and passive properties of the resulting foams. In this manner, enhancement of mechanical properties or fire retardancy, together with biocompatibility, adsorption behavior, and other specific functionalities derived from the clay and biopolymer synergy, make this type of foam materials of interest for applications as diverse as scaffolds in tissue engineering or bioreactor accommodating enzymatic systems.

9.2 ROLE OF CLAYS DURING FOAM FORMATION

The development of stable solid foams is closely related to the synthesis of materials featuring a mixed behavior between their mechanics (foams are self-supported materials, thus showing some degree of rigidity) and their low apparent density. Hence, one of the limiting factors for the preparation of foams resides in the elastic properties of the bulk. If the bulk component is not sufficiently rigid, the thin walls and/or beams surrounding and supporting the pores (required for the

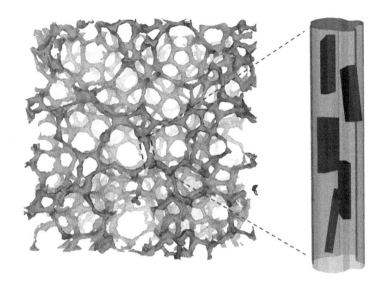

FIGURE 9.2 A polymer–clay nanocomposite foam is a polymer foam reinforced with nano-sized clay fillers. The dimensions of the reinforcing agent should allow for their inclusion within the structural elements of the foams. On the left is the general aspect of the foam displaying open porosity. On the right is the schematic representation of the dispersion of individual clay lamellae within the foam struts.

low apparent density of the foam) will be insufficient to hold the foam together and the material will collapse. The insertion of clay nanoparticles in the foam composition aims at strengthening the cell walls and beams rendering the material *foamable* as is depicted in Figure 9.2. Moreover, the enhancement of the elastic properties of the structural element of the foam may allow for larger pore volumes. This effect is due to a reduction of the amount of matter needed to support the final material. This strengthening effect (actually using a composite material to form the foam instead of a single phase component) is especially relevant in the case of biopolymers since their mechanical properties are often quite modest in comparison to technical polymers (Mark 1999). Besides the obvious reinforcing effect, nanofillers such as clay minerals have other functionalities in the context of biopolymer composite foams. The subsequent paragraphs aim at providing the reader with an overview of how clay minerals impact on the formation and structure of foams.

One of the major effects induced by the introduction of clay minerals in the structure of polymer foams has been described for polyurethane foams reinforced with organically modified MMT. According to the authors, the inclusion of layered nanofillers in the polyurethane formulation led to a visible reduction of cell size due to increased nucleation (Pardo-Alonso et al. 2012). Moreover, the authors have described a more homogeneous cell distribution throughout the foam. Finally, the authors have reported that the apparent density, initially around 52 kg m^{-3} of the material, remained unchanged with the addition of the clay moiety. In this chapter, we focus on these systems and discuss their impact on the future of biopolymer–clay nanocomposite foams.

Naturally, the effect of clay nanoparticles dispersed inside the biopolymer matrix cannot be completely dissociated from the processing techniques employed to generate the required porosity. Depending on the foam formation, for example, by supercritical carbon dioxide (scCO$_2$), reactive foaming, ice-templating, or high-speed stirring, clay nanoparticles play diverse roles in the formation and stabilization of foams:

Nucleation effect and stability against coalescence—It has been observed that even low clay contents (1 wt%) significantly reduce the final cell size of polyurethane foams (Pardo-Alonso et al. 2012). General explanations for these phenomena rely on increased nucleation sites. This effect is due to the considerable enhancement of gas bubble nucleation sites in presence clay particles dispersed within the matrix. Nucleation of gas bubbles in a polymer matrix is considered to be as a homogeneous process characterized by a nucleation site limit determined by the gas solubility in the polymer matrix. When an additional phase is added to the matrix, to the same gas bubble nucleation sites, new nucleation sites located at the interface between the continuous and the disperse matrix are added. These energetically favorable nucleation sites on the clay surface account for the noticeable augmentation of pore number and smaller pore size (Colton and Suh 1987a, b; Harikrishnan et al. 2011; Mitsunaga et al. 2003; Pardo-Alonso et al. 2012). On the other hand, added clays increase the suspension viscosity and also hamper gas diffusion within the foaming suspension. These processes can convey stability against cell coalescence and cell growth during the foaming stage. From results obtained with high-resolution microfocus X-ray radioscopy, Pardo-Alonso et al. (2012) suggested the increasing number of nucleation sites with added clay as the major cell size reduction mechanism in reactive polyurethane foaming, while no evidence for coalescence was observed. However, this may only hold for this particular polymer and foaming system and may have to be confirmed in the individual case. For instance, kaolin was observed to cause an increase in cell size in starch foams due to suspected bubble coalescence (Kaewtatip et al. 2013).

Supporting the nucleation site hypothesis, Zeng et al. (2003) observed reduced cell size and increased cell densities by comparing intercalated with exfoliated MMT during $scCO_2$ foaming of polystyrene (PS) and poly(methyl methacrylate) (Figure 9.3). By exfoliating the clay lamella, the effective particle concentration is increased, which can provide more nucleation sites, and consequently, a higher cell density was observed. Similar conclusions were also drawn for $scCO_2$ foaming of a poly(lactic acid)–clay nanocomposite showing reduced cell size and increased cell density as consequence of the presence of organo-clay (Ema et al. 2006).

Cell size and shape—While the cell size is rather influenced by nucleation and growth, the cell shape is determined by contact and deformation of initially spherical bubbles that develop polyhedral geometries as a result of volumetric confinement and mechanical balances. Another possible influence on foam morphology can arise from liquid drainage from the foam where clay platelets can have an influence on the wetting properties (Harikrishnan et al. 2011).

Finely dispersed organo-clay platelets at the polymer–gas (CO_2 or air) interface may have various effects, such as reduction of surface tension (Zeng et al. 2003; Ema et al. 2006) and reduction of gas permeability (Harikrishnan et al. 2012). The first effect can alter the cell nucleation rate (Colton and Suh 1987c), as well as the liquid drainage, while the latter can reduce cell coarsening and cell coalescence (References within [Pardo-Alonso et al. 2012]).

Another way of generating foams is through the creation of air bubbles suspended in aqueous clay-polymer dispersion via high-speed stirring and subsequent freeze-drying (Viggiano et al. 2013).

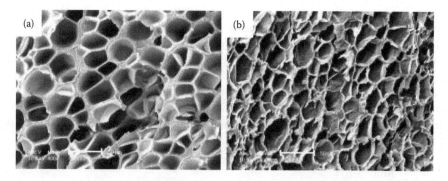

FIGURE 9.3 The scanning electron micrographs demonstrate the influence of nanoclay on the pore density of PS foams: without clay (a) and with 5% of clay (b). Scale bar corresponds to 50 μm. (From Zeng, C. et al.: Polymer–clay nanocomposite foams prepared using carbon dioxide, *Adv. Mater.* 2003. 15. 1743–1747. Copyright Wiley-VCH Verlag GmbH & Co. KGaA. Reproduced with permission.)

The addition of xanthan gum as a viscosity control agent helped to entrap the formed air bubbles, while the suspended clay improved the strength of the cell walls between the bubbles and thus, possibly preventing the migration of the bubbles during freezing.

Another crucial technique associated with the processing of nanocomposite foams is freeze-drying. This technique, vastly used in the food processing industry (soluble coffee, powder milk, dehydrated soups, etc.) consists of the controlled freezing of a solution or suspension and subsequent sublimation of the frozen solvent to retrieve a dried porous material. The resulting foam is often a negative reproduction of the frozen solvent structure. Given the hydrophilic nature of most clay minerals and biopolymers, water is the most commonly used solvent in freeze-drying foam processing. Since the foam internal morphology is highly dependent on the frozen solvent domains, the freezing procedure is critical for morphological control purpose. The faster the freezing step occurs, the smaller the ice domains and consequently smaller and more numerous pores will be formed. Conversely, slow freezing steps tend to allow the development of extensive ice crystals that systematically render larger, less numerous and ill-defined pores. Finally, one important criterion that has recently boosted the use of freeze-based techniques to generate nanocomposite foams regards the possibility to orient porosity by mastering the temperature gradient during the freezing step (Darder et al. 2011). Depending on the processing conditions, the unidirectional freezing techniques allow to control pore size, pore connectivity, pore orientation, as well as the orientation of the inorganic filler within the biopolymer matrix. These characteristics permit a fine control over mechanical properties of the prepared foams since the anisotropic filler are in general oriented during the unidirectional freezing step.

9.3 STRUCTURAL PROPERTIES OF CLAY–BIONANOCOMPOSITE FOAMS

9.3.1 Mechanical Properties

One of the most challenging features in biopolymer–clay nanocomposite foams regards the structural complexity induced by the generated porosity. If composite models such as Halpin-Tsai tend to explain (with variable degrees of reliability) the effects of a reinforcing phase on the mechanical properties (namely, the elastic modulus) of nonporous materials, modeling porous composite materials is a far more complex problem (Halpin and Kardos 1976). From a strictly conceptual point of view, the pores in a foam structure can be modeled as inclusions displaying mechanical properties as low as that of air. However, some key aspects block the direct application of standard composite theories to model the mechanical properties of bionanocomposite foams. The first relates to the pore (and wall) architecture and orientation. The second relates to the impact of the clay minerals on the mechanical properties of the biopolymer, as seen in the case of sepiolite-gelatin nanocomposite materials. This effect implies some variation of the mechanical properties of the biopolymer-clay matrix (a composite material itself) related to the ratio between biopolymer and clay mineral (Fernandes et al. 2011a, b). Finally, a third aspect relates to the important pore volume, which places these materials outside of most composite theory limits. These difficulties in establishing a rational approach to the design of nanocomposite foams with mechanical properties *a la carte* have led to a more experimental approach. The Schiraldi group developed over the recent years a large body of work concerned with the improvement of mechanical properties of clay–biopolymer foams. MMT was used as reinforcing nanofiller in matrices as diverse as casein, (Gawryla et al. 2008) natural rubber, (Pojanavaraphan et al. 2011) pectin, (Chen et al. 2013) alginate, (Chen et al. 2012) polyelectrolytes, (Viggiano et al. 2013) milk, (Wang and Schiraldi 2012) or nanocellulose (Gawryla et al. 2009). Employing different cross-linking strategies such as ionic chelating with Na^+, Ca^{2+}, or Al^{3+} (Chen et al. 2012); covalent cross-linking with S_2Cl_2 to generate disulfide bridges (Pojanavaraphan et al. 2011); and enzymatic cross-linking with transglutaminase (Wang and Schiraldi 2012). The resulting nanocomposite foams and aerogels (Figure 9.4), produced from *green* ice-templating methods, generally showed superior mechanical properties with respect to both the pure clay and the pure biopolymer

FIGURE 9.4   Photo of pectin-MMT aerogel monoliths. (Reprinted with permission from Chen, H.-B. et al., *ACS Appl. Mater. Interf.*, 5, 1715–1721, 2013. Copyright 2013 American Chemical Society.)

foams, thus clearly manifesting a clear reinforcing effect. The compressive moduli of these aerogels are in some cases as high as 100 MPa and specific moduli of over 600 MPag^{-1} cm^{-3} (Gawryla et al. 2008).

In a different approach, nanocomposite foams based on a mixture of starch/poly(lactic acid)/organoclay (Cloisite 30B) were melt compounded (Lee and Hanna 2009). Using the microfibrous clay sepiolite as nanofiller, Frydrych et al. prepared via a freeze-drying method mechanically improved gelatin nanocomposite foams. The addition of sepiolite had a significant influence on the foam microstructure such as mean cell size and apparent density, which was reported to be responsible for a 300% increase in strength and modulus (Frydrych et al. 2011b). The presence of sepiolite, similar to other suspended particles, (Kurz and Fischer 1992; Scherer 1993) is believed to enhance the nucleation of ice crystals and thus, helps to reduce the cell size.

In summary, though only few results are available to date on the elaboration of bionanocomposite foams from biopolymers and clay minerals, this strategy seems to establish as a solid trend in the development of self-supported, mechanically resistant porous nanocomposite materials. The following points aim at unveiling the applicative potential behind these structures by reviewing the relevant reports on the subject of insulation, fire retardancy, and electromagnetic shielding.

9.3.2   THERMAL AND ACOUSTIC INSULATION

The use of macroporous materials for applications in the domain of insulation is far from being a novelty. However, the introduction of reinforced biomaterials in this context might play a significant role regarding the sustainability of insulating materials, but also regarding their efficiency.

Historically, the porosity in inorganic materials aiming at high-temperature thermal insulation has been templated by polymers and subsequently calcined to burn off the organic part and anneal the percolating inorganic matrix. Recently, biopolymers such as methylcellulose have been mixed together with kaolin and subsequently calcined to achieve effective all-inorganic insulating materials displaying thermal conductivities as low as 0.054 Wm^{-1} K^{-1} (Bourret et al. 2013). While these materials can hardly be considered as biopolymer–clay hybrids (the thermal treatment is set to 1100°C), their intermediate steps do represent a close association between the biopolymer and a clay fraction. The elaboration of bionanocomposite foams to produce insulating materials adopts the same intermediate step, usually in form of a suspension of inorganic particles in a polymer solution, but generates the porosity according to above mentioned techniques that do not imply the degradation of the biopolymer (Ruiz-Hizky et al. 2010a). Naturally, since the thermal stability range of biopolymers is reduced with respect to that of all-inorganic foams, the range of applicable temperatures for thermal insulating bionanocomposite foams is limited.

Many reports addressing the elaboration of bionanocomposite foams mention the possibility to apply these materials in insulating applications. However, while the mechanical properties of such foams are often explored in some detail, the experimental determination of their insulating properties is often relegated to a secondary plane. In our opinion, a stronger effort in fully characterizing the sound abating and thermal insulating properties of clay based bionanocomposite foams should be pursued. The tunable porosity, low cost raw materials, as well as green processing technologies (Ojijo and Sinha Ray 2013), are strong arguments to further the application of bionanocomposite foams to low-temperature commercial applications.

9.3.3 FIRE RETARDANT FOAMS

The employment of petroleum and more recently bio-based foams in buildings and vehicles often requires the additional quality of fire or flame retardancy since the safety requisites for insulation materials are constantly increasing. The use of nanoclays such as microfibrous sepiolite, exfoliated bentonites, and aluminum containing metal hydroxide nanoparticles (e.g., layered double hydroxides) as nontoxic and eco-friendly alternatives to questionable brominated compounds as flame retardancy agents in polymers has been recognized (Sanchez et al. 2005) and is currently expanded to the field of biopolymer foams. The great advantage of clays in polymer nanocomposites is their ability to create intumescent systems (Laoutid et al. 2009), that is, induction of char formation in the event of fire, migration of nanoclay particles to the external surface, and creation of a ceramic-like layer, which is often sufficiently resistant to protect the underlying polymer matrix. Due to the open structure of foams as compared to dense nanocomposites, layer-forming flame retardancy mechanism in foams is more difficult. Nevertheless, Chen et al. (2012) reported the successful preparation of low flammability ammonium alginate-sodium MMT aerogels, which resist fire due to the creation of a MMT-based surface barrier for heat and mass transport (Figure 9.5). Attributed to Ca^{2+} cross-linking, these foam-like materials exhibited superior mechanical properties (100 MPa modulus) as compared to rigid polyurethane foams (up to 25 MPa modulus), being a benchmark material in the field. Furthermore, the addition of clay to the alginate aerogel effectively increased the time to ignition and decreased the total heat release as compared to expanded polystyrene.

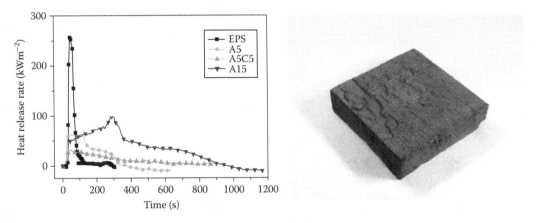

FIGURE 9.5 Cone calorimetry data of various alginate-MMT foams together with expanded polystyrene (left). Photo of an alginate-MMT sample after burning in cone calorimeter (right). (Reprinted from *Polymer*, 53, Chen et al., Low flammability, foam-like materials based on ammonium alginate and sodium montmorillonite clay, 5825–5831, Copyright 2012, with permission from Elsevier.)

9.4 FUNCTIONAL PROPERTIES OF CLAY–BIONANOCOMPOSITE FOAMS

9.4.1 SUPPORT OF BIOCATALYTIC ENTITIES

One of the most relevant domains of application for clay-based bionanocomposite foams is that of the development of scaffolds for biological entities. Other than the paradigmatic example of trabecular bone (addressed in Chapters 12 and 13), recent works show the potential of biopolyelectrolyte-based hydrogels (i.e., charged polysaccharides) to act as viable scaffolds for the immobilization and functional conservation of biological species such as enzymes and microorganisms. Incorporated nanoclay particles can be useful to adjust the structural properties of these hydrogels and thus, influence, for instance, release properties. In this way, Liu et al. (2011b) modified chitosan-xanthan gum hydrogels with MMT to slow down the controlled release of encapsulated enzymes from these freeze-dried polyelectrolyte beads.

Another example for enzyme encapsulation using in this case poly(vinyl alcohol) (PVA) was reported by Wicklein et al. (2013), who constructed in a bottom-up approach a multifunctional PVA-sepiolite foam suitable for biocatalysis and biosensing. Following the nanoarchitectonics concept (Ariga et al. 2012), a lipid bilayer was deposited on the surface of sepiolite fibers, which served to accommodate urease enzymes (Figure 9.6a). This biohybrid was doped into a PVA-sepiolite suspension and ice templating rendered meso- and macroporous foams whose pore architecture could be tailored by borax cross-linking. The different pore scales proved to be crucial for rapid diffusion of substrate and product molecules through the foam and thus, to ensure a pseudo first order catalysis reaction rate. Additional doping with carbon black (Figure 9.6b) rendered the foams electrically conductive (0.3 mS cm^{-1}), which allows for the *in situ* recording of catalytical reactions (Figure 9.6c), a useful quality that could be in the future explored for bioelectrocatalysis and biosensing devices.

Materials displaying dual porosity can also be achieved by the incorporation of microporous MMT into macroporous chitosan aerogel beads (Ennajih et al. 2012). Chitosan intercalates into

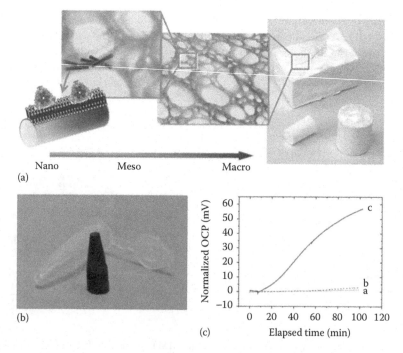

FIGURE 9.6 Nanoarchitectural construction of enzyme active PVA-sepiolite foam (a), photo of carbon black doped foam (b), open circuit potential curves of urease reaction recorded *in situ* (c). (Reproduced from Wicklein, B. et al., *J. Mater. Chem. B*, 1, 2911, 2013. Reproduced by permission of The Royal Society of Chemistry.)

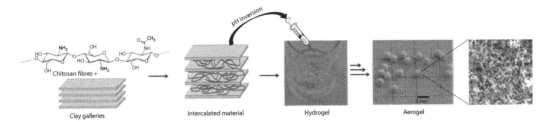

FIGURE 9.7 A schematic illustration of the preparation of aerogel beads. (Reprinted from *Micropor. Mesopor. Mater.*, 152, Ennajih, H. et al., Chitosan–montmorillonite bio-based aerogel hybrid microspheres, 208–213, Copyright 2012, with permission from Elsevier.)

the MMT galleries in acidic conditions, while during pH inversion, the free chitosan chains rapidly gel and form hydrogel beads (Figure 9.7). Through a series of solvent exchanges and eventually scCO$_2$ drying, chitosan-MMT aerogel beads were prepared. The large specific surface area of about 320 m^2 g^{-1} of these materials can be utile for effective adsorption of organic contaminants in environmental remediation processes or for heterogeneous catalysis through site-selective functionalization of either the chitosan matrix or the microporous clay filler.

9.4.2 ADSORBENTS: WATER REMEDIATION

Bionanocomposites based on clays are being investigated for environmental applications. The use of foams instead of other powdered materials offers the advantage of better adsorbency and easier recovering from aqueous media. Zein, an abundant protein from corn containing hydrophobic amino acid residues, can form foams with ligninic compounds (Oliviero et al. 2012), being also able to assemble lamellar and fibrous clays giving rise to bionanocomposites that can be conformed as foams (Alcântara 2013; Ruiz-Hitzky et al. 2010). Removal of pollutants such as pesticides in water is efficiently carried out due to the good mechanical properties and the improved resistance in water of these foams (Alcântara 2013). Clay-based bionanocomposites using polysaccharides such as chitosan and starch have been studied for preparation of efficient adsorbents useful for the removal of pollutants from water, but their conformation as foams present stability problems in water. Chitosan can easily intercalate smectites such as MMT due to its positively charged chains in acidic media (Darder et al. 2003). The immobilized biopolymer exhibit charged sites able to adsorb anions such as nitrate and sulfate. Starch is a neutral biopolymer that must be functionalized by introducing diverse groups, such as carboxylate, acrylonitrile, acrylamide, and phosphate that can act as active centers for pollutants removal (Abdel-Aal et al. 2006; Guo et al. 2006; Khalil and Abdel-Halim 2001; Xie et al. 2011; Xing et al. 2006). In this way, bionanocomposites based on polysaccharides such as alginate (Lezehari et al. 2010), cellulose acetate (Zhou et al. 2012), and cationized starch (Koriche et al. 2013) combined with smectites of diverse nature have been used in the removal of pollutants such as dyes. The use of sepiolite and palygorskite fibrous clay instead of smectites does not require modification of the polysaccharide resulting in systems highly stable in water that allows their use as films in the removal of diverse type of pollutants (e.g., heavy metal ions and dyes) (Alcântara et al. 2014). However, what appears of greater interest is the preparation of stable foams using polysaccharide clay–bionanocomposites, which is a topic still open for research. In this way, it has been reported that the formation of macroporous foams from chitosan-xanthan gum colloidal suspension incorporating MMT, which afford enhanced mechanical properties to the resulting systems (Liu et al. 2011a). Ternary systems based also on the combination of two polysaccharides and sepiolite fibrous clay drive to foams through freeze-dried processes (Ruiz-Hizky et al. 2010b), resulting in foams that show relatively good stability in water and adsorption properties of interest for environmental applications, for example, in the removal of dyes (Darder et al. 2014).

9.4.3 CLAY–BIONANOCOMPOSITES AS SOURCE OF CONDUCTING FOAMS

An interesting property of carbohydrates is their use as precursors of carbonaceous materials by thermal treatment under controlled conditions. For instance, the use of clays combined with sucrose yields supported graphenes with interesting characteristics (electrical conductivity and large specific surface area), useful for applications in energy storage (Darder and Ruiz-Hitzky 2005; Gomez-Aviles et al. 2007; Ruiz-García et al. 2014; Ruiz-Hitzky et al. 2011). Biopolymers, such as gelatin, assembled to lamellar or fibrous clays are excellent precursors of this type of carbonaceous materials giving rise in this case to supported *N*-doped graphenes of good conductivity. The most interesting materials are derived from sepiolite-gelatin bionanocomposites submitted to a carbonization process by treatment at 700°C–800°C under nitrogen flow. This is an approach in which gelatin adsorbed on sepiolite fibers drives to the formation of electrically conducting silicate-carbon fibers (Figure 9.8) (Fernandes 2011; Ruiz-Hitzky and Fernandes 2012; Ruiz-Hitzky et al. 2011).

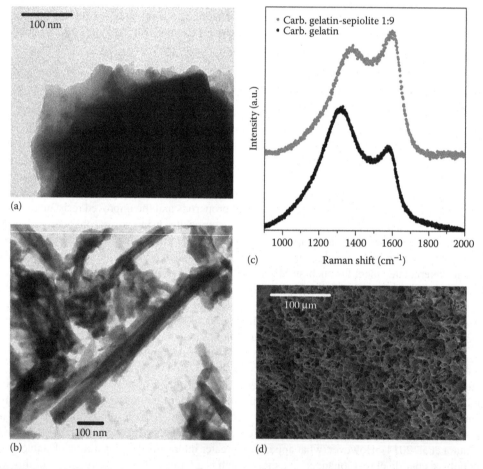

FIGURE 9.8 (a) Transmission electron micrograph of a stack of graphene layers obtained by carbonization of gelatin in presence of sepiolite. (b) Transmission electron micrograph of graphene deposited over sepiolite fibers. (c) Raman spectra of carbonized gelatin in the presence (gray dots) and absence (black dots) of sepiolite showing the inversion of the G to D ratio. (d) Macroporous structure of conducting graphene-like foam obtained from the carbonization of gelatin-sepiolite colloid after freeze-drying. (From Ruiz-Hitzky et al.: Supported graphene from natural resources: Easy preparation and applications. *Adv. Mater.* 2011. 23. 5250–5255. Copyright Wiley-VCH Verlag GmbH & Co. KGaA. Reproduced with permission.)

Gelatin-sepiolite bionanocomposites processed as foams were used in the preparation of conducting nanocomposite carbon-sepiolite foams adopting different conditioning temperatures and freeze-drying conditions (Figure 9.8d) (Fernandes 2011). The use of a preconditioning temperature below the renaturation temperature of gelatin (e.g., 4°C) might induce an enhancement of the foam structural integrity with improved mechanical properties. The second factor is the influence of the freezing temperature (and thus the velocity at which the ice forms) on the pore dimensions, form and number (Fernandes 2011). After freeze-drying, four different foams presented a foam-like aspect and variable handling properties. The samples that had been preconditioned at 4°C presented much stronger mechanical properties and were not deformed by manual handling, while samples preconditioned at 60°C were easily deformable. Besides, samples frozen in liquid nitrogen present considerably reduced pore size compared with samples frozen at −20°C. Interestingly, gelatin-sepiolite foams that had been conditioned at 4°C present a stable structure upon carbonization, leading to sepiolite-carbon foams that maintain its structural integrity. Raman spectroscopy of these materials proved the existence of considerable amount of graphitic domains with spectra comparable to those corresponding to supported-graphenes on sepiolite using sucrose or caramel as carbon precursors (Ruiz-Hitzky et al. 2011). The conductivity values for the materials derived from gelatin-sepiolite foams, measured after grinding and pellets conformation, ranges between 1×10^{-5} and 1×10^{-4} S/cm (Fernandes 2011). These characteristics are appropriate regarding potential applications in fields such as antistatic materials (Kupke et al. 1998; Pokhmurs'kyi et al. 2005), sensors (Davis and Weber 1990), and electrocatalysts and biosensing (Wu et al. 2007), among other applications. For instance, the prepared carbonized gelatin-sepiolite foams were used to produce conductive composites based on thermosetting epoxy resin by filling of the foam pores by the resin (Fernandes 2011).

9.4.4 ELECTROMAGNETIC INTERFERENCE SHIELDING

The diversity of electromagnetic devices, and thus of electromagnetic radiation to which we are daily exposed to, along with the large possibilities of interference between a myriad of different signals, have led to the development of electromagnetic interference shielding technologies. Foams and more specifically conductive foams are especially suited in this specific application, since they can easily combine the conductive character that derives from the carbonization of the biopolymers with the ability to host nano-sized magnetic entities. Though few works report the stabilization of biopolymers and magnetic nanoparticle-containing clays under the form of rigid foams, these aspects of electromagnetic interference shielding materials have been addressed separately. González-Alfaro et al. (2011) have developed a clay-based conductive and paramagnetic system by associating carbon nanotubes and sepiolite clay microfibers decorated with magnetite by impregnation of the material with a ferrofluid. The resulting materials display the paramagnetic effect associated with the magnetite ferrofluid while presenting a pronounced electronic conductivity due to the presence of carbon nanotubes. The impregnation of bionanocomposite foams by ferrofluids has also been addressed to achieve lightweight magnetic materials (Ruiz-Hizky et al. 2011). In summary, the advances in both fields, conductive foams derived from bionanocomposite foams, along with the ability to impregnate such foams with ferrofluids, opens a clear pathway to the development of electromagnetic interference shielding materials from bionanocomposite foams.

9.5 FUTURE AND PERSPECTIVES IN CLAY–BIONANOCOMPOSITE FOAMS

The previous sections illustrate that clay–bionanocomposite foams are an emerging and attractive field of research not only focused to biomedical applications as widely reported elsewhere, but also for different applications of bionanocomposite foams.

A large and elaborate body of works on structural and especially on mechanical properties of clay–bionanocomposite foams has been created over the last decade. It becomes evident now that the functional part of such multicomponent foams clearly lags behind in terms of diversity, robustness, and viable demonstrators.

Development of new and improved methodologies for foaming clay-based bionanocomposites is still a task of primordial importance, particularly in view to scale up for mass production. Multicomponent composites including different types of both biopolymers and inorganic particles, including clay mineral, offer novel perspectives toward multifunctional foams.

ACKNOWLEDGMENTS

This work was supported by the CICYT, Spain (projects MAT2012-31759), and the EU COST Action MP1202.

REFERENCES

Abdel-Aal, S.E., Gad, Y.H., Dessouki, A.M., 2006. Use of rice straw and radiation-modified maize starch/acrylonitrile in the treatment of wastewater. *J. Hazard. Mater.* 129, 204–215.

Alcântara, A.C.S., 2013. *Biohybrid Materials Based on Zein: Synthetic Approaches, Characterization and Exploration of Properties*. PhD Dissertation, Universidad Autonoma de Madrid.

Alcântara, A.C.S., Darder, M., Aranda, P., Ruiz-Hitzky, E., 2014. Polysaccharide–fibrous clay bionanocomposites. *Appl. Clay Sci.* 96, 2–8.

Alexandre, M., Dubois, P., 2000. Polymer-layered silicate nanocomposites: Preparation, properties and uses of a new class of materials. *Mater. Sci. Eng., R*. 28, 1–63.

Aranda, P., Fernandes, F.M., Wicklein, B., Ruiz-Hitzky, E., Hill, J., Ariga, K., 2011. Bioinspired materials chemistry I: Organic-inorganic nanocomposites, in: Swiegers, G.F. (Ed.), *Bioinspiration and Biomimicry in Chemistry*. John Wiley & Sons, Hoboken, NJ, pp. 121–138.

Ariga, K., Ji, Q., Hill, J.P., Bando, Y., Aono, M., 2012. Forming nanomaterials as layered functional structures toward materials nanoarchitectonics. *NPG Asia Mater.* 4, e17.

Avérous, L., Pollet, E., 2012. *Environmental Silicate Nano-Biocomposites, Environmental Silicate Nano-Biocomposites, Green Energy and Technology*. Springer, London.

Bordes, P., Pollet, E., Averous, L., 2009. Nano-biocomposites: Biodegradable polyester/nanoclay systems. *Prog. Polym. Sci.* 34, 125–155.

Bourret, J., Tessier-Doyen, N., Naït-Ali, B., Pennec, F., Alzina, A., Peyratout, C.S., Smith, D.S., 2013. Effect of the pore volume fraction on the thermal conductivity and mechanical properties of kaolin-based foams. *J. Eur. Ceram. Soc.* 33, 1487–1495.

Chen, H.-B., Chiou, B.-S., Wang, Y.-Z., Schiraldi, D.A., 2013. Biodegradable pectin/clay aerogels. *ACS Appl. Mater. Interf.* 5, 1715–1721.

Chen, H.-B., Wang, Y.-Z., Sánchez-Soto, M., Schiraldi, D.A., 2012. Low flammability, foam-like materials based on ammonium alginate and sodium montmorillonite clay. *Polymer* 53, 5825–5831.

Colton, J.S., Suh, N.P., 1987a. The nucleation of microcellular thermoplastic foam with additives: Part II: Experimental results and discussion. *Polym. Eng. Sci.* 27, 493–499.

Colton, J.S., Suh, N.P., 1987b. The nucleation of microcellular thermoplastic foam with additives: Part I: Theoretical considerations. *Polym. Eng. Sci.* 27, 485–492.

Colton, J.S., Suh, N.P., 1987c. The nucleation of microcellular thermoplastic foam with additives: Part I: Theoretical considerations. *Polym. Eng. Sci.* 27, 485–792.

Darder, M., Aranda, P., Ferrer, M.L., Gutiérrez, M.C., del Monte, F., Ruiz-Hitzky, E., 2011. Progress in bionanocomposite and bioinspired foams. *Adv. Mater.* 23, 5262–5267.

Darder, M., Aranda, P., Ruiz-Hitzky, E., 2007. Bionanocomposites: A new concept of ecological, bioinspired, and functional hybrid materials. *Adv. Mater.* 19, 1309–1319.

Darder, M., Colilla, M., Ruiz-Hitzky, E., 2003. Biopolymer-clay nanocomposites based on chitosan intercalated in montmorillonite. *Chem. Mater.* 15, 3774–3780.

Darder, M., Ruiz-Hitzky, E.R., 2005. Caramel-clay nanocomposites. *J. Mater. Chem.* 15, 3913–3918.

Darder, M., Tolosana, P., Aranda, P., Ruiz-Hitzky, E., 2014. Application of bionanocomposite materials in contaminated wastewater dye adsorption (unpublished results).

Davis, B.K., Weber, S.G., 1990. Electrochemical characterization of a microcellular carbon foam/epoxy composite electrode. *Anal. Chem.* 62, 1000–1003.

Detellier, C., Letaief, S., 2013. Chapter 13.2—Kaolinite–polymer nanocomposites, in: Bergaya, F., Lagaly, G. (Eds.), *Developments in Clay Science*. Elsevier, Amsterdam, the Netherlands, pp. 707–719.

Ema, Y., Ikeya, M., Okamoto, M., 2006. Foam processing and cellular structure of polylactide-based nanocomposites. *Polymer (Guildf)*. 47, 5350–5359.

Ennajih, H., Bouhfid, R., Essassi, E.M., Bousmina, M., El Kadib, A., 2012. Chitosan–montmorillonite bio-based aerogel hybrid microspheres. *Micropor. Mesopor. Mater*. 152, 208–213.

Fernandes, F.M., 2011. *On the Structural and Functional Properties of Sepiolite in Polymer-Clay Nanocomposites and Materials Derived Thereof*. Autonomous University of Madrid, Madrid, Spain.

Fernandes, F.M., Darder, M., Ruiz, A.I., Aranda, P., Ruiz-Hitzky, E., 2011a. Gelatine-based bio-nanocomposites, in: Mittal, V. (Ed.), *Nanocomposites with Biodegradable Polymers. Synthesis, Properties, and Future Perspectives*. Oxford University Press, New York.

Fernandes, F.M., Manjubala, I., Ruiz-Hitzky, E., 2011b. Gelatin renaturation and the interfacial role of fillers in bionanocomposites. *Phys. Chem. Chem. Phys*. 13, 4901–4910.

Frydrych, M., Wan, C., Stengler, R., O'Kelly, K.U., Chen, B., 2011a. Structure and mechanical properties of gelatin/sepiolite nanocomposite foams. *J. Mater. Chem*. 21, 9103.

Frydrych, M., Wan, C., Stengler, R., O'Kelly, K.U., Chen, B., 2011b. Structure and mechanical properties of gelatin/sepiolite nanocomposite foams. *J. Mater. Chem*. 21, 9103.

Gawryla, M.D., Nezamzadeh, M., Schiraldi, D.A., 2008. Foam-like materials produced from abundant natural resources. *Green Chem*. 10, 1078.

Gawryla, M.D., van den Berg, O., Weder, C., Schiraldi, D.A., 2009. Clay aerogel/cellulose whisker nanocomposites: A nanoscale wattle and daub. *J. Mater. Chem*. 19, 2118.

Gomez-Aviles, A., Darder, M., Aranda, P., Ruiz-Hitzky, E., 2007. Functionalized carbon-silicates from caramel-sepiolite nanocomposites. *Angew. Chemie-Int Ed*. 46, 923–925.

González-Alfaro, Y., Aranda, P., Fernandes, F.M., Wicklein, B., Darder, M., Ruiz-Hitzky, E., 2011. Multifunctional porous materials through ferrofluids. *Adv. Mater*. 23, 5224–5228.

Guo, L., Zhang, S., Ju, B., Yang, J., 2006. Study on adsorption of Cu(II) by water-insoluble starch phosphate carbamate. *Carbohydr. Polym*. 63, 487–492.

Halpin, J.C., Kardos, J.L., 1976. The Halpin-Tsai equations: A review. *Polym. Eng. Sci*. 16, 344–352.

Harikrishnan, G., Patro, T.U., Unni, A.R., Khakhar, D.V., 2011. Clay nanoplatelet induced morphological evolutions during polymeric foaming. *Soft Matter* 7, 6801.

Harikrishnan, G., Singh, S.N., Lindsay, C.I., Macosko, C.W., 2012. An aqueous pathway to polymeric foaming with nanoclay. *Green Chem*. 14, 766.

Kaewtatip, K., Tanrattanakul, V., Phetrat, W., 2013. Preparation and characterization of kaolin/starch foam. *Appl. Clay Sci*. 80–81, 413–416.

Khalil, M.I., Abdel-Halim, M.G., 2001. Preparation of anionic starch containing carboxyl groups and its utilization as chelating agent research paper. *Starch* 53, 35–41.

Koriche, Y., Darder, M., Aranda, P., Semsari, S., Ruiz-Hitzky, E., 2013. Efficient and ecological removal of anionic pollutants by cationic starch-clay bionanocomposites. *Sci. Adv. Mater*. 5, 994–1005.

Kupke, M., Wentze, H.-P., Schulte, K., 1998. Electrically conductive glass fibre reinforced epoxy resin. *Mater. Res. Innov*. 2, 164–169.

Kurz, W., Fischer, D.J., 1992. *Fundamentals of Solidification*, 3rd ed. Trans Tech Publications, Aedermannsdorf, Switzerland.

Lambert, J.-F., Bergaya, F., 2013. Chapter 13.1—Smectite–polymer nanocomposites, in: Bergaya, F., Lagaly, G. (Eds.), *Developments in Clay Science*. Elsevier, Amsterdam, the Netherlands, pp. 679–706.

Laoutid, F., Bonnaud, L., Alexandre, M., Lopez-Cuesta, J.-M., Dubois, P., 2009. New prospects in flame retardant polymer materials: From fundamentals to nanocomposites. *Mater. Sci. Eng., R* 63, 100–125.

Lee, S.Y., Hanna, M.A., 2009. *Tapioca Starch-Poly (lactic acid) -Cloisite 30B Nanocomposite Foams. Polym. Compos*. 30, 665–672.

Lezehari, M., Basly, J.-P., Baudu, M., Bouras, O., 2010. Alginate encapsulated pillared clays: Removal of a neutral/anionic biocide (pentachlorophenol) and a cationic dye (safranine) from aqueous solutions. *Colloids Surf. A Physicochem. Eng. Asp*. 366, 88–94.

Liu, H., Nakagawa, K., Chaudhary, D., Asakuma, Y., Tadé, M.O., 2011a. Freeze-dried macroporous foam prepared from chitosan/xanthan gum/montmorillonite nanocomposites. *Chem. Eng. Res. Des*. 89, 2356–2364.

Liu, H., Nakagawa, K., Kato, D., Chaudhary, D., Tadé, M.O., 2011b. Enzyme encapsulation in freeze-dried bionanocomposites prepared from chitosan and xanthan gum blend. *Mater. Chem. Phys*. 129, 488–494.

Lvov, Y., Abdullayev, E., 2013. Functional polymer–clay nanotube composites with sustained release of chemical agents. *Prog. Polym. Sci*. 38, 1690–1719.

Mark, J.E., 1999. *Polymer Data Handbook*. Oxford University Press, Oxford.

Mitsunaga, M., Ito, Y., Ray, S.S., Okamoto, M., Hironaka, K., 2003. Intercalated polycarbonate/clay nanocomposites: Nanostructure control and foam processing. *Macromol. Mater. Eng.* 288, 543–548.

Ojijo, V., Sinha Ray, S., 2013. Processing strategies in bionanocomposites. *Prog. Polym. Sci.* 38, 1543–1589.

Oliviero, M., Verdolotti, L., Nedi, I., Docimo, F., Di Maio, E., Iannace, S., 2012. Effect of two kinds of lignins, alkaline lignin and sodium lignosulfonate, on the foamability of thermoplastic zein-based bionanocomposites. *J. Cell. Plast.* 48, 516–525.

Pardo-Alonso, S., Solórzano, E., Estravís, S., Rodríguez-Perez, M.A., de Saja, J.A., 2012. In situ evidence of the nanoparticle nucleating effect in polyurethane–nanoclay foamed systems. *Soft Matter* 8, 11262.

Pojanavaraphan, T., Liu, L., Ceylan, D., Okay, O., Magaraphan, R., Schiraldi, D.A., 2011. Solution cross-linked natural rubber (NR)/clay aerogel composites. *Macromolecules* 44, 923–931.

Pokhmurs'kyi, V.I., Piddubnyi, V.K., Zin', I.M., Lavryshyn, B.M., Bilyi, L.M., Voloshyn, M.P., 2005. Influence of surface-modified conducting fillers on the properties of epoxy coatings. *Mater. Sci.* 41, 495–500.

Ray, S.S., Okamoto, M., 2003. Polymer/layered silicate nanocomposites: A review from preparation to processing. *Prog. Polym. Sci.* 28, 1539–1641.

Ruiz-García, C., Darder, M., Aranda, P., Ruiz-Hitzky, E., 2014. Toward a green way for the chemical production of supported graphenes using porous solids. *J. Mater. Chem. A* 2, 2009.

Ruiz-Hitzky, E., Aranda, P., Darder, M., 2008. Bionanocomposites, in: Kirk-Othmer *Enciclopedia of Chemical Technology*. John Wiley & Sons, Hoboken, NJ, pp. 1–28.

Ruiz-Hitzky, E., Aranda, P., Darder, M., 2009. Polymer and biopolymer-layered solid nanocomposites: Organic-inorganic assembling in two-dimensional hybrid systems, in: Ariga, K., Nalwa, H.S. (Eds.), *Bottom Up Nanofabrication*. American Scientific, Stevenson Ranch, CA, pp. 1–38.

Ruiz-Hitzky, E., Aranda, P., Darder, M., Alcântara, A.C.S., 2010. Composite material based on zeine-clay biohybrids, the method for production thereof and uses of these materials. WO2010146216 (A1).

Ruiz-Hitzky, E., Aranda, P., Darder, M., Fernandes, F.M., 2013a. Chapter 13.3—Fibrous clay mineral–polymer nanocomposites, in: Bergaya, F., Lagaly, G. (Eds.), *Developments in Clay Science*. Elsevier, Amsterdam, the Netherlands, pp. 721–741.

Ruiz-Hitzky, E., Aranda, P., Darder, M., Fernandes, F.M., Santos, M.C.R., 2010a. Rigid foams of composite type based on biopolymers combined with fibrous clays and method for the preparation thereof. WO 2010/081918 (A1).

Ruiz-Hitzky, E., Aranda, P., Fernandes, F.M., Matos, C.R.S., 2010b. Rigid foams of composite type based on biopolymers combined with fibrous clays and method for the preparation thereof. WO2010081918 (A1).

Ruiz-Hitzky, E., Aranda, P., González-Alfaro, Y., 2011. Method for obtaining materials with superparamagnetic properties. WO2011110711 (A1).

Ruiz-Hitzky, E., Ariga, K., Lvov, Y.M., 2007. *Bio-Inorganic Hybrid Nanomaterials, Strategies, Syntheses, Characterization and Applications*. Wiley-VCH, Weinheim, Germany.

Ruiz-Hitzky, E., Darder, M., Aranda, P., 2005. Functional biopolymer nanocomposites based on layered solids. *J. Mater. Chem.* 15, 3650–3662.

Ruiz-Hitzky, E., Darder, M., Aranda, P., 2010. Progress in bionanocomposite materials, in: Zhang, Q., Brinker, C.J. (Eds.), *Annual Review of Nanoresearch*. World Scientific Publishing, Singapore, pp. 149–189.

Ruiz-Hitzky, E., Darder, M., Fernandes, F.M., Wicklein, B., Alcântara, A.C.S., Aranda, P., 2013b. Fibrous clays based bionanocomposites. *Prog. Polym. Sci.* 38, 1392–1414.

Ruiz-Hitzky, E., Darder, M., Fernandes, F.M., Zatile, E., Palomares, F.J., Aranda, P., 2011. Supported graphene from natural resources: Easy preparation and applications. *Adv. Mater.* 23, 5250–5255.

Ruiz-Hitzky, E., Fernandes, F.M., 2012. Composition of carbonaceous materials that can be obtained by carbonisation of a clay-supported biopolymer. WO2012160229 (A1).

Ruiz-Hitzky, E., Fernandes, F.M., 2013. Progress in bionanocomposites: From green plastics to biomedical applications. *Prog. Polym. Sci.* 38, 1391.

Ruiz-Hitzky, E., Van Meerbeek, A., 2006. Clay mineral- and organoclay-polymer nanocomposite, in: Bergaya, F., Theng, B.K.G., Lagaly, G. (Eds.), *Handbook of Clay Science*. Elsevier, Amsterdam, the Netherlands.

Sanchez, C., Julián, B., Belleville, P., Popall, M., 2005. Applications of hybrid organic–inorganic nanocomposites. *J. Mater. Chem.* 15, 3559.

Scherer, G.W., 1993. Freezing gels. *J. Non. Cryst. Solids* 155, 1–25.

Viggiano, R.P., Gawryla, M.D., Schiraldi, D.A., 2013. Foam-like polymer/clay aerogels which incorporate air bubbles. *J. Appl. Polym. Sci.* 131, 39546–39551.

Wang, Y., Schiraldi, D., 2012. Foam-like materials produced from milk and sodium montmorillonite clay using a freeze-drying process. *Green Mater.* 1, 11–15.

Wicklein, B., Aranda, P., Ruiz-Hitzky, E., Darder, M., 2013. Hierarchically structured bioactive foams based on polyvinyl alcohol–sepiolite nanocomposites. *J. Mater. Chem. B* 1, 2911.

Wicklein, B., Darder, M., Aranda, P., Ruiz-Hitzky, E., 2010. Bio-organoclays based on phospholipids as immobilization hosts for biological species. *Langmuir* 26, 5217–5225.

Wu, S., Ju, H.X., Liu, Y., 2007. Conductive mesocellular silica–Carbon nanocomposite foams for immobilization, direct electrochemistry, and biosensing of proteins. *Adv. Funct. Mater.* 17, 585–592.

Xie, G., Shang, X., Liu, R., Hu, J., Liao, S., 2011. Synthesis and characterization of a novel amino modified starch and its adsorption properties for Cd(II) ions from aqueous solution. *Carbohydr. Polym.* 84, 430–438.

Xing, G., Zhang, S., Ju, B., Yang, J., 2006. Study on adsorption behavior of crosslinked cationic starch maleate for chromium(VI). *Carbohydr. Polym.* 66, 246–251.

Zeng, C., Han, X., Lee, L.J., Koelling, K.W., Tomasko, D.L., 2003. Polymer–clay nanocomposite foams prepared using carbon dioxide. *Adv. Mater.* 15, 1743–1747.

Zhou, C.-H., Zhang, D., Tong, D.-S., Wu, L.-M., Yu, W.-H., Ismadji, S., 2012. Paper-like composites of cellulose acetate–organo-montmorillonite for removal of hazardous anionic dye in water. *Chem. Eng. J.* 209, 223–234.

10 Bio-Based Polyurethane Foams

Aleksander Prociak

CONTENTS

10.1 INTRODUCTION

Polyurethanes (PUs) are very interesting part of polymers, which are manufactured as solids and porous materials. Their development is dated from the first patent application of Otto Bayer in 1937. PUs are synthesized in the polyaddition reaction (10.1) of isocyanates with components containing hydroxyl groups, that is, polyols.

$$\sim R-N=C=O \quad + \quad HO-R'\sim \quad \longrightarrow \quad \sim R-\underset{\underset{H}{|}}{N}-\overset{\overset{O}{||}}{C}-O-R'\sim \qquad (10.1)$$

The commercialization of the first PU products, such as foams, elastomers, coatings, fibers, and adhesives, took place in the 1940s (Randall and Lee 2002; Król 2009).

PU's properties can be modified in a broad range and tailored to match the demands of various applications. Rigid or flexible foams and microcellular elastomers or solids with apparent density from 10 to 1100 kg m^{-3} are produced using different ratios of various polyols and isocyanates, as well as additional components, such as catalysts, blowing agents, surfactants, and fillers (Prociak and Michałowski 2012a; Szycher 2013).

PUs are used in different branches of the industry, such as furniture, automotive, buildings, heat insulating, and constructions. The structure of PU products in the world market is shown in Figure 10.1. In 2010, 65 wt% of PU systems were used to obtain foams of different types. The most significant share of PU foams (28 wt%) are flexible products (http://marketsandmarkets.com, 2011). The prediction for the future of the PU foams market in the nearest years is rather optimistic. The use of PU foams will be increased, especially as rigid heat insulating foams due to the energy saving requirements, imposed by the increasing costs of raw materials, as well as environmental aspects and legislation solutions, for example, in the European Union (Prociak 2008a; Peacock 2009).

Foams have a special position among PU products (Prociak 2008a; Paciorek-Sadowska et al. 2010). Low weight and apparent density, as well as excellent mechanical properties, make them very attractive polymeric materials (Yan 2002). In order to obtain porous PUs, chemical and physical blowing agents are used. The most popular method is foaming PUs with carbon dioxide, which is generated in the reaction (10.2) of water with isocyanate groups.

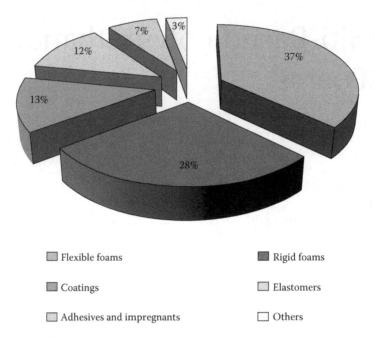

Flexible foams   Rigid foams

Coatings   Elastomers

Adhesives and impregnants   Others

FIGURE 10.1   Global market of PUs by material type in 2010.

$$\sim R - N = C = O \quad + \quad H_2O \quad \longrightarrow \quad \sim R - NH_2 \quad + \quad CO_2 \uparrow \qquad (10.2)$$

Foams can be produced in blocks or can be molded in different forms and sizes (Veronese et al. 2011). Nowadays, flexible foams cover the major uses of PUs in the market (Petrovic 2008; Dumont and Narine 2010). The major applications of flexible PU foams are cushioning materials. They are used in furniture, mattresses, and automotive applications. They are also applied in transportation, textiles, packaging, medical supplies, sporting goods, or as sound absorbents. PU industry is interested in low-density flexible foams due to lower costs of such materials. Recently, technologies that use liquid carbon dioxide as an auxiliary blowing agent have emerged and have gained wide attention. The apparent densities of flexible foams can be varied in the range of 10 to 30 kg·m^{-3}, based on the market requirements (Landrock 1995; Rojek and Prociak 2012).

Among the flexible foams, viscoelastic products are especially interesting due to their specific properties. The characteristic behaviors like slow return to the previous shape after deformation and high ability for energy absorption made them interesting for astronautic applications many years ago. Nowadays, viscoelastic foams (VEFs) are widely used in the mattresses' production, especially for rehabilitation, as well as to produce various elements of body protection, which are able to absorb energy (Vaughan et al. 2011; Huo et al. 2012).

Rigid PU foams are widely used as thermal insulation in refrigerators and constructions, as well as for acoustic insulation, because of their combination of low density, low thermal conductivity, and good mechanical properties (Modesti et al. 2002; Branca et al. 2003; Veronese et al. 2011). Nowadays, rigid PU foams are the most effective heat insulating materials among commercial products used in refrigerators and in building industry. Their thermal conductivity mainly depends on the type of gas closed in cells, as well as on dimensions and shape of cells. The important challenge is to obtain foams with nano-sized cells (Prociak 2007b; Kurańska and Prociak 2012).

The synthesis of PU foams with different properties, tailored for specific applications, is possible due to the broad base of raw materials, especially polyols. Taking into account the need of sustainable development, new PU formulations are developed with consideration of aspects such as product quality, components and energy saving, low emission of volatile organic compounds, waste reduction, and life cycle assessment (Petrovic 2008; Prociak 2008a).

An introduction of renewable raw materials to the PU systems is one of the solutions that allow the synthesis of sustainable PU foams. Oils, fats, and plant fillers are gaining popularity due to certain attractive properties related to their specific structures, as well as due to general concern for the environment and sustainability (Shogren et al. 2004; Veenendaal 2007). The derivatives of vegetable oils can potentially replace petrochemical polyols in the synthesis of PUs; moreover, they exhibit capacity to biological degradation (Helling and Russell 2009; David et al. 2010; Palanisamy et al. 2011).

A special type of oil is castor oil, which contains about 90% of ricinoleic acid, having three hydroxyl groups in each molecule. Pricewise, oils such as soybean, palm, and rapeseed are the most attractive ones for large-scale industrial products. Soybean oil is the most attractive raw material in North and South America since it combines both stable and low price with a relatively high degree of unsaturation, while Europe is more oriented toward rapeseed oil and Asia toward palm oil (Hill 2000; Guo et al. 2002; Petrovic et al. 2005).

10.2 BIO-BASED COMPONENTS FOR PU SYNTHESIS

The intensive development of components from renewable raw materials is reflected in their use in the world production of PU materials (Hill 2000; Prociak 2008a). Petrochemical components are replaced with bio-based raw materials in the synthesis of different type PU foams and elastomers (Veenendaal 2007; Petrovic 2008). Vegetable oil-based polyols are mainly used to produce flexible foams for automotive industry sector. Nevertheless, such polyols are also used in the synthesis of VEF for furniture industry and other specific applications (Prociak et al. 2014).

The main components of the PUs are polyols and isocyanates. Nowadays, efforts are being made to replace petrochemical polyols with those based on renewable raw materials in order to obtain bio-based and low cost PU materials (Yeganeh and Mehdizadeh 2004; Meier et al. 2007; Prociak et al. 2014). Typical polyols contain 2–8 reactive hydroxyl groups and have average molecular weight from 300 to 10,000 g·mol^{-1}. The two types of polyols, that is, polyether and polyester polyols, are most often used in the synthesis of foams. In the PU industry, renewable resources always played an important role in the development of polyols. For example, glycerol is the most important starter in the synthesis of polyether polyols for flexible PU foams, while polyether polyols for rigid foams are produced via the hydrolysis of natural triglycerides from vegetable or animal resources (Ionescu 2005). Sucrose, the most important starter of polyether polyols for rigid foams, is produced exclusively by extraction from natural resources. Tetrahydrofuran, a cyclic monomer used in the synthesis of polytetramethyleneglycols, can be produced from furfurol, which is in turn obtained by the reaction of acid hydrolysis of pentosanes existing in many agricultural wastes, such as corn on the cob and straw. A very interesting natural starting material to prepare polyols for rigid foams is lignin, available in large quantities from the wood and cellulose industry (Ionescu 2005).

Recently, research efforts also are aimed at application of natural raw materials, especially vegetable oil derivatives, in the formulations of PU systems (Prociak 2007a; Rojek and Prociak 2012).

Vegetable oils are triglycerides of predominantly unsaturated fatty acids. They are considered to be relatively unreactive chemically. In order to apply them for formation of the urethane groups —NH—COO—, most of vegetable oils have to be transformed into compounds containing at least two hydroxyl groups capable of reacting with isocyanates (Prociak 2008a).

A unique vegetable oil is castor oil, extracted from the seeds of the plant *Ricinus communis*. This triglyceride of ricinoleic acid contains three secondary hydroxyl groups. Castor oil has played an important role, especially in the earlier stages, of the PU industry for more than 50 years. However, castor oil has certain major disadvantages: low functionality, low hydroxyl number, and low reactivity due to secondary hydroxyl groups. Thus, castor oil as sole polyol is generally used for the synthesis of semi-rigid PU foams (Hu et al. 2002; Ionescu 2005; Prociak et al. 2014).

The unsaturated vegetable oils, having double bonds but without hydroxyl groups, can be transformed into polyols using various chemical reactions. Over the last 10 years, two basic types of reactions have been developed:

1. Transesterification and transamidation involving esteric groups
2. Converting double bonds into hydroxyl groups

In the first case, monoglicerydes or other hydroxyl derivatives of different functionalities are obtained in the transesterification reaction of oils with glycols, glycerol, sorbitol, and sucrose or transamidization reaction usually with, for example, diethanolamine (Ionescu 2005; Petrovic 2008; Kirpluks et al. 2013).

This method leads to polyols with molecular weight usually below 1000 g mol^{-1} and a functionality of approximately 2. In the glycelorysis process of vegetable oils, a mixture of mono- and di-glycerides is generally obtained. In order to obtain mixture with approximately 70 wt% participation of monoester, a large excess of glycerol is necessary (Nowicki 2007; Prociak et al. 2014).

In case of conversion of double bonds, the following solutions were implemented: hydroformylation and hydrogenation, epoxidation followed by oxirane ring-opening, microbial conversion, thermal polymerization followed by transesterification, ozonolysis and hydrogenation or halogen addition, and nucleophilic substitution (Zhang et al. 2007; Lubguban et al. 2009; Garrett and Du 2010; Rojek and Prociak 2012).

Vegetable oil-based polyol structure depends not only on the method of the insertion of hydroxyl groups but also on the type of triglycerides. Structure of vegetable oils varies with the distribution and position of the double bonds in the fatty acid chain. Edible oils have three dominating unsaturated fatty acids: oleic, linoleic, and linolenic. All these unsaturated fatty acids have *cis*-double bonds. The position of the first double bond is between the ninth and 10th carbons, the second between the 12th and 13th, and the third between the 15th and 16th (Petrovic 2008). Generic structure of vegetable oil containing oleic, linoleic, and linolenic acid chains is shown in Figure 10.2 (Stirna et al. 2011).

The composition of the fatty acids in the selected vegetable oils and the potential functionalities of polyols based on them are shown in Table 10.1 (Petrovic 2008; Prociak et al. 2014).

Vegetable oil-based polyols with hydroxyl groups introduced at the positions of double bonds usually have molecular weights from 600 to 6000 g·mol^{-1}. Due to specific triglyceride structure, functional groups in the oil-based polyols are not at the end of fatty acid chains. In case of oleic fatty acid, the hydroxyl groups are located exactly in the middle of the chain. When such bio-polyols are cross-linked, a part of chains (most often from the 10th carbon) creates so-called dangling chains, which significantly affects the properties of the produced foams (Petrovic 2008).

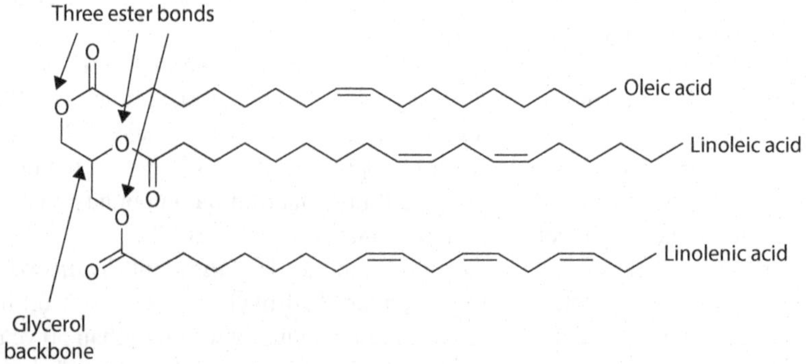

FIGURE 10.2 Generic structure of vegetable oil containing oleic, linoleic, and linolenic acid chains. (Data from Stirna, U. et al., Sci. *J. Riga Tech. Univ.*, 6, 85–90, 2011.)

TABLE 10.1

Composition of the Fatty Acids in the Selected Vegetable Oils

	Type of Fatty Acid, Number of Double Bonds				
Type of Oil	Oleic, 1	Linoleic, 2	Linolenic, 3	Others	Functionality
Palm	39	11	0.3	50.7	1.8
Rapeseed	61	21	10	8	3.9
Corn	28	58	1	13	4.4
Soybean	25	53	6	16	4.5
Sunflower	23	65	0.5	11.5	4.6
Linseed	18	16	56.5	9.5	6.4

Source: Prociak, A., Materiały Poliuretanowe, Państwowe Wydawnictwa Naukowe PWN, Warszawa, Poland, 2014.

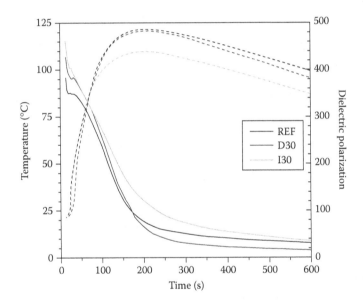

FIGURE 10.3 The profiles of the temperature and dielectric polarization of the PU systems. (Data from Rojek, P. and Prociak, A., Foaming process and cellular structure analyses of bio-based flexible polyurethane foams, *Blowing Agents and Foaming Processes*, Mainz, Germany, 2013a.)

In order to analyze the reactivity of the bio-polyols in the preparation of PU foams, different derivatives based on rapeseed oil (ROP) were synthesized using the two-step method of epoxidation and opening oxirane rings. In the first step rapeseed oil with the same epoxy numbers was obtained. In the second step, two types of agents, namely isopropanol and diethylene glycol, were used to obtain polyols called rapeseed oil-based polyol (RIso) and diethylene glycol (RDEG), respectively, with various types of hydroxyl groups and different reactivity. The reactivity of PU systems with such polyols were measured using the Foamat device, which allows to measure the profiles of temperature, pressure, and dielectric polarization of the reaction mixture, as well as the rise height of foams. In Figure 10.3 (Rojek and Prociak 2013a), profiles of the temperature and dielectric polarization of the PU systems modified with two types of ROPs in comparison to the reference PU system based only on petrochemical polyol are shown.

The highest reaction temperatures occur for formulation D30 modified with 30 wt% of RDEG polyol, because this polyol has primary and secondary hydroxyl groups and also the highest hydroxyl value (OHV). The highest reactivity of D30 systems is also confirmed by the highest slope of dielectric polarization curves, which provide information about the yield of curing. The lowest reactivity was observed for I30 systems modified with 30 wt% of RIso polyol, because of nearly twice lower OHV of RIso polyol in comparison to RDEG polyol, and presence of only secondary hydroxyl groups in RIso. These systems have the lowest slope of dielectric polarization curves (Rojek and Prociak 2013a).

The viscosity of polyols has also an appreciable effect on the foaming time, liquidity of the reaction mixture, and appropriate filling of the mold. It was found that the dynamic viscosity of rapeseed oil-based polyols decreases with higher shear rate and temperature. Generally, such polyols are non-Newtonian fluids. The viscosity values of polyols depends on the ROPs molecular weight and most often is in the range of 500–5000 MPa·s (Prociak et al. 2011; Dworakowska et al. 2012).

The high viscosity soy-polyols (13,000–31,000 MPa·s) can be made from fully epoxidized soybean oil by alcoholysis reaction. In the alcoholysis reaction, epoxidized soybean oil is combined with ethylene glycol using p-toluenesulfonic acid as catalyst to promote the oxirane ring opening and form hydroxyl groups (Fan et al. 2013).

Different types of polyols can also be prepared based on the lignocellulose from bark, wood, starch, and waste paper. Such materials can be liquefied into polyols with the addition of an acid catalyst. The lignocellulose-containing polyols possess a large amount of phenolic and alcoholic hydroxyl groups with high reactivity and are suitable for preparing PU foams. Studies have also shown that tannin and bark could act as effective cross-linking agents in PU foams synthesis (Lee et al. 2002; Wang et al. 2008; Zhao et al. 2012).

The second main isocyanate component for the PU synthesis may also be prepared using plant oils. An easy and efficient synthesis of unsaturated plant oil triglycerides having isocyanate groups was reported. In the first step of the synthesis, the triglyceride is brominated at the allylic positions in a reaction with N-bromosuccinimide. Such brominated derivatives are then reacted with AgNCO to convert them into isocyanate-containing triglycerides with efficiency of ca. 60%–70%.

The double bonds of the triglycerides are not consumed in this method (Cayli and Kusefoglu 2008). Aliphatic diisocyanates from dimerized fatty acids are also known; however, their reactivity is insufficient for application in foams. In the PU foam, formulations isocyanates ought to be aromatic and therefore isocyanate prepolymers based on bio-polyols and aromatic isocyanates (toluene diisocyanate and methylene diphenyl diisocyanate) have been developed (Petrovic 2008; Prociak et al. 2014).

10.3 FLEXIBLE AND VISCOELASTIC BIO-BASED PU FOAMS

Flexible foams have the largest share in the market of PU products, thus use of natural renewable polyols for flexible foams is highly desirable. Moreover, life cycle assessment of vegetable oil-based polyols shows clear environmental benefits of their application in the area of flexible PU foams. Compared to petrochemical-based polyols, the vegetable polyols most often contain from 33% to 64% of the fossil resources and generate very low greenhouse gas emissions (Helling and Russell 2009).

A direct approach is to use the natural polyol, for example, castor oil. Foams made solely from this natural polyol have a temperature dependent modulus. The materials made from this natural polyol are limited to low resiliency foams, which means only partial recovery can be attained immediately after deformation. (Nozawa et al. 2005). The narrow range of achievable foam properties and the relatively high cost of castor oil have turned researchers to transform other vegetable oils for polyols (Zhang et al. 2007).

Nowadays, soy oil-based polyols are the most often used natural oil-based polyols (NOPs) in PU materials (Dawe et al. 2007; Rojek and Prociak 2012). However, flexible PU foams modified with palm and rapeseed oil-based polyols were also examined (Shaari et al. 2006; Rojek and Prociak 2012).

The main field of PU application is the furniture industry and ca. 30 wt% of the total worldwide PUs is used for the production of mattresses from flexible slabstock foams. Automotive industry is the second important consumer of flexible and semiflexible PUs (seat cushioning, bumpers, sound insulation, and so forth) (Ionescu 2005).

There are two basic methods of flexible foams synthesis. Slabstock foams are produced continuously by pouring the formulation mixture on a conveyor belt, and the foaming process is carried out in open air to form a cuboid. The foams are cut to desired shapes such as mattresses (Petrovic 2008). Molded foams are formed in closed molds and have higher technical requirements. The final shape of product depends on the mold shape. These foams should have low apparent density and good elastic recovery and support properties without large deformation (Carnicer 2005; Prociak et al. 2014).

Flexible foam characteristics depend on the degree of cell opening, which should be complete for desirable properties. Polyols for slabstock flexible foams are usually polyethers triols with secondary OH groups but the reactivity of those for molded application must be higher. That is achieved by capping with ethylene oxide, which also improves miscibility with water. Addition of copolymer polyols containing 20–30 wt% of dispersed micron-sized particles of usually styrene/acrylonitrile copolymer helps cell opening in case of flexible foams (Petrovic 2008; Prociak et al. 2014). A similar function is played by oil-based polyols with primary hydroxyl groups, which can be obtained by oil epoxydation and subsequent opening of oxirane rings with diethylene glycols. The reactivity of such polyols may be higher in comparison to the typical polyether polyols used in the synthesis of flexible foams (Prociak et al. 2011; Rojek and Prociak 2013b).

In case of foams synthesized using entirely soybean, rapeseed or palm oil-derived polyols, the materials were not of good quality and improvements in different aspects, such as surfactant efficiency and NOPs reactivity were needed. Partial incorporation of vegetable oil-derived polyols has been much more successful (John et al. 2002; Zhang et al 2007; Prociak et al. 2011).

The effects of hydroxyl groups positions on the PU system reactivity and foaming process were investigated using soybean oil-based polyols (SOPs) and ROPs. Flexible PU foams from three SOPs with different functionalities and positions of OH-groups (primary or secondary OH-groups) were synthesized. It was found that the SOP containing primary OH-groups was the most reactive and allowed to obtain foams with smaller and more uniform cells (Campanella et al. 2009).

Soybean oil-derived polyols synthesized by epoxidation, followed by an oxirane ring-opening reaction using a mixture of water and methanol, are less reactive than conventional polyols because of its secondary hydroxyl groups (Zhang et al. 2007). Similar results have been found for rapeseed oil-based polyols (RIso) obtained with isopropanol as oxirane ring opening agent. The introduction of primary hydroxyl groups by using diethylene glycol (RDEG) as oxirane ring opening agent allows to increase the maximal temperature of reaction mixture and to accelerate the cross-linking process (lower values of dielectric polarization in Figure 10.4b). However, such polyols, practically,

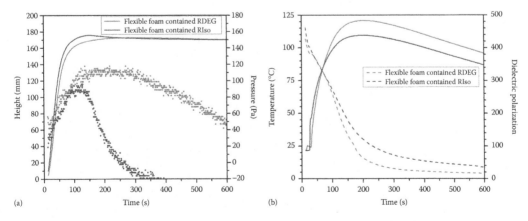

FIGURE 10.4 Foaming process parameters of flexible foam formulations with different structure of hydroxyl groups: (a) rise height and pressure and (b) temperature and dielectric polarization of reaction mixture.

do not influence the rise height of the reaction mixture and show lower tendency for cell opening (higher pressure values in Figure 10.4a (Rojek and Prociak 2013b).

Recently, it was found that the addition of 20–30 wt% vegetable oil-based polyols into a foam formulation allow to achieve similar mechanical, insulating, and other physical properties as those foams containing solely 100% petroleum-based polyols (Tu et al. 2007; Zhang et al. 2007).

For all foams with petrochemical polyols partially replaced with NOPs higher hardness and compressive modulus are observed. Moreover, a significant load-bearing increase measured by indentation force deflection test was reported. The experimental results confirmed that the SOPs are capable of improving mechanical properties, namely compressive properties (Herrington and Malsam 2006; Zhang et al. 2007).

Compression properties were also examined and compared via determination of compression values and compression stress—strain characteristics, as well as tensile strength and elongation at break for flexible PU foams modified by partial substitution of petrochemical polyether triol with ROPs and obtained with different isocyanate index. Bio-polyols differed in functionalities—2.5 (Polyol I) and 5.2 (Polyol II)—and in OHV 114 and 196 mg KOH g^{-1}, respectively. Foams modified with Polyol I had higher values of resilience and elongation at break than those with Polyol II, while higher tensile and compression strength and superior cell structure were observed in case of foams modified with Polyol II (Rojek and Prociak 2012).

Compression stress of flexible foams modified with ROPs depends on the characteristic of polyols, including the type and number of hydroxyl groups (OHV). The compression stress at 25% of strain increases with the increasing content of RDEG (with primary and secondary hydroxyl groups; OHV = 156 mg KOH g^{-1}) while this property does not change significantly with the increasing content of RIso (with secondary hydroxyl group only, OHV = 80 mg KOH g^{-1}). It is caused by higher functionality of RDEG and thus increased hard segment content in modified foams (Rojek and Prociak 2013a).

It was also confirmed (differential scanning calorimetry and dynamic mechanical analysis results) that the improved PU modulus in the foams modified with SOPs is attributed to a combination of factors: a high T_g of SOP-rich phase, high hard segment concentration, and improved hard domain ordering. In SOP-modified foams, an SOP-rich region, which has a higher T_g than the polyether polyol-based soft domains, may exist. Atomic force micrographs are allowed to confirm the possibility of presence of two types of soft domains in such foams. Moreover, atomic force micrographs have shown that addition of 30% SOP gave smaller hard domains with a distribution of interdomain spacings. The Fourier transform infrared spectroscopy results confirmed that SOP-containing foams have the most ordered hard domain structures, implying a well-separated hard phase (Zhang et al. 2007).

As shown in Figure 10.5, in case of a large majority of investigated foams, the resilience by ball rebound test of the foams was reduced with increasing isocyanate index. However, unexpected effect of considerable increase of resilience versus NCO index for foams modified with 30 wt% of Polyol I has also been noticed. This effect corresponded to the changes of compression properties of the analyzed foams. The modification of PU systems with Polyol II also influenced the resilience changes in case of products with 20 wt%.

Generally, foams with petrochemical polyols partially replaced with Polyol I had better resilience than those modified with Polyol II. It can be explained as an effect of higher hard segment content in the foams with Polyol II, due to its higher functionality and OHV. As a consequence, the foams modified with Polyol II had lower elasticity. In case of relatively high resilience Polyol I-derived foams, the plasticizing effect prevails (Rojek and Prociak 2012).

NOPs with hydroxyl groups introduced at the positions of double bonds have no terminal hydroxyl groups and dangling chains in such structure play the role of a plasticizer (Petrovic 2008).

The replacement of petrochemical polyol with NOPs significantly affects the foam's behavior upon stretching, as shown in Figure 10.6. The tensile strengths of the reference foam (REF) and foams modified with up to 50wt% with ROP are on a similar level (Figure 10.6a), while the tensile strength of foams modified with palm oil polyol (POP) is higher by ca. 10% in comparison to

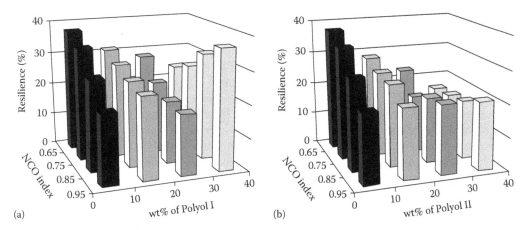

(a)   (b)

FIGURE 10.5   Resilience of flexible polyurethane foams versus isocyanate index and percentage of ROP: (a) foams modified with Polyol I and (b) foams modified with Polyol II. (Data from Rojek, P. and Prociak, A., *J. Appl. Polym. Sci.* 125: 2936–2945, 2012.)

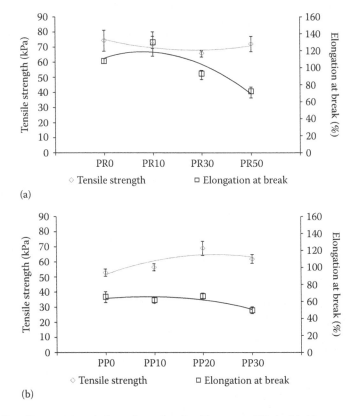

(a)

(b)

FIGURE 10.6   Tensile strength and elongation at break of foams modified with (a) rapeseed oil-based polyol PR and (b) palm oil-based polyol PP. (Data from Prociak, A. et al., *J. Cell Plast.*, 48, 489–499, 2012b.)

the reference material (Figure 10.6b). In case of foams containing ROP or POP, the decrease of elongation at break is observed and this is an effect of shortening of the soft segment length, when NOPs content in the foam formulation is increased (Prociak et al. 2012b).

In the twenty-first century, a dynamic development of so-called VEF is observed. Research effort allowed to obtain modified VEF materials with different properties, as well as to broaden their applications in various industrial branches (Zhang and Yu 2005). Due to their ability to change the elasticity as an effect of human body temperature, VEF allows to increase the comfort of their use as cushions, mattresses, and parts of furniture of high standards. Low resilience makes such materials able to damp mechanical vibrations and to absorb up to 90% of energy. Therefore, VEFs are increasingly often used in automotive and sport applications (Dai et al. 2007; Vaughan et al. 2011).

In the synthesis of VEF, castor oil, as well as other various types of vegetable oil-based polyols, can be used to replace up to 40 wt% of petrochemical polyols (Andries and Macken 2009; Abraham et al. 2010). The type and amount of applied NOPs considerably influences the chemical structure and physical–mechanical properties of the final products (Obi et al. 2007). The application of NOPs may cause elongation of gel and tack-free times of foams, what is unbeneficial from technological point of view (Singh and Bhattacharya 2004). Moreover, the dangling chains of fatty acids cause plasticization of the PU matrix (Petrovic 2008; Vaughan et al. 2011).

Generally, the addition of NOPs to VEF formulations causes the following:

* Decreased surface tension (Prociak et al. 2014)
* Increased cell number, however, and difficulty of their cracking (Ionescu 2005; Singh and Bhattacharya 2004)
* Increased apparent density and changes of mechanical properties (Tanaka et al. 2008)
* Increased hardness and compression strength (Prociak et al. 2014)
* Decreased tensile strength and elongation at break (Sharma and Kundu 2006; Pawlik and Prociak 2012)

The comparison of typical properties of VEFs modified with castor oil , rapeseed oil-based polyol (RP) and palm oil-based polyol with the REF prepared basing petrochemical polyols is shown in Table 10.2.

The analysis showed that the properties of the foams modified with NOPs (castor oil, PP, and RP) are comparable with the REF. However, replacement of the petrochemical polyol (30 wt%) with PP and RP causes the apparent density of modified foams to increase, while the foam CO30 has similar density to REF. It was probably caused by higher viscosity of PP and RP, than of the replaced petrochemical polyol. The increased content of closed cells can also be explained by the effect of higher initial viscosity of PU systems modified with rapeseed and palm oil-based polyols. In case of modified foams, resilience and tensile strength are higher, while the elongation at break is on the similar level. It can be correlated with

TABLE 10.2
Selected Properties of Viscoelastic Foams

Properties	REF	CO30	RP30	PP30
Apparent density (kg/m^3)	40.0	38.6	45.1	48.4
Closed cells content (%)	4.8	5	6.8	10.3
Compressive stress at 40% (kPa)	1.6	2.4	9.0	11.5
Resilience (%)	5.7	7.7	7.4	9.6
Tensile strength (kPa)	58	104	153	159
Elongation at break (%)	152	186	156	141
T_g (°C)	13.1	8.6	25.1	22.1
				43.1

Source: Pawlik, H. and Prociak, A., Influence of natural oil-based polyols on selected properties of viscoelastic polyurethane foams, modern polymeric materials for environmental applications, *Proceedings of the 5th International Seminar*, vol. 5, 101–106, Kraków, Poland, 2013.

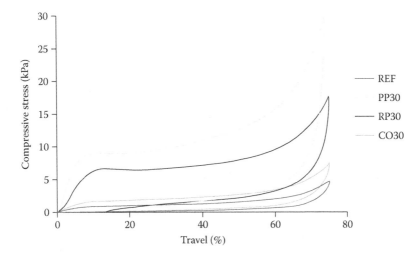

FIGURE 10.7 The hysteresis of foams modified with PP, RP, and castor oil polyols. (Data from Pawlik, H. and Prociak, A., Influence of natural oil-based polyols on selected properties of viscoelastic polyurethane foams, modern polymeric materials for environmental applications, *Proceedings of the 5th International Seminar*, vol. 5, 101–106, Kraków, Poland, 2013.)

higher cross-linking density due to the presence of oligomers fraction in PP and RP polyols, and it also can be an effect of shorter chains between branching points. For this reason, also an increase of compression stress is observed (Pawlik and Prociak 2013).

The compression behavior of the foams during loading and unloading is shown in Figure 10.7. The highest compressive stress was noticed when loading the foams containing PP and RP polyols, while the unloading curves represent similar shape and level for all foams. Hence, different energy is dissipated or lost as heat in case of investigated materials. Therefore, the greatest value of the hysteresis loss occurs in case of PP30 foam, which means that this material has the best energy damping properties.

The area under the loading curves and the hysteresis loop were used to determine the ability of the foam to absorb energy. The hysteresis loss was calculated as the ratio of the loop area and the area below the loading curve. The increase of hysteresis loss of foams modified with castor oil, PR, and PP in comparison to REF was ca. 5%, 10%, and 20%, respectively (Pawlik and Prociak 2013).

Currently, the introduction of new type NOPs to the PU systems for flexible and VEF is further developed. Different types (soybean, rapeseed, and palm) of NOPs are investigated and several of them are produced on an industrial scale. In order to obtain good quality and economically acceptable foams, the new PU systems with appropriate additives have been elaborated. One of the most important challenges in the nearest future is the selection of the so-called low emission and no emission catalysts due to the environmental aspects and lower reactivity of NOPs in comparison to typical petrochemical polyols (Prociak et al. 2014).

10.4 RIGID BIO-BASED PU FOAMS

Rigid PU foams are most often used for thermal insulation of buildings and refrigerators, cold stores, pipes, refrigerated transport, thermal insulation in chemical and food industries (Ionescu 2005; Prociak et al. 2014).

Different types of vegetable oils, such as palm, rapeseed, soybean, and linseed oil, have been used to synthesize rigid foams. The results of the reaction kinetics of PU foams made from modified SOPs confirm that PU foams can be successfully made from NOPs. The properties of rigid foams can be changed by controlling variables such as water content, isocyanate index, and catalysts (John et al. 2002).

TABLE 10.3

**Influence of ROP on Cell Structure and Selected Properties
of Rigid PU Foams Synthesized in Different Molds**

Properties	Foam Symbol			
	RF1	R1DG3	R1DG4	R1DG5
Content of ROP in polyol premix	0	30	40	50
Foams from Mold I				
Apparent density (kg/m³)	27.5	25.2	28.6	38.9
Closed cell content (%)	83	86	72	62
Thermal conductivity (mW/m·K)	25.8	24.0	25.4	26.7
Foams from Mold II				
Apparent density (kg/m³)	35.6	40.9	41.0	42.7
Closed cell content (%)	95	94	92	89
Thermal conductivity (mW/m·K)	20.7	20.3	20.7	20.9

Source: Prociak, A., *Polimery*, 3, 195–200, 2008b.

Generally, the resulting products exhibit improved thermal conductivity and good dimensional stability (Prociak 2007a; Zhang et al. 2007). The absence of the ether linkage is the key to improved thermal and oxidation properties (Guo et al. 2000). The foams modified with NOPs have mechanical properties comparable to the foams based only on petroleum raw materials (Prociak 2008a; Kurańska and Prociak 2012). Mechanical testing of rigid foams synthesized from a number of soybean oil-derived polyols showed that they exhibit similar compressive properties to the petroleum-based foams (Tu et al. 2007, 2008).

Thermal conductivity of all tested foams showed no significant changes and was independent of polyol properties. Rigid foams can also be made from rapeseed oil-derived polyols; however, researchers recommended that both polyol reactivity and long-term stability of the polyols should be further understood (Hu et al. 2002). The low reactivity of oil-derived polyols can cause difficulties in foam density control. The increase of apparent density of rigid foams modified with ROP content is shown in Table 10.3 (Prociak 2008b).

Cell structure of rigid foams depends on a large number of factors. The application of NOPs as replacements of petrochemical polyols up to 30 wt% in polyol premix allows to reduce considerably the cell size; however, it can also increase the anisotropy (Prociak 2007c, 2008a).

The influence of initial viscosity of NOPs on foaming process and cell structure is very significant. Generally, the POPs, ROPs, and SOPs described in literature have viscosity in the range from 1000 to 10,000 MPa·s, what is typical for the petrochemical polyols (Prociak et al. 2014). However, high viscosity (21,000–31,000 MPa·s) SOPs can be also used in the synthesis of rigid foams. The structure micrographs of such foams showed that they have numerous cells in the shape of spheres or polyhedra. With increasing soy-polyol percentage, the cell size decreased while their number increased (Fan et al. 2013).

The foaming process and cell structure depend also on the presence of other components including nanofillers and the dispersion quality of polyol premix (Woo et al. 2006, Semenzato et al. 2009). The 3D-struture reconstruction of rigid foams modified with SOP and nanofillers (cellulose microfibers and nanoclays) has shown that the application of such components in PU system makes it possible to decrease the cell size and improve the cell uniformity (Zhu et al. 2012).

In case of rigid foams, which are applied as heat insulating material, thermal conductivity is one of the most important properties. The cell structure has a considerable influence on thermal conductivity due to the heat transport through the PU matrix and radiation. Moreover, the thermal conductivity of foams depends on the closed cell content and type of gas filling them. The effect of closed cells content and their anisotropy on the thermal conductivity of the foams modified with different amount of ROP is shown in the Table 10.3 (Prociak 2008b; Prociak et al. 2014).

The foams prepared in mold II had cells elongated perpendicularly to the heat transport direction during the thermal conductivity measurements, and therefore the values of this property are lower when compared with results for foams from mold I (cells are elongated in parallel to the heat transport direction), although their apparent density was higher (Prociak 2008b).

Rigid foams made with SOPs of high viscosity had slightly higher values than the REFs based only on petrochemical polyols. This was caused by a lower closed cell content of PU foams made from SOPs than that in the reference material. However, micrographs showed that with increasing SOP percentage, the foam cell size decreased and cell number increased. This change was attributed to the different chemical structure and the initial viscosity of compared PU systems. With increasing SOP content, the viscosity of the reaction mixture increased reducing the tendency for coalescence of cells (Fan et al. 2013).

The application of NOPs in rigid foam formulation considerably influenced mechanical properties of final products. The investigation of foams made with castor oil, canola oil-based polyol and SOP showed that PU system performance depends on the number and position of OH-groups and dangling chains in the starting materials, which also influences the cell structure and as a consequence also the mechanical and thermal properties of rigid PU foams. The foams based on canola oil showed better compressive properties than commercially available soybean polyol, but not as good as castor oil foam (Narine et al. 2007).

Foams made with high viscosity (21,000–31,000 MPa·s) SOPs exhibited similar or superior density-compressive strength properties to the control foam made from only petrochemical polyol. Based on the analysis of isocyanate content and compressive strength of foams, it was concluded that rigid PU foams could be made by replacing 50% petroleum-based polyol with a high viscosity soy-polyol resulting in a 30% reduction in the isocyanate content (Fan et al. 2013).

Compressive strength of rigid foams with modified NOPs depends on different factors, such as hydroxyl number and functionality of the polyol, the type of blowing agent, apparent density, and chemical structure of foams (Prociak et al. 2014). The comparison of physical–mechanical properties of rigid PU foams modified with different types of NOPs allows to state that applied formulations make it possible to obtain products with permanent dimensional long-term stability by replacing 30 wt% of petrochemical polyol with rapeseed and sunflower and 50 wt% with linseed oil-derived polyols. Among the compared foams, those modified with linseed oil-based polyol have the lowest apparent density, which results in relatively worse mechanical properties measured in parallel to the foam rise direction. However, values of compressive strength measured in the most critical direction (perpendicular to the foam rise) were comparable for all three foams (Prociak 2007a). Another work (Veronese et al. 2011) focused on the properties of rigid PU foams made from soybean oil-based polyol (OHV = 477 mg KOH g^{-1}) and castor oil-based polyols (OHV = 393 and 441 mg KOH g^{-1}). The foams made from SOP with a higher OHVr had higher T_g because of higher cross-linking density. All, bio-based PU foams showed slightly worse mechanical properties than the REF based on petrochemical polyol.

The effects of various factors such as catalyst, blowing agent, functionality, and viscosity of SOPs on the rigidity of rigid PU foams were also evaluated. The results showed that higher content of catalyst or higher functionality of SOP allow to obtain a more rigid foam structure. The application of SOPs with similar OHV but with different viscosity, influences mechanical properties of final products. The foams made from SOP with higher viscosity had higher compressive strength (Banik and Sain 2008).

Dimensional stability of foamed materials depends on their cell structure, the type of used blowing agents and mechanical properties of the PU matrix. The chemical structure of polyols has a considerable effect on mechanical properties of rigid PU foams. Plasticization effect of increased content of ROP on compressive strength of foams is shown in the Table 10.4 for different directions of measurements (Prociak 2008a).

The opposite effect, that is, increased compressive strength was noticed for rigid foams modified using SOP with functionality 4.4 (Tan et al. 2011). However, commercially available petroleum-derived polypropylene-based polyol with functionality 3.0 was chosen as the reference polyol. The foam density increased with increasing SOP content but only significantly at 100% substitution of reference polyol. Compressive strength increased with SOP content probably due to the decreased foam cell size. In Figure 10.8, the results of dynamic mechanical properties of these foams are shown. In the glassy state, all samples with different content of SOP have comparable plateau G'

TABLE 10.4
Selected Properties of Rigid Foams Modified with Rapeseed Oil-Based Polyols

	Foam Symbol			
Polyol	SPPUR	SPPUR30	SPPUR40	SPPUR50
Alfapol RF-551 (% mas.)	100	70	60	50
ROP (% mas.)	0	30	40	50
Foam Properties				
Apparent density (kg/m³)	36.0	42.0	42.5	43.0
Compressive strength at 10% (kPa)				
• Parallel	188	180	148	102
• Perpendicularly 1	142	126	113	77
• Perpendicularly 2	96	86	57	48

Source: Prociak, A., Poliuretanowe materiały termoizolacyjne nowej generacji, Wydawnictwo Politechniki Krakowskiej, Kraków, Poland, 2008a.

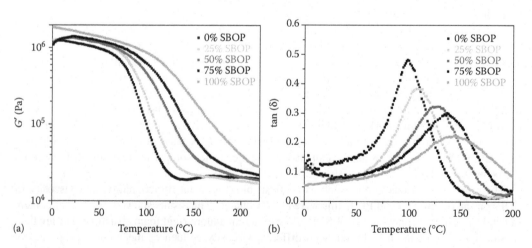

FIGURE 10.8 Dynamic elastic modulus (a) and tan δ (b) of PU foams modified with SOP versus temperature. (Data from Tan, S. et al., *Polymer*, 52, 2840–2846, 2011.)

values (Figure 10.8a). The higher G' for the 100% SOP sample is probably caused by the higher foam apparent density. Glass transition temperature T_g was determined from the peak in tan δ and increased from 98°C to 142°C with increasing SOP content (Figure 10.8b) (Tan et al. 2011).

The improvement of dimensional stability and mechanical strength, as well as decrease of brittleness of rigid PU foams, can be achieved by application of different fibers, for example, glass, cellulose, and nanofillers. Nowadays, natural fillers such as wood, flax, and kenaf are used in PU formulations in order to increase the content of renewable raw materials and to improve the selected properties of final foams. However, the introduction of such fillers considerably increased the initial viscosity of the reaction mixture (Zhu et al. 2012; Prociak et al. 2014).

The properties of rigid PU foams depend on the quality of applied fibers and their size and water content. The application of flax fibers up to 5 wt% in PU systems does not cause difficulties in foaming process and can slightly improve the compressive strength, especially in anisotropic materials (Xu et al. 2007; Prociak et al. 2014). In case of wood fibers, good compatibility with PU matrix is observed. The break of composite takes place in the middle of fiber, which confirms very good adhesion of fibers to PU matrix (Casado et al. 2009).

The beneficial increase of compressive strength of rigid foams modified by addition of cellulose microfibers is shown in Figure 10.9. Lignocellulose fibers were applied in the synthesis of rigid foams based only on petrochemical polyols and those modified with SOP. Results show that incorporation of cellulose microfibers of 0.5, 1.0, and 2.0 php (per hundred polyol) into foam did not have a significant influence on the density. This indicates that addition of microfibers at these levels does not change the behavior of foaming process. Compressive strengths increased significantly in the presence of microfibers of 1.0 and 2.0 php (Zhu et al. 2012).

Over the last few years, many researchers have analyzed the influence of renewable raw materials on rigid PU foams and composites of low density. The effect of cellulosic fiber from commercially available writing paper, bleached kraft pulp, and commercially available rice flour on the properties of PU foams was studied. Generally, the compressive strength of PU foams modified with such natural fillers was higher than that of the reference materials (Banik and Sain 2009). In case of wood flours, the addition of this filler to rigid PU foams improved the compressive properties of porous composites, whereas flexural and tensile properties were worsened (Yuan and Shi 2009).

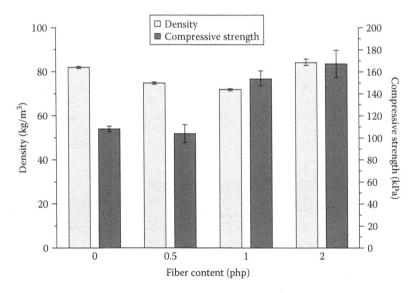

FIGURE 10.9 Effects of cellulose microfibers on density and compressive strength of rigid PU foams. (Data from Zhu, M., *J. Appl. Polym. Sci.* 124, 4702–4710, 2012.)

The investigations have shown that rapeseed oil-based polyols are a raw material, which can be successfully used in the synthesis of porous PU composites with high apparent density ca. 300 kg m^{-3}. In such materials, petrochemical polyols can be replaced with 80 wt% of rapeseed oil derivative, beneficially influencing the heat-insulating properties of such bio-composites. The modification of PU systems by polyol obtained in an epoxidation process and using flax fibers (5 php) increased the compressive strength of PU composites by ca. 40% in comparison to the reference material. Moreover, partial replacement of petrochemical polyol with ROP significantly decreases the water absorption of modified rigid cellular composites (Kurańska et al. 2013).

The development of rigid PU-PIR (PU-polyisocyanurate) foams for heat insulating applications is underway. The results of research on the water-blown PIR foams prepared from sunflower seed, flaxseed, and rapeseed oils, which are synthesized by way of transesterification of vegetable oils with triethanolamine or amidization with diethanolamine, are presented. A detailed experimental analysis has proved that the water-blown PIR foams from vegetable oil polyols possess competitive physical and mechanical properties to those exhibited by the traditional petrochemical origin foams. The optimal physical and mechanical properties suitable for practical applications are achieved at the isocyanate index values 150 < II < 200 (Stirna et al. 2008; Kirpluks et al. 2013).

In the near future, decreasing the flammability of bio-based rigid PU and PU-PIR foams will be one of the key challenges taking into account the tendency to use halogen-free flame retardants in such materials (Prociak et al. 2014).

REFERENCES

Abraham T.W., Dai D.G., De Genova R., Malsam R. 2010. Viscoelastic polyurethane foams comprising amidated or transesterified oligomeric natural oil polyols. Patent Application US 2010/0087561.

Andries K., Macken J.A.S. 2009. Process for making visco-elastic foams. Patent Application US 2009/0286897.

Banik I., Sain M.J. 2008. Water blown soy polyol-based polyurethane foams of different rigidities. *J Reinf Plast Compos* 27: 357–373.

Banik I., Sain M.M. 2009. Water-blown soy polyol based polyurethane foams modified by cellulosic materials obtained from different sources. *J Appl Polym Sci* 112: 1974–1987.

Branca C., Di Blasi C., Casu A., Morone V., Costa C. 2003. Reaction kinetics and morphological changes of a rigid polyurethane foam during combustion. *Thermochim Acta* 399: 127–137.

Campanella A., Bonnaillie L., Wool R. 2009. Polyurethane foams from soyoil-based polyols. *J Appl Polym Sci* 112: 2567–2578.

Carnicer J. November 23–24, 2005. Global flexible polyurethane markets: Present and future trends, *FSK Congress*, Heidelberg, Germany.

Casado U., Marcovich N.E., Aranguren M.I., Mosiewicki M.A. 2009. High-strength composites based on tung oil polyurethane and wood flour: Effect of the filler concentration on the mechanical properties. *Polym Eng Sci* 49: 713–721.

Cayli G., Kusefoglu S. 2008. Biobased polyisocyanates from plant oil triglycerides: Synthesis, polymerization, and characterization. *J Appl Polym Sci* 109: 2948–2955.

Dai J., Genova R., Simpson D. September 24–26, 2007. Recent developments in natural oil-based polyols for the production of viscoelastic foams, *Polyurethanes 2007 Technical Conference*, Orlando, FL.

David J., Vojtova L., Bednarik K., Kucerık J., Vavrova M., Janca J. 2010. Development of novel environmental friendly polyurethane foams. *Environ Chem Lett* 8: 381–385.

Dawe B., Casati F., Fregni S., Miyazaki Y. 2007. Natural oil polyols: Applications in molded polyurethane foams, *Proceedings of UTECH 2007 Conference*, Orlando, FL.

Dumont M.-J., Narine S.S. 2010. Physical properties of new polyurethanes foams from benzene polyols synthesized from erucic acid. *J Appl Polym Sci* 118: 3211–3217.

Dworakowska S., Bogdal D., Prociak A. 2012. Microwave-assisted synthesis of polyols from rapeseed oil and properties of flexible polyurethane foams. *Polymers* 4(3): 1462–1477.

Fan H., Tekeei A., Suppes G.J., Hsieh F.-H. 2013. Rigid polyurethane foams made from high viscosity soy-polyols. *J Appl Polym Sci* 127: 1623–1629.

Garrett T.M., Du X.X. 2010. Polyols from plant oils and methods of conversion. US Patent 7,674,925 B2.

Guo A., Demydov D., Zhang W., Petrovic Z. 2002. Polyols and polyurethanes from hydroformylation of soybean oil. *J Polym Environ* 10: 49–52.

Guo A., Javni I., Petrovic Z.S. 2000. Rigid polyurethane foams based on soybean oil. *J Appl Polym Sci* 77: 467–473.

Helling R., Russell D. 2009. Use of life cycle assessment to characterize the environmental impacts of polyol production options. *Green Chem* 11: 380–389.

Herrington R., Malsam J. 2006. Flexible polyurethane foams prepared using modified vegetable oil-based polyols. US Patent Application 2005/0070620.

Hill K. 2000. Fats and oils as oleochemical raw materials. *Pure Appl Chem* 72: 1255–1264.

Hu Y.H., Gao Y., Wang D.N., Hu C.P., Zu S., Vanoverloop, L., Randall, D. 2002. Rigid polyurethane foam prepared from a rape seed oil based polyol. *J Appl Polym Sci* 84: 591–597.

Huo S., Nie M., Kong Z., Wu G., Chen J. 2012. Crosslinking kinetics of the formation of lignin-aminated polyol-based polyurethane foam. *J Appl Polym Sci* 125: 152–157.

Ionescu M. 2005. *Chemistry and Technology of Polyols for Polyurethanes*, Rapra Technology, Shawbury.

John J., Bhattacharya M., Turner R.B. 2002. Characterization of polyurethane foams from soybean oil. *J Appl Polym Sci* 86: 3097–3107.

Kirpluks M., Cabulis U., Kurańska M., Prociak A. 2013. Three different approaches for polyol synthesis from rapeseed oil. *Key Eng Mater* 559: 69–74.

Król P. 2009. Polyurethanes—A review of 60 years of their syntheses and applications. *Polimery* 54(7–8): 489–500.

Kurańska M., Prociak A., 2012. Porous polyurethane composites with natural fibres. *Compos Sci Technol* 72: 299–304.

Kurańska M., Prociak A., Kirpluks M., Cabulis U. 2013. Porous polyurethane composites based on bio-components. *Compos Sci Technol* 75: 70–76.

Landrock A.H. 1995. *Handbook of Plastic Foams*, Noyes Publications, Park Ridge, NJ.

Lee S.H., Teramoto Y., Shiraishi N. 2002. Biodegradable polyurethane foam from liquefied waste paper and its thermal stability, biodegradability, and genotoxicity. *J Appl Polym Sci* 83: 1482–1489.

Lubguban A.A., Tu Y.-C., Lozada Z.R., Hsieh F.-H., Suppes G.J. 2009. Functionalization via glycerol transesterification of polymerized soybean oil. *J Appl Polym Sci* 112: 19–27.

Markets and Markets. 2011. Methylene Diphenyl Diisocyanate (MDI), Toluene Diisocyanate (TDI) and Polyurethane Market (2011–2016), http://marketsandmarkets.com.

Meier M.A.R., Metzgerb J.O., Schubert U.S. 2007. Plant oil renewable resources as green alternatives in polymer science. *Chem Soc Rev* 36: 1788–1802.

Modesti M., Lorenzetti A., Simioni F., Camino G. 2002. Expandable graphite as an intumescent flame retardant in polyisocyanurate-polyurethane foam. *Polym Degrad Stab* 77: 195–202.

Narine S., Kong X., Bouzidi L., Sporns P. 2007. Physical properties of polyurethanes produced from polyols from seed oils. *J Am Oil Chem Soc* 84: 65–72.

Nowicki J. 2007. Nowoczesna kataliza heterogeniczna w procesach estryfikacji i transestryfikacji oleochemikaliów. *Przem Chem* 86: 901–904.

Nozawa K., Sasaki M., Okubo K. October 17–19, 2005. Novel vegetable oil-based based polyol. *Proceedings of Polyurethane 2005: Technical Conference and Trade Fair*, Houston, TX.

Obi B., Butler D., Babb D., Larre A. September 24–26, 2007. Recent advances in TDI 80/20 with stannous octoate catalyst viscoelastic (VE) foams produced from both hydrocarbon based polyols, as well as natural oil derived polyol (NOP), *Polyurethanes 2007 Technical Conference*, Orlando, FL.

Paciorek-Sadowska J., Czupryński B., Liszkowska J., Jaskółowski W. 2010. Nowy poliol boroorganiczny do produkcji sztywnych pianek poliuretanowo-poliizocyjanurowych. Cz. II. Otrzymywanie sztywnych pianek poliuretanowo-poliizocyjanurowych z zastosowaniem nowego poliolu boroorganicznego. *Polimery* 55: 99–105.

Palanisamy A., Karuna M., Satyavani T., Kumar D. 2011. Development and characterization of water-blown polyurethane foams from diethanolamides of karanja oil. *J Am Oil Chem Soc* 88: 541–549.

Pawlik H., Prociak A. 2012. Influence of palm oil-based polyol on the properties of flexible polyurethane foams. *J Polym Environ* 20: 438–445.

Pawlik H., Prociak A. May 15–17, 2013. Influence of natural oil-based polyols on selected properties of viscoelastic polyurethane foams, modern polymeric materials for environmental applications, *Proceedings of the 5th International Seminar*, vol. 5, 101–106, Kraków, Poland.

Peacock R. 2009. Reviewed and revised—The global polyurethane industry 2005–2012: Future prospects in an uncertain world, *UTECH Europe, Conference Papers on CD*, Maastricht, the Netherlands.

Petrovic Z., Cevallos M., Javni I., Schaffer D., Justice R. 2005. Soy-oil-based segmented polyurethanes. *J Polym Sci Part: B Polym Phys* 43: 3178–3190.

Petrovic Z.S. 2008. Polyurethanes from vegetable oils. *Polym Rev* 48: 109–155.

Prociak A. 2007a. Porous polyurethane composites based on bio-components. *Cell Polym* 26: 381–392.

Prociak A. 2007b. Spienione poliuretany. Kierunki rozwoju w XXI w. *Przem Chem* 9: 918–920.

Prociak A. September 24–26, 2007c. Rigid polyurethane foams modified with recycled and natural oil-based polyols, *Polyurethanes 2007 Technical Conference*, Orlando, FL.

Prociak A. 2008a. Poliuretanowe materiały termoizolacyjne nowej generacji, Wydawnictwo Politechniki Krakowskiej, Kraków, Poland.

Prociak A. 2008b. Heat-insulating properties of rigid polyuretane foams synthesized with use of vegetable oils: Based polyols. *Polimery* 3: 195–200.

Prociak A., Michałowski S., 2012a. Porowate kompozyty poliuretanowe z udziałem surowców odnawialnych in Biokompozyty z Surowców Odnawialnych, Colegium Columbinum.

Prociak A., Rojek P., Pawlik H. 2012b. Flexible polyurethane foams modified with natural oil-based polyols. *J Cell Plast* 48: 489–499.

Prociak A., Rojek P., Pawlik H., Kurańska M. 2011. Synteza poliuretanów z udziałem surowców odnawialnych. *Przem Chem* 90: 1000–1005.

Prociak A., Rokicki G., Ryszkowska J. (editors) 2014. Materiały Poliuretanowe, Państwowe Wydawnictwa Naukowe PWN, Warszawa, Poland.

Randall D., Lee S. 2002. *The Polyurethanes Book*, John Wiley & Sons, New York.

Rojek P., Prociak A. 2012. Effect of different rapeseed-oil-based polyols on mechanical properties of flexible polyurethane foams. *J Appl Polym Sci* 125: 2936–2945.

Rojek P., Prociak A. May 14–15, 2013a. Foaming process and cellular structure analyses of bio-based flexible polyurethane foams, *Blowing Agents and Foaming Processes*, Mainz, Germany.

Rojek P., Prociak A. 2013b. Structure-property relationship of polyurethane flexible foams modified with different rapeseed oil-based polyols. *Modern Polym Mater Environ Appl* 5: 187–192. 5th International Seminar, May 15–17, 2013, Kraków, Poland.

Semenzato S., Lorenzetti A., Modesti M., Ugel E., Hrelja D., Besco S., Michelin R.A. et al. 2009. A novel phosphorus polyurethane FOAM/montmorillonite nanocomposite: Preparation, characterization and thermal behaviour. *Appl Clay Sci* 44: 35–42.

Shaari N.Z.K., Lye O.T., Ahmad S. 2006. Production of moulded palm-based flexible polyurethane foams. *J Oil Palm Res* 18: 198–203.

Sharma V., Kundu P.P. 2006. Addition polymers from natural oils. A review. *Prog Polym Sci* 31: 983–1008.

Shogren R.L., Petrovic Z.S., Liu Z., Erhan S.Z. 2004. Biodegradation behavior of some vegetable oil-based polymers. *J Polym Environ* 12: 173–178.

Singh P., Bhattacharya M. 2004. Viscoelastic changes and cell opening of reacting polyurethane foams from soy oil. *Polym Eng and Sci* 44(10): 1977–1986.

Stirna U., Cabulis U., Beverte I. 2008. Structure and properties of the polyurethane and polyurethane foam synthesized from castor oil polyols. *J Cell Plast* 44: 139–160.

Stirna U., Fridrihsone A., Misane M., Vlsone Dz. 2011. Rapeseed oil as renewable resource for polyol synthesis. *Sci J Riga Tech Univ* 6: 85–90.

Szycher M. 2013. *Szycher's Handbook of Polyurethanes*, CRC Press, Boca Raton, FL.

Tan S., Abraham T., Ference D., Macosko C.W. 2011. Rigid polyurethane foams from a soybean oil-based polyol. *Polymer* 52: 2840–2846.

Tanaka R., Hirose S., Hatakeyama H. 2008. Preparation and characterization of polyurethane foams using a palm oil-based polyol. *Bioresour Technol* 99: 3810–3816.

Tu Y.-C., Kiatsimkul P., Suppes G., Hsieh F.-H. 2007. Physical properties of water-blown rigid polyurethane foams from vegetable oil-based polyols. *J Appl Polym Sci* 105: 453–459.

Tu Y.-C., Suppes G.J., Hsieh F.-H. 2008. Water-blown rigid and flexible polyurethane foams containing epoxidized soybean oil triglycerides. *J Appl Polym Sci* 109: 537–544.

Vaughan B.R., Wilkes G.L., Dounis D.V., McLaughlin C. 2011. Effect of vegetable-based polyols in unimodal glass-transition polyurethane slabstock viscoelastic foams and some guidance for the control of their structure-property behavior. I. *J Appl Polym Sci* 119: 2683–2697.

Veenendaal B. 2007. Renewable content in the manufacture of polyurethane polyols-An opportunity for natural oils. *PU Magaz Int* 4(6): 352–359.

Veronese V.B., Menger R.K., de C. Forte M.M., Petzhold C.L. 2011. Rigid polyurethane foam based on modified vegetable oil. *J Appl Polym Sci* 120: 530–537.

Wang T.P. Zhang L.H., Li D., Yin, J., Wu S., Mao Z.H. 2008. Mechanical properties of polyurethane foams prepared from liquefied corn stover with PAPI. *Bioresour Technol* 99: 2265–2268.

Woo T., Halley P., Martin D., Kim D.S. 2006. Effect of different preparation routes on the structure and properties of rigid polyurethane-layered silicate nanocomposites. *J Appl Polym Sci* 102: 2894–2903.

Xu Z., Tang X., Gu A., Fang Z. 2007. Novel preparation and mechanical properties of rigid polyurethane foam/organoclay nanocomposites. *J Appl Polym Sci* 106: 439–447.

Yan H.H., Yun G., De N.W., Chun P.H., Zu S., Vanoverloop L., Randall D., 2002. Rigid polyurethane foam prepared from a rape seed oil based polyol. *J Appl Polym Sci* 84: 591–597.

Yeganeh H., Mehdizadeh M.R. 2004. Synthesis and properties of isocyanate curable millable polyurethane elastomers based on castor oil as a renewable resource polyol. *Eur Polym J* 40: 1233–1238.

Yuan J, Shi S.O. 2009. Effect of the addition of wood flours on the properties of rigid polyurethane foam. *J Appl Polym Sci* 113: 2902–2909.

Zhang L., Jeon H.K., Malsam J., Herrington R., Macosko C.W. 2007. Substituting soybean oil-based polyol into polyurethane flexible foams. *Polymer* 48: 6656–6667.

Zhang Q., Yu J. 2005. High performance all MDI viscoelastic foams, *Polyurethanes 2005, Technical Conference and Trade Fair*, Houston, TX.

Zhao Y., Yan N., Feng M. 2012. Polyurethane foams derived from liquefied mountain pine beetle-infested barks. *J Appl Polym Sci* 123: 2849–2858.

Zhu M., Bandyopadhyay-Ghosh S., Khazabi M., Cai H., Correa C., Sain M. 2012. Reinforcement of soy polyol-based rigid polyurethane foams by cellulose microfibers and nanoclays. *J Appl Polym Sci* 124: 4702–4710.

11 Foaming Technologies for Thermoplastics

Luigi Sorrentino, Salvatore Iannace, S.T. Lee, and Roberto Pantani

CONTENTS

11.1 INTRODUCTION

Many different thermoplastic polymer foam processing technologies exist, and these can be grouped into two large classes: continuous and discontinuous. Continuous techniques produce foamed products or semifinished products using a time-continuous process, while discontinuous techniques are characterized by time-recurring operations.

The technology choice for thermoplastic polymer expansion is determined by a variety of parameters. Some of the most important of these are the thermal transitions, specifically the glass transition and melting temperatures. Both depend on the polymer selected and on the following conditions: the article geometry (either complex in three dimensions, or regular along one direction), the process productivity (the number of articles needed to produce in a time unit), the surface quality and precision, the apparent density of the product, which is heavily influenced by the rheological and sorption properties of the specific polymer. Based on the article's design and the various parameters, a specific production technology is chosen.

An object's geometry is one of the most important parameter that influences the choice of technology. In fact, to make large products with a constant cross section and an arbitrary length, extrusion foaming is the best solution. It is used to produce both finished and semifinished products for use in subsequent processes such as thermoforming, where the possibility for very complex sections, with different thicknesses and shapes, exists. The product's length can be determined by cutting a continuous strand either in-line or off-line. The largest variety of polymers can be processed using extrusion foaming and density values from solid (no porosity) to extremely low (<20 kg m^{-3}) can be obtained. There is a great advantage to the extrusion process. In this, as in extrusion foaming, although with some restrictions in the latter, recycled polymers can be used to produce foams. Recycled materials from municipal waste recycling depots or from manufacturing scraps may be treated with additives to improve their polymer properties to facilitate foaming. A continuous, steady output is another characteristic of the extrusion process, which supports consistent processing conditions and constant product quality.

If objects with three-dimensional complex shapes have to be produced, injection molding foaming technology may be chosen. It is a well-established technology, which can be automated and has strong productivity rates. Surface finishing can be very high, depending on the expanded article's density. Injection molding's most common drawback is with density, since it is not easy to obtain foams with a density below 500 kg m^{-3}. Flow conditions, which are determined by the mold shape, have to be managed according to the desirable final geometry, and have to be coupled with the nonuniform temperature profiles in the various cavities. The temperature is lower at the mold wall and increases toward the center of the shape. An advantage here relates to the potential to make small, very complex, and precise three-dimensional objects in a single production step. Also, there is the added potential of coupling different materials (co-molding), to vary aesthetic or tactile characteristics. Since injection molding technology generally uses an extruder to melt and meter the compound, it also allows for the use of recycled raw materials or scraps, within appropriate percentages, and which are treated with additives to improve their polymer properties.

Where very low densities coupled with complex shapes are needed, the chest molding technology may be the answer. It consists of pre-expanded particles sintering into a mold. The final piece's degree of expansion may be very low (20 kg m^{-3}) and will be determined by the density of the pre-expanded particles. The surface quality is not very high, and strongly relies on the size of the expanded particles. It is a very simple and consolidated technology, which has become highly automated. Consequently, it is not very expensive to implement. But chest molding's two disadvantages are the need to use raw materials with very stringent requirements and its inability to recycle manufacturing scraps or to use recovered raw materials.

In addition to the most popular technologies, other techniques are also used to expand polymers in ways not practically obtainable. For example, batch foaming is a niche technology wherein polymers are expanded in a high-pressure vessel. Solid semifinished products are saturated with a physical blowing agent in the high-pressure vessel. After a time interval, depending on the sorption temperature and the sample's geometry, particularly its thickness, the article is expanded by means of thermodynamic instability, which means either a temperature increase or a pressure release. After expansion, the object is cooled below the glass transition temperature to consolidate its cellular structure. This technique allows for the polymer's expansion in all three directions (i.e., free foaming), if a specific shape is required, a closed mold can be used. The main drawback here relates to the high pressures that are typically used to solubilize the physical blowing agent both within the polymer and for the duration of this processing stage.

Each technology requires that the polymers exhibit some rheological and chemicophysical properties. Polymers cannot be expanded indifferently with the various expansion technologies. Their properties have to demonstrate improvement. The polymer's viscosity is central to this issue. Specifically, its elongational viscosity affects its capacity to develop the desired cellular morphology, especially if micro-sized cells are required. In addition, it consolidates the cellular structure. To improve the polymers' ability to bear extensional loads in the melt state, a preliminary treatment can be applied to increase their molecular weight, and thus also their rheological properties. Additives are the main method used to increase polymer viscosity. These act as chain extenders by connecting the ends of macromolecules to increase the molecular weight. Another way to modify the rheological behavior of polymers, especially their melt strength, is via the dispersion of nanoparticles, which can affect molecular mobility. If properly managed, the presence of particles comparable in size with macromolecules can increase the cell walls' strength (strain hardening) during the increase in bubble diameters. This facilitates the formation of a large amount of small size cells by hindering the coalescence phenomenon; that is, the increase in bubble diameters due to breakage of adjacent walls between two or more bubbles.

Rheological behavior is not the only parameter responsible for the development of good cellular morphology. In fact, crystallinity plays a major role in semicrystalline polymers. The polymer's ability to develop the crystalline phase can be exploited to affect the cellular morphology (closed or open cells), and to consolidate the cellular morphology to avoid coalescence. The crystallization rate should be carefully managed. Polymers with a low crystallization rate will develop crystals too late to stabilize cells. Nucleating agents can be used in this case.

The foaming technologies of thermoplastics are very broad, and it is not the purpose of this chapter to cover all of them. But some basic aspects related to the processing technologies of bio-based thermoplastic polymers are offered here to provide a background for those unfamiliar with foaming technologies. While many specific and detailed books are available on this topic, most conventional foam extrusion technologies are covered in a recent book by Lee and Park (2014).

11.2 BLOWING AGENTS

Thermoplastic foams can be produced with porosities ranging from 50% to 98%, depending on the polymer used and the end-use application. Higher density foams typically use chemical blowing agents (CBAs) that decompose to release gas that dissolves in the melt. A higher porosity can only be achieved by using physical blowing agents (PBAs), which dissolve in the polymer melt and expand the polymer after a pressure release.

11.2.1 CHEMICAL BLOWING AGENTS

CBAs are substances that decompose at processing temperatures, thus, releasing gases such as carbon dioxide and/or nitrogen (Nema et al. 2008). Solid organic and inorganic substances, such as azodicarbonamide and sodium bicarbonate, are used as CBAs. In general, CBAs are divided by the enthalpy of the reaction that turns the gas into two groups, including exothermic and endothermic foaming agents (Thompson et al. 2006). In the present day, a combination of exothermic and endothermic CBAs is also used for foaming. High-density foams based on bio-based polymers find their applications in competition with products traditionally made from wood. The CBA can be roughly mixed with polymer and additives. Care must be taken to avoid premature decomposition. If the decomposition temperature is too low, the gas will be prematurely released, eventually being vented through the feed throat. Conversely, if the decomposition temperature is too high, with respect to the processing temperatures, the gas will not be released. The proper CBA selection depends on the polymer's processing temperatures. The decomposition temperature can be adjusted by using activating additives. CBA benefits include the possibility of adopting conventional injection molding machines. This is because CBAs are mixed with the polymer pellets inside the hopper, like a masterbatch. Temperature is critical, since the decomposition temperature must be reached inside the injection chamber for foaming to occur only after the pressure decrease during the melt injection into the mold.

11.2.2 Physical Blowing Agents

Physical blowing agents are compounds that release gases as a result of the physical processes of evaporation and desorption, at elevated temperatures or reduced pressures. This results in foam densities as low as 20 kg m^{-3}, when liquid or gaseous blowing agents are used. Commonly used PBAs include hydrochlorofluorocarbons 141b and 142b; hydrocarbons propane, butane, and pentane; and nitrogen or carbon dioxide gas. Physical foaming agents that have been reported (Sun and Turng 2014) to have been used in microcellular processing include water (Peng 2012), argon, nitrogen, and carbon dioxide. PBAs are both inexpensive and environmental friendly. They produce high degrees of foaming and thus a lower density in the parts. No residue of the blowing agent remains in the part, which means that physically blown parts are suitable for applications such as food packaging and in medical products. Unlike CBAs, however, PBAs generally need special equipment such as a gas dosage unit and a gas injector.

PBA must be soluble in the polymer at the processing pressure and temperature, and must separate from the polymer when the pressure is reduced below the partial pressure of the blowing agent. The most common PBAs in biopolymers are nitrogen (N_2) and carbon dioxide (CO_2). N_2 has a lower solubility with respect to CO_2, but a higher diffusivity inside the polymer melts (Durril and Griskey 1966). After PBAs have dissolved in the melt at the concentrations used to obtain low-density products, they strongly plasticize the polymer. Thus, viscosity is reduced, and this affects transition temperatures such as the glass transition, crystallization, and melting temperatures. The lower solubility allows the N_2 to reach a higher degree of supersaturation as compared to CO_2 for the same gas amount. Therefore, it induces a stronger nucleation. However, the higher solubility allows the CO_2 to produce higher amounts of dissolved gas, which means greater plasticization. The literature suggests that a blend of CO_2 and N_2 could be a better blowing agent than the two gases alone (Di Maio et al. 2005).

11.2.3 Other Foaming Systems

Another potential way to foam using a conventional extruder/injection molding machine is to saturate the pellets with a PBA under high pressure, and then use them for further processing. The pellets can be prepared in an autoclave under high pressure at room temperature (Pfannschmidt and Michaeli 1999). In this case, sorption times are long and the pellets must be used shortly after preparation. Another method is to prepare the *gas-laden pellets* using an extruder (Lee et al. 2011a). In this case, the PBA is injected and mixed into the polymer during extrusion. The extruded material is quickly cooled down to prevent the gas from foaming within the polymer. Then, it is pelletized, and the pellets can be used in a conventional injection molding machine. Similar systems are also reported for CBAs. For instance, structural foams have been obtained by adopting the so-named expandable thermoplastic microspheres (Peng et al. 2011b). These microscaled spheres consist of an outer polymer shell, which encapsulates a hydrocarbon core. When heated, the expanded hydrocarbon swells the microspheres. Compared to conventionally foamed parts, these systems are reported to produce parts with a better surface quality (Peng et al. 2011a). This is because the outer shell of the polymer prevents the gas from leaving swirl marks on the part's surface.

11.3 EXTRUSION PROCESS

Extrusion is the most used technology in the plastics industry, and this also applies to polymer foaming. The foaming extrusion process is continuous, very flexible, and allows high throughput. The widest range of thermoplastic polymers can be processed in all industrial foaming technologies. In addition, it can achieve the widest density ranges. The extrusion process aims to shape a polymeric compound, and this is done by the extruder. It acts as a pump, which uses one or two screws to convey the polymer along the barrel through a die, which is specifically designed to take into account the polymer's viscoelasticity. The die is responsible for the forming of the desired shape of the polymer after cooling. The extruder can produce finished or semifinished products, but it is also used in combination with other industrial machines, such as in the injection molding process, or as a part of a multiextruder line, that is, a tandem configuration.

One major extruder design parameter is the number of screws. In general, the single-screw extruder has a one-piece screw, which can limit the torque applied to the polymer due to the deep-cut feed section that leaves a smaller screw-root cross section available. Single-screw extruders are usually preferred over twin-screw types because they cost roughly half the price of them. They are also easier to understand and to maintain. On the other hand, twin-screw extruders are preferred for their following performance: greater and more steady material solids feeding, higher stress transmission to the melted compound, higher mixing efficiency, better dispersion of additives, nucleating agents and even sorption of blowing agents, better temperature control and homogenization, and their relatively strong and stable dynamic seals before gas injection points. All of these mean that mixing can be performed over shorter transport distances and in less time.

The twin-screw or single-screw extruder can sequentially perform subprocesses along its length. These can be performed by designing the screw profile, and by eventually using a segmented design. Segmented screws are highly desirable because elements can be strung onto screw shafts to optimize subprocesses. They can be changed according to specific needs and polymers, without having to replace the entire screw.

Both single- and twin-screw extruders can be used to produce bio-based foamed products. In addition to melting, pumping, and forming subprocesses, foaming extruders carry out compounding while delivering the melt in a process window suitable for uniform bubble nucleation and growth. This affects several parameters, such as the choice of the screw types and segmenting, formulation modifications, and operating conditions. In particular, the mixing elements are essential to the quick development of an acceptable dispersion of additives before the gas is introduced. A specific extrusion foaming requirement is to prevent the flow-back of injected gas. It is a challenging task since high pressures are usually used, and short disc stacks and special seals can replace longer, torque-consuming sealing sections. Since the gas is soluble in the polymer, no dispersive mixing is usually needed after the gas has been injected. Otherwise, open meshing elements are used for pumping as the pressure rise is distributed over a larger length of the barrel. These are also preferred for heat transfer and thermal homogenization to static mixers as they consume low energy and the morphology results are similar. Temperature homogenization is needed to avoid excessive expansion, cell rupture, coalescence, or collapse of the cellular structure after the die exit. This requires a reduction of the melt temperature, which could be difficult to do with a single extruder, since it conflicts with the plasticating function of the extruder, thus limiting the production rate. This issue can be overcome by using a second extruder operating in tandem with the plasticating extruder. The primary function of the second extruder is to cool the melt to a temperature range where good quality foam may be formed.

11.3.1 EXTRUSION LINES FOR BIO-BASED FOAMS

The main components of an extrusion line can be schematized in three conceptual zones: polymer feeding, plasticating, and feeding through the shaping die (Figure 11.1). Usually, more complex designs are used to perform the process, especially if compounds or foams have to be produced and a peculiar pressure profile is generated (Figure 11.2). To produce a foam, the following extruder zones should be present:

1. Materials feeding
2. Melting and compounding of additives
3. Dynamic sealing to prevent the gas blow-back
4. Gas injection (in case a physical blowing agent is used)
5. Gas solubilization
6. Melt homogenization and cooling
7. Metering through the die
8. Die and post-die forming

Foam processes differ. This depends upon the target density, the polymers used, the physical blowing agent versus a CBA, the finished product shape and form (including whether part of a coextrusion), and any special performance properties.

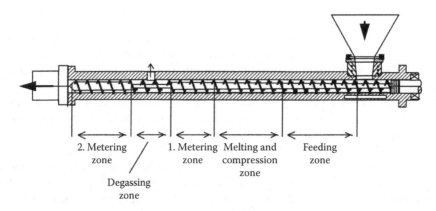

FIGURE 11.1 Schematics of an extruder showing the different zones. (Data from Scholz, D., Development of endothermic chemical foaming/nucleation agents and its processes. In *Polymeric Foams: Technology and Developments in Regulation, Process, and Products*, Eds. S.T. Lee, Chapter 2, CRC Press, Boca Raton, FL, 2009. With permission.)

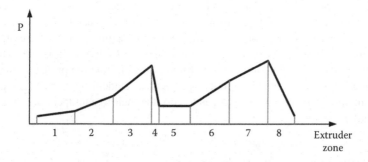

FIGURE 11.2 Pressure profile in the different zones of an extrusion line: (1) materials feeding, (2) melting and compounding of additives, (3) dynamic sealing to prevent the gas blow-back, (4) gas injection, (5) gas solubilization, (6) melt homogenization and cooling, (7) metering through the die, and (8) die and post-die forming.

11.3.1.1 Materials Feeding

The feed throat is used to feed the polymer and the additives into the screw. This section can be cooled for low-melting-point polymers, which may melt before entering the barrel and stick to the inside surface, if the feed throat is allowed to reach a high temperature by heat conduction from the extruder barrel. Screw and barrel temperatures, bulk density, and pellet temperature affect the feeding of solids. Feeding is optimized with low friction on the screw and high friction on the barrel to maximize the material transport and the applied stresses. On the other hand, high-melting-point polymers may best be fed by allowing the feed-throat section to reach a higher temperature. In fact, cold pellets may occur in a lower feeding rate due to the reduction of their friction coefficient on the barrel surface. To avoid this, a preheating treatment on pellets using a continuous hopper dryer is often used to stabilize the polymer flow.

Twin-screw extruders are generally equipped with volumetric or, preferably, gravimetric feeders to precisely control the flow rate of all the components of the formulation. The need to produce a cellular structure can render the dispersion of additives in the polymer, such as nucleating agents and colorants, very challenging if a single extruder is used. This is because gravimetrically fed components, such as nanoparticles or low percentage components, do not deliver accurately at low rates. Metering a premixed formulation could allow for the use of only one feeder, but additives must not segregate, and preferably powders should be used to make stable premixes.

11.3.1.2 Melting and Compounding

If only melting and light mixing are needed, a single-screw extruder would be adequate. However, in most cases, different components have to be intimately dispersed in the polymer, and their agglomeration has to be carefully avoided. This generally is addressed by using a twin-screw extruder, which may deagglomerate and disperse the nucleating agents, as well as the colorants, and the fillers in a controlled way. This would assure good bubble nucleation and growth. Further, if blending two polymers is part of the compounding process, the twin-screw design is mandatory. In such cases, dispersive mixing may be required, and the screw design to perform this subprocess can include dispersive mixers and/or newer generation geometries. Also, the compounding/melting process in the extruder must consider the limited flow rate, since screw speeds of as low as 30 rpm are used. This strongly reduces the shear stress applied to the melt and renders the dispersive process most difficult.

If a CBA is used, great care must be taken during melting and compounding to avoid a premature decomposition of the blowing agent, which will result in a loss of gas through the feed throat and in increased foam density. However, the temperature should not be too low because it may result in poor mixing and poor extrudate quality.

Screws, which produce good mixing, and are increased if an intermeshing/counter-rotating design is used, and which control the melt temperature for a given polymer will produce good foam extrusion. Some adjustment of the barrel zone temperatures may be required to initiate the blowing agent's decomposition at the right moment. This will also provide an optimum melt temperature as the polymer enters the die. Excessively low operating pressure will result in poor mixing, large temperature gradients in the melt stream, and large density gradients in the product. Conversely, excessively high pressure will cause a high melt temperature and result in overblowing and cell collapse.

11.3.1.3 Venting and Blowing Agent Sealing

The melting or compounding of all constituents could lead to the release of low-molecular-weight compounds. These should be extracted from the melt to avoid possible degradations or unwanted results along the barrel or after the foam exits the die. This task can be performed by reducing the screw root diameter, thus lowering the pressure and allowing the desorption of volatiles. A venting port can only be employed when physical blowing agents are used, and can help to evacuate these substances from the internal barrel volume before the gas injection. If a venting port is employed, care must be taken to raise the melt pressure before the gas injection zone. At low gas injection pressures of 5–20 bar, this operation can usually be performed by allowing the complete filling of the screw channel with the compounded melt. For high pressures, a dedicated seal section has to be added to avoid the melt or a gas backflow. To avoid excessive heat generation and power consumption, large sections equipped with kneading blocks or other high power dissipating elements are not used. They are usually replaced with cascaded discs, special flights, or other appropriate barriers mounted on the screw shaft.

11.3.1.4 Blowing Agent Injection

For CBAs, the melt temperature should reach the decomposition temperature in the compression section before it does so in the metering section. This allows time for the released gas to dissolve and be dispersed in the melt under high pressure. Failure to reach the decomposition temperature in time results in higher density foams. Die temperature can be increased to compensate for late CBA decomposition, but care has to be taken because this could lead to a different density between the skin and the core of the foamed stream, or to an overblow on the surface material, causing cell rupture and a very rough appearance.

Metering of gaseous blowing agents requires instrumentation to measure pressure, temperature, and volume flow rate to control the mass flow rate. A barrier-type screw with an intensive mixing section downstream from the blowing agent injection point is recommended. An length/dismeter (L/D) ratio of at least 32:1 is advised in order to provide sufficient time and shear to obtain a good dispersion of the blowing agent.

The gas may come from a liquid or a gaseous phase. In either state, it is critical to meter the physical blowing agent precisely. Pressure and temperature monitoring in the extruder barrel near the injector is important to monitor the process stability. The main reason for monitoring the pressure is to have information about gas escaping through the dynamic seals. It is also a way to verify that the gas metering is constant. Some designs use dual PBAs as complementary blowing agents to nucleate and further grow gas bubbles. A special design aid is a short and stiff PBA metering channel. This is needed to avoid large gas density fluctuations. If the metering line is long, temporary pressure variations in the barrel translate into delayed variations of pressure in the injection channel, which in turn can heavily affect the pressure in the melt. Gas delivery in short circuits is faster and readily reacts to pressure fluctuations in the barrel.

11.3.1.5 Blowing Agent Sorption

The gas should be soluble in the polymer in the amount needed to reach the target foam density. When the gas has dissolved, the rheology of the material is then uniform. Before that has been achieved, a liquid or gas phase exists concurrent with a polymer melt phase. It is critical to prevent mass ponding of the gas or liquid. To minimize this, the injection should occur over very intensive rate distributive mixers, with substantial pressure under the injector. In this way, the gas is rapidly introduced and distributed in small isolated domains in the polymer melt and is ready to dissolve by diffusion. Large gas or liquid droplets must be avoided since they need a long time to dissolve in the polymer and can give rise to either long residence times or irregular cellular morphologies.

The PBA dissolution rate is in inverse proportion to its drop size, and efficient mixing helps to quickly solubilize the gas in the polymers. For this reason, twin-screw extruders are largely preferred for thermoplastic foaming due to their highly efficient use of energy to homogenize the gas/polymer solution. In tandem extrusion lines, compounding, dynamic sealing, and gas injection and dissolution are almost always performed in the first twin-screw extruder. This occurs while the thermal treatment of the gas/polymer solution and its metering through the die are performed in the second extruder, usually a single-screw type.

11.3.1.6 Metering

Metering causes a major pressure buildup, depending on the polymer viscosity and the die shape. This is particularly so if thin parts have to be produced. It would be better to extrude the polymer/gas solution through the forming die using twin-screw extruders, but with the tandem configuration it can be performed by a single-screw extruder, which is responsible for temperature homogenization and lowering. Pumping could alternately be done with a gear melt pump. Like other unit operations on the extruder, pumping should be done as efficiently as possible to place the least torque on the screw shafts. An exception to this rule would be elements sufficiently severe to cause frictionally heated spots on the surface to become hot enough to cause premature foaming.

11.3.1.7   Temperature Management

To minimize the amount of heat that must be extracted in the cooling extruder, the downstream zones should be held at a temperature below the melt temperature and higher than the crystallization temperature. In this way, the viscoelastic properties of the melt can be improved to bear the elongational flow that occurs during bubble expansion. When a tandem configuration is used, care must be taken to make the connection between the plasticating extruder and the cooling extruder very short. This is to keep the pressure drop low. Higher pressure requires a higher screw speed to maintain the same output rate. Since the polymer has already been melted and mixed prior to entering the cooling extruder, the screw design of the cooling extruder is different from that of the plasticating one, and the only objective is to reduce the melt temperature. This implies that shear stresses should be kept very low to avoid heat generation for the viscous flow.

The blowing agent dissolved into the polymer acts as a plasticizer, and the melt viscosity drops as does its softening point. This affects the foaming process because the actual extrusion temperature is much higher than the neat polymer glass transition temperature, and careful cooling has to be applied to raise the viscosity to an adequate value. If the melt is too warm, early nucleation and bubble growth can occur before the die exit, thus causing large cells and coalescence. On the other hand, if the melt temperature is too low, nucleation and bubble growth will occur too late and bubbles may not reach the expected size. Specific screw sections are used to meter the melt with low friction. The barrel end section can be liquid cooled to reduce the cooling time. High distribution rate mixers can be used at low energy settings to homogenize the melt.

11.3.1.8   Die Forming and Post-Die Treatments

Die designs vary greatly with the final shape, melt rheological properties, and the products' density. As foam densities reduce, their shape must change to account for increasing and nonuniform foaming. The die temperature control must be absolutely accurate to keep bubble nucleation and growth even across the stream section. Premature and uneven pressure drops inside the die should be avoided because they can result in premature bubble nucleation and, potentially, in cell coalescence due to excessive elongational stresses applied to cell walls. In fact, when the pressure within the die drops below the partial pressure of the blowing agent, then the nucleation process starts. If the extrusion rate is too low or the melt temperature is too high, foam will form within the die, cells will be ruptured by the shearing action, and the structure will collapse.

If the die shape is complex, a localized temperature or shear compensation may be needed. The temperature has to be lowered and homogenized before the melt reaches the die, thus improving gas solubilization and reducing premature foaming before the die exit. The concentration of the blowing agent and the melt temperature determine the amount of density reduction, but the final density is heavily influenced by the strong plasticizing action of the blowing agent. This results in a further viscosity reduction. The reduced gas pressure and the increased viscosity of the polymer terminate the expansion process.

Other key points in die-forming design concern the control of both the pressure drop across the section and of the shear stress distribution due to the melt's history. They can affect the final shape and dimensions of the foamed extrudate, and a bad design can result in profile deformations. Thus, a calculation of the pressure drops in each region of the die, along with a consideration of each section's geometry and an application of the appropriate flow analysis, is involved. Further, the pressure drop rate is of great importance for the cellular morphology, since high depressurization rates permit higher nucleation and smaller cell sizes.

11.3.1.9 Tandem Extruder Configuration

In some applications, the high melt temperature and intensive shear required to properly and rapidly disperse the blowing agents prevent direct extrusion into a die. This is because the high partial pressure of the blowing agent and the low viscosity will cause overblowing, cell rupture, and collapse of the extrudate. Although a screw can be designed to minimize work input in the down-stream region, conflicting requirements for cooling versus intensive mixing and a limited surface area for heat extraction severely limit the production rate capability. For this reason, commercial systems often employ a second extruder in tandem with the plasticating extruder (Figure 11.3). The second extruder is typically one size larger than the plasticating extruder and functions as a heat exchanger designed to optimize heat extraction in order to reduce the melt temperature into a range where a satisfactory foam can be formed.

11.3.2 EXAMPLES OF EXTRUSION FOAMING LINES

11.3.2.1 Starch-Based Foams

Food extrusion is more advanced than plastics extrusion. It seems to be a predecessor of plastics processing. In the 1930s, extrusion was a popular process in pasta making in Italy. Raw materials were first ground to the right particle size, usually with the course flour, which was then mixed with other ingredients such as liquid sugar, fats, dyes, and water. This *slurry* possessed unique rheological properties out of intra-starch and inter-component hydrogen bonding and could then be processed through an extruder to make pasta products. In our day, the plastics extrusion process is far more sophisticated than it was over three quarters of a century ago. But the same principle is applied to starch foam processing, which can generate highly expanded beads, foam sheet, and foam board with unique degradable characteristics.

Starch, a natural —OH-rich material and —OH-based foams, has gained a significant amount of extrusion attention in the last two decades (Boonstra and Berkhout 1975; Lacourse and Altieri 1989; Loudrin et al. 1995; Narayan 2006). Starch is hydrophilic in nature and readily absorbs moisture from the environment. It is rapidly biodegraded by enzymes. Caution has to be exercised between degradability, durability, and serviceability though. In general, starch contains about 20%–30% water-insoluble amylose and 70%–80% of water-soluble amylopectin. Both are hydrophobic and have the same base unit of $C_6H_{10}O_5$, as shown in Figure 11.4. Yet the chain length and structure exhibit vastly different water solubilization. In the mid 1970s, water-insoluble hydrophobic starch was a successfully expanded cellular structure with 8.5% moisture (Boonstra and Berkhout 1975).

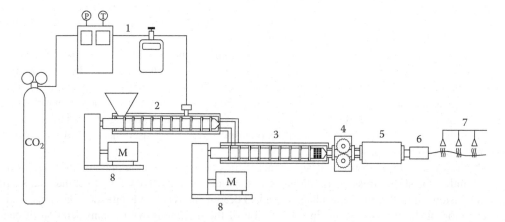

FIGURE 11.3 Schematics of a tandem extrusion line. 1, Gas injection system; 2, primary single-screw extruder; 3, secondary single-screw extruder; 4, melt pump; 5, static mixer; 6, extrusion die; 7, air cooling system; 8, motor. (Data from Ma, P. et al., *J. Cell. Plast.*, 48, 191, 2012. With permission.)

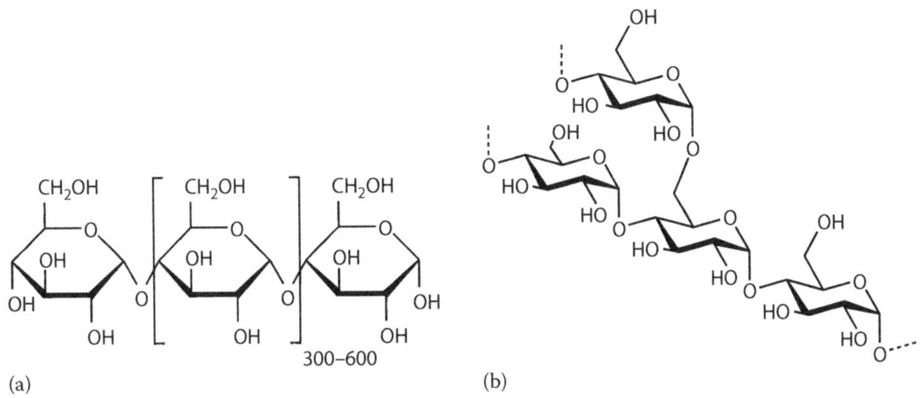

FIGURE 11.4 Formulations: (a) amylose and (b) amylopectin.

TABLE 11.1
Starch Foam Advantage and Disadvantage Summary

	Advantage	Disadvantage
Product	High expansion (i.e., over 30X) with water blowing agent	Brittle and low in mechanical properties, and poor water resistance
Raw material	Raw material from plant or food source	A concern in food shortage area
Processing	Extrudable	Need processing aids
Environment	Degradable	–
Chemical modification	Readily react with —OH	Economics
Recycle	Recyclable via slurry process	Specially designed system
Reuse	–	Not as good as polystyrene foam

It was also found that the amylose content played a critical role in extrusion with moisture. Over 45% amylose was the preferred amount (Lacourse and Altieri 1989). Because it lacks sufficient water-insolubilized content, its viscosity strongly depends on temperature decrease, or on residence time. It tends to gel up and cause extruder pressure to shut down, and often it is necessary to dissemble the die and screw to clean up the *baked* residual. Additionally, a high amylose content starch raises economic concerns. For 20%–30% general starch, adding poly-glycol or poly(vinyl alcohol) (PVOH) was deemed necessary to make it more suitable for thermoplastic processing (Neumann and Seib 1993; Loudrin et al. 1995; Glenn et al. 2006; Lee 2014). Also, starch can be grafted with methyl acrylate to form a starch copolymer to enlarge its processing latitude and to enhance its product elongation (Fisk 1998). Table 11.1 presents a brief summary of starch foam.

It is known that hydrogen bonding forms relatively quickly when —OH groups meet. It opens a big processing door for extrusion of starch and —OH-based polymers with organic solvents such as methanol, ethanol, and propanol, and inorganic moisture as foaming agents. PVOH is a good example of synthetic polymers containing a large amount of OH groups in the chemical structure. Its inherent melting point and decomposition point is too close to process through most plasticators. After modification, its processing window significantly increases. Thus, extrusion is a suitable system for PVOH foaming with liquid alcohols (Malwitz and Lee 1992). Conventional foam extruders; single- or twin-based, can be easily adapted to make low density PVOH foam sheets, boards, and rods.

In foam-extrusion applications, an organic solvent with a hydroxyl group is generally classified as a flammable blowing agent. Safety is inherent in the processing and handling of such products. Alternatively, water is a unique inorganic liquid substance. It has a low molecular weight and a 100°C boiling point. It is commonly available and without toxic or flammable issues. Thus, it is a preferred foaming agent. Water blown foam could be extremely dynamic at the extrusion exit. Also, it tends to condense as it cools to lower than 100°C. If the product cannot build up sufficient material strength by that point, then dimensional stability becomes a big issue in curing. It was reported that a high amylose content in starch is necessary for extrusion with a good processing window. But the most commonly available starch has a low amylose content. Normally, melting starch is not a problem. Yet, the melt viscosity is very sensitive to cooling. The sharp increase in viscosity at a decreased temperature could easily cause high pressure to shut down the process and, worse yet, open up the extrusion to clean up the x-linked mass. A common extrusion process involves feeding starch and water slurry into the extruder, which is not desirable from a material handling perspective. Often devolatilization is required to remove excessive water. However, a short extrusion process was developed to process the starch and moisture to make high expansion peanut foam. This process is more food processing than it is plastics processing. A short and shallow channel depth, with near-paralleled flights, is adequate to break the seeds for melting and mixing. A typical loose-fill screw is presented in Figure 11.5. In general, a high rpm and a low residence time are necessary to break the granules by mechanical force, so that moisture solubilizes the internal hydrogen bond. This means there is not too much time for gelatinization to stop extrusion. This product is ready to be packaged. Sorghum seed is rich in starch. Adding PVOH as a coupling agent makes extrusion a suitable technique to produce *loose-fill* foamed peanuts for packaging (Markham and Martin 2005).

Another way to increase melt strength is to add fiber to starch. With a proper coupling agent, plastics can be added for plasticization and to enlarge processing window. For instance, using PVOH as a coupling agent, paper fiber, starch, and polyolefin could be processed into a cellular structure while using moisture as the sole blowing agent. When strand die is incorporated, the strands can be fused into thick board for insulation applications, as shown in Figure 11.6 (Matsushita 2013). When a single strand is made, it has a strain-induced crystallization in the machine direction. This offers a unique cross-machine direction modulus, which can be used in items such as handbags netting, and so forth. It should be noted that even this extrusion requires more residence time than that of the loose-fill beads. The L/D is much less relevant to conventional foam extrusion, in which cooling makes up nearly 60% of the process. A general screw for melting and mixing is adequate for this process. Either a co-rotating or counter-rotating twin screw or a single screw would work.

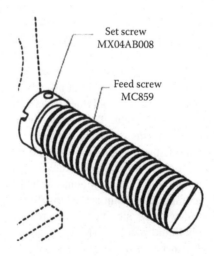

Set screw
MX04AB008

Feed screw
MC859

FIGURE 11.5   Screw of loose-fill extruder. (Courtesy of Maddox Metal Works, Dallas, TX.)

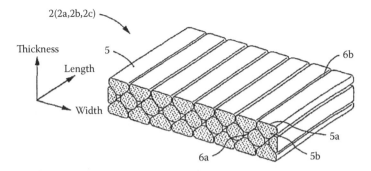

FIGURE 11.6 Foam plank made with fused strands for insulation; a plate-like foamed product (2) is formed by connecting a plurality of rod-like elements (5) extending in an axial direction; there are voids (6) between the rod-like elements (5); US Patent 8,414,998.

Foaming thermoplastic starch is a unique process used to fully capture the degradability and light weight benefits of starch. These benefits are especially useful in packaging. Thus, blending polymers and fibers enhance the properties of starch in a wide range of applications. But the high open cell content associated with water foaming decreases the foamed starch product's mechanical properties. Along with its poor water resistance, its application window is quite narrow. Chemical modification to its crystalline structure seems necessary to widen its processing window and to improve its mechanical properties and moisture resistance. Oxidation is an interesting approach in which carboxyl and carbonyl groups are introduced to starch chains. When the hydroxyl groups in the glucose ring are oxidized to aldehyde or ketone groups, the hydrogen bonds between the starch chains are reduced to exhibit hydrophobic thermoplastic behavior. In this way, crystallization in the starch is decreased, or an increased mechanical property is established (Zhang 2007; Zhang et al. 2010). Figure 11.7 shows the chemistry involved. This could offer a viable solution for starch foam to retain its good mechanical properties and moisture resistance, without fully sacrificing its degradability status.

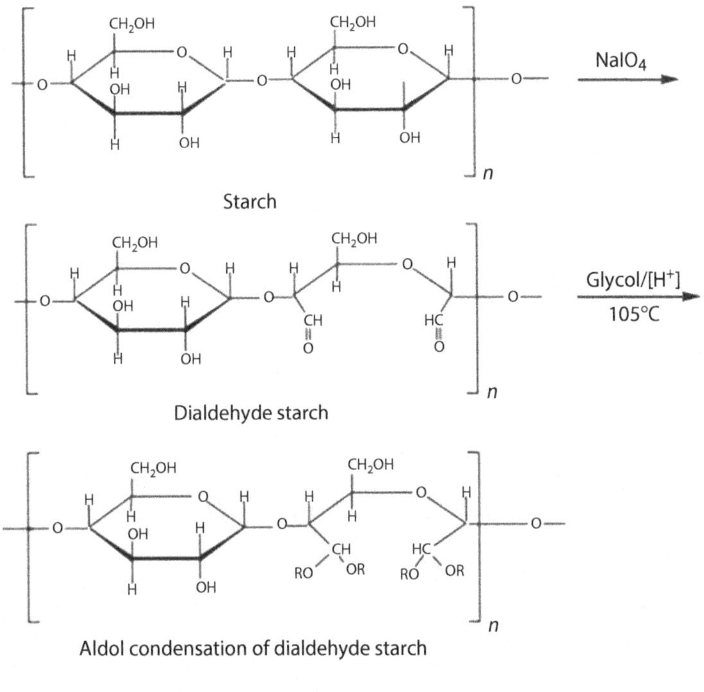

FIGURE 11.7 The preparation steps for dialdehyde starch and *g*-dialdehyde starch.

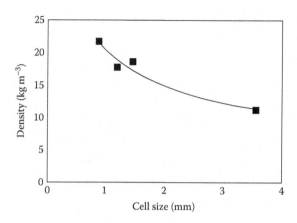

FIGURE 11.8 Cell size effects on starch foam with 7.2% water as the blowing agent. (Data from Yang, Z. et al., *Polym. Eng. Sci.*, 53, 857–867, 2013. With permission.)

To improve moisture resistance, polyhydroxyl ether was added to starch on the low density sheet extrusion line while using water as the blowing agent to relax the hydrogen-bond rich starch (Yang et al. 2013). A co-rotating twin with L/D at 16 was used to make around 20 kg m^{-3} foam sheets from an annular die. Figure 11.8 shows that the smaller the cell the more expansion there was. As cell size increases, cell wall thickness also increases to mitigate cell wall weakness, or open cell content, to make lower density possible. In addition to water resistance, it also exhibits some cushioning characteristics at a load of 1.4 kPa, or 0.2 psi. Yet deceleration curves shift up from the second drop to the fifth drop. This suggests that damage exists in the internal morphology following each drop, which is common for brittle cellular products. It is reported that alpha-crystalline emulsifiers are capable of stabilizing starch swelling and granule rupture up to 90°C. In food processing, this is critical to create a wider processing window. Starch gelatinization is better controlled and thus improves cellular morphology (Richardson 2003). It basically affects amylose aggregation, which can assist the hydrogen bond dependency of the granule breakdown into a homogenized melt for a smooth foam formation. It can do this in the pilot mode, as well as on a commercial scale. But there is a practical consideration here. Even though crystallization and water resistance improve starch-based foam, one must ask if it is economically viable. It certainly requires the combined efforts of various groups to make it a commercial success.

11.3.2.2 Polyesters Foams

Among polyesters, poly(lactic acid) (PLA) is widely recognized as the most promising compostable polymer. This is due to its low cost as a biodegradable polymer and its good mechanical properties, which are close to those of several synthetic and nonbiodegradable polymers (Garlotta 2001; Queiroz and Collares-Queiroz 2009; Parker et al. 2011b) and its biodegradability. It is mainly used in food packaging, but other applications in medicine or short life cycle commodity applications are widening.

Considerable effort has gone into the development of PLA based extrusion foams. This is because PLA has a low thermal stability and a very low melt strength, which hinders the production of other foams that use conventional extrusion foaming lines. A low shear rate technology has been developed by Plastic Engineering Associates Licensing, Inc., Boca Raton, Florida. Their Turbo-Screws® technology has proved to be effective in foaming amorphous and semicrystalline PLA resins. The Turbo-Screws technology has rectangular holes through the screw flights at the root. These move the melt from the root to the barrel wall for faster cooling and a higher output, and the technology appears to works best within a tandem foam extrusion system (Parker et al. 2011). Sheets with a thickness of between 0.5 and 5 mm and densities of ~50 kg m^{-3} can be obtained with a morphology mainly characterized by closed cells. This technology can use several PBAs, but the main industrial results have been achieved by using hydrocarbons. In high temperature food applications, a high degree of crystallinity is needed, and the production process must allow for the formation of a large amount of crystals before the complete cooling of the extrudate.

If a standard extrusion line is available, PLA foaming is quite challenging. The foaming window is often very narrow because PLA has a very low elongational viscosity and a poor melt strength. The main method used to improve the PLA's melt performance is to increase its molecular weight. This goal has been obtained with chain-extenders or light cross-linkers (Inata and Matsumura 1987; Villalobos et al. 2006). Chain extenders react with PLA macromolecules by joining chain ends, thanks to their two or more reactive functionalities. Several reactive species can be used, such as hydroxyl, amine, epoxy, carboxylic acid/anhydride, ionomer, and isocyanate, but usually epoxy and/or acid/anhydride or ionomeric containing reactants are preferred (Di et al. 2005a; Pilla et al. 2009; Mihai et al. 2010). Recently, specific commercial additives have been developed from major polymer suppliers. Other approaches to improve the melt rheology of PLA involves the use of nanofillers (Di et al. 2003, Di et al. 2005b, Ema et al. 2006) that are able to improve the elongational viscosity of the melt. In carefully controlled situations, peroxides can also be used to introduce branching or light cross-linking.

Reignier et al. (2007) reported that PBAs proved to be the best foaming agents for PLA. Using a lower than 10 wt% content of CO_2 as a blowing agent, they were able to prepare extruded PLA foams with densities of 20–25 kg m^{-3}. CBAs can be used in PLA but they result in high density foams, and their processing window is influenced by the need to reach the activation temperatures needed for gas releasing reactants (Matuana et al. 2009). Blending, on the other hand, can control the cellular morphology and produce low-density foams with good cell size distribution. In particular, the PLA-starch blends have been studied in extrusion foaming by a number of research groups (Preechawong et al. 2005; Mihai et al. 2007). Good-quality foams were obtained with a CO_2 blowing agent.

11.4 INJECTION MOLDING

Injection molding is a major processing technique for converting thermoplastic materials, and it accounts for about 30% of the world resin production. Injection molding relies on the ability of a thermoplastic material to be softened by heating, formed under pressure, and solidified by cooling. It is a cyclic process, and the number of operations that take place in an injection molding machine between two consecutive moldings is called the *molding cycle* (Figure 11.9). In a typical injection molding machine, granular material (the plastic resin) is fed from the hopper (a feeding device) into one end of the cylinder (the melting device). It is heated, melted (plasticization), and conveyed by a rotating screw into the injection chamber. When the accumulated plastic melt reaches the desired volume, the screw is stopped. Then, the accumulated plastic melt is forced into the mold cavity either by moving the screw forward or by moving a separate accumulator ram forward (injection step). Enough melt must be injected into the mold cavity to fill it and to compensate for the plastic's shrinkage as it cools. The melt must, therefore, be maintained under relatively high pressure until the material in the mold cavity is cooled sufficiently to prevent backflow (packing/holding step). If the cavity is not packed, the part will exhibit poor surface quality, sink marks, voids in thick sections, and differential distortion. Since the mold is cold, the resin will start to solidify at the mold surface immediately after the first contact with (namely, during the injection step). The solidification proceeds from the surface toward the mid-plane of the plastic part, according to the energy balance. Conventionally, however, the so-named cooling step starts at the end of the holding step. This is when the thin section at the cavity inlet, called the *gate*, solidifies or is closed by an external control, thus creating a barrier between the molded part and the series of channels known as the *runners*. These convey the molten material from the injection nozzle to the cavity. While the cooling takes place, the screw rotates, thus preparing the following shot for the *batching*

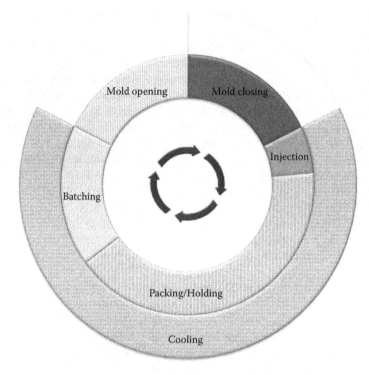

FIGURE 11.9 Schematization of a molding cycle in its basic steps.

or *metering* step. During this operation, the screw moves back while pushing the material toward the injection chamber. The hydraulic system exerts back pressure to control the screw's backwards movement. The cooling step should last until the part is completely solidified, although in common practice it lasts until the part is sufficiently solidified it to retain its shape at demolding. Following the cooling step, the mold opens and the part is ejected.

11.4.1 FOAM INJECTION MOLDING

The objective of foam injection molding is twofold. First, the traditional injection molding technique is used to produce a plastic part with a well-defined shape. Second, this is to be done by consuming less material and without sacrificing mechanical properties. Material is saved by creating voids by means of a foaming agent. If the plastic resin is mixed with a volatile material and is forced into the mold with a volume lower that what is needed to obtain the part's final geometry, the volatile material forms gas bubbles in the melt and expands until the solution completely fills the cavity. The extent of foaming has been found to be a function of the amount of volatile material added to the melt, the temperature of the melt and the mold cavity, the speed of injection, the uniformity of dispersion of the volatile material, and the pressure applied to the melt during various stages of transfer from the accumulator to the mold cavity.

While replacing plastics with gas creates a significant weight reduction in the part, the voids also serve as crack arrestors by blunting crack tips, and thereby greatly enhance the part's toughness. Since the gas fills the interstitial sites between polymer molecules, it effectively reduces the viscosity and the glass transition temperature (Di Maio et al. 2006). Therefore, the part can be injection molded at lower temperatures and pressures, leading to a significant reduction in the clamp tonnage requirement and cycle time. Further, the expansion forces the material to completely fill the cavity, thus canceling the need for a packing phase.

Basically, the foam injection molding process involves the following four distinctive steps:

- Gas dissolution—the foaming agent is added into the cylinder to form a solution with the polymer for processing.
- Nucleation—a large number of nucleation sites are formed by a rapid and substantial pressure drop when material is being pushed into the cavity through the nozzle.
- Cell growth—cell growth and cell coalescence take place during mold filling and postfilling stages and is controlled by such processing conditions as melt pressure and temperature.
- Shaping—the shaping of the part takes place inside the mold via solidification.

11.4.1.1 Structural Foam

The typical result of foam injection molding is described as *structural foam*. It is characterized as a plastic structure with a foamed core and unfoamed skins (Figure 11.10), usually with a 15–40 wt% density reduction.

Sometimes, the term *integral foam* is also found in the literature (Shutov et al. 1986). Structural foam is conventionally divided into three zones along the thickness direction, going from the surface of the part toward the mid-plane (Bledzki et al. 2012). The first zone is a compact skin layer. It is the result of the quick solidification of the polymer/gas solution, which comes into contact with the cold mold, and does not have the time to foam before solidifying. This layer ends when the first cells are found to be going toward the sample's center. By increasing the distance from the mold, the cooling rate is decreased, and thus the foaming process can take place. The second layer, called the *transition layer*, is the area that begins at the end of the skin layer and ends where the cell diameter increases. Since growth takes place when the polymer is still under shear, as shown in Figure 11.2, the cells in the transition layer are normally elongated in the flow direction. The third area lies in the middle of the sample and is known as the *core layer*. In this layer, the cooling rate and the deformation rate are low. As a result, the gas bubbles have the time to grow and the possibility to coalesce, thus creating large cells. A structural foam is typically characterized by closed cells. However, different methods to induce a significant amount of open cells have been explored, as reported in the scientific literature. For instance, open-cell thermoplastic foams have been created by leaching a soluble filler from a polymer matrix (Kramschuster and Turng 2010).

11.4.1.1.1 Surface Appearance

One of the main defects of structural foams is their surface finishing (Chen et al. 2010). Light-scattering or a strip-type flow marks, that is, bright sliver streaks or swirl marks, appear on the microcellular foam-injected molded part's surface (Figure 11.11).

This affects the part's clarity and its aesthetics, and limits this technology to applications involving interior products. The cause of the flow marks generated on the surface of the foamed injection

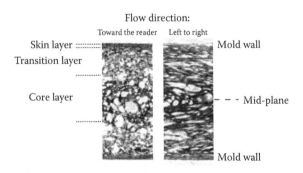

FIGURE 11.10 Tomographic image of a sample of polystyrene foamed with supercritical N_2. The thickness of the sample is 5 mm.

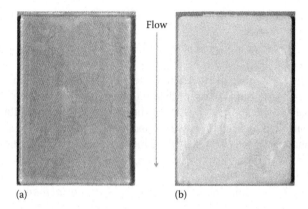

FIGURE 11.11 Difference in the appearance between an unfoamed (a) and a foamed (b) sample made of PLA. The unfoamed sample is transparent. The shown samples are 100 mm long, 70 mm wide, and 5 mm thick.

molded parts is not completely clear. Cha and Yoon (2005) pointed out that swirl marks are caused by the gas that is trapped on the mold surface as the plastic/gas solution begins to solidify. The disturbances that the trapped gas causes form grooves on the part's surface. Silver streaks are oriented in the direction of the flow path. Michaeli and Cramer (2006) reported that silver streaks result from macroscopic gas bubbles, which are sheared on the cavity wall during the filling phase. Their fragmentation leads to many small microscopic gas bubbles that cause dents in the surfaces of foamed parts. The different light reflection properties of those dents appear as shining silver streaks on the part's surface.

Improving surface appearance is one of the main concerns in foam injection molding. Special processes such as in-mold decoration or co-injection (Turng and Kharbas 2004), and secondary operations such as surface polishing allow one to bypass the problem. The control of cell growth during the early filling by such means as breathing molds and gas-counter pressure, can effectively improve the surface quality (Bledzki et al. 2008). Cha and Yoon (2005) found that maintaining a high mold temperature above the melting point during the filling process significantly reduced the swirl marks. However, maintaining a high mold temperature during the filling process can unacceptably increase the cycle time. Subsequently, several methods have been suggested to obtain a high polymer temperature at the first contact with the mold surface without needing to heat the entire mold (Wang and Hu 2011). These methods include induction heating (Chen et al. 2008c), or the addition of an insulating layer on the cavity surface, which induces a heat transfer delay that makes for a higher contact temperature (Chen et al. 2008a,b). The possibility of applying these methods to mass production is still questionable. Mold surface heating means significant additional costs, and the long-time stability of a coating layer must be verified.

11.4.1.2 Methods for Achieving Structural Foams

Foam injection molding can be schematized in several ways. One possibility is to distinguish between low pressure and high pressure molding.

11.4.1.2.1 Low-Pressure Process

Low-pressure or short-shot conventional foam processing methods are the most commonly used. A controlled melt solution of plastic and the blowing agent is injected into the cavity to only partially fill the mold, thus realizing a so-named short shot. Due to the sudden pressure reduction, the blowing agent expands and foam fills the cavity.

The main process variables are the amount of melt injected, that is, the percentage of the cavity filled with the short shot, the mold and melt temperatures, the type and concentration of blowing agent, and the capabilities of the molding machine, particularly the imposed flow rate.

This process is hindered by the fact that the low pressure at the melt front leads to foaming, and the bubbles are brought to the part's surface.

In order to control foaming during the early stages of filling, the gas counter pressure technique was developed (Michaeli and Cramer 2006; Bledzki et al. 2008). This method uses a sealed mold pressurized with an inert gas. The pressure is sufficient to suppress foaming as the plastic mix enters the mold cavity. After the measured shot is injected, the mold pressure is released, allowing instantaneous foaming to form the core between the already formed solid skins. The gas counter pressure technique has been reported to effectively improve the part's surface appearance (Chen et al. 2012). The pressure levels needed are between 40 and 100 bar depending on the polymer/gas solution and on the injection temperature.

Another possible way to improve the surface appearance is the so-named low pressure with coinjection technique. This involves the injection of two compatible plastics that are coinjected using two injection cylinders (Chien et al. 2004; Turng and Kharbas 2004). A solid plastic is injected from one cylinder to form a solid, smooth skin against the surfaces of the mold cavity. Simultaneously, a second material, a measured short shot containing a blowing agent, is injected to form the foamed core.

11.4.1.2.2 *High-Pressure Process*
In high-pressure foaming the melt solution is injected into the mold to completely fill the cavity, creating a cavity pressure higher than the blowing-agent gas pressure. The high pressure prevents any foaming from occurring while the skin portion starts solidifying against the mold surfaces. As soon as the skin surface hardens to a desired thickness, the cavity mold pressure is reduced to allow the remaining melt to foam between the skins. Depending on the type of equipment, this pressure reduction is obtained either by withdrawing cores or by special press motions that partially open the mold halves, the so-named mold breathing action (Bledzki et al. 2008; Stumpf et al. 2011).

11.4.1.3 Machine Requirements
Conventional injection molding machines can be successfully adopted for foaming. The basic system (Figure 11.12) needs a gas injection control unit with one or more gas nozzles directing the gas inlet into the plasticating cylinder.

However, some elements are essential for carrying out the foaming process. These elements are listed as follows:

11.4.1.3.1 *Control of Back Pressure*
After the batching stage, a standard injection molding machine enters an idle period in which the pressure on the screw is released. The screw can thus be pushed backward by gas pressure in

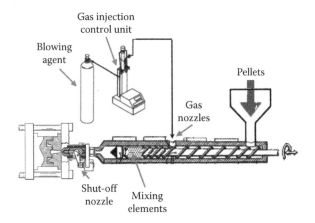

FIGURE 11.12 Schematic of an injection machine for foaming with physical blowing agents.

the solution accumulated in front of the screw tip. This can induce a phase separation between the gas and the melt.

The simplest way to maintain the screw position while, at the same time, keeping a constant back pressure during the screw idle period, is to introduce a mechanical brake on the back of the screw to maintain its position after the batching stage. Another way of doing this is to introduce a hydraulic control, so that the servo valve controls the final screw position. This latter method has the advantage of directly keeping the solution's pressure in the injection chamber at a value higher than the saturation point of the injection temperature.

11.4.1.3.2 Screw Modification

Usually the injection machine does not have a long enough mixing section, and thus the introduction of one is necessary to maintain the quality of the microcellular injection molding screw (Bravo and Hrymak 2005). Rauwendaal (1998) has reported successful gas dispersion by using special mixing elements at the end of a single-screw extruder to increase stretching and eventual breakup of the liquid additives' bubbles and droplets. Park (Lee et al. 2008) has used a static mixer in the nozzle, which created not only a shearing effect but also an elongational effect. This is a much more effective way to promote bubble breakup and hence better gas bubble dispersion.

A special middle check valve or restrictive element is also useful for maintaining the melt pressure inside the screw wherever the gas existed. In other words, once the gas is added into the screw, it must be kept at a certain level of pressure all the while until it is injected into the mold. The middle check valve must be chosen to match the screw tip performance to avoid either losing the melt pressure in the single-phase solution or of having a built-up pressure surge during the injection.

11.4.1.3.3 Shut-Off Nozzle

A shut-off nozzle, or a valve gate, is a critical component to make a good foamed injection molded part. Without the shut-off nozzle, foaming takes place before injection, leading to a very poor product.

11.4.1.3.4 Special Modification to the Injection System

Most of the equipment designed for foam injection molding is based on single-screw extrusion, with mixing or other elements aimed at promoting bubble breakup. Modifications to the injection system have been presented in the literature to improve the mixing between the gas and the melt. These systems often present an additional accumulator after the screw where the solution is kept before injection to allow for a better gas dispersion inside the polymer. For example, Xu et al (2008) presented a technology that decoupled the gas dissolution step from the injection and molding operations. This was based on the introduction of a gear pump and an additional accumulator to make the polymer/gas solution formation step continuous, in spite of the intrinsic discontinuous nature of the injection molding process (Xu et al. 2008).

11.4.1.4 Process Variables

The major challenge of foam injection molding is in controlling the process variables to create fine and uniform microcells throughout the part. Once the polymer and the foaming agent are selected, the number of variables involved is large and can be divided into two groups: those involved in the preparation of the solution to be injected and those involved in the injection. The following are examples of variables belonging to the first group: back pressure, screw rotation speed, amount of gas injected into the cylinder, melt temperature, and residence time of the solution inside the injection chamber. All of these are aimed at reaching the correct degree of gas supersaturation inside the injection chamber. We note (Lee et al. 2011b) that when the gas amount dissolved in the polymer melt is increased, the molded parts have a smaller cell size and a higher cell density, and maintain an acceptable surface quality until a certain degree of supersaturation or gas is reached. When this value is reached, the cell nucleation rate becomes so high that swirl marks appear on the surface.

The solubility of a gas decreases with an increase in melt temperature. Thus, for the same amount of gas injected, the supersaturation level decreases with decreased temperature. A high and constant screw rotation speed is useful for mixing gas inside the polymer melt (Lee et al. 2008). A high back pressure allows not only a longer mixing time but also a higher concentration of gas inside the polymer. Eventually, in order to promote a homogeneous distribution of gas inside the shot to be injected, it is useful to allow the solution a sufficient residence time inside the injection chamber. This interval is obviously inversely dependent on the diffusivity of the gas inside the polymer. As an example, for N_2 in polystyrene, a time in the order of 100 s has been found to be effective for stabilizing the cell density in the foamed parts (Rizvi and Bhatnagar 2009).

Examples of the variables belonging to the second group are gate geometry, injection flow rate, ratio between injected volume (short shot) and final volume, and mold temperature. The gate design is critical for foaming. This is because the pressure before it should be sufficiently high to prevent premature foaming, and the pressure after it should be lower than the solubility so as to induce immediate cell nucleation inside the cavity. This normally requires a very thin gate. The injection flow rate is probably the most critical parameter. In general, a high injection flow rate induces smaller cells and a more homogeneous distribution of the morphology inside the molded part (Lee et al. 2008). However, beyond a certain value, no further reduction of cell dimension is found, and even a reduction in cell density has been reported (Rizvi and Bhatnagar 2009). This phenomenon could be the result of an extremely high injection speed causing high pressure at locations closer to the gate, and which inhibits foaming during filling. When pressure decreases to a suitable value, the local temperature can be too low for foaming (Pantani et al. 2013). The effect of mold temperature has been described in detail above.

11.4.1.5 Foam Injection Molding of Biodegradable Polymers

As previously discussed, foam injection molding's main advantage is the potential to produce parts with a reduced weight and a stiffness-to-weight ratio comparable to their unfoamed counterparts. Further, as noted above, the presence of gas inside the polymer can reduce its viscosity (Areerat et al. 2002; Choudhary et al. 2005), and can thus enable the polymer's processability at lower temperatures and pressures. This is of particular advantage for biodegradable polymers, which are thermally sensitive and have narrow processing windows (Speranza et al. 2014). The increasing interest in biodegradable polymers has, therefore, further boosted the appeal of foam injection (Jeon et al. 2013). Among the biodegradable polymers, PLA has received most of the research attention. However, it is very difficult to control PLA foaming by injection molding, because of its low melt strength and slow crystallization kinetics (Pantani et al. 2010). Also, for this reason, the literature studies on PLA foam injection molding are few in number and quite recent (Kramschuster et al. 2007; Kunimune et al. 2010; Pilla et al. 2010; Ameli et al. 2013).

In 2013 (Pantani et al. 2013), the foam injection molding of PLA was analyzed by pointing out the effects of the injection rate and the mold temperature. The essential results of this work are presented subsequently as an example of the foam injection molding of biodegradable polymers. The material adopted was a commercial grade supplied by NatureWorks, 4032D. The cavity was rectangular, with the dimensions reported in Figure 11.11, and a film gate 1 mm thick. When dealing with PLA, the injection temperature is limited to a narrow range between a lower value, at which the polymer presents too great a viscosity, and a high value at which it degrades very quickly. The mold temperature is upwardly limited by the material's relatively low glass transition temperature of about 55°C–60°C. The injection temperature was, therefore, set at 200°C. Two mold temperatures were considered: 25°C and 40°C. The injection point was located at 35 mm from the gate. A low-pressure foam injection molding was adopted, keeping constant the shot size in such a way as to fill 80% of the cavity. The density reduction was calculated as follows:

$$d_r = \frac{\rho_{unfoamed} - \rho_{foamed}}{\rho_{unfoamed}}$$

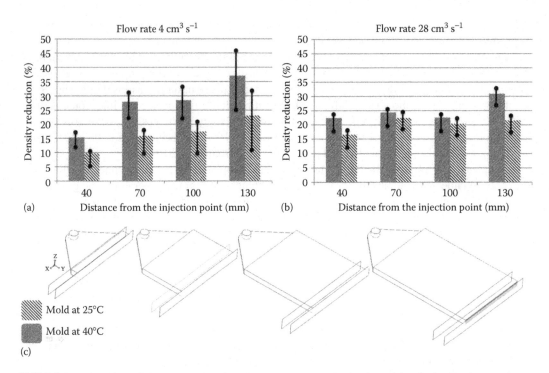

FIGURE 11.13   Effect of flow rate and mold temperature on the distribution of density inside the samples.

Figure 11.13 shows the effect of the flow rate on density reduction. At low flow rates, the samples were quite inhomogeneous, with more consistent foaming occurring at the cavity's tip. The sections closer to the gate had vastly variable results, as indicated by the large error bars. This effect was more evident at the higher mold temperature, where the density reductions at the tip were very large. With an increased flow rate, more homogeneous samples resulted, and the density reduction was larger for the highest applied mold temperature.

Figure 11.14 shows the morphology distribution inside the samples, as well as some optical images of sample sections taken at different distances from the mold. Despite the homogeneous density reduction, in both samples, the morphology changes with increased density from the gate (Volpe 2013). In particular, the core section, which is indicated by dotted lines in Figure 11.14, becomes thinner. The resulting morphology is more homogeneous when a higher mold temperature is adopted, and there is a reduced core section. Further, the cell sizes in the core are smaller for the hotter mold. This is more clearly evidenced by the scanning electron micrographs taken at 100 mm from the injection point, which Figure 11.14 also shows.

In conclusion, PLA foams with a relatively high void fraction and an acceptably uniform structure can be obtained by selecting suitable molding conditions. In particular, a high injection flow rate and a high mold temperature significantly improve the homogeneity of cell dimensions. Adding a nucleating agent such as talc (Pantani et al. 2013, Ameli et al. 2013) can further improve the structure's uniformity.

11.5   CHEST MOLDING

Chest molding is a technology originally developed for expanded polystyrene and is widely used to produce very low-density foams with complex three-dimensional shapes starting with pre-expanded beads. This technology is used in many applications from packaging, including food packaging, to thermal insulating boards for buildings. Since expanded polystyrene is based on polystyrene, it is a nonrenewable synthetic resource and is not biodegradable. Several efforts have been recently made

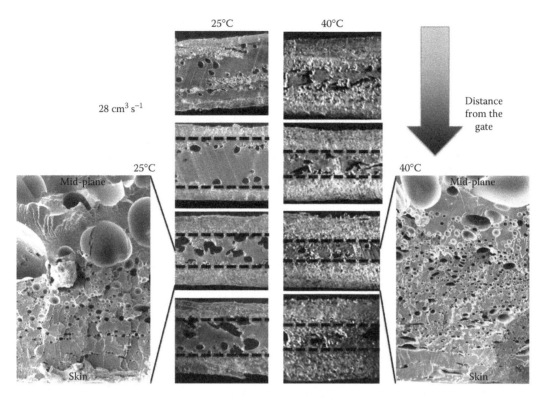

FIGURE 11.14 Distribution of morphology inside the samples obtained by imposing a high flow rate. The flow is directed toward the reader. The dotted lines identify the core layer of the samples.

to apply chest molding to biodegradable or bio-based polymers to reduce expanded polystyrene manufacturing waste. This effort has also been driven by the recent increase in oil prices. Large initiatives have been undertaken to develop bio-based polymeric formulations, which could allow for the use of the standard chest molding facilities (Parker et al. 2011b).

11.5.1 Unit Operations in Chest Molding

Chest molding technology consists of the postexpansion of pre-expanded beads in a closed mold through the use of a high temperature gaseous flow, which is usually water steam at temperatures of between 110°C and 125°C. The gas flows through the pre-expanded beads, thus softening and then bonding their external surfaces as in the sintering process of particles.

Spherical particles having a diameter ranging from 0.5 to 1.3 mm are first saturated by using liquid PBAs, such as pentane. This blowing agent can be retained for a long time under controlled stocking conditions, and usually at a temperature far below the room temperature. The first step in the production of beads foams is the pre-expansion phase, during which the small bead freely swells to almost 50 times its original size. During the pre-expansion phase, the temperature is raised to a specific value at which the polymer/blowing agent is sufficiently softened to allow the blowing agent to nucleate bubbles and to migrate from the polymer to the growing bubble. The process ends when the desired average bead expansion has been reached. After that, a pressure equalization phase must be performed to allow the internal pressure of the bubbles to equilibrate with the atmospheric pressure and to avoid cell collapse in the subsequent processing phases. This *aging* can take from 12 h to two days, depending on the target density of the expanded beads. When the aging is complete, the beads are ready to be formed in the molds.

We note that if water steam is used a subsequent drying phase should be applied to avoid humidity accumulation between the expanded beads.

The molding process consists of filling of the mold, which is kept slightly open, with the pre-expanded beads, and then letting a dry water steam at a temperature of between 110°C and 125°C flow through the bead interstices. The beads are briefly exposed to steam, which flows alternatively in two opposite directions, to pre-heat their surfaces, through the open mold. Then, the mold is completely closed and another steam flow is applied. This second steam flow, applied during the compaction of the beads into the final shape, is responsible for the sintering of almost the entire external surface of the beads. The latent heat from the steam induces a pressure increase and further expands the beads. Since the mold is a confined environment, the only way the beads can expand is to fill up any voids between them, causing the soft surfaces to fuse together into a polyhedral-type solid structure. The molding process must be carefully performed to prevent moisture from forming within the beads, which can result in a low welding degree and, consequently, a low performance level.

The final products obtained by using the chest molding technique usually have excellent thermal insulation capacities and high energy absorption per unit weight properties, which is also lower than 20 kg m^{-3}.

11.5.2 PLA Bead Foaming

The chest molding process has been adapted to the requirements of specific biodegradable or bio-based polymers. PLA, in particular, has been investigated as a viable replacement for the polystyrene for beads used in the sintering process. Different processing routes have been developed by research groups or companies to get a bead foamed PLA part. These focus on the polymer viscoelastic properties or the deep control of specific phases of the molding process.

The Biopolymer Network Ltd., New Zealand, has developed a PLA particle foam technology (Figure 11.15). This uses commercial PLA polymer, without the addition of nucleating agents (Witt and Shah 2007). Their expanded PLA process is based on liquid carbon dioxide as a blowing agent, which they claim is faster, with respect to gas impregnation, and cheaper, with respect to supercritical CO_2. PLA polymer beads are impregnated with CO_2 at 60 bar for up to four h and reach a blowing agent content of between 18 and 35 wt%. The beads prefoaming is performed at temperatures of 50°C to 70°C while the claimed density of the final objects ranges between 30 and 60 kg m^{-3}. BPN have stated that they worked on a process that could be applied to different PLAs, which were either modified to improve its elongational viscosity and hence, its foamability, or standard grades. They also claimed that their production process could be integrated into standard expanded polystyrene bead chest molding lines without needing special beads treatments or having polymer rheological properties.

Several companies have investigated the bead foaming process, such as JSP Corporation (Japan) (Shinohara et al. 2002, Haraguchi and Ohta 2005). They developed a specific thermal treatment, which is applied during the pre-expansion phase. They used nucleating agents such as talc to increase the number of cells and to reduce their dimensions. PLA beads are impregnated with gaseous CO_2 at pressures of between 30 and 40 bar for up to 3.5 hours, obtaining gas absorption up to 14 wt%. A steam flow is applied to the beads for five s (the expansion temperature was reported to be 64°C to 70°C) giving the expanded beads an apparent density of 50–60 kg m^{-3}. Before being transferred into the mold, the pre-expanded beads are impregnated again to effect further expansion through steam, and they are sintered at a temperature of around 120°C.

Another route to obtaining sintered PLA beads is to surface coat pre-expanded beads to improve their stitching, such as in the Synbra Technology, the Netherlands (Noordegraaf et al. 2007). They use semicrystalline PLA beads loaded with small percentages of nucleating agent. They can use different biodegradable coatings, which result in a relative amount of between 0.5 and 15 wt%, with respect to the weight of the PLA beads. The beads are then impregnated with gaseous CO_2 at 20 bar

FIGURE 11.15 Loose PLA foam beads (a) and PLA foam-molded particle foam products (b, c, and d). (Courtesy of Biopolymer Network Ltd., New Zealand, and partners; Reprinted from Parker, K. et al., Polylactic acid [PLA] foams for packaging applications, in *Handbook of Bioplastics and Biocomposites Engineering Applications*, Ed. S. Pilla., Scrivener Publishing, Salem, MA, 2011b. With permission.)

for sorption times from 20 min to 16 h to get a final blowing agent content of between 5 and 8 wt%. Then, they are pre-expanded or directly expanded in the mold.

Sekisui Plastics Company Limited (Japan) has developed another method to prepare pre-expanded beads (Hirai 2008). This uses a conventional extrusion foaming process based on a CO_2 blowing agent to mix the polymer with the blowing agent. This is then cut into small pellets, which can expand at a final density of higher than 100 kg m^{-3}. These foamed pellets, after being impregnated for a second time, can be used in place of pre-expanded beads in the chest molding process. A final part density of around 50 kg m^{-3} can be achieved.

REFERENCES

Ameli, A., Jahani, D., Nofar, M., Jung, P. U., and Park, C. B. 2013. Processing and characterization of solid and foamed injection-molded polylactide with talc. *J Cell Plast* 49(4), 351–374.

Areerat, S., Nagata, T., and Ohshima, M. 2002. Measurement and prediction of LDPE/CO_2 solution viscosity. *Polym Eng Sci* 42, 2234–2245.

Bledzki, A. K., Kirschling, H., Rohleder, M., and Chate, A. 2012. Correlation between injection moulding processing parameters and mechanical properties of microcellular polycarbonate. *J Cell Plast* 48(4), 301–340.

Bledzki, A. K., Kuehn, J., Kirschling, H., and Pitscheneder, W. 2008. Microcellular injection molding of PP and PC/ABS with precision mold opening and gas counterpressure. *Cell Polym* 27, 91–100.

Boonstra, D. J. and Berkhout, F. 1975. Preparation of Redispersible Hydrophobic Starch Derivatives, US Patent: 3,891,624.

Bravo, V. and Hrymak, A. 2005. Nozzle injection of physical blowing agents in the injection molding of microcellular foams. *Int Polym Proc* 20, 149–156.

Cha, S. W., and Yoon, J. D. 2005. The relationship of mold temperatures and swirl marks on the surface of microcellular plastics. *Polym Plast Technol Eng* 44(5), 795–803.

Chen, H.-L., Chien, R.-D., and Chen, S.-C. 2008a. Variable mold temperature to improve surface quality of microcellular injection molded parts using induction heating technology. *Int Commun Heat Mass* 35, 991–994.

Chen, S.-C., Hsu, P.-S., and Hwang, S.-S. 2012. The effects of gas counter pressure and mold temperature variation on the surface quality and morphology of the microcellular polystyrene foams. *J Appl Polym Sci* 127 (6), 4769–4776.

Chen, S. C., Li, H. M., Huang, S. T., and Wang, Y. C. 2010. Effect of decoration film on mold surface temperature during in-mold decoration injection molding process. *Int Commun Heat Mass Transf* 37(5), 501–505.

Chen, S.-C., Li, H.-M., Hwang, S.-S., and Wang, H.-H. 2008b. Passive mold temperature control by a hybrid filming-microcellular injection molding processing. *Int Commun Heat Mass* 35, 822–827.

Chen, S.-C., Lin, Y.-W., Chien, R.-D., and Li, H.-M. 2008c. Using thermally insulated polymer film for mold temperature control to improve surface quality of microcellular injection molded parts. *Adv Polym Tech* 27, 224–232.

Chien, R. D., Chen, S.-C., Lee, P.-H., and Huang, J.-S. 2004. Study on the molding characteristics and mechanical properties of injection-molded foaming polypropylene parts. *J Reinf Plast Comp* 23, 429–444.

Choudhary, M., Delaviz, Y., Loh, R., Polasky, M., Wan, C., Todd, D. B., Hyun, K. S., Dey, S., and Wu, F. 2005. Measurement of shear viscosity and solubility of polystyrene melts containing various blowing agents. *J Cell Plast* 41, 589–599.

Di, Y., Iannace, S., Di Maio, E., and Nicolais, L. 2003. Nanocomposites by melt intercalation based on polycaprolactone and organoclay. *J Polym Sci Part B: Polym Phys* 41(7), 670–678.

Di, Y., Iannace, S., Di Maio, E., and Nicolais, L. 2005a. Reactively modified poly (lactic acid): Properties and foam processing. *Macromol Mater Eng* 290(11), 1083–1090.

Di, Y., Iannace, S., Di Maio, E., and Nicolais, L. 2005b. Poly(lactic acid)/organoclay nanocomposites: Thermal, rheological properties and foam processing. *J Polym Sci Part B: Polym Phys* 43(6), 689–698.

Di Maio, E., Iannace, S., Mensitieri, G., and Nicolais, L. 2006. A predictive approach based on the Simha–Somcynsky free-volume theory for the effect of dissolved gas on viscosity and glass transition temperature of polymeric mixtures. *J Polym Sci Pol Phys* 44, 1863–1873.

Di Maio, E., Mensitieri, G., Iannace, S., Nicolais, L., Li, W., and Flumerfelt, R. 2005. Structure optimization of polycaprolactone foams by using mixtures of CO_2 and N_2 as blowing agents. *Polym Eng Sci* 45, 432–441.

Durril, P. L., and Griskey, R. G. 1966. *AICHE J* 12, 1147.

Ema, Y., Ikeya, M., and Okamoto, M. 2006. Foam processing and cellular structure of polylactide-based nanocomposites. *Polymer* 47, 5350–5359.

Fisk, D. 1998. Polymer Composition Containing Prime Starch, US Patent: 5,739,244.

Garlotta, D. 2001. A literature review of poly(lactic acid). *J Polym Environ* 9(2), 63–84.

Glenn, G., Orts, W., Klamczynski, A., Ludvik, C., Chiou, B.-S., Iman, S., and Wood, D. 2006. Heat-expanded starch-based compositions. *Annual BEPS International Meeting*, Chicago, IL.

Haraguchi, K., and Ohta, H. 2005. Expandable polylactic acid resin particles, European Patent application EP 1 683 828 A2.

Hirai, T., Nishijama, K., and Ochiai, T. 2008. Polylactic acid resin foam particle for in-mold foam forming, process for producing the same, and process for producing polylactic acid resin foam moulding. Sekisui Plastics, Japan, European Patent Office EP 2 135 724 A l.

Inata, H., and Matsumura, S. 1987. *J Appl Polym Sci* 33(8), 3069–3079.

Jeon, B., Kim, H. K., Cha, S. W., Lee, S. J., Han, M.-S., and Lee, K. S. 2013. *Int. J. Precis. Eng. Manuf.* 14, 679–690.

Kramschuster, A., Gong, S., Turng, L.-S., Li, T., and Lil, T. J. 2007. *Biobased Mater Bioenerg* 1, 37–45.

Kramschuster, A., and Turng, L. S. 2010. An injection moulding process for manufacturing highly porous and interconnected biodegradable polymer matrices for use as tissue engineering scaffolds. *J Biomed Mater Res Part B: Appl Biomater* 92(2), 366–376.

Kunimune, N., Kunimune, T., Nagasawd, T., Tamada, S., Yamada, K., Leong, Y. W., and Hamada, H. 2010. *Annu Tech Conf Proc* 1, 683–686.

Lacourse, N.L. and Altieri, P.A. 1989. Biodegradable Packaging Material and the Method of Preparation Thereof, US Patent: 4,863,655.

Lee, J., Turng, L.-S., Dougherty, E., and Gorton, P. 2011a. *Polym Eng Sci* 51, 2295–2303.

Lee, J., Turng, L.-S., Dougherty, E., and Gorton, P. 2011b. *Polymer* 52, 1436–1446.

Lee, J. W. S., Wang, J., Yoon, J. D., and Park, C. B. 2008. *Ind Eng Chem Res* 47, 9457–9464.

Lee, S. T. 2014. Starch Foam, editorial in Foam Update, March 2014.

Lee, S.T., and Park, C.B. 2014. *Foam Extrusion: Principles and Practice* (Second Edition), Eds. Lee S.T. and Park C.B., Boca Raton, FL: CRC Press.

Loudrin, D., Valle, G. D., and Colonna, P. 1995. Influence of amylose content on starch films and foams. *Carbohyd Polym* 27, 261–270.

Ma, P., Wang, X., Liu, B., Li, Y., Chen, S., Zhang, Y., and Xu, G. 2012. Preparation and foaming extrusion behavior of polylactide acid/polybutylene, succinate/montmorillonoid nanocomposite. *J Cell Plast* 48, 191.

Malwitz, N. and Lee, S. T. 1992. Thermoplastic Compositions for Water Soluble Foams, US Patent: 5,089,535.

Markham, J. P. and Martin, T. K. 2005. Mould Inhibitor Integrated within a Matrix and Method of Making same, US Patent 6,894,136.

Matsushita, T. 2013. Heat Insulator, US Patent: 8,414,998.

Matuana, L. M., Faruk, O., and Diaz, C. A. 2009. Cell morphology of extrusion foamed poly (lactic acid) using endothermic chemical foaming agent. *Bioresour Technol* 100(23), 5947–5954.

Michaeli, W., and Cramer, A. 2006. *Annu Tech Conf Proc* 3, 1210–1214.

Mihai, M., Huneault, M.A., and Favis, B.D. 2010. Rheology and extrusion foaming of chain-branched poly(lactic acid). *Polym Eng Sci* 50(3): 629–642.

Mihai, M., Huneault, M. A., Favis, B. D., and Li, H. 2007. Extrusion foaming of semi-crystalline PLA and PLA/thermoplastic starch blends. *Macromol Biosci* 7(7), 907–920.

Narayan, R. 2006. Biobased & biodegradable polymer materials: Rationale, drivers, and technology examplars. *ACS Symp Ser* 939, 282.

Nema, A. K., Deshmukh, A. V., Palanivelu, K., Sharma, S. K., and Malik, T. 2008. *J Cell Plast* 44, 277–292.

Neumann, P. E., and Seib, P. A. 1993. Starch-Based Biodegradable Packing Filler and Method of Preparing Same, US Patent: 5,208,267.

Noordegraaf, J., Britton, R.N., Van, D.F.A.H.C., Molenveld, K., and Schennink, G.G.J. 2007. Particulate expandable polylactic acid, a method for producing the same, a foamed moulded product based on particulate expandable polylactic acid, as well as a method for producing the same. WO 2008/130226 A2.

Pantani, R., De Santis, F., Sorrentino, A., De Maio, F., and Titomanlio, G. 2010. *Polym Degrad Stabil* 95, 1148–1159.

Pantani, R., Sorrentino, A., Volpe, V., and Titomanlio, G. 2014. Foam injection molding of poly(lactic acid) with physical blowing agents. *AIP Conf. Proc* 1593, 397.

Parker, K., Garancher, J.-P., Shah, S., and Fernyhough, A. 2011b. Expanded polylactic acid—An eco-friendly alternative to polystyrene foam. *J Cell Plast* 47, 233.

Parker, K., Garancher, J.-P., Shah, S., Weal, S. and Fernyhough, A. 2011a. Polylactic acid (PLA) foams for packaging applications. In *Handbook of Bioplastics and Biocomposites Engineering Applications*, Ed. S. Pilla. Salem, MA: Scrivener Publishing.

Peng, J., Turng, L.-S., and Peng, X.-F. 2012. *Polym Eng Sci* 52: 1464–1473.

Peng, J., Wang, J., Li, K., Turng, L. S., and Peng, X. F. 2011a. *Annu Tech Conf ANTEC Conf Proc* 3, 2643–2648.

Peng, J., Yu, E., Sun, S., Turng, L. S., and Peng, X. F. 2011b. Study of microcellular injection moulding with expandable thermoplastic microsphere. *Int Polym Process* 26(3), 249–255.

Pfannschmidt, O., and Michaeli, W. 1999. Foam injection molding of thermplastics loaded with carbon dioxide prior to processing. *SPE ANTEC Tech Papers*, 2100-2103.

Pilla, S., Kim, S.G., Auer, G.K., Gong, S., Park, C.B. 2009. Microcellular extrusion-foaming of polylactide with chain-extender. *Pol. Eng. Sci.* 49(8), 1653–1660.

Pilla, S., Kramschuster, A., Lee, J., Clemons, C., Gong, S., and Turng, L.-S. 2010. *J Mater Sci* 45, 2732–2746.

Preechawong, D., Peesan, M., Supaphol, P., and Rujiravanit, R. 2005. Preparation and characterization of starch/poly(L-lactic acid) hybrid foams. *Carbohyd polym* 59(3), 329–337.

Queiroz, A.U.B. and Collares-Queiroz, F.P. 2009. Innovation and industrial trends in bioplastics. *Polym. Rev.* 49(2): 65–78.

Rauwendaal, C., Osswald, T., Gramann, P., and Davis, B. 1998. New dispersive mixers for single screw extruders. *SPE ANTEC Proc* 44, 277.

Reignier, J., Gendron, R., and Champagne, M. F. 2007. Extrusion foaming of poly (lactic acid) blown with CO_2: Toward 100% green material. *Cell polym* 26(2), 83–115.

Richardson, G. 2013. Foam formation and starch gelatinization with alpha-crystalline emulsifiers. PhD thesis, Lund University, Lund, Sweden

Rizvi, S. J. A., and Bhatnagar, N. 2009. *Int Polym Proc* 24, 399–405.

Scholz, D. 2009. Development of endothermic chemical foaming/nucleation agents and its processes (Chapter 2). In *Polymeric Foams: Technology and Developments in Regulation, Process, and Products*, Eds. S.T. Lee. Boca Raton, FL: CRC Press.

Shinohara, M., Tokiwa, T., and Sasaki, H. 2002. Expanded polylactic acid resin beads and foamed moulding obtained therefrom, in European patent office, JSP Corporation, Japan.

Shutov, F. A., Henrici-Olivé, G., and Olivé, S. 1986. *Integral/Structural Polymer Foams: Technology, Properties, and Applications*, pp. 1–6. Berlin, Germany: Springer-Verlag.

Speranza, V., De Meo, A., and Pantani, R. 2014. *Polym Degrad Stabil* 100(1), 37–41.

Stumpf, M., Spoerrer, A., Schmidt, H.-W., and Altstaedt, V. 2011. *J Cell Plast* 47, 519–534.

Sun, X., and Turng, L.-S. 2014. *Polym Eng Sci* 54, 899–913.

Thompson, M. R., Qin, X., Zhang, G., and Hrymak, A. N. J. 2006. *Appl Polym Sci* 102: 4696–4706.

Turng, L. S., and Kharbas, H. 2004. *Int Polym Proc* 19: 77–86.

Villalobos, M., Awojulu, A., Greeley, T., and Turco, G. 2006. Oligomeric chain extenders for economic reprocessing and recycling of condensation plastics. *Energy* 31(15), 3227–3234.

Volpe, V. 2013. *Foam Injection Moulding with Physical Blowing Agents*, PhD thesis, University of Salerno, Italy.

Wang, Y., and Hu, G. 2011. *Appl Mech Mater* 66, 2010–2016.

Witt, M.R.J., and Shah, S. 2007. Methods of manufacture of polylactic acid foams, WO 2008/093284 A 1.

Xu, X., Park, C. B., Lee, J. W. S., and Zhu, X. 2008. *J Appl Polym Sci* 109: 2855–2861.

Yang, Z., Gravier, D., and Narayan, R. 2013. Extrusion of humidity-resistant starch foam sheet. *Polym Eng Sci* 53(4): 857–867.

Zhang, S. D. 2007. *The Preparation of Oxidized Starch as Biodegradable Materials and the Relationship between Its Structure and Properties*. PhD thesis, Sichuan University, China.

Zhang, S. D., Wang, X. L., Zhang, Y. R., Yang, K. K., and Wang, Y. Z. 2010. Preparation of a new dialdehyde starch derivative and investigation of its thermoplastic properties. *J Polym Res* 17(3), 439–446.

12 Fabrication of Bio-Based Cellular and Porous Materials for Tissue Engineering Scaffolds

Hao-Yang Mi, Xin Jing, and Lih-Sheng Turng

CONTENTS

12.1 TISSUE ENGINEERING AND TISSUE ENGINEERING SCAFFOLDS

12.1.1 DEFINITION OF TISSUE ENGINEERING

The loss or failure of an organ or tissue is one of the most common, devastating, and costly problems in human health care. The increasing number of transplantation surgeries has led to high donor organ demand, which far exceeds supply. Each year, about 15% of potential candidates for liver or heart transplantation in the United States die while on the waiting list. Although they have proven successful with regard to patient recover, autografts (autotransplantations) and allografts (allotransplantations) have obvious shortcomings in that the organ from another part of the patient (autograft) or from the donor (allograft) must be sacrificed or damaged. Xenograft (xenotransplantation), which uses organ sources from animals (e.g., pigs), also has drawbacks due to issues surrounding the immunological barrier, potential microbiological hazards, and long-term incompatibility (Bach 1998). Alternatively, instead of transplanting living organs, other substitutes—such as biocompatible metals and polymers—can be used for transplantation. This is referred to as *tissue engineering* (also sometimes called *regenerative therapeutics*).

The term *tissue engineering* was coined at a U.S. National Science Foundation workshop in 1988 to describe the applications of the principles and methods of engineering and life sciences toward the fundamental understanding of structure–function relationships in normal and pathological mammalian tissues, and the development of biological substitutes that restore, maintain, or improve tissue functions. One of the most well-known definitions of tissue engineering was provided by Robert Langer in 1993. He defined tissue engineering as "an interdisciplinary field that applies the principles of biology and engineering to the development of functional substitutes for damaged tissue" (Langer and Vacanti 1993, p. 920).

In tissue engineering, knowledge of various subjects is needed to produce artificial scaffolds that fulfill the multiple requirements of living tissues. To date, biological progress has been made in the areas of cell and stem cell biology, biochemistry, and molecular biology, and their application to the research of cell conduct and behavior, as well as to stimulate or guide different types of cells to proliferate, differentiate, and migrate. Likewise, advances in materials science, chemical engineering, mechanical engineering, and computer-assisted engineering have been applied to artificial tissue modeling and fabrication with regard to the physical, chemical, and geometric demands of particular tissue types. In addition, the clinical applicability of artificial tissue substitutes must be verified with help from the fields of medical science and human therapy as applied by surgeons and physicians. Therefore, the rise of tissue engineering promotes the development of those subjects, as well as stimulating collaboration among different disciplines.

12.1.2 Fundamentals of Tissue Engineering

Due to the high cost of failed tissue and organ transplants, as well as the limited number of compatible donors, tissue engineering aims to restore defective or damaged tissue and organ function by employing biological and engineering strategies to these clinical issues. Tissue engineering uses a set of tools from the biomedical and engineering sciences to produce artificial scaffolds that can mimic the human extracellular cell matrix and support the growth of living cells, or stimulate endogenous cells to aid in tissue formation and regeneration, with the help of external biological factors.

The general paradigm of tissue engineering is illustrated in Figure 12.1. A typical tissue engineering process contains five steps: (1) obtain original cells, (2) reproduce cells, (3) seed cells on a proper scaffold, (4) culture cells on the scaffold, and (5) transplant the scaffold (Mi et al. 2014d). As shown in the figure, the fundamental components of tissue engineering include a bioresorbable scaffold, cells, and biological factors. All of these components are related to various areas that could affect the final engineered tissue or organ performance. For example, the fabrication of scaffolds may involve the synthesis of materials and the production of scaffolds with specific geometries and suitable properties. Furthermore, the materials selected and the properties the scaffolds possess are crucial for the applicability of the engineered tissue.

In tissue engineering, cells are combined with scaffolds to construct the bulk artificial tissue that will be transplanted into the human body. To facilitate tissue growth, scaffolds can be transplanted with or without cells. Generally, tissue engineering strategies can be classified into two categories regarding the cell source: (1) *in vitro* tissue engineering followed by transplantation, as shown in Figure 12.2a and (2) *in vivo* tissue engineering by transplantation of scaffold and recruitment and reorganization of host cells, as shown in Figure 12.2b. In the first strategy, cells are obtained from an external source or isolated from the patient. Cells are cultured and reproduced and then seeded on a suitable scaffold. Then the whole scaffold is further cultured to allow the cells to fully attach, followed by transplantation into the patient. This method is sometimes called *cells with scaffold*

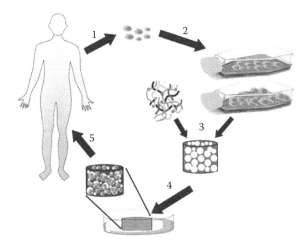

FIGURE 12.1   A typical tissue engineering cycle. (1) A small amount of cells are isolated from the human body. (2) The cells are cultured to produce more cells. (3) These cells are seeded onto porous scaffolds together with growth factors. (4) The seeded scaffolds are further cultured to increase the number of cells. (5) Finally, the regenerated tissue is implanted into the site of the defect to integrate with the natural tissue. (Adapted from Mi, H. Y. et al., *J. Cell. Plast.*, 2014d, with permission from SAGE.)

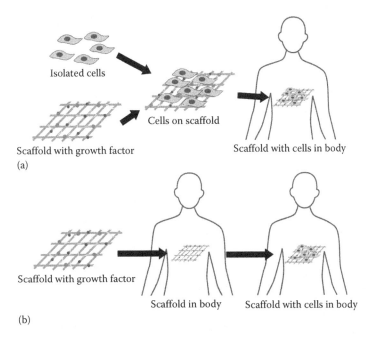

FIGURE 12.2   Illustration of two tissue engineering approaches. (a) Cells are seeded on the scaffold containing growth factors and transplanted into the patient after cell reproduction. (b) A scaffold containing growth factors is directly transplanted into the patient's body and the cells reproduce *in vivo*.

transplantation. The disadvantage of the first strategy is obvious, as cell isolation, reproduction, and adhesion take a long time. Most patients do not have time to wait for the artificial tissue to be ready for use. Hence, the second strategy, where the scaffold can be transplanted with growth factors but without cell seeding, offers another useful option in some clinical setting. This strategy requires *in situ* tissue regeneration by providing an instructive scaffold that can guide and control the regeneration process *in vivo* (Zhao and Karp 2009).

The three main elements of tissue engineering include cells, signaling (biological factors), and scaffolds, as described in details below.

12.1.2.1 Cells

Cells are the building blocks of tissue and play a critical role in tissue healing and regeneration. Cells used in tissue engineering can be obtained from various sources, including autologous cells from the patient, allogeneic cells from other human beings, and xenogeneic cells from different species. However, allogeneic and xenogeneic cells always suffer from immune rejection. The common cell types that have been used in tissue engineering *in vivo* and *in vitro* consist of stem cells, differentiated mature cells, or a mixture of differentiated cells. Among them, stem cells are the most often investigated cell type due to their ability to be stimulated and differentiated into different cell types and their relatively high proliferation ability. In tissue engineering, cells seeded on scaffolds further reproduce to generate more cells on the scaffold.

Ultimately, the whole scaffold degrades and is replaced by cells to form a new functional tissue.

12.1.2.2 Biological Factors

Biological factors, including hormones, cytokines, growth factors, extracellular matrix molecules, cell surface molecules, and nucleic acids, can influence the function and behavior of cells in the scaffold significantly. The success of scaffold transplantation largely depends on the appropriate presentation of signals that mediate host cell mobilization and coordination of cell behavior, such as cell adhesion, migration, and proliferation. The employment of different growth factors can guide cell behavior and stimulate cell conduct, thus leading to cells that perform their normal functions during the tissue engineering process.

12.1.2.3 Scaffold

The scaffold acts as a synthetic analog of the natural extracellular cell matrix. The role of the scaffold is to recapitulate the normal tissue development process by allowing cells to formulate the desired microenvironment to fulfill the requirements of the defective tissue (Lee et al. 2008). As illustrated in Figure 12.2, the scaffold plays a pivotal role in the tissue engineering process in both of these areas. It provides necessary support for cells to attach, migrate, proliferate, and maintain their differentiated function for the subsequent regeneration of new tissue. Eventually, the scaffold will be degraded by the cells in the media or in the *in vivo* environment when the new tissue is completely formed by the cells. Therefore, producing feasible scaffolds is a crucial step in tissue engineering. This chapter will mainly focus on introducing different methods to fabricate scaffolds for tissue engineering applications. Both traditional and recently developed methods will be covered in the following sections.

12.1.3 Tissue Engineering Scaffold Requirements

Biomaterials-based scaffolds play a pivotal role in tissue engineering as a three-dimensional template that provides mechanical stability, delivers therapeutic agents, and facilitates cell growth in tissue repair. In order to fulfill the requirements of the tissue engineering application, the ideal scaffold should have the following characteristics: (1) a three-dimensional, highly porous structure with an interconnected network to facilitate cell migration; reproduction; and diffusion of nutrients, metabolic waste, and paracrine factors (Hutmacher 2001); (2) be biocompatible and biodegradable with controllable degradation rates to match cell growth *in vitro* and *in vivo* (Cima et al. 1991; Kweon et al. 2003); (3) a suitable surface chemistry for cell attachment, migration, proliferation, and differentiation (Mandal and Kundu 2009a, b); (4) ideal mechanical properties that match the surrounding tissue of the implantation site and provide cues to control the structure and function of the newly formed tissue (Hutmacher 2000, 2001); and (5) be able to be easily processed into a variety of shapes and sizes that fit the implantation site and are able to maintain that shape for a specific period of time (Hutmacher 2001). These scaffold characteristics vary according to different tissue application needs.

12.1.3.1 Scaffold Structure

A bulk artificial tissue with a specific geometry will be transplanted during an implantation operation, thus a three-dimensional scaffold is necessary for cells to live and thrive. A challenge in scaffold fabrication is that the scaffold has to be a porous structure with interconnected microchannels (also referred to as an *open porous structure*). A three-dimensional porous structure that possesses a high surface-to-volume ratio is favorable for a large number of cells to attach and live. This structure will facilitate cell migration and reproduction into the inner part of the scaffold and support the free transport of nutrients and waste. However, fabricating a uniform, open porous structure scaffold with a desirable pore size, pore density, porosity, and interconnectivity is still a challenge. A number of methods (e.g., solvent casing and particulate leaching, thermally induced phase separation [TIPS], electrospinning, rapid prototyping, and gas foaming) have been developed to produce the ideal scaffold structure while improving upon practical applications and promoting commercialization. These methods will be introduced in detail in the subsequent sections.

It is important to note that the structure of the scaffold has a significant impact on cell behavior. Different cell types also react differently to different scaffold structures. For example, a 75% void fraction is not high enough for most cells. However, it has been reported that void fractions higher than 90% cause micro-vascular epithelial cells to form disconnected webs (Zeltinger et al. 2001). The optimal pore size also varies for different cell types, with pore sizes ranging from 50 to 70 μm suitable for cardiac cells, while human osteoblast cells typically require pore sizes of around 200 μm to achieve good performance (Liu and Ma 2004; Odedra et al. 2011). It has also been suggested that cells cannot migrate through pores smaller than 10 μm (Verreck et al. 2003). Thus, it is necessary to understand the microenvironment that is best suited for the target cell type before producing the scaffold. Generally, the pore size and porosity can be controlled by adjusting multiple processing parameters during the scaffold fabrication process. Nevertheless, good interconnectivity is necessary for a favorable scaffold. Interconnection channels allow cells to migrate into the inner part of the scaffold and ensure the transport of nutrition factors and cellular waste. Moreover, the three-dimensional structure of the scaffold leads to potential cell seeding challenges.

In the simplest tissue engineering process, as mentioned above, the cells are seeded and cultured on the surface of the scaffold. Then the whole scaffold, which contains the cells, is further cultured for longer periods of time to form a functional tissue, with the expectation that the cells will grow into the inner part of the scaffold. However, in reality, cellular growth is usually slow and cell penetration is limited. Sometimes, cells prefer to grow on top of each other and form clusters instead of migrating deep into the scaffold (Vunjak-Novakovic et al. 1998). Multiple methods have been applied to solve these issues and achieved promising results, including directly seeding cells into the inner area of the scaffold after sufficiently soaking the scaffold in culture media, applying mild centrifugation to force the cells into deeper pores, suffusing culture media from top to bottom or using oscillation perfusion to guide cells to migrate in the direction of the media flow, and controlling the culture media flow by using a functionalized bioreactor to form a dynamic culture environment (i.e., spinner flask bioreactors, rotating-wall vessels, and hollow-fiber bioreactors) (Jasmund and Bader 2002; Li et al. 2001; Vunjak-Novakovic et al. 1996; Wendt et al. 2003).

12.1.3.2 Scaffold Surface Properties

Scaffold surface properties, including both chemical and topographical characteristics, are important for cellular adhesion, migration, and proliferation since the scaffold surface is the initial and primary site of interaction with surrounding cells and tissue (Boyan et al. 1996). In *in vivo* native tissue environments, cells attached to the extracellular cell matrix are mediated by binding between integrins (receptors on the cell surface) and extracellular cell matrix adhesion proteins such as fibronectin, vitronectin, laminin, and collagen. The same principal applies to biomaterial surfaces

TABLE 12.1

General Methods for Polymer Surface Modification

Methods	Examples
Roughening	Sand blasting and etching
Oxidation	Alkaline treatment, chromium treatment, and plasma treatment
Coating	Casting, lamination, and plasma polymerization
Blending	Surfactant addition and block and graft co-polymer addition
Ion implantation	High-energy argon and nitrogen injection
Graft polymerization	Ultraviolet, ionizing radiation, and low-temperature plasma

Source: *Biomaterials*, 15, Ikada, Y., Surface modification of polymers for medical applications, 725–736, Copyright 1994, with permission from Elsevier.

and cell interactions. Several methods have been used for polymer surface modification as listed in Table 12.1 (Ikada 1994).

Among these, grafting is a widely used method for surface modification by coupling reaction of existing polymer chains or graft polymerization of monomers, which can introduce multiple functional groups on the material surface and consequently affect surface properties. Researchers have focused on several goals of surface modification, such as improvement of hydrophilicity, protein immobilization, and surface micro- and nano-patterning (Ma et al. 2007). While hydrophilicity will affect cell attachment significantly, neither a highly hydrophilic nor a highly hydrophobic surface is favorable for cells. The optimal hydrophilicity depends on the different cell types and surface modification methods. For example, it was reported that the best water contact angle for endothelial cell attachment and proliferation was 70°, while for chondrocytes, it was 76° (Ma 2003; Zhu et al. 2002a).

To covalently immobilize protein molecules on chemically inert polymeric biomaterials, reactive groups such as hydroxyl, carboxyl, and amino groups must first be introduced as coupling sites and then be reacted with other coupling agents and proteins. Typically, the scaffold and cell interactions improve significantly after introducing proteins onto the scaffold surface. Figure 12.3 gives an example of the schematic of the protein immobilization procedure, in which Y.B. Zhu et al. reported several approaches to grafting collagen onto a polyester surface (Zhu et al. 2002b, 2004a, b, c).

Surface micro- and nano-patterning method could also effectively manipulate cell behavior in terms of cell shape and migration, protein synthesis, and gene expression on biomaterial surfaces by patterning different materials into various shapes that are favorable or unfavorable for cells (Ito et al. 1998; Stevens and George 2005). Since this method is mainly based on lithographic technology, however, it has limited application on three-dimensional scaffolds so far.

In addition to surface chemistry, scaffold surface roughness and surface charge also have influences on cell behavior. It has been reported that high poly(methyl methacrylate) (PMMA) surface roughness improved the hydrophobicity, cell adhesion, and cell migration area, thus indicating that rougher scaffold surfaces are better for cell adhesion (Lampin et al. 1997). For titanium scaffolds, an average surface roughness of 3 μm (macro-sandblasted titanium) is more suitable than an average surface roughness of 0.5 μm (micro-sandblasted titanium) in favoring osteoblast differentiation *in vitro* (Marinucci et al. 2006). Cell behavior can also be modulated by surface charging the scaffold. It has been reported that a positively charged poly(ε-caprolactone) (PCL) electrospun fiber can promote cell adhesion, and that fibroblasts are more likely to form spherical shapes in the horizontal direction and flat shapes in the cross-section direction (Kim et al. 2012). In addition, positively charged polyethylene (PE) sheets also showed better platelet adhesion than negatively charged sheets (Lee et al. 1998).

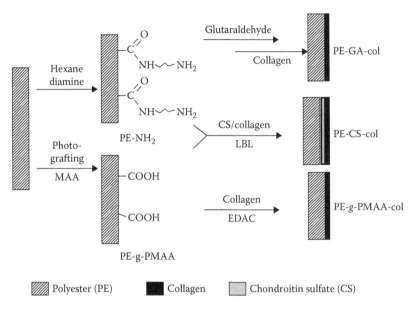

FIGURE 12.3 Schematic representation of methods used to immobilize biomacromolecules on a polyester surface. (Reprinted from *Colloid. Surf. B*, 60, Ma, Z.W. et al., Surface modification and property analysis of biomedical polymers used for tissue engineering, 137–157, Copyright 2007, with permission from Elsevier.)

12.1.3.3 Scaffold Physical Properties

Using the proper mechanical properties for a scaffold in a tissue engineering application is critical to the success of the implantation. The scaffold should possess mechanical properties close to the targeted surrounding tissue in order to fulfill the supporting function of the normal tissue. It is even more essential to retain the mechanical strength of the scaffold after implantation for the reconstruction of hard, load-bearing tissues such as bones and cartilage. Furthermore, it is important to evaluate one or more of the mechanical properties to ensure the suitability of the scaffold. Several important parameters are usually measured to characterize tissue engineering scaffolds including the following: (1) elastic modulus—defined as the slope of the stress–strain curve in the elastic deformation region in response to a given tensile or compressive test; (2) flexural modulus—measured relationship between a bending stress and the resulting strain in response to a load perpendicular to the specimen; (3) tensile strength—maximum stress that the material can withstand before it breaks; and (4) maximum strain—ductility of a material or total strain exhibited prior to fracture (Dhandayuthapani et al. 2011).

The scaffold's physical properties depend not only on the scaffold material used but also can be attributed to scaffold structure and morphology. Generally, higher porosities, larger pore sizes, and greater pore densities lead to lower mechanical properties, especially compressive properties (Borden et al. 2003; Lin et al. 2003). Therefore, researchers sometimes have to make a tradeoff between high porosity and better mechanical performance for a given material. Moreover, tailoring scaffolds' physical properties is also required in order to fulfill various tissue applications. For example, the Young's modulus for cancellous bone is about 50–100 MPa, it is 65–541 MPa for ligament, 143–2,310 MPa for tendon and 0.7–15.3 MPa for cartilage (Baji et al. 2006; Kramschuster and Turng 2010; Yang et al. 2001). Typically, tailoring scaffolds' physical properties can be achieved by adjusting and optimizing scaffold processing parameters for different fabrication methods or by combing with various materials (e.g., natural biomaterials, other biocompatible synthetic polymers, or biocompatible additives) into scaffolds to produce scaffolds with desirable properties. For example, Li et al. (2013) have used poly (β-hydroxybutyrateco-β-hydroxyvalerate) (PHBV) microspheres to enhance poly (L-lactide-*co*-caprolactone) (PLCL) scaffolds to promote their application in cartilage regeneration. Zhao et al. (2011a) found that the composite scaffold showed improved mechanical properties and cell compatibility after combining chitosan with PLA. Likewise, Wang et al. (2010) reported that addition of bioceramic hydroxyapatite (HA) improved the mechanical properties of PCL scaffolds significantly.

Besides providing the necessary physical support, the physical properties of scaffolds influence cell behavior as well. For example, polyacrylamide (PA) collagen-coated gels with stiffness similar to that of muscle tissue can enhance myotubular actin/myosin striations generated by fused skeletal muscle cells (Engler et al. 2004); gels that have stiffness similar to that of muscle, brain, or bone tissues lead to cell differentiation to myoblasts, neurons, or osteoblasts, respectively (Engler et al. 2006). Furthermore, the mechanical properties of poly(L-lactide acid) (PLLA)/poly(lactic-*co*-glycolic acid) (PLGA) scaffolds have been shown to influence the organization, differentiation, and viability of skeletal muscle cells, namely—rigid scaffolds supported cell differentiation toward the formation of aligned myotubes and improved cell viability during the culture period (Levy-Mishali et al. 2009). It was also reported that substrate compliance can influence communication between neighboring endothelial cells as well (Dado and Levenberg 2009). In a word briefly, the closer the scaffold mechanical properties to the target tissue, the better the cells perform, assuming that other conditions are held constant.

12.1.3.4 Scaffold Materials

The ultimate goal of tissue engineering is to transplant the artificial tissue (on scaffold) into the human body to cure damage to the native tissue. The transplanted scaffold, which is viewed as alien by the human body, should be compatible with the native tissue and should be replaced by newly generated tissue in the long term. Therefore, the selected polymeric materials for tissue engineering have to be biocompatible and biodegradable. Biocompatibility refers to the ability of a biomaterial to perform its desired function with respect to a medical therapy without eliciting any undesirable local or systemic effects in the recipient or beneficiary of that therapy while generating the most appropriate beneficial cellular or tissue response in that specific situation and optimizing the clinically relevant performance of that therapy (Williams 2003, 2008).

Biodegradation, on the other hand, is the chemical dissolution of materials by bacteria or other biological means. Biodegradable materials have been promoted all over the world in recent decades because of their renewability and environmental-friendly advantages. In tissue engineering, *biodegradability* typically refers to the biodegradation of materials in the biological environment, a mild, aqueous salt solution at 37°C, which can lead to rapid or gradual breakdown of many materials. Biodegradation generally happens in two modes, including surface erosion and bulk erosion, depending on the nature of the material. Surface erosion mainly occurs in hydrophobic polymers, in which the bond cleavage rate is faster than the water diffusion speed. Hence, the bulk polymer molecular structure would not change significantly during degradation, thus guaranteeing the physical properties during degradation, although the volume of the polymer would be gradually reduced. In contrast, bulk degradation occurs gradually with the molecular bonds being cleaved slowly throughout the whole polymeric sample. This mode applies to hydrophilic materials that absorb water relatively faster than their bonds cleave. The direct consequence of bulk degradation is a change in the density of the sample and the molecular weight of the polymer; moreover, the mechanical properties decrease as the molecular weight decreases. (Gopferich 1996). Thus the degradation and restoration rates of the scaffolds should match the rate of tissue growth *in vitro* and *in vivo*, so that the scaffolds can continue providing the necessary physical support during new tissue formation and guide the new tissue to form the same shape as the implanted scaffold. Sometimes nonbiodegradable polymeric scaffolds are also used in tissue engineering because they are biologically stable and can provide permanent support, ideally for the life of the patient. For example, PMMA has been used as bone cement for hip and knee replacements, and high density PE can be used to form the articulating surfaces of hip and knee joints (Ramakrishna et al. 2001).

To date, many types of biocompatible and biodegradable polymers have been used to produce tissue engineering scaffolds. In general, these polymers can be classified as natural and synthetic polymers. Natural polymers can be considered to be the first biodegradable biomaterials used clinically (Nair and Laurencin 2007). Owing to their excellent bioactive properties, natural materials have good compatibility and interactions with cells, which allow them to enhance cell performance in biological systems. Natural polymers can be classified as proteins (silk, collagen, gelatin, fibrinogen, elastin, keratin, actin, and myosin), polysaccharides (cellulose, amylose, dextran, chitin, and glycosaminoglycans), or polynucleotides (DNA and RNA) (Presteich and Atzet 2013). Synthetic polymers are another kind of widely used and researched biomaterial that could potentially be used for tissue engineering scaffolds. In the design of biodegradable biomaterials, many important factors must be considered. These materials must (1) not evoke a sustained inflammatory response, (2) possess a degradation time coinciding with their function, (3) have appropriate mechanical properties for their intended use, (4) produce nontoxic degradation products that can be readily resorbed or excreted, and (5) include appropriate permeability and processability for the desired application (Lloyd 2002; Ulery et al. 2011). Many kinds of polymers have been investigated for their ability to be used in tissue engineering scaffold applications. Table 12.2 lists several kinds of commonly used polymers in tissue scaffold applications.

12.2 TISSUE ENGINEERING SCAFFOLD FABRICATION

As mentioned before, the fabrication of scaffolds is a critical step in tissue engineering. It is worth putting some effort into understanding the most commonly used and recently developed scaffold fabrication methods. Several polymeric scaffold fabrication methods have been developed in recent decades to fulfill the requirements of tissue scaffolds. Based on whether solvent is needed in the scaffold preparation procedure, these methods can be roughly classified into wet methods and solvent-free methods. Wet methods typically include solvent casting and particle leaching, TIPS, and electrospinning. Solvent-free methods include blending and particle leaching, gas foaming, and solid freeform fabrication (SFF). Recently, researchers have been trying to take advantage of different scaffold fabrication methods by combining these methods as well. Various basic scaffold fabrication methods will be introduced in the sections that follow.

12.2.1 WET METHODS

In this category, polymers need to be dissolved in some type of solvent to form a solution. Then the polymer solution is further processed and the solvent is subsequently removed to form a porous polymeric scaffold. Solvent casting and particle leaching, TIPS, and electrospinning methods fit into this category. Since most organic solvents are detrimental to cells, residual solvent in the scaffolds is the main concern of these methods.

12.2.1.1 Solvent Casting and Particle Leaching

Solvent casting/particle leaching (SC/PL) is one of the most popular and traditional techniques used to fabricate scaffolds for tissue engineering applications. In the typical SC/PL process, the homogeneous polymer solution is cast into the porogen-filled mold. When the solvent evaporates, the composite material created contains the polymer and the porogens. Then the composite is put into a bath to dissolve the porogens and the porous structure is obtained. In this method, salt, sugar, and paraffin are mainly used as porogens to create pores in the polymer matrix (Ma and Choi 2001; Vaquette et al. 2008; Wan et al. 2008). In addition, polymer particles such as PMMA also serve as porogens to create a porous structure. Dinu et al. (2013) successfully prepared novel chitosan hydrogels with polyhedral porous structures tailored by an ice-templating and PMMA-porogen leaching method.

The SC/PL technique is very simple to carry out and easy to manipulate. The macro geometry of the scaffold is determined by the casting mold cavity. The porous structure of the scaffold can be

TABLE 12.2
Properties of Polymers Used as Scaffolds

Polymers	Melting Temperature (°C)	Glass Transition Temperature (°C)	Degradation Time (Months)	Applications	Polymer Repeat Unit Structure Example (Can Vary from Different Synthesis Formulas)	References
Polylactic acid (PLA)	173–178	60–65	12–18	Fracture fixation, interference screws, meniscus repair, and bone grafts		Koegler and Griffith (2004), Shin et al. (2003)
Polyglycolic acid (PGA)	225–230	35–40	Three to four	Suture anchors, meniscus repair, and bone grafts		Koegler and Griffith (2004), Shin et al. (2003), Wen and Tresco (2006)
Polylactic-co-glycolic acid (PLGA)	Amorphous	50–55	Three to six	Sutures and artificial skin		Bendix (1998), Shin et al. (2003), Wen and Tresco (2006)
Poly(ε-caprolactone) (PCL)	58–63	–60	>24	Suture coating, orthopedic and dental implants, and bone grafts		Lepoittevin et al. (2002), Li et al. (2002)
Poly(propylene fumarate) (PPF)	30–50	–60	>24, depends on formula	Orthopedic and dental implants, and bone grafts		Shi et al. (2005), Sitharaman et al. (2007)
Poly(glycolide-co-caprolactone) (PGCL)	Amorphous	19	50% mass loss within two months	Smooth muscle tissue applications		Lee et al. (2003)

(Continued)

TABLE 12.2 (Continued)
Properties of Polymers Used as Scaffolds

Polymers	Melting Temperature (°C)	Glass Transition Temperature (°C)	Degradation Time (Months)	Applications	Polymer Repeat Unit Structure Example (Can Vary from Different Synthesis Formulas)	References
Polyurethane (PU)	180–210	−60 to 30	>24, depends on formula	Soft tissue engineering and vascular grafts		Guan et al. (2005)
Poly(hydroxybutyrate-cohydroxyvalerate) (PHBV)	168–171	−15 to 0	10–22	Bone pins and plates and vascular grafts		Adamus et al. (2012), Arcos-Hernandez et al. (2012)
Poly(propylene carbonate) (PPC)	25–45	15–200	Two to three months, depends on test environment	Nerve tissues and vascular grafts		Luinstra and Borchardt (2012), Nagiah et al. (2012), Wang et al. (2011)

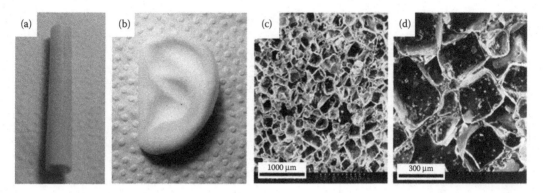

FIGURE 12.4 Photographs of (a) tubular and (b) ear-shaped PLGA porous scaffolds. Scanning electron microscopy (SEM) images of the ear-shaped PLGA scaffold cross-sectional morphologies in (c) low magnification, and (d) high magnification. (Reprinted from *Biomaterials*, 27, Wu, L.B. et al., A "room-temperature" injection molding/particulate leaching approach for fabrication of biodegradable three-dimensional porous scaffolds, 185–191, Copyright 2006, with permission from Elsevier.)

controlled by the content of the porogens added and the size and shape of the porogens (Plikk et al. 2009; Zhang et al. 2005). Highly porous scaffolds with porosities of up to 93% have been prepared using this technique (Mikos et al. 1994). Researchers have made efforts to further improve the interconnectivity of the scaffolds. For example, centrifuging the polymer solution containing the porogen helps porogen particles connect to each other and further improves the porosity and interconnectivity; pre-fusing the porogen particles slightly improves the connected areas between the particles, which leads to larger interconnected channels in the scaffold; adding another kind of water soluble polymer besides salt into the polymer solution further enhances scaffold porosity (Murphy et al. 2002; Sin et al. 2010).

In addition, it has been reported that scaffold mechanical properties can be improved by incorporating other additives in the polymer solution, such as β-tricalcium phosphate (TCP), magnesium phosphate, zein, and wollastonite, and that these composite scaffolds have the potential to be used in bone tissue engineering applications (Wei et al. 2009; Wu et al. 2012a, b).

Different scaffold geometries can be achieved using different mold cavities. For instance, Wu et al. (2006) used a combined *room temperature* injection molding approach using the salt particle leaching method to fabricate three-dimensional PLGA tubular scaffolds and ear-shaped porous scaffolds. The porosities of the scaffolds reached 94%. Photographs and scanning electron microscopy morphologies of these scaffolds are shown in Figure 12.4 (Wu et al. 2006). Although the SC/PL technique is very convenient, the solvent involved in the process greatly limits its application in tissue engineering due to the toxicity of the residual solvent on the cells (Yang et al. 2001).

12.2.1.2 Thermally Induced Phase Separation

Phase separation is a versatile technique, which includes solvent-induced phase separation (SIPS) and TIPS. The difficulty of finding a suitable solvent to dissolve the crystallized polymers in limits the application of SIPS. Therefore, TIPS has been used more often in recent years. TIPS is often combined with the freeze-drying technique to fabricate scaffolds. The most attractive characteristic of TIPS over other approaches is that it can form an intrinsically interconnected polymer network in one simple step just by manipulating processing parameters and system compositions.

The principle of TIPS is that a homogeneous solution of polymer–solvent or polymer–solvent–nonsolvent at elevated temperatures is quenched to a lower temperature by cooling to induce the uniform solution into polymer-rich and polymer-lean phases. After demixing is induced, the solvent is removed by extraction, evaporation, or freeze-drying, and pores form in the scaffold due to the removal of the solvent (Lo et al. 1995). There are two mechanisms of phase separation.

(1) Solid–liquid phase separation is where the solvent forms crystals at the lower temperature before phase separation occurs between the solvent and the polymer. Then sublimation of the solvent gives rise to a ladder-like porous structure in the scaffold. (2) Liquid–liquid phase separation, on the other hand, generates polymer-rich and -lean liquid phases. The further development of polymer-lean phases in the spinodal phase separation creates a highly interconnected porous structure in the scaffolds by the sublimation of the solution (Hua et al. 2003). The typical schematic representation of the temperature-concentration phase diagram for a binary polymer–solvent system with an upper critical solution temperature for TIPS is shown in Figure 12.5 (Mi et al. 2014d). The solid binodal curve represents the thermodynamic equilibrium of liquid–liquid (L–L) demixing, which is often used to approximate the cloud point at which the clear solution becomes turbid. In the L–L demixing region, polymer-rich and -lean phases coexist (Vandeweerdt et al. 1991; Williams and Moore 1987). Meanwhile, on the dashed spinodal curve, the second derivative of the Gibbs free energy of mixing is equal to zero. The critical point (maximum point) is where the binodal and spinodal curves merge. The area between the binodal and spinodal curves is the metastable region where the polymer concentration is lower or higher than the concentration at the critical point, while the area below the spinodal curve is the unstable region. L–L demixing located in the metastable region leads to bead-like or poorly connected closed pores for polymer concentrations lower or higher than the critical point, respectively (Matsuyama et al. 2000; Tsai and Torkelson 1990). When L–L demixing is in the unstable region, an open porous and well-interconnected structure can be obtained (Schugens et al. 1996b). If the temperature of the solution is decreased to or beyond the freezing point of the solvent prior L–L phase separation, solid–liquid phase separation induced by the solidification of the solvent occurs and the anisotropic ladder-like structure with relatively small pores forms (Schugens et al. 1996a).

Recently, the TIPS method has been used for fabricating biodegradable scaffolds. The first prepared scaffold with a ladder-like structure was reported in 1996. In this study, Schugens and coworkers found that solid–liquid phase separation was the main reason for the special structure without L–L demixing (Schugens et al. 1996a). Some researchers have also employed the principle of solid–liquid phase separation to fabricate a highly oriented porous structure (Goh and Ooi 2008; Hu et al. 2008; Yang et al. 2006). To obtain a more porous and interconnected structure, many researchers have introduced nonsolvents, such as water and ethanol, to induce the occurrence of liquid–liquid phase separation using a polymer–solvent–nonsolvent ternary system. It was found

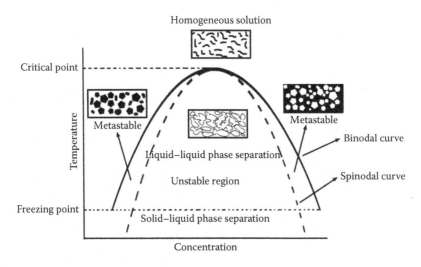

FIGURE 12.5 Schematic diagram of a binary polymer–solvent system for TIPS. (Adapted from Mi, H.Y. et al., *J. Cell. Plast.*, 2014d, with the permission from SAGE.)

that the introduction of a nonsolvent can increase the L–L demixing temperature, which is helpful for creating a porous structure with interconnected pores (He et al. 2009; Hua et al. 2002, 2003; Molladavoodi et al. 2013).

The fabrication of macro porous scaffolds by L–L TIPS is a dynamic process. The morphology of the scaffolds is controlled by parameters such as quenching temperature, quenching rate, quenching period, coarsening time, polymer molecular weight, polymer concentration, and solvent-to-nonsolvent ratio. Chen et al. (2010) investigated the effects of fabrication parameters and the composition of the ternary system on the macro- and microscopic morphologies of PLA scaffolds with a PLA–dioxane–water ternary system. It was found that the cloud-point temperature increased with an increase of water in the system. In addition, a fast cooling rate introduced smaller pores, while a higher polymer concentration increased the cloud point, which resulted in an increased coarsening time and pore size (Chen et al. 2010). In recent years, to fabricate nanofibrous (NF) scaffolds, which are better for cell adhesion and proliferation, via the TIPS process, researchers have made modifications to the process, such as using porogens to improve porosity, or combining polymers with other natural biomaterials (Mao et al. 2012; Wei and Ma 2006; Zhao et al. 2012). Liu and Ma combined TIPS with a porogen leaching method to fabricate NF gelatin scaffolds. They obtained NF-gelatin scaffolds that had high surface areas, high porosities, well-connected macro pores, and NF pore wall structures, as shown in Figure 12.6 (Liu and Ma 2009).

To better mimic the mineral components and microstructures of natural bone, as well as to improve cell adhesion on the scaffolds, HA has been commonly used in TIPS to fabricate composite scaffolds. For example, PLLA/HA, PLGA/HA, PHBV/HA, and PA/HA composite scaffolds have been produced using the TIPS method (Huang et al. 2008; Jack et al. 2009; Wang et al. 2007; Wei and Ma 2004). In one of the authors' studies, thermoplastic polyurethane (TPU)/HA composite scaffolds were fabricated. The effects of solvent and HA size were investigated. It was found that a ladder-like structure was obtained when dioxane was used as the solvent, while an interconnected porous structure was achieved after adding a small amount of water as a nonsolvent into the dioxane. It was also found that nano-HA could efficiently stimulate mineralization behavior in the scaffolds in simulated body fluid (Mi et al. 2014a). Some of the results are shown in Figure 12.7. Other additives, such as carbon nanotubes, bioactive glass, and calcium phosphate, have also been used to produce composite scaffolds using the TIPS technique to improve scaffold properties and performance. It was found that chitosan-multiwall carbon nanotubes composites promoted osteoblast cell proliferation, and calcium phosphate crystals, the mineralization products of osteoblast cells, were found on the surface of scaffolds after just a few days of culture (Olivas-Armendariz et al. 2010). The addition of carbon nanotubes improved the compression properties and enhanced the surface roughness of polyurethane scaffolds fabricated via TIPS (Jell et al. 2008). In addition, PLGA/bioglass and

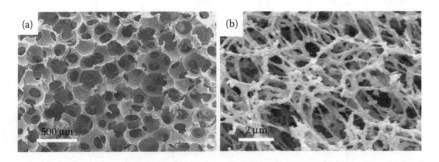

FIGURE 12.6 Scanning electron microscopy (SEM) images of NF-gelatin scaffolds with highly porous structures: (a) a uniform interconnected porous structure and (b) enlarged image of the pore wall structure. (Reprinted from *Biomaterials*, 30, Liu, X.H. and Ma, P.X., Phase separation, pore structure, and properties of nanofibrous gelatin scaffolds, 4094–4103, Copyright 2009, with permission from Elsevier.)

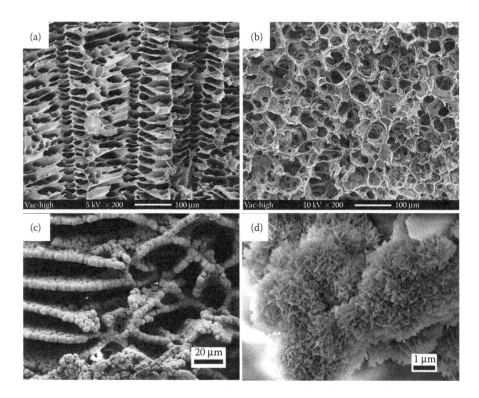

FIGURE 12.7 Scanning electron microscopy (SEM) images of TPU/HA scaffolds (a) using only dioxane as a solvent, (b) using dioxane and water (8.5:1.5) as a solvent, (c) TPU/nanoHA scaffold after mineralization in simulated body fluid for three weeks, and (d) the mineralized HA crystals at the three-week mineralization test. (With kind permission from Springer Science+Business Media: *J. Mater. Sci.*, Morphology, mechanical properties, and mineralization of rigid thermoplastic polyurethane/hydroxyapatite scaffolds for bone tissue applications: Effects of fabrication approaches and hydroxyapatite size, 49, 2014a, 2324–2337, Mi, H.Y. et al.)

PLA/calcium phosphate biodegradable scaffolds have also been reported as suitable substitutes for bone tissue engineering applications (Boccaccini and Maquet 2003; Charles-Harris et al. 2008).

12.2.1.3 Electrospinning

Electrospinning was first observed and recorded by William Gilbert in the late sixteenth century. The most attractive characteristic of electrospinning is that it allows for the production of polymer fibers with diameters in the range of 15 nm to 10 μm or greater (Lannutti et al. 2007). Moreover, the setup for electrospinning is very simple and inexpensive, consisting of a syringe pump, collector, and high voltage supply (shown in Figure 12.8). During the electrospinning process, a polymer droplet is held at the tip of a needle by surface tension. When a sufficiently high voltage is applied to the droplet, it becomes charged and an electrostatic repulsion force is formed. When the electrostatic force caused by the high-voltage source counteracts the surface tension, an electrically charged jet of polymer solution erupts. As the jet moves forward, but before reaching the collector, the solvent evaporates and the jet solidifies into thin fibers (Fong et al. 1999). The electrospinning technique has been widely used in tissue engineering scaffold fabrication so far.

The structures of the obtained fibers can be manipulated by various parameters involved in the deposition process, which can then be divided into three categories: solution parameters, processing parameters, and ambient parameters. Solution parameters include the nature of the solvent used, the polymer molecular weight, the molecular structure, the solution viscosity, the solution concentration, and the surface tension (Ki et al. 2005; Koski et al. 2004; Pant et al. 2011; Sukigara et al. 2003;

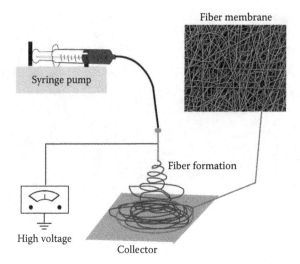

FIGURE 12.8 Schematic illustration of the electrospinning setup. A syringe is filled with a polymer solution. It is then placed on a syringe pump and connected with a blunt needle. An electric field is created between the needle tip and the ground collector by high voltage supply. The solution is then pushed out consistently and stretched into fibers in the electric field. The fibers are obtained from the ground collector.

Zhao et al. 2005). Processing parameters include the applied voltage, the distance between the needle tip and the collector (working distance), and the solution flow rate (Demir et al. 2002; Geng et al. 2005; Wannatong et al. 2004). Ambient parameters include the temperature and humidity (Casper et al. 2004; Mit-uppatham et al. 2004). Typically, high material molecular weight, high solution concentration, high solution flow rate, low applied voltage, small working distance, and applied air flow will lead to a smaller electrospun fiber diameter. Moreover, typical electrospinning fibers are nonwoven and randomly dispersed (shown in Figure 12.8 inset image), but the fiber orientation can be adjusted using different collection devices. As shown in Figure 12.9, aligned fibers can be obtained using a conductive rotation drum or two parallel conductive plates (Figure 12.9a,b). Woven net structure fibers can be produced using four orthogonal conductive plates (Figure 12.9c) (Mi et al. 2014d).

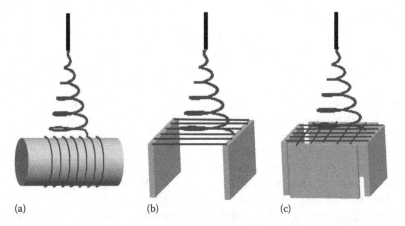

(a)	(b)	(c)

FIGURE 12.9 Schematic illustration of electrospinning process collection devices for the fabrication of tissue scaffolds with directed fiber orientations: (a) using a rotation drum to collect the aligned fibers, (b) using two parallel plates to collect the aligned fibers, and (c) using four orthogonally arranged plates to collect the net structured fibers. (Adapted from Mi, H.Y. et al., *J Cell. Plast.*, 2014d, with the permission from SAGE.)

Materials used in electrospinning can be either natural materials or synthetic polymers. Natural materials are used in electrospinning due to their enhanced biocompatibility and biofunctional motifs (Almany and Seliktar 2005). A number of natural materials have been used to fabricate tissue engineering scaffolds via electrospinning such as gelatin, hyaluronic acid, collagen, silk, fibroin, and chitosan (Lovett et al. 2007, 2008, 2010; Tchemtchoua et al. 2011). There has been a lot of research using synthetic polymers to fabricate fibrous scaffolds for tissue engineering as well. Compared with natural materials, synthetic polymers have a relatively low cost and higher availability.

Polylactide (PLA), PLGA, PCL, and their copolymers are the most extensively used polymers for biodegradable scaffolds. For example, electrospun PLGA mats have been successfully used as wound dressings for skin wounds and fibroblast behavior was evaluated in terms of proliferation, morphology, and gene expression on PLGA electrospun scaffolds with fiber diameters ranging from 150 to 6,000 nm (Kumbar et al. 2008). Li et al. (2005) reported that multilineage differentiation of mesenchymal stem cells was fully supported within the NF scaffolds, indicating that PCL-based NF scaffolds are a promising candidate for tissue engineering scaffolds. PCL fibers in the nanometer and submicron range closely resemble the size scale of extracellular components (Li et al. 2003). A significant disadvantage of these polymers is their brittleness and that they all have relatively high moduli that are typically suitable only for bone tissue applications without any modification.

There are several types of soft polymers that can be fabricated into scaffolds via electrospinning that have the potential to be used in soft tissue applications. Polyurethane, a segmented block copolymer composed of soft and hard segments, has been extensively used in soft tissue engineering applications. Theron et al. (2010) reported using the electrospinning method to fabricate cross-linked TPU vascular grafts. polypropylene carbonate (PPC), a recently reported polymer, was synthesized from propylene oxide and carbon dioxide (CO_2). PPC has high affinity to biological organisms and good biodegradability (Kim et al. 2008). Wang et al. (2011) prepared PPC-electrospun aligned fibers, and they showed significantly enhanced peripheral nerve regeneration *in vitro*.

To further mimic the extracellular cell matrix, many investigators combine the advantages of different polymers by dissolving them into a solvent or co-solvent and electrospinning the blended solution. For example, Yin et al. (2013) fabricated a poly (L-lactic acid-*co*-caprolactone) (P[LLA-CL])/collagen/chitosan scaffold by electrospinning. The endothelial cells showed better proliferation and more adhesion on the blended scaffolds than on the P(LLA-CL) scaffold alone (Yin et al. 2013). The hydrophilicity of PCL electrospun fibers can be improved when blended with gelatin (Gautam et al. 2013).

Instead of dissolving various polymers into the same solvent, different polymer solutions can be spun concurrently to produce mixed fiber mats. This method is known as *co-electrospinning*. Du et al. (2012) fabricated chitosan/PCL NF vessel scaffolds by sequential co-electrospinning. They also used heparin to modify the hybrid fibers to enhance endothelial cell proliferation and anticoagulation (Du et al. 2012). Another study prepared fiber mats using two polymers with different degradation speeds, thus the scaffold porosity increased gradually during tissue in-growth (Kidoaki et al. 2005). Similarly, by co-spinning solutions of PCL and gelatin, bigger pore sizes were achieved. Furthermore, gelatin's degradation facilitated rabbit bone marrow cell migration (Zhang et al. 2005). Moreover, co-axial electrospun scaffolds have been fabricated to produce scaffolds with core-shell morphology. Gulfam et al. (2011) fabricated core–shell fiber scaffolds using PCL as the core material and gelatin as the shell material to improve biocompatibility.

Synthetic polymer fibers can be modified to mimic the topography of natural materials. It has been found that PCL fibers could be further modified through self-induced crystallization of PCL on the surface of the fiber by soaking the as-spun fiber in a low concentration of a PCL/acetic acid solution (Wang et al. 2013). PCL molecules in the solution slowly crystallized onto the PCL electrospun fibers to form a hybrid nanoscale shish-kebab structure which showed a similar topography to natural collagen fibers, as can be seen in Figure 12.10. It was further demonstrated that it was

FIGURE 12.10 Scanning electron microscopy (SEM) images of (a) as-spun electrospun PCL fibers and (b) self-induced crystallized shish-kebab structure fibers. Scale bars are 1 μm in (a) and 2 μm in (b). (Reprinted with permission from Wang, X.F. et al., *Biomacromolecules*, 14: 3557–3569. Copyright 2013 American Chemical Society.)

easier for cells to adhere to the rougher surface of the shish-kebab structure. Several biocompatible additives, such as HA, TCP, bioglass, and carbon nanotubes, have been combined with polymers to fabricate scaffolds with modified properties. Meng et al. (2010) found that a very small loading level of carbon nanotubes (e.g., 0.5wt%) in PCL fibers could enhance the tensile strength of the nanofibers by 46%, versus pure PCL fibers, and that the degradation rate of the composite fiber scaffolds were accelerated. In another study, PLA/HA composite electrospun scaffolds showed favorable mechanical strengths and good osteoconductive effects on cultured osteoblasts and mesenchymal stem cells (Jeong et al. 2008).

Moreover, electrospinning has been employed on polymer melts. The mechanism is the same as solution electrospinning, with the difference being that the polymer is heated above the melt point to achieve suitable viscosity for electrospinning. The advantage of melt electrospinning is that the solvent is absent, while the degradation of polymer may occur in the process. Karchin et al. (2011) evaluated the melt electrospinnability of several kinds of synthesized PU and successfully produced nontoxic electrospun fibers suitable for tissue engineering scaffold applications.

12.2.2 Solvent-Free Methods

Besides the wet methods introduced above, some other techniques have been developed to produce porous tissue engineering scaffolds without solvents. These methods can be roughly classified as blending and particle leaching, gas foaming, and SFF methods. Gas foaming methods generate porous structures by the expansion of gas and include batch foaming, microcellular injection molding, and melt extrusion foaming. SFF typically includes three-dimensional printing, selective laser sintering (SLS), fused deposition modeling (FDM), and laser stereolithography.

12.2.2.1 Blending and Particle Leaching (Particulate Leaching)

Similar to the solvent casting and particle leaching method, the blending and particle leaching method is a simple method that incorporates polymers with a porogen and creates pore structures by leaching out the porogen after scaffold molding. The main advantage of blending and particle leaching is that no solvent is needed in the process. There are generally two ways to blend polymers with a porogen. The polymer can be melt blended with the porogen via a twin-screw extruder or mixing machine and then molded into a specific geometry via compression or injection molding. In this method, the dispersion of the porogen in the polymer matrix is uniform, while the size of the porogen is always reduced by the melt blending leading to a smaller pore size. Alternatively, polymer pellets or powder can be physically blended with porogen first and then added to the mold, followed by hot compression. In this way, the porogen will maintain its size in the resulting sample, although it is usually hard to achieve perfect dispersion. The porous scaffolds are obtained after leaching out the porogen. In this method, the porosity and pore size of the scaffold is solely determined by the porogen content and size (Leung et al. 2008).

Among these approaches, compression/particle leaching is the most commonly used method to fabricate three-dimensional scaffolds. As an example of this technique, a mixture of PLGA powder and gelatin microparticles was loaded into a Teflon mold. Next the mixture was heated to above the glass transition temperature of PLGA and compressed into a mold under constant force. The composite was then removed from the mold and the gelatin particles were leached out using water (Thomson et al. 1995). To further improve the porosity and interconnectivity of the scaffold, a second porogen, typically a water soluble polymer, can be added to the blend. Other examples include highly porous PCL/nHA (nanoscale HA) scaffolds that have been fabricated using both salt and polyethylene glycol (PEG) as porogens via compression and particle leaching methods (Liu et al. 2012), and porous poly(urethane urea)-based scaffolds for bone tissue fabricated by Shokrolahi et al. (2011) using this method. In Shokrolahi's study, PEG and salt, which were used as co-porogens, were melt mixed with poly(urethane urea). The porogens were eventually leached out to generate an interconnected porous structure (Shokrolahi et al. 2011). In addition, the mechanical properties of scaffolds can be modified by combining the matrix polymer with other polymers or additives. For example, PCL/PLGA (Barbanti et al. 2011), PLGA/HA (Lee and Lee 2006), and PLA/TCP (Zhao et al. 2011b) scaffolds with improved mechanical properties have been produced using this approach.

Besides hot compression, extrusion and injection molding are alternative and efficient continuous methods to produce polymer/porogen blends. Highly porous PCL scaffolds with fully interconnected pores were fabricated by Reignier and Huneault (2006). They used twin-screw extrusion-blended PCL with polyethylene oxide (PEO) and salt as porogens, which were subsequently leached out to generate pores and interconnected channels (Reignier and Huneault 2006). The advantage of injection molding is that the molded scaffold has the same geometry as the mold cavity. Teng et al. (2013) fabricated L-shaped nasal scaffolds using injection molding combined with salt leaching. Ghosh et al. (2008) prepared PLLA/PEO blends using injection molding, where the PEO phase was leached out to a create porous scaffold.

12.2.2.2   Gas Foaming

Gas foaming is a method that generates porous structures by the expansion of gas (CO_2 or N_2). In this method, the gas is dissolved in the polymer or polymer melt at a certain temperature and pressure, with the gas in the supercritical state where the supercritical fluid (SCF) has good miscibility with the polymer. Bubbles are created by gas expansion when the high pressure is released and the gas turns from the supercritical state into the gaseous state. Tissue engineering scaffold fabrication methods employing gas foaming mechanisms include batch foaming, microcellular injection molding, and extrusion foaming.

12.2.2.2.1   Batch Foaming

Batch foaming is commonly used to produce polymer foams for various applications, such as heat or sound insulation, shock absorption materials, packaging, automotive parts, and biomedical devices. Carbon dioxide (CO_2) has been most widely used in the batch foaming process as it is inexpensive, nontoxic, nonflammable, recoverable, and reusable. In batch foaming, the polymer is first saturated with CO_2 at a high pressure, and sometimes high temperature, in a vessel or specific cylinder to allow the CO_2 to turn into the supercritical state, followed by depressurization to ambient levels at room temperature or in a hot oil bath to trigger the nucleation of CO_2. A porous structure forms due to the expansion of CO_2 when it turns from the supercritical state into the gaseous state. The nature of the polymer, the miscibility of the polymer and CO_2, the foam temperature, the amount of dissolved

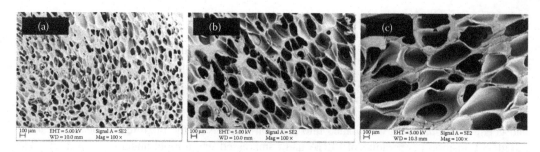

FIGURE 12.11 Scanning electron microscopy (SEM) cross sections along the foaming direction of foams prepared via the batch foaming method at different temperatures: (a) 40°C, (b) 45°C, and (c) 50°C. All other conditions were held constant. The higher the foaming temperature, the larger the bubbles that were obtained. (From Floren, M. et al.: Porous poly(D,L-lactic acid) foams with tunable structure and mechanical anisotropy prepared by supercritical carbon dioxide. *J Biomed Mater Res B*. 2011. 99B. 338–349. Copyright Wiley-VCH Verlag GmbH & Co. KGaA. Reproduced with permission.)

CO_2, the rate and type of gas nucleation, and the rate of gas diffusion to pore nuclei are factors that primarily determine the pore structure and porosity of the resulting foam.

It was found that batch foaming at near-critical pressure (5.5 MPa) can produce highly porous scaffolds (>90% porosity with around 100 μm pore size) from preprocessed solid PLGA discs (Mooney et al. 1996). Floren et al. (2011) investigated the process conditions, such as pressure, temperature, and venting time, to prepare poly(D,L-lactide acid) (PDLLA) scaffolds with an extensive range of scaffold morphologies, which are crucial to the success of engineered tissue replacements. It was found that the porosity, pore shape anisotropy, and pore wall thickness play a role in the mechanical performance of the foams. Figure 12.11 shows the typical structure of batch foaming. The pore morphology was affected by the foaming temperature (Floren et al. 2011). PCL foams, which have high porosities and the potential to be used as tissue scaffolds, have also been prepared using supercritical CO_2 as a foaming agent via the batch foaming process (Leonard et al. 2008). The batch foaming method has been applied to polyurethane as well to produce porous scaffolds (Yeh et al. 2013).

Although batch foaming is a very simple way to make porous structures without the use of organic solvents, two defects need to be considered. (1) Batch foaming is not an efficient way to mass produce scaffolds because, typically, it takes at least a few hours to saturate the polymer with enough CO_2 to produce a high porosity scaffold. (2) The pores created by the gas foaming procedure yield a mostly closed pore structure because the gas expansion force is not high enough to break the pore walls between the pores. Regarding the second defect, researchers have tried different methods to improve the interconnectivity of the scaffold. Salerno et al. (2011) fabricated porous PCL and PCL/HA scaffolds with bimodal pore size distributions using a two-step depressurization solid-state supercritical CO_2 foaming process. As shown in Figure 12.12, the bimodal porous structure helped to improve the interconnections among the pores. The biocompatibilities of the bimodal scaffolds were finally verified by *in vitro* culture of human mesenchymal stem cells (Salerno et al. 2011).

Adding another porogen (i.e., sodium chloride [NaCl]) to be leached out is a common way to improve both scaffold porosity and interconnectivity. Osteochondral biomimetic PCL and nanometric HA scaffolds for bone tissue engineering were fabricated via a CO_2 batch foaming and micronized NaCl particle leaching method (Salerno et al. 2012). Annabi et al. (2011) fabricated three-dimensional porous PCL/elastin composites that impregnated elastin into PCL matrix. In their research, high pressure CO_2 and salt particles were used as foaming agents and porogen to create interconnected porous scaffold (Annabi et al. 2011). Salerno et al. (2009) combined the gas foaming technique with selective polymer extraction from co-continuous blends to prepare μ-bimodal PCL scaffolds. First, they produced PCL/thermoplastic gelatin co-continuous blends and used an N_2/CO_2 mixture blowing agent to foam the blends. Then thermoplastic gelatin was selectively extracted from the foamed blend by soaking the sample in deionized H_2O. The results showed that the combined

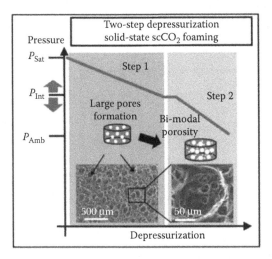

FIGURE 12.12 Illustration of the two-step depressurization method used to produce a bimodal porous structure. Materials were saturated with CO_2 at 37°C and 20 MPa. The blowing agent was released in two steps: (1) from 20 to 10 MPa at a slow depressurization rate to form large pores and (2) from 10 MPa to ambient pressure at a fast depressurization rate to form small pores. (From Salerno, A. et al.: Design of bimodal PCL and PCL-HA nanocomposite scaffolds by two step depressurization during solid-state supercritical CO2 foaming. *Macromol Rapid Comm*. 2011. 32. 1150–1156. Copyright Wiley-VCH Verlag GmbH & Co. KGaA. Reproduced with permission.)

gas foaming–polymer extraction technique allowed for the preparation of PCL scaffolds with a unique multiscaled and highly interconnected micro-architecture (Salerno et al. 2009).

12.2.2.2.2 Microcellular Injection Molding

Similar to batch foaming, microcellular injection molding is a relatively new method used to produce porous parts with a specific geometry by gas foaming via an injection molding machine. It was recently introduced to tissue engineering scaffold fabrication and is an ideal manufacturing method to prepare three-dimensional scaffolds owing to its low cost, design flexibility, and high productivity. Microcellular injection molding combines the advantages of traditional injection molding and batch foaming to create foamed polymers by using a physical blowing agent (N_2 or CO_2). A schematic of the microcellular injection molding process is shown in Figure 12.13 (Mi et al. 2014d).

In this method, SCF is mixed with the polymer melt in the injection barrel by the rotating screw to form a single-phase polymer/gas solution, which is then injected into the mold cavity. The sudden pressure drop as the material passes through the gate and cavity induces thermodynamic instability. As a result, dissolved gas starts emerging from the solution and generates bubbles inside of the molded part. Microcellular injection molding is capable of producing parts with a microcellular structure using lower injection pressures, shorter cycle times, and less material. It eliminates the need for packing pressure and improves the dimensional stability of the molded parts. Furthermore, this method can be easily industrialized.

Compared with batch foaming, there are more parameters in microcellular injection molding that affect the final foamed structure due to the complexity of the process. These parameters can be material properties like the polymer's chemical structure, molecular weight, miscibility with the foaming agent, and viscosity; or process conditions such as the processing temperature, gas content, injection speed, shot dosage, mold temperature, and cooling time. In addition, the mold design also affects the foaming structure (Barzegari and Rodrigue 2009; Leicher et al. 2005; Sporrer and Altstadt 2007). Researchers have made efforts to promote this method in tissue engineering scaffold fabrication. Wintermantel and coworkers successfully fabricated polyether urethane scaffolds

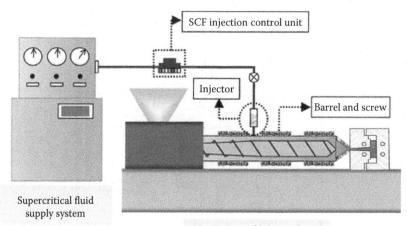

FIGURE 12.13 Schematic illustration of the microcellular injection molding process. The supercritical blowing agent is provided by a supercritical fluid (SCF) supply system. An SCF control unit is used to adjust the SCF flow rate. The SCF is injected into the middle section of the barrel through an injector and mixed with polymer melt with the help of screw rotation. Finally, the mixed polymer/gas solution is injected and foamed in the mold cavity. (Adapted from Mi, H.Y. et al., *J. Cell. Plast.*, 2014d, with the permission from SAGE.)

using microcellular injection molding with CO_2 as the foaming agent. They even obtained open pore structure polyurethane scaffolds with 71% porosity by carefully adjusting the processing parameters (Leicher et al. 2005; Wu et al. 2012c). In another study, they applied water as the foaming agent and introduced NaCl as an additional porogen to produce porous tissue scaffolds via the microcellular injection molding method. Positive interactions between the resulting scaffold and human fibroblasts were observed (Haugen et al. 2006a, b).

Scaffold properties can also be tuned by blending polymers that have different attributes. For example, TPU/PLA blended scaffolds have been produced via microcellular injection molding. The mechanical properties of the scaffolds can be adjusted by varying the ratio between the TPU and PLA in the blends (Mi et al. 2013b). The morphology of the TPU/PLA scaffold is shown in Figure 12.14.

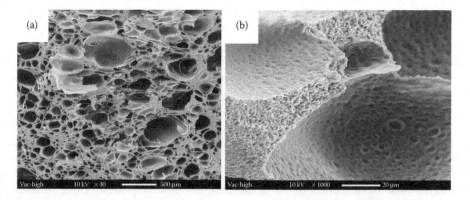

FIGURE 12.14 Scanning electron microscopy (SEM) images of a TPU/PLA microcellular injection molded scaffold with 75 wt% PLA at (a) low magnification showing the porous structure and (b) high magnification showing the surface topography. Scale bars are 500 μm in (a) and 20 μm in (b). (Reprinted from *Mater. Sci. Eng. C-Mater.*, 33, Mi, H.Y. et al., Characterization of thermoplastic polyurethane/polylactic acid [TPU/PLA] tissue engineering scaffolds fabricated by microcellular injection molding, 4767–4776, Copyright 2013b, with permission from Elsevier.)

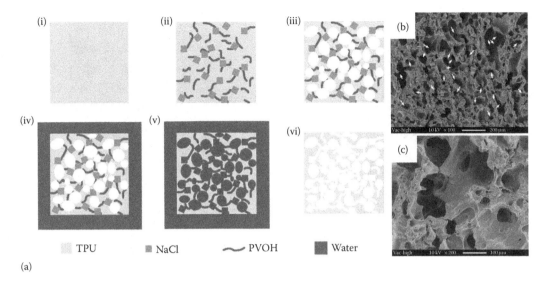

(a)

FIGURE 12.15   (a) Schematic illustration of the scaffold fabrication method combining microcellular injection molding and particle leaching: (i) TPU matrix, (ii) TPU mixed with PVOH and NaCl, (iii) microcellular injection molded TPU/PVOH/NaCl components with bubbles, (iv) beginning of leaching process, (v) end of leaching process, and (vi) final porous and interconnected structure. SEM images of fabricated scaffolds at (b) low magnification and (c) high magnification. Arrows indicate interconnecting tunnels inside of the pores.

Similar to batch foaming, microcellular injection molding also has a closed porous structure that limits its direct application to tissue engineering. To further improve the interconnectivity of the scaffolds, Turng and coworkers modified the method by introducing two additional porogens. In this study, highly porous PLA scaffolds were fabricated using CO_2 as the foaming agent and NaCl and polyvinyl alcohol (PVOH) as additional porogens, which were subsequently leached out after foaming (Kramschuster and Turng 2010). The authors have produced TPU scaffolds using this method as well. The schematic illustration of this method, and the interconnected TPU scaffold, are shown in Figure 12.15 (Mi et al. 2014c). In another study, PCL/HA scaffolds were successfully produced using N_2 as the blowing agent. The porosity and interconnectivity were improved by leaching out the NaCl and PEO which were pre-blended with PCL. The scaffold then proved suitable for osteoblast-like cell growth (Cui et al. 2012).

However, some problems still exist in this method that restricts its application and commercialization. For example, typically, the microcellular injection molded foam structure will be affected by mold cavity conditions and melt fountain flow behavior, which leads to a nonuniform pore size distribution—namely, larger pores in the center and smaller pores close to the mold cavity wall. Also, usually a solid skin layer will form on the surface of the sample due to the fast surface cooling, as shown in Figure 12.16a (Mi et al. 2013a). Prior to cell culture, the skin layer often has to be removed to expose the porous structure. It was found that using water and CO_2 as co-blowing agents in microcellular injection molding increased the foaming area and eliminated the skin layer as shown in Figure 12.16b (Mi et al. 2014b).

12.2.2.2.3 Extrusion Foaming

Extrusion is a well-known process applied to form objects of a predefined, fixed cross section. It has been used to process thermoplastic bioresorbable polymers for three-dimensional scaffold fabrication with macroscale cross-sectional areas. When combined with SCF, a foamed polymer stream can be produced continuously by extrusion foaming. The productivity of extrusion foaming is much higher than that of batch foaming, and the foaming agent can be better mixed with polymer melt

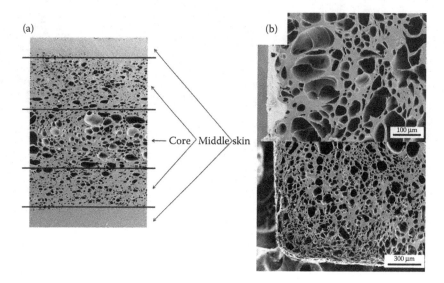

FIGURE 12.16 (a) Traditional microcellular injection molded TPU sample morphology: large pores in the core area, smaller pores in the middle sections, and a solid skin layer. (b) SEM images of a microcellular injection molded sample using water and CO_2 as co-blowing agents showing the elimination of the skin layer. (From Mi, H.Y. et al.: A novel thermoplastic polyurethane scaffold fabrication method based on injection foaming with water and supercritical carbon dioxide as coblowing agents, *Polym. Eng. Sci.*, 2014b. Copyright Wiley-VCH Verlag GmbH & Co. KGaA. Reproduced with permission.)

with the help of screw rotation. Compared with microcellular injection foaming, extrusion foaming has many fewer variables during processing and the foam structure is more uniform and easier to control. Typically, the processing parameters of extrusion foaming include processing temperature, screw rotation speed, gas injection flow rate, die temperature, and die design. The final foamed morphology is typically affected by the gas flow rate and pressure drop. Higher gas flow rates and pressure drops usually lead to higher porosities (Matuana and Diaz 2010). The extrusion foaming method has been used in continuous starch scaffold fabrication as CO_2 is widely used as a blowing agent for starch processing (Gomes et al. 2002).

PLA scaffolds were successfully produced by melt extrusion foaming and it was found that a proper pore structure could be maintained with a gas content of less than 10wt% (Lee and Kang 2005). However, the defects of extrusion foaming are obvious. In extrusion foaming, only the cross-sectional structure can be controlled, which is determined by the die design. Furthermore, like batch foaming, the pore structure of the extrusion foamed sample does not always have interconnections. As with microcellular injection molding, scaffold interconnectivity can be improved by adding porogen particles or water soluble polymers.

12.2.2.3 Solid Freeform Fabrication

The utilization of computer-aided technologies in tissue engineering has evolved into an emerging field called *computer-aided tissue engineering* (Sun and Lal 2002). SFF, also known as *rapid prototyping* (RP), is a group of advanced manufacturing processes that can produce custom made objects layer by layer in an additive manner, directly from computer data such as computer-aided design, computed tomography, and magnetic resonance imaging data. Generally, the fabrication process starts with a three-dimensional design of the scaffold. Next, the design is transferred to a .STL (stereo lithography) file format, where it is virtually sliced into thin, horizontal cross sections and transferred to the RP machine, where the scaffold is directly fabricated layer by layer

(Sodian et al. 2005). These techniques include three-dimensional printing (3DP) (powder with binder or liquid resins), SLS (polymer powder, usually nylon or ceramic powder), FDM (thermoplastics), or laser stereolithography (SL) (photosensitive polymers) (Lantada and Morgado 2012). Due to the ability of these technologies to incorporate advanced computer-aided design technologies to produce complex and precise parts, SFF technologies are fast becoming the technologies of choice for fabricating scaffolds due to their reliability, high degree of reproducibility, and potential to overcome the limitations of conventional manual-based fabrication techniques (Leong et al. 2003).

12.2.2.3.1   Three-Dimensional Printing

3DP was the first RP technique to be used for biomedical and tissue engineering purposes (Salgado et al. 2004). In 3DP, the first step is spreading powder onto a platform with a roller, followed by an inkjet print head printing a two-dimensional pattern onto the powder layer. After that, the next powder layer is spread and then overlapped by another layer of inkjet printing until the part is finished. The unused powder is brushed or blown off afterward. The process is usually followed by a temperature treatment to burn the binder off and a final sintering step. Recent research on three-dimensional printed scaffolds focuses on evaluating the mechanical properties of the scaffolds both *in vivo* and *in vitro*. Ge et al. (2009) proved that the mechanical properties of three-dimensional printed PLGA scaffolds were close to trabecular bone and that human fetal osteoblasts seeded on the scaffolds could proliferate and differentiate normally. Serra et al. (2013) fabricated three-dimensional biodegradable PLA/PEG/bioactive glass composite scaffolds with an interconnected porous structure and improved surface roughness and hydrophilicity due to the addition of PEG and bioglass.

12.2.2.3.2   Selective Laser Sintering

In SLS, three-dimensional objects are fabricated by sintering the powder material in the cross-sectional area of each layer on a powder bed using a CO_2 laser. The powder is brought to its glass transition temperature (T_g) by heaters in the machine prior to fusing by the laser. The nonsintered particles act as a support for any hollow sections, overhangs, or undercuts in the part. Finally, after the completed part is removed from the chamber, the loose powder can be removed (Chua et al. 2004). Yeong et al. (2010) fabricated PCL scaffolds with 85% porosity and 40–100 μm pore diameters by SLS and tested them with myoblast cells to investigate their feasibility for cardiac tissue engineering. Duan et al. (2010 and 2011) conducted extensive research on fabricating polymer/bioceramic composite scaffolds using the SLS method. They successfully produced calcium phosphate (CP)/poly(hydroxybutyrate-co hydroxyvalerate) (PHBV) and carbonated hydroxyapatite (CHAp)/PLLA nanocomposite scaffolds that have potential application in bone tissue scaffold applications. Examples of the composite scaffolds are shown in Figure 12.17 (Duan et al. 2010, 2011; Duan and Wang 2010; Zhou et al. 2009).

12.2.2.3.3   Fused Deposition Modeling

The FDM method, which is also known as *extrusion-based RP*, forms three-dimensional objects from computer-generated solid or surface models like in a typical RP process. Models can also be derived from computer tomography scans, magnetic resonance imaging scans, or model data created from three-dimensional object digitizing systems. FDM uses a small temperature-controlled extruder to force out a thermoplastic filament material and deposit the semi-molten polymer onto a platform in a layer-by-layer process (Zein et al. 2002). For example, Korpela et al. (2013) prepared biodegradable and bioactive porous scaffolds using different polymers and additives including PLA, PCL, their copolymer PLCL (L-lactide/ε-caprolactone 75/25), and bioactive glass. The biocompatibility of the scaffolds was verified by fibroblast cell culture (Korpela et al. 2013). Alternative methods have been proposed, such as three-dimensional fiber deposition (3DFD) (Woodfield et al. 2004), bioplotting (Pfister et al. 2004), and precision extruding deposition (PED) to overcome the limitations of traditional FDM like the high heat effect on raw material (Wang et al. 2004). Additionally, researchers

FIGURE 12.17 Three-dimensional scaffolds produced by the SLS method: (a) PHBV, (b) CP/PHBV, (c) PLLA, and (d) CHAp/PLLA. (Reprinted from *Acta Biomater.*, 6, Duan, B. et al., Three-dimensional nanocomposite scaffolds fabricated via selective laser sintering for bone tissue engineering, 4495–4505, Copyright 2010, with permission from Elsevier.)

have tried to combine the FDM method with electrospinning to further improve the morphological properties and scaffold performance by introducing nanofibers into the molded three-dimensional scaffolds (Centola et al. 2010; Lee et al. 2011; Owida et al. 2011).

12.2.2.3.4 Laser Stereolithography

Like most SFF techniques, SL, is an additive fabrication process that allows the fabrication of parts from a computer-aided design file. It is based on the spatially controlled solidification of a photo-curable liquid polymer. Using a computer-controlled laser beam, or a digital light projector with a computer-driven building stage, a pattern is illuminated on the surface of a resin. After the first layer solidifies and attaches to the platform, successive layers are cured by lowering the platform and recoating it with an exact thickness of polymer liquid (Hollister 2005). The main limitation for SL is that the number of resins that are commercially available for processing are few in that the resin must be a liquid that rapidly solidifies upon illumination with light (Melchels et al. 2010b). The first report using SL to fabricate biodegradable structures with ε-caprolactone (CL) and trimethylene carbonate (TMC) appeared in 2000 (Matsuda et al. 2000). Melchels et al. made a great effort in developing the SL method. They fabricated PEG/PLA hydrogels, poly(D,L-lactide-*co*-ε-caprolactone)-based resins, and PLA scaffolds via SL methods with mechanical properties suitable for tissue engineering applications (Melchels et al. 2009, 2010a; Seck et al. 2010). To further improve the bioactivity of the scaffolds in bone tissue engineering, Ronca et al. (2013) fabricated composite structures containing HA. Natural materials (such as gelatin) have also been used in SL to extend its usability in tissue engineering applications (Schuster et al. 2009).

REFERENCES

Adamus, G., Sikorska, W., Janeczek, H., Kwiecien, M., Sobota, M., Kowalczuk, M. (2012) Novel block copolymers of atactic PHB with natural PHA for cardiovascular engineering: Synthesis and characterization. *Eur Polym J* 48: 621–631.
Almany, L., Seliktar, D. (2005) Biosynthetic hydrogel scaffolds made from fibrinogen and polyethylene glycol for 3D cell cultures. *Biomaterials* 26: 2467–2477.

Annabi, N., Fathi, A., Mithieux, S.M., Weiss, A.S., Dehghani, F. (2011) Fabrication of porous PCL/elastin composite scaffolds for tissue engineering applications. *J Supercrit Fluid* 59: 157–167.

Arcos-Hernandez, M.V., Laycock, B., Pratt, S., Donose, B.C., Nikolic, M.A.L., Luckman, P., Werker, A., Lant, P.A. (2012) Biodegradation in a soil environment of activated sludge derived polyhydroxyalkanoate (PHBV). *Polym Degrad Stab* 97: 2301–2312.

Bach, F.H. (1998) Xenotransplantation: Problems and prospects. *Annu Rev Med* 49: 301–310.

Baji, A., Wong, S.C., Srivatsan, T.S., Njus, G.O., Mathur, G. (2006) Processing methodologies for polycaprolactone-hydroxyapatite composites: A review. *Mater Manuf Process* 21: 211–218.

Barbanti, S.H., Santos, A.R., Zavaglia, C.A.C., Duek, E.A.R. (2011) Poly(epsilon-caprolactone) and poly(D,L-lactic acid-co-glycolic acid) scaffolds used in bone tissue engineering prepared by melt compression-particulate leaching method. *J Mater Sci-Mater M* 22: 2377–2385.

Barzegari, M.R., Rodrigue, D. (2009) The effect of injection molding conditions on the morphology of polymer structural foams. *Polym Eng Sci* 49: 949–959.

Bendix, D. (1998) Chemical synthesis of polylactide and its copolymers for medical applications. *Polym Degrad Stabil* 59: 129–135.

Boccaccini, A.R., Maquet, V. (2003) Bioresorbable and bioactive polymer/bioglass (R) composites with tailored pore structure for tissue engineering applications. *Compos Sci Technol* 63: 2417–2429.

Borden, M., El-Amin, S.F., Attawia, M., Laurencin, C.T. (2003) Structural and human cellular assessment of a novel microsphere-based tissue engineered scaffold for bone repair. *Biomaterials* 24: 597–609.

Boyan, B.D., Hummert, T.W., Dean, D.D., Schwartz, Z. (1996) Role of material surfaces in regulating bone and cartilage cell response. *Biomaterials* 17: 137–146.

Casper, C.L., Stephens, J.S., Tassi, N.G., Chase, D.B., Rabolt, J.F. (2004) Controlling surface morphology of electrospun polystyrene fibers: Effect of humidity and molecular weight in the electrospinning process. *Macromolecules* 37: 573–578.

Centola, M., Rainer, A., Spadaccio, C., De Porcellinis, S., Genovese, J.A., Trombetta, M. (2010) Combining electrospinning and fused deposition modeling for the fabrication of a hybrid vascular graft. *Biofabrication* 2: 014102.

Charles-Harris, M., Koch, M.A., Navarro, M., Lacroix, D., Engel, E., Planell, J.A. (2008) A PLA/calcium phosphate degradable composite material for bone tissue engineering: An in vitro study. *J Mater Sci-Mater M* 19: 1503–1513.

Chen, J.S., Tu, S.L., Tsay, R.Y. (2010) A morphological study of porous polylactide scaffolds prepared by thermally induced phase separation. *J Taiwan Inst Chem E* 41: 229–238.

Chua, C.K., Leong, K.F., Tan, K.H., Wiria, F.E., Cheah, C.M. (2004) Development of tissue scaffolds using selective laser sintering of polyvinyl alcohol/hydroxyapatite biocomposite for craniofacial and joint defects. *J Mater Sci-Mater M* 15: 1113–1121.

Cima, L.G., Vacanti, J.P., Vacanti, C., Ingber, D., Mooney, D., Langer, R. (1991) Tissue engineering by cell transplantation using degradable polymer substrates. *J Biomech Eng-T Asme* 113: 143–151.

Cui, Z.X., Zhao, H.B., Peng, Y.Y., Kaland, M., Turng, L.S., Shen, C.Y. (2012) Morphology and properties of porous and interconnected poly(epsilon-caprolactone) matrices using solid and microcellular injection molding. *J Biobased Mater Bioenerg* 6: 259–268.

Dado, D., Levenberg, S. (2009) Cell-scaffold mechanical interplay within engineered tissue. *Semin Cell Dev Biol* 20: 656–664.

Demir, M.M., Yilgor, I., Yilgor, E., Erman, B. (2002) Electrospinning of polyurethane fibers. *Polymer* 43: 3303–3309.

Dhandayuthapani, B., Yoshida, Y., Maekawa, T., Kumar, D.S. (2011) Polymeric scaffolds in tissue engineering application: A review. *Int J Polym Sci* 2011: 1–19.

Dinu, M.V., Pradny, M., Dragan, E.S., Michalek, J. (2013) Ice-templated hydrogels based on chitosan with tailored porous morphology. *Carbohyd Polym* 94: 170–178.

Du, F.Y., Wang, H., Zhao, W., Li, D., Kong, D.L., Yang, J., Zhang, Y.Y. (2012) Gradient nanofibrous chitosan/poly epsilon-caprolactone scaffolds as extracellular microenvironments for vascular tissue engineering. *Biomaterials* 33: 762–770.

Duan, B., Cheung, W.L., Wang, M. (2011) Optimized fabrication of Ca-P/PHBV nanocomposite scaffolds via selective laser sintering for bone tissue engineering. *Biofabrication* 3: 1–13.

Duan, B., Wang, M. (2010) Customized Ca-P/PHBV nanocomposite scaffolds for bone tissue engineering: Design, fabrication, surface modification and sustained release of growth factor. *J R Soc Interf* 7: S615–S629.

Duan, B., Wang, M., Zhou, W.Y., Cheung, W.L., Li, Z.Y., Lu, W.W. (2010) Three-dimensional nanocomposite scaffolds fabricated via selective laser sintering for bone tissue engineering. *Acta Biomater* 6: 4495–4505.

Engler, A.J., Griffin, M.A., Sen, S., Bonnetnann, C.G., Sweeney, H.L., Discher, D.E. (2004) Myotubes differentiate optimally on substrates with tissue-like stiffness: Pathological implications for soft or stiff microenvironments. *J Cell Biol* 166: 877–887.

Engler, A.J., Sen, S., Sweeney, H.L., Discher, D.E. (2006) Matrix elasticity directs stem cell lineage specification. *Cell* 126: 677–689.

Floren, M., Spilimbergo, S., Motta, A., Migliaresi, C. (2011) Porous poly(D,L-lactic acid) foams with tunable structure and mechanical anisotropy prepared by supercritical carbon dioxide. *J Biomed Mater Res B* 99B: 338–349.

Fong, H., Chun, I., Reneker, D.H. (1999) Beaded nanofibers formed during electrospinning. *Polymer* 40: 4585–4592.

Gautam, S., Dinda, A.K., Mishra, N.C. (2013) Fabrication and characterization of PCL/gelatin composite nano-fibrous scaffold for tissue engineering applications by electrospinning method. *Mat Sci Eng C-Mater* 33: 1228–1235.

Ge, Z.G., Wang, L.S., Heng, B.C., Tian, X.F., Lu, K., Fan, V.T.W., Yeo, J.F., Cao, T., Tan, E. (2009) Proliferation and differentiation of human osteoblasts within 3D printed poly-lactic-co-glycolic acid scaffolds. *J Biomater Appl* 23: 533–547.

Geng, X.Y., Kwon, O.H., Jang, J.H. (2005) Electrospinning of chitosan dissolved in concentrated acetic acid solution. *Biomaterials* 26: 5427–5432.

Ghosh, S., Vianac, J.C., Reis, R.L., Mano, J.F. (2008) Development of porous lamellar poly(L-lactic acid) scaffolds by conventional injection molding process. *Acta Biomater* 4: 887–896.

Goh, Y.Q., Ooi, C.P. (2008) Fabrication and characterization of porous poly(L-lactide) scaffolds using solid-liquid phase separation. *J Mater Sci-Mater M* 19: 2445–2452.

Gomes, M.E., Godinho, J.S., Tchalamov, D., Cunha, A.M., Reis, R.L. (2002) Alternative tissue engineering scaffolds based on starch: Processing methodologies, morphology, degradation and mechanical properties. *Mat Sci Eng C-Bio S* 20: 19–26.

Gopferich, A. (1996) Mechanisms of polymer degradation and erosion. *Biomaterials* 17: 103–114.

Guan, J.J., Fujimoto, K.L., Sacks, M.S., Wagner, W.R. (2005) Preparation and characterization of highly porous, biodegradable polyurethane scaffolds for soft tissue applications. *Biomaterials* 26: 3961–3971.

Gulfam, M., Lee, J.M., Kim, J.E., Lim, D.W., Lee, E.K., Chung, B.G. (2011) Highly porous core-shell polymeric fiber network. *Langmuir* 27: 10993–10999.

Haugen, H., Aigner, J., Brunner, M., Wintermantel, E. (2006a) A novel processing method for injection-molded polyether-urethane scaffolds. Part 2: Cellular interactions. *J Biomed Mater Res B* 77B: 73–78.

Haugen, H., Will, J., Fuchs, W., Wintermantel, E. (2006b) A novel processing method for injection-molded polyether-urethane scaffolds. Part 1: Processing. *J Biomed Mater Res B* 77B: 65–72.

He, L.M., Zhang, Y.Q., Zeng, X., Quan, D.P., Liao, S., Zeng, Y.S., Lu, J., Ramakrishna, S. (2009) Fabrication and characterization of poly(L-lactic acid) 3D nanofibrous scaffolds with controlled architecture by liquid-liquid phase separation from a ternary polymer-solvent system. *Polymer* 50: 4128–4138.

Hollister, S.J. (2005) Porous scaffold design for tissue engineering. *Nat Mater* 4: 518–524.

Hu, X.X., Shen, H., Yang, F., Bei, J.Z., Wang, S.G. (2008) Preparation and cell affinity of microtubular orientation-structured PLGA(70/30) blood vessel scaffold. *Biomaterials* 29: 3128–3136.

Hua, F.J., Kim, G.E., Lee, J.D., Son, Y.K., Lee, D.S. (2002) Macroporous poly(L-lactide) scaffold 1. Preparation of a macroporous scaffold by liquid-liquid phase separation of a PLLA-dioxane-water system. *J Biomed Mater Res* 63: 161–167.

Hua, F.J., Park, T.G., Lee, D.S. (2003) A facile preparation of highly interconnected macroporous poly(D,L-lactic acid-co-glycolic acid) (PLGA) scaffolds by liquid-liquid phase separation of a PLGA-dioxane-water ternary system. *Polymer* 44: 1911–1920.

Huang, Y.X., Ren, J., Chen, C., Ren, T.B., Zhou, X.Y. (2008) Preparation and properties of poly(lactide-co-glycolide) (PLGA)/nano-hydroxyapatite (NHA) scaffolds by thermally induced phase separation and rabbit MSCs culture on scaffolds. *J Biomater Appl* 22: 409–432.

Hutmacher, D.W. (2000) Scaffolds in tissue engineering bone and cartilage. *Biomaterials* 21: 2529–2543.

Hutmacher, D.W. (2001) Scaffold design and fabrication technologies for engineering tissues—State of the art and future perspectives. *J Biomat Sci-Polym E* 12: 107–124.

Ikada, Y. (1994) Surface modification of polymers for medical applications. *Biomaterials* 15: 725–736.

Ito, Y., Chen, G.P., Imanishi, Y. (1998) Micropatterned immobilization of epidermal growth factor to regulate cell function. *Bioconjugate Chem* 9: 277–282.

Jack, K.S., Velayudhan, S., Luckman, P., Trau, M., Grondahl, L., Cooper-White, J. (2009) The fabrication and characterization of biodegradable HA/PHBV nanoparticle-polymer composite scaffolds. *Acta Biomater* 5: 2657–2667.

Jasmund, I., Bader, A. (2002) Bioreactor developments for tissue engineering applications by the example of the bioartificial liver. *Adv Biochem Eng/Biotechnol* 74: 99–109.

Jell, G., Verdejo, R., Safinia, L., Shaffer, M.S.P., Stevens, M.M., Bismarck, A. (2008) Carbon nanotube-enhanced polyurethane scaffolds fabricated by thermally induced phase separation. *J Mater Chem* 18: 1865–1872.

Jeong, S.I., Ko, E.K., Yum, J., Jung, C.H., Lee, Y.M., Shin, H. (2008) Nanofibrous poly(lactic acid)/hydroxyapatite composite scaffolds for guided tissue regeneration. *Macromol Biosci* 8: 328–338.

Karchin, A., Simonovsky, F.I., Ratner, B.D., Sanders, J.E. (2011) Melt electrospinning of biodegradable polyurethane scaffolds. *Acta Biomater* 7: 3277–3284.

Ki, C.S., Baek, D.H., Gang, K.D., Lee, K.H., Um, I.C., Park, Y.H. (2005) Characterization of gelatin nanofiber prepared from gelatin-formic acid solution. *Polymer* 46: 5094–5102.

Kidoaki, S., Kwon, I.K., Matsuda, T. (2005) Mesoscopic spatial designs of nano- and microfiber meshes for tissue-engineering matrix and scaffold based on newly devised multilayering and mixing electrospinning techniques. *Biomaterials* 26: 37–46.

Kim, G., Ree, M., Kim, H., Kim, I.J., Kim, J.R., Lee, J.I. (2008) Biological affinity and biodegradability of poly(propylene carbonate) prepared from copolymerization of carbon dioxide with propylene oxide. *Macromol Res* 16: 473–480.

Kim, J., Kim, D.H., Lim, K.T., Seonwoo, H., Park, S.H., Kim, Y.R., Kim, Y., Choung, Y.H., Choung, P.H., Chung, J.H. (2012) Charged nanomatrices as efficient platforms for modulating cell adhesion and shape. *Tissue Eng Part C Method* 18: 913–923.

Koegler, W.S., Griffith, L.G. (2004) Osteoblast response to PLGA tissue engineering scaffolds with PEO modified surface chemistries and demonstration of patterned cell response. *Biomaterials* 25: 2819–2830.

Korpela, J., Kokkari, A., Korhonen, H., Malin, M., Narhi, T., Seppala, J. (2013) Biodegradable and bioactive porous scaffold structures prepared using fused deposition modeling. *J Biomed Mater Res B* 101B: 610–619.

Koski, A., Yim, K., Shivkumar, S. (2004) Effect of molecular weight on fibrous PVA produced by electrospinning. *Mater Lett* 58: 493–497.

Kramschuster, A., Turng, L.S. (2010) An injection molding process for manufacturing highly porous and interconnected biodegradable polymer matrices for use as tissue engineering scaffolds. *J Biomed Mater Res B* 92B: 366–376.

Kumbar, S.G., Nukavarapu, S.P., James, R., Nair, L.S., Laurencin, C.T. (2008) Electrospun poly(lactic acid-co-glycolic acid) scaffolds for skin tissue engineering. *Biomaterials* 29: 4100–4107.

Kweon, H., Yoo, M.K., Park, I.K., Kim, T.H., Lee, H.C., Lee, H.S., Oh, J.S., Akaike, T., Cho, C.S. (2003) A novel degradable polycaprolactone networks for tissue engineering. *Biomaterials* 24: 801–808.

Lampin, M., WarocquierClerout, R., Legris, C., Degrange, M., SigotLuizard, M.F. (1997) Correlation between substratum roughness and wettability, cell adhesion, and cell migration. *J Biomed Mater Res* 36: 99–108.

Langer, R., Vacanti, J.P. (1993) Tissue Engineering. *Science* 260: 920–926.

Lannutti, J., Reneker, D., Ma, T., Tomasko, D., Farson, D.F. (2007) Electrospinning for tissue engineering scaffolds. *Mater Sci Eng C-Bio S* 27: 504–509.

Lantada, A.D., Morgado, P.L. (2012) Rapid prototyping for biomedical engineering: Current capabilities and challenges. *Annu Rev Biomed Eng* 14: 73–96.

Lee, C.T., Lee, Y.D. (2006) Preparation of porous biodegradable poly(lactide-co-glycolide)/hyaluronic acid blend scaffolds: Characterization, in vitro cells culture and degradation behaviors. *J Mater Sci Mater M* 17: 1411–1420.

Lee, H., Yeo, M., Ahn, S., Kang, D.O., Jang, C.H., Lee, H., Park, G.M., Kim, G.H. (2011) Designed hybrid scaffolds consisting of polycaprolactone microstrands and electrospun collagen-nanofibers for bone tissue regeneration. *J Biomed Mater Res B* 97B: 263–270.

Lee, J., Cuddihy, M.J., Kotov, N.A. (2008) Three-dimensional cell culture matrices: State of the art. *Tissue Eng Pt B-Rev* 14: 61–86.

Lee, J.H., Khang, G., Lee, J.W., Lee, H.B. (1998) Platelet adhesion onto chargeable functional group gradient surfaces. *J Biomed Mater Res* 40: 180–186.

Lee, J.R., Kang, H.J. (2005) Preparation of poly(L-lactic acid) scaffolds by melt extrusion foaming. *Polym-Korea* 29: 198–203.

Lee, S.H., Kim, B.S., Kim, S.H., Choi, S.W., Jeong, S.I., Kwon, I.K., Kang, S.W. et al. (2003) Elastic biodegradable poly(glycolide-co-caprolactone) scaffold for tissue engineering. *J Biomed Mater Res A* 66A: 29–37.

Leicher, S., Will, J., Haugen, H., Wintermantel, E. (2005) MuCell (R) technology for injection molding: A processing method for polyether-urethane scaffolds. *J Mater Sci* 40: 4613–4618.

Leonard, A., Calberg, C., Kerckhofs, G., Wevers, M., Jerome, R., Pirard, J.P., Germain, A., Blacher, S. (2008) Characterization of the porous structure of biodegradable scaffolds obtained with supercritical CO(2) as foaming agent. *J Porous Mat* 15: 397–403.

Leong, K.F., Cheah, C.M., Chua, C.K. (2003) Solid freeform fabrication of three-dimensional scaffolds for engineering replacement tissues and organs. *Biomaterials* 24: 2363–2378.

Lepoittevin, B., Devalckenaere, M., Pantoustier, N., Alexandre, M., Kubies, D., Calberg, C., Jerome, R., Dubois, P. (2002) Poly(epsilon-caprolactone)/clay nanocomposites prepared by melt intercalation: Mechanical, thermal and rheological properties. *Polymer* 43: 4017–4023.

Leung, L., Chan, C., Song, J., Tam, B., Naguib, H. (2008) A parametric study on the processing and physical characterization of PLGA 50/50 bioscaffolds. *J Cell Plast* 44: 189–202.

Levy-Mishali, M., Zoldan, J., Levenberg, S. (2009) Effect of scaffold stiffness on myoblast differentiation. *Tissue Eng Pt A* 15: 935–944.

Li, C., Zhang, J.J., Li, Y.J., Moran, S., Khang, G., Ge, Z.G. (2013) Poly (L-lactide-co-caprolactone) scaffolds enhanced with poly (beta-hydroxybutyrateco-beta-hydroxyvalerate) microspheres for cartilage regeneration. *Biomed Mater* 8: 1–13.

Li, G., Gill, T.J., DeFrate, L.E., Zayontz, S., Glatt, V., Zarins, B. (2002) Biomechanical consequences of PCL deficiency in the knee under simulated muscle loads—An in vitro experimental study. *J Orthop Res* 20: 887–892.

Li, W.J., Danielson, K.G., Alexander, P.G., Tuan, R.S. (2003) Biological response of chondrocytes cultured in three-dimensional nanofibrous poly(epsilon-caprolactone) scaffolds. *J Biomed Mater Res A* 67A: 1105–1114.

Li, W.J., Tuli, R., Huang, X.X., Laquerriere, P., Tuan, R.S. (2005) Multilineage differentiation of human mesenchymal stem cells in a three-dimensional nanofibrous scaffold. *Biomaterials* 26: 5158–5166.

Li, Y., Ma, T., Kniss, D.A., Lasky, L.C., Yang, S.T. (2001) Effects of filtration seeding on cell density, spatial distribution, and proliferation in nonwoven fibrous matrices. *Biotechnol Prog* 17: 935–944.

Lin, A.S.P., Barrows, T.H., Cartmell, S.H., Guldberg, R.E. (2003) Microarchitectural and mechanical characterization of oriented porous polymer scaffolds. *Biomaterials* 24: 481–489.

Liu, L., Wang, Y.Y., Guo, S.R., Wang, Z.Y., Wang, W. (2012) Porous polycaprolactone/nanohydroxyapatite tissue engineering scaffolds fabricated by combining NaCl and PEG as co-porogens: Structure, property, and chondrocyte-scaffold interaction in vitro. *J Biomed Mater Res B* 100B: 956–966.

Liu, X.H., Ma, P.X. (2004) Polymeric scaffolds for bone tissue engineering. *Ann Biomed Eng* 32: 477–486.

Liu, X.H., Ma, P.X. (2009) Phase separation, pore structure, and properties of nanofibrous gelatin scaffolds. *Biomaterials* 30: 4094–4103.

Lloyd, A.W. (2002) Interfacial bioengineering to enhance surface biocompatibility. *Med Device Technol.* 13: 18–21.

Lo, H., Ponticiello, M.S., Leong, K.W. (1995) Fabrication of controlled release biodegradable foams by phase separation. *Tissue Eng* 1: 15–28.

Lovett, M., Cannizzaro, C., Daheron, L., Messmer, B., Vunjak-Novakovic, G., Kaplan, D.L. (2007) Silk fibroin microtubes for blood vessel engineering. *Biomaterials* 28: 5271–5279.

Lovett, M., Eng, G., Kluge, J.A., Cannizzaro, C., Vunjak-Novakovic, G., Kaplan, D.L. (2010) Tubular silk scaffolds for small diameter vascular grafts. *Organogenesis* 6: 217–224.

Lovett, M.L., Cannizzaro, C.M., Vunjak-Novakovic, G., Kaplan, D.L. (2008) Gel spinning of silk tubes for tissue engineering. *Biomaterials* 29: 4650–4657.

Luinstra, G.A., Borchardt, E. (2012) Material properties of poly(propylene carbonates). *Syn Biodegrad Polym* 245: 29–48.

Ma, P.X., Choi, J.W. (2001) Biodegradable polymer scaffolds with well-defined interconnected spherical pore network. *Tissue Eng* 7: 23–33.

Ma, Z.W., 2003. *Modification and Cytocompatibility of Poly-Lactic Acid (PLLA) Scaffold for Cartilage Tissue Engineering.* Zhejiang University, Hangzhou, China.

Ma, Z.W., Mao, Z.W., Gao, C.Y. (2007) Surface modification and property analysis of biomedical polymers used for tissue engineering. *Colloid Surf B* 60: 137–157.

Mandal, B.B., Kundu, S.C. (2009a) Cell proliferation and migration in silk fibroin 3D scaffolds. *Biomaterials* 30: 2956–2965.

Mandal, B.B., Kundu, S.C. (2009b) Osteogenic and adipogenic differentiation of rat bone marrow cells on non-mulberry and mulberry silk gland fibroin 3D scaffolds. *Biomaterials* 30: 5019–5030.

Mao, J.F., Duan, S., Song, A.N., Cai, Q., Deng, X.L., Yang, X.P. (2012) Macroporous and nanofibrous poly(lactide-co-glycolide)(50/50) scaffolds via phase separation combined with particle-leaching. *Mat Sci Eng C-Mater* 32: 1407–1414.

Marinucci, L., Balloni, S., Becchetti, E., Belcastro, S., Guerra, M., Calvitti, M., Lilli, C., Calvi, E.M., Locci, P. (2006) Effect of titanium surface roughness on human osteoblast proliferation and gene expression in vitro. *Int J Oral Max Impl* 21: 719–725.

Matsuda, T., Mizutani, M., Arnold, S.C. (2000) Molecular design of photocurable liquid biodegradable copolymers. 1. Synthesis and photocuring characteristics. *Macromolecules* 33: 795–800.

Matsuyama, H., Teramoto, M., Kuwana, M., Kitamura, Y. (2000) Formation of polypropylene particles via thermally induced phase separation. *Polymer* 41: 8673–8679.

Matuana, L.M., Diaz, C.A. (2010) Study of cell nucleation in microcellular poly(lactic acid) foamed with supercritical CO_2 through a continuous-extrusion process. *Ind Eng Chem Res* 49: 2186–2193.

Melchels, F.P.W., Bertoldi, K., Gabbrielli, R., Velders, A.H., Feijen, J., Grijpma, D.W. (2010a) Mathematically defined tissue engineering scaffold architectures prepared by stereolithography. *Biomaterials* 31: 6909–6916.

Melchels, F.P.W., Feijen, J., Grijpma, D.W. (2009) A poly(D,L-lactide) resin for the preparation of tissue engineering scaffolds by stereolithography. *Biomaterials* 30: 3801–3809.

Melchels, F.P.W., Feijen, J., Grijpma, D.W. (2010b) A review on stereolithography and its applications in biomedical engineering. *Biomaterials* 31: 6121–6130.

Meng, Z.X., Zheng, W., Li, L., Zheng, Y.F. (2010) Fabrication and characterization of three-dimensional nano-fiber membrane of PCL-MWCNTs by electrospinning. *Mater Sci Eng C-Mater* 30: 1014–1021.

Mi, H.Y., Jing, X., Peng, J., Turng, L.S., Peng, X.F. (2013a) Influence and prediction of processing parameters on the properties of microcellular injection molded thermoplastic polyurethane based on an orthogonal array test. *J Cell Plast* 49: 439–458.

Mi, H.Y., Jing, X., Salick, M.R., Cordie, T.M., Peng, X.F., Turng, L.S. (2014a) Morphology, mechanical properties, and mineralization of rigid thermoplastic polyurethane/hydroxyapatite scaffolds for bone tissue applications: Effects of fabrication approaches and hydroxyapatite size. *J Mater Sci* 49: 2324–2337.

Mi, H.Y., Jing, X., Salick, M.R., Peng, X.F., Turng, L.S. (2014b) A novel thermoplastic polyurethane scaffold fabrication method based on injection foaming with water and supercritical carbon dioxide as coblowing agents. *Polym Eng Sci.* 54: 2947–2957.

Mi, H.Y., Jing, X., Salick, M.R., Turng, L.S., Peng, X.F. (2014c) Fabrication of thermoplastic polyurethane (TPU) tissue engineering scaffold by combining microcellular injection molding and particle leaching. *J Mater Res* 29: 911–922.

Mi, H.Y., Jing, X., Turng, L.S. (2014d) Fabrication of porous synthetic polymer scaffolds for tissue engineering. *J Cell Plast* 51: 165–196.

Mi, H.Y., Salick, M.R., Jing, X., Jacques, B.R., Crone, W.C., Peng, X.F., Turng, L.S. (2013b) Characterization of thermoplastic polyurethane/polylactic acid (TPU/PLA) tissue engineering scaffolds fabricated by microcellular injection molding. *Mater Sci Eng C-Mater* 33: 4767–4776.

Mikos, A.G., Thorsen, A.J., Czerwonka, L.A., Bao, Y., Langer, R., Winslow, D.N., Vacanti, J.P. (1994) Preparation and characterization of poly(L-lactic acid) foams. *Polymer* 35: 1068–1077.

Mit-uppatham, C., Nithitanakul, M., Supaphol, P. (2004) Ultratine electrospun polyamide-6 fibers: Effect of solution conditions on morphology and average fiber diameter. *Macromol Chem Phys* 205: 2327–2338.

Molladavoodi, S., Gorbet, M., Medley, J., Kwon, H.J. (2013) Investigation of microstructure, mechanical properties and cellular viability of poly(L-lactic acid) tissue engineering scaffolds prepared by different thermally induced phase separation protocols. *J Mech Behav Biomed* 17: 186–197.

Mooney, D.J., Baldwin, D.F., Suh, N.P., Vacanti, L.P., Langer, R. (1996) Novel approach to fabricate porous sponges of poly(D,L-lactic-co-glycolic acid) without the use of organic solvents. *Biomaterials* 17: 1417–1422.

Murphy, W.L., Dennis, R.G., Kileny, J.L., Mooney, D.J. (2002) Salt fusion: An approach to improve pore interconnectivity within tissue engineering scaffolds. *Tissue Eng* 8: 43–52.

Nagiah, N., Sivagnanam, U.T., Mohan, R., Srinivasan, N.T., Sehgal, P.K. (2012) Development and characterization of electropsun poly(propylene carbonate) ultrathin fibers as tissue engineering scaffolds. *Adv Eng Mater* 14: B138–B148.

Nair, L.S., Laurencin, C.T. (2007) Biodegradable polymers as biomaterials. *Prog Polym Sci* 32: 762–798.

Odedra, D., Chiu, L., Reis, L., Rask, F., Chiang, K., Radisic, M. (2011) Cardiac Tissue Engineering. Burdick, J.A. and Mauck, R.L. (eds.). *Biomaterials for Tissue Engineering Applications: A Review of the Past and Future Trends*, 421–456.

Olivas-Armendariz, I., Garcia-Casillas, P., Martinez-Sanchez, R., Martinez-Villafane, A., Martinez-Perez, C.A. (2010) Chitosan/MWCNT composites prepared by thermal induced phase separation. *J Alloy Compd* 495: 592–595.

Owida, A., Chen, R., Patel, S., Morsi, Y., Mo, X.M. (2011) Artery vessel fabrication using the combined fused deposition modeling and electrospinning techniques. *Rapid Prototyp J* 17: 37–44.

Pant, H.R., Nam, K.T., Oh, H.J., Panthi, G., Kim, H.D., Kim, B.I., Kim, H.Y. (2011) Effect of polymer molecular weight on the fiber morphology of electrospun mats. *J Colloid Interf Sci* 364: 107–111.

Pfister, A., Landers, R., Laib, A., Hubner, U., Schmelzeisen, R., Mulhaupt, R. (2004) Biofunctional rapid prototyping for tissue-engineering applications: 3D bioplotting versus 3D printing. *J Polym Sci Pol Chem* 42: 624–638.

Plikk, P., Malberg, S., Albertsson, A.C. (2009) Design of resorbable porous tubular copolyester scaffolds for use in nerve regeneration. *Biomacromolecules* 10: 1259–1264.

Presteich, G.D., Atzet, S., 2013. *Biomaterials Science: An Introduction to Materials in Medicine*, 3rd ed. Elsevier, Waltham, MA.

Ramakrishna, S., Mayer, J., Wintermantel, E., Leong, K.W. (2001) Biomedical applications of polymer-composite materials: A review. *Compos Sci Technol* 61: 1189–1224.

Reignier, J., Huneault, M.A. (2006) Preparation of interconnected poly(epsilon-caprolactone) porous scaffolds by a combination of polymer and salt particulate leaching. *Polymer* 47: 4703–4717.

Ronca, A., Ambrosio, L., Grijpma, D.W. (2013) Preparation of designed poly(D,L-lactide)/nanosized hydroxyapatite composite structures by stereolithography. *Acta Biomater* 9: 5989–5996.

Salerno, A., Guarnieri, D., Iannone, M., Zeppetelli, S., Di Maio, E., Iannace, S., Netti, P.A. (2009) Engineered mu-bimodal poly(epsilon-caprolactone) porous scaffold for enhanced hMSC colonization and proliferation. *Acta Biomater* 5: 1082–1093.

Salerno, A., Iannace, S., Netti, P.A. (2012) Graded biomimetic osteochondral scaffold prepared via CO_2 foaming and micronized NaCl leaching. *Mater Lett* 82: 137–140.

Salerno, A., Zeppetelli, S., Di Maio, E., Iannace, S., Netti, P.A. (2011) Design of bimodal PCL and PCL-HA nanocomposite scaffolds by two step depressurization during solid-state supercritical CO_2 foaming. *Macromol Rapid Comm* 32: 1150–1156.

Salgado, A.J., Coutinho, O.P., Reis, R.L. (2004) Bone tissue engineering: State of the art and future trends. *Macromol Biosci* 4: 743–765.

Schugens, C., Maquet, V., Grandfils, C., Jerome, R., Teyssie, P. (1996a) Biodegradable and macroporous polylactide implants for cell transplantation. 1. Preparation of macroporous polylactide supports by solid-liquid phase separation. *Polymer* 37: 1027–1038.

Schugens, C., Maquet, V., Grandfils, C., Jerome, R., Teyssie, P. (1996b) Polylactide macroporous biodegradable implants for cell transplantation. 2. Preparation of polylactide foams by liquid-liquid phase separation. *J Biomed Mater Res* 30: 449–461.

Schuster, M., Turecek, C., Weigel, G., Saf, R., Stampfl, J., Varga, F., Liska, R. (2009) Gelatin-based photopolymers for bone replacement materials. *J Polym Sci Pol Chem* 47: 7078–7089.

Seck, T.M., Melchels, F.P.W., Feijen, J., Grijpma, D.W. (2010) Designed biodegradable hydrogel structures prepared by stereolithography using poly(ethylene glycol)/poly(D,L-lactide)-based resins. *J Control Release* 148: 34–41.

Serra, T., Planell, J.A., Navarro, M. (2013) High-resolution PLA-based composite scaffolds via 3-D printing technology. *Acta Biomater* 9: 5521–5530.

Shi, X.F., Hudson, J.L., Spicer, P.P., Tour, J.M., Krishnamoorti, R., Mikos, A.G. (2005) Rheological behaviour and mechanical characterization of injectable poly(propylene fumarate)/single-walled carbon nanotube composites for bone tissue engineering. *Nanotechnology* 16: S531–S538.

Shin, H., Jo, S., Mikos, A.G. (2003) Biomimetic materials for tissue engineering. *Biomaterials* 24: 4353–4364.

Shokrolahi, F., Mirzadeh, H., Yeganeh, H., Daliri, M. (2011) Fabrication of poly(urethane urea)-based scaffolds for bone tissue engineering by a combined strategy of using compression moulding and particulate leaching methods. *Iran Polym J* 20: 645–658.

Sin, D., Miao, X.G., Liu, G., Wei, F., Chadwick, G., Yan, C., Friis, T. (2010) Polyurethane (PU) scaffolds prepared by solvent casting/particulate leaching (SCPL) combined with centrifugation. *Mater Sci Eng C-Mater* 30: 78–85.

Sitharaman, B., Shi, X., Tran, L.A., Spicer, P.P., Rusakova, I., Wilson, L.J., Mikos, A.G. (2007) Injectable in situ cross-linkable nanocomposites of biodegradable polymers and carbon nanostructures for bone tissue engineering. *J Biomater Sci Polym Ed* 18: 655–671.

Sodian, R., Fu, P., Lueders, C., Szymanski, D., Fritsche, C., Gutberlet, M., Hoerstrup, S.P., Hausmann, H., Lueth, T., Hetzer, R. (2005) Tissue engineering of vascular conduits: Fabrication of custom-made scaffolds using rapid prototyping techniques. *Thorac Cardiov Surg* 53: 144–149.

Sporrer, A.N.J., Altstadt, V. (2007) Controlling morphology of injection molded structural foams by mold design and processing parameters. *J Cell Plast* 43: 313–330.

Stevens, M.M., George, J.H. (2005) Exploring and engineering the cell surface interface. *Science* 310: 1135–1138.

Sukigara, S., Gandhi, M., Ayutsede, J., Micklus, M., Ko, F. (2003) Regeneration of Bombyx mori silk by electrospinning—Part 1: Processing parameters and geometric properties. *Polymer* 44: 5721–5727.

Sun, W., Lal, P. (2002) Recent development on computer aided tissue engineering—A review. *Comput Meth Prog Bio* 67: 85–103.

Tchemtchoua, V.T., Atanasova, G., Aqil, A., Filee, P., Garbacki, N., Vanhooteghem, O., Deroanne, C. et al. (2011) Development of a chitosan nanofibrillar scaffold for skin repair and regeneration. *Biomacromolecules* 12: 3194–3204.

Teng, P.T., Chern, M.J., Shen, Y.K., Chiang, Y.C. (2013) Development of novel porous nasal scaffold using injection molding. *Polym Eng Sci* 53: 762–769.

Theron, J.P., Knoetze, J.H., Sanderson, R.D., Hunter, R., Mequanint, K., Franz, T., Zilla, P., Bezuidenhout, D. (2010) Modification, crosslinking and reactive electrospinning of a thermoplastic medical polyurethane for vascular graft applications. *Acta Biomater* 6: 2434–2447.

Thomson, R.C., Yaszemski, M.J., Powers, J.M., Mikos, A.G. (1995) Fabrication of biodegradable polymer scaffolds to engineer trabecular bone. *J Biomat Sci-Polym E* 7: 23–38.

Tsai, F.J., Torkelson, J.M. (1990) Microporous poly(methyl methacrylate) membranes—Effect of a low-viscosity solvent on the formation mechanism. *Macromolecules* 23: 4983–4989.

Ulery, B.D., Nair, L.S., Laurencin, C.T. (2011) Biomedical applications of biodegradable polymers. *J Polym Sci Pol Phys* 49: 832–864.

Vandeweerdt, P., Berghmans, H., Tervoort, Y. (1991) Temperature concentration behavior of solutions of polydisperse, atactic poly(methyl methacrylate) and its influence on the formation of amorphous, micro-porous membranes. *Macromolecules* 24: 3547–3552.

Vaquette, C., Frochot, C., Rahouadj, R., Wang, X. (2008) An innovative method to obtain porous PLLA scaf-folds with highly spherical and interconnected pores. *J Biomed Mater Res B* 86B: 9–17.

Verreck, G., Chun, I., Rosenblatt, J., Peeters, J., Van Dijck, A., Mensch, J., Noppe, M., Brewster, M.E. (2003) Incorporation of drugs in an amorphous state into electrospun nanofibers composed of a water-insoluble, nonbiodegradable polymer. *J Control Release* 92: 349–360.

Vunjak-Novakovic, G., Freed, L.E., Biron, R.J., Langer, R. (1996) Effects of mixing on the composition and morphology of tissue-engineered cartilage. *Aiche J* 42: 850–860.

Vunjak-Novakovic, G., Obradovic, B., Martin, I., Bursac, P.M., Langer, R., Freed, L.E. (1998) Dynamic cell seeding of polymer scaffolds for cartilage tissue engineering. *Biotechnol Prog* 14: 193–202.

Wan, Y., Wu, H., Cao, X.Y., Dalai, S. (2008) Compressive mechanical properties and biodegradability of porous poly(caprolactone)/chitosan scaffolds. *Polym Degrad Stabil* 93: 1736–1741.

Wang, F., Shor, L., Darling, A., Khalil, S., Sun, W., Guceri, S., Lau, A. (2004) Precision extruding deposition and characterization of cellular poly-epsilon-caprolactone tissue scaffolds. *Rapid Prototyping J* 10: 42–49.

Wang, H.N., Li, Y.B., Zuo, Y., Li, J.H., Ma, S.S., Cheng, L. (2007) Biocompatibility and osteogenesis of bio-mimetic nano-hydroxyapatite/polyamide composite scaffolds for bone tissue engineering. *Biomaterials* 28: 3338–3348.

Wang, X.F., Salick, M., Wang, X.D., Cordie, T., Han, W.J., Peng, Y.Y., Turng, L.S., Li, Q. (2013) Poly (epsilon-caprolactone) nanofibers with a self-induced nanohybrid shish-kebab structure mimicking collagen fibrils. *Biomacromolecules* 14: 3557–3569.

Wang, Y., Dai, J., Zhang, Q.C., Xiao, Y., Lang, M.D. (2010) Improved mechanical properties of hydroxyapa-tite/poly(epsilon-caprolactone) scaffolds by surface modification of hydroxyapatite. *Appl Surf Sci* 256: 6107–6112.

Wang, Y., Zhao, Z., Zhao, B., Qi, H.X., Peng, J., Zhang, L., Xu, W.J., Hu, P., Lu, S.B. (2011) Biocompatibility evaluation of electrospun aligned poly(propylene carbonate) nanofibrous scaffolds with peripheral nerve tissues and cells in vitro. *Chinese Med J-Peking* 124: 2361–2366.

Wannatong, L., Sirivat, A., Supaphol, P. (2004) Effects of solvents on electrospun polymeric fibers: Preliminary study on polystyrene. *Polym Int* 53: 1851–1859.

Wei, G.B., Ma, P.X. (2004) Structure and properties of nano-hydroxyapatite/polymer composite scaffolds for bone tissue engineering. *Biomaterials* 25: 4749–4757.

Wei, G.B., Ma, P.X. (2006) Macroporous and nanofibrous polymer scaffolds and polymer/bone-like apatite composite scaffolds generated by sugar spheres. *J Biomed Mater Res A* 78A: 306–315.

Wei, J., Chen, F.P., Shin, J.W., Hong, H., Dai, C.L., Su, J.C., Liu, C.S. (2009) Preparation and character-ization of bioactive mesoporous wollastonite—Polycaprolactone composite scaffold. *Biomaterials* 30: 1080–1088.

Wen, X.J., Tresco, P.A. (2006) Fabrication and characterization of permeable degradable poly(DL-lactide-co-glycolide) (PLGA) hollow fiber phase inversion membranes for use as nerve tract guidance channels. *Biomaterials* 27: 3800–3809.

Wendt, D., Marsano, A., Jakob, M., Heberer, M., Martin, I. (2003) Oscillating perfusion of cell suspensions through three-dimensional scaffolds enhances cell seeding efficiency and uniformity. *Biotechnol Bioeng* 84: 205–214.

Williams, D. (2003) Revisiting the definition of biocompatibility. *Med Device Technol* 14: 10.

Williams, D.F. (2008) On the mechanisms of biocompatibility. *Biomaterials* 29: 2941–2953.

Williams, J.M., Moore, J.E. (1987) Microcellular foams—phase-behavior of poly(4-methyl-1-pentene) in diisopropylbenzene. *Polymer* 28: 1950–1958.

Woodfield, T.B.F., Malda, J., de Wijn, J., Peters, F., Riesle, J., van Blitterswijk, C.A. (2004) Design of porous scaffolds for cartilage tissue engineering using a three-dimensional fiber-deposition technique. *Biomaterials* 25: 4149–4161.

Wu, F., Liu, C.S., O'Neill, B., Wei, J., Ngothai, Y. (2012a) Fabrication and properties of porous scaffold of magnesium phosphate/polycaprolactone biocomposite for bone tissue engineering. *Appl Surf Sci* 258: 7589–7595.

Wu, F., Wei, J., Liu, C.S., O'Neill, B., Ngothai, Y. (2012b) Fabrication and properties of porous scaffold of zein/PCL biocomposite for bone tissue engineering. *Compos Part B-Eng* 43: 2192–2197.

Wu, H.B., Haugen, H.J., Wintermantel, E. (2012c) Supercritical CO_2 in injection molding can produce open porous polyurethane scaffolds—A parameter study. *J Cell Plast* 48: 141–159.

Wu, L.B., Jing, D.Y., Ding, J.D. (2006) A "room-temperature" injection molding/particulate leaching approach for fabrication of biodegradable three-dimensional porous scaffolds. *Biomaterials* 27: 185–191.

Yang, F., Qu, X., Cui, W.J., Bei, J.Z., Yu, F.Y., Lu, S.B., Wang, S.G. (2006) Manufacturing and morphology structure of polylactide-type microtubules orientation-structured scaffolds. *Biomaterials* 27: 4923–4933.

Yang, S.F., Leong, K.F., Du, Z.H., Chua, C.K. (2001) The design of scaffolds for use in tissue engineering. Part 1. Traditional factors. *Tissue Eng* 7: 679–689.

Yeh, S.K., Liu, Y.C., Wu, W.Z., Chang, K.C., Guo, W.J., Wang, S.F. (2013) Thermoplastic polyurethane/clay nanocomposite foam made by batch foaming. *J Cell Plast* 49: 119–130.

Yeong, W.Y., Sudarmadji, N., Yu, H.Y., Chua, C.K., Leong, K.F., Venkatraman, S.S., Boey, Y.C.F., Tan, L.P. (2010) Porous polycaprolactone scaffold for cardiac tissue engineering fabricated by selective laser sintering. *Acta Biomater* 6: 2028–2034.

Yin, A.L., Zhang, K.H., McClure, M.J., Huang, C., Wu, J.L., Fang, J., Mo, X.M., Bowlin, G.L., Al-Deyab, S.S., El-Newehy, M. (2013) Electrospinning collagen/chitosan/poly(L-lactic acid-co-epsilon-caprolactone) to form a vascular graft: Mechanical and biological characterization. *J Biomed Mater Res A* 101A: 1292–1301.

Zein, I., Hutmacher, D.W., Tan, K.C., Teoh, S.H. (2002) Fused deposition modeling of novel scaffold architectures for tissue engineering applications. *Biomaterials* 23: 1169–1185.

Zeltinger, J., Sherwood, J.K., Graham, D.A., Mueller, R., Griffith, L.G. (2001) Effect of pore size and void fraction on cellular adhesion, proliferation, and matrix deposition. *Tissue Eng* 7: 557–572.

Zhang, J.C., Wu, L.B., Jing, D.Y., Ding, J.D. (2005) A comparative study of porous scaffolds with cubic and spherical macropores. *Polymer* 46: 4979–4985.

Zhang, Y.Z., Ouyang, H.W., Lim, C.T., Ramakrishna, S., Huang, Z.M. (2005) Electrospinning of gelatin fibers and gelatin/PCL composite fibrous scaffolds. *J Biomed Mater Res B* 72B: 156–165.

Zhao, J.H., Han, W.Q., Tu, M., Huan, S.W., Zeng, R., Wu, H., Cha, Z.G., Zhou, C.R. (2012) Preparation and properties of biomimetic porous nanofibrous poly(l-lactide) scaffold with chitosan nanofiber network by a dual thermally induced phase separation technique. *Mater Sci Eng C-Mater* 32: 1496–1502.

Zhao, M.Y., Li, L.H., Li, X., Zhou, C.R., Li, B. (2011a) Three-dimensional honeycomb-patterned chitosan/poly(L-lactic acid) scaffolds with improved mechanical and cell compatibility. *J Biomed Mater Res A* 98A: 434–441.

Zhao, W.A., Karp, J.M. (2009) Controlling cell fate in vivo. *Chembiochem* 10: 2308–2310.

Zhao, X.F., Li, X.D., Kang, Y.Q., Yuan, Q. (2011b) Improved biocompatibility of novel poly(L-lactic acid)/beta-tricalcium phosphate scaffolds prepared by an organic solvent-free method. *Int J Nanomed* 6: 1385–1390.

Zhao, Y.Y., Yang, Q.B., Lu, X.F., Wang, C., Wei, Y. (2005) Study on correlation of morphology of electrospun products of polyacrylamide with ultrahigh molecular weight. *J Polym Sci Pol Phys* 43: 2190–2195.

Zhou, W.Y., Duan, B., Wang, M., Cheung, W.L. (2009) Crystallization kinetics of poly(L-lactide)/carbonated hydroxyapatite nanocomposite microspheres. *J Appl Polym Sci* 113: 4100–4115.

Zhu, H.G., Ji, J., Lin, R.Y., Gao, C.Y., Feng, L.X., Shen, J.C. (2002a) Surface engineering of poly(DL-lactic acid) by entrapment of alginate-amino acid derivatives for promotion of chondrogenesis. *Biomaterials* 23: 3141–3148.

Zhu, Y.B., Gao, C.Y., He, T., Shen, J.C. (2004a) Endothelium regeneration on luminal surface of polyurethane vascular scaffold modified with diamine and covalently grafted with gelatin. *Biomaterials* 25: 423–430.

Zhu, Y.B., Gao, C.Y., Liu, X.Y., He, T., Shen, J.C. (2004b) Immobilization of biomacromolecules onto amino-lyzed poly(L-lactic acid) toward acceleration of endothelium regeneration. *Tissue Eng* 10: 53–61.

Zhu, Y.B., Gao, C.Y., Liu, X.Y., Shen, J.C. (2002b) Surface modification of polycaprolactone membrane via aminolysis and biomacromolecule immobilization for promoting cytocompatibility of human endothelial cells. *Biomacromolecules* 3: 1312–1319.

Zhu, Y.B., Gao, C.Y., Liu, Y.X., Shen, J.C. (2004c) Endothelial cell functions in vitro cultured on poly(L-lactic acid) membranes modified with different methods. *J Biomed Mater Res A* 69A: 436–443.

13 Composite and Hybrid Porous Structures for Regenerative Medicine

Vincenzo Guarino, Maria Grazia Raucci, and Luigi Ambrosio

CONTENTS

ABSTRACT: The success of scaffolds for clinical applications is strictly related to the validation of key features such as biocompatibility, degradability, mechanical integrity, and bioactivity. For these reasons, the material design to replace lost structures, or to improve existing structures by the promotion of new tissue formation, moves toward the choice of materials, which show a wide survey of physical and chemical properties able to satisfy the heterogeneous demand in tissue engineering and regenerative medicine. In this chapter, a review of main strategies used to fabricate polymer-based composite and hybrid scaffolds in regenerative medicine is reported. The state of the art of porous scaffold and process technique are illustrated in the first paragraph. Composite scaffolds are introduced in two subsections dealing with degradation properties of matrices and bioactivity of calcium phosphate fillers. Finally, the second section is referred to synthesis approaches based on the assembly hybrid materials as scaffolds, empathizing the use of solgel reaction for the fabrication of porous nanocomposites.

13.1 INTRODUCTION

The main challenge of the material scientists currently consist of the engineering of *bioinspired* scaffolds, which are able to ideally mimic the living tissue from a mechanical, but also, from the chemical, biological, and functional points of view. Starting from the example of the best materials scientist, which is the nature, it is strictly mandatory to provide the design of smart structural components that respond, *in situ*, to exterior stimulus adapting the microstructure and corresponding properties, taking into account the occurring phenomena due to specific mechanisms involved in the tissue formation—that is, vascularization and osteogenesis. In this context, the demand of materials compatibility with natural tissues historically leads to consider traditional materials classes such as metals and ceramics alone, less suitable than polymers for the applications in regenerative medicine and tissue repair. This is mainly due to the mismatch with properties of natural tissues like bone.

In this case, metals may be preferred for high strength, ductility, and their wear resistance but may offer some problems in terms of low biocompatibility, corrosion, too high stiffness compared to the natural tissue, high densities, and release of metal ions with possible allergic tissue reactions. Otherwise, unreinforced polymers are typically more ductile but not stiff enough to be used to replace hard tissues in load-bearing applications and transmit the mechanical stimuli to the native microenvironment. In this context, polymer-based composites represent a very convenient solution for tissue repair and regeneration, providing a wider set of options and possibilities in implant design with tailored chemical physical and biological features.

In this chapter, a wide range of recent studies concerning the design of polymer-based composites and hybrid scaffolds will be examined, emphasizing the crucial role of chemical/physical properties of constituent materials, that is, degradation, mechanical properties, and bioactivity—on the biological function of porous scaffolds at micro, submicro, or nanometric scale.

13.2 POROUS SCAFFOLDS: STATE OF THE ART

Scaffolds for tissue regeneration occupy a fundamental role in tissue regeneration processes, because they must support the proliferation and differentiation of cells as they mature into a functional tissue. It should act as a template for *in vitro* and *in vivo* bone growth in three-dimensional (3D) able to resorb at the same rate as new bone tissue is formed, to exhibit mechanical properties matching that of the host tissue and, meanwhile, leaving waste products, which do not alter cell adhesion and activity but promote a strong bond to the host tissue without scar tissue formation. The presence of an open porous architecture generally assures an efficient cell penetration, tissue ingrowth and eventually vascularization on implantation. Indeed, the scaffold-assisted regeneration of specific tissues has been shown to be strongly dependent on morphological parameters such as surface-to-volume ratio and pore size and interconnectivity. Indeed, these micro-architectural features not only significantly influence cell morphology, cell binding, and phenotypic expression but also control the extent and nature of nutrient diffusion and tissue ingrowth (Yang et al. 2001). Furthermore, it has been suggested that the pore dimensions may directly affect some biological events; as a result, different tissues require optimal pore sizes for their regeneration (Zeltinger et al. 2001). Therefore, scaffolds with bimodal micron scale (l-bimodal) porosities may often be necessary for the regeneration of highly structured biological tissues, such as bone and cartilage (Salerno et al. 2009). On the other hand, transport issues, 3D cell colonization, and tissue ingrowth would be inhibited if the pores are not well interconnected, even if the porosity of the scaffolds is high (Moore et al. 2004).

Recent works have underlined the differences between 3D structures with single macropore size and multiscale porosity (i.e., micro- and macro-porosity) to influence mechanical properties and biocompatibility after *in vitro* culture. A finely structured design of porosity, pore size, and interconnectivity at different dimensional levels is recommended to provide a morphologically friendly environment to cells. This may be reached either macroscopically, by controlling scaffold rod spacing, or microscopically, by varying rod porosity. Macroporosity—that is, pores over 50 μm—is thought to contribute to osteogenesis by facilitating cell and ion transport (Bignon et al. 2003). Likewise, other studies suggest that microporosity—that is, pores less than 20 μm—improves bone growth into scaffolds by increasing surface area for protein adsorption (Hing et al. 2005), increasing ionic solubility in the microenvironment (Le Nihouannen et al. 2005), and providing attachment points for cells (Bignon et al. 2003). Finally, pore interconnectivity is mandatory in order to drive bone deposition rate and depth of infiltration *in vitro* and *in vivo*. Regular, interconnected pores provide spacing for the vasculature required to nourish new bone and to remove waste products (Hing et al. 2002). The combination of pore size and interconnectivity required for optimal osteoconductivity has yet to be determined, making the ability to adjust these parameters an important capability for scaffold fabrication. In this context, several techniques have been used to fabricate 3D porous scaffolds, each characterized by its own advantages and limitations. Conventional methods allow construction of scaffolds, which are characterized by a continuous and uninterrupted pore structure. However, their use does not provide any long range channeling microarchitecture.

To date, a plethora of processing techniques are available in the literature for producing 3D scaffolds from various biodegradable polymers. These include fiber bonding (Mikos et al. 1993a), solvent casting, and particulate leaching/porogen leaching (Murphy et al. 2002), membrane lamination (Mikos et al. 1993), phase inversion/particulate leaching (Guarino et al. 1993b), melt molding (Hollister 2005), solvent (Athanasiou et al. 1995) or gel casting (Coombes and Heckman 1992), thermally induced phase separation—sublimation—lyophilization—emulsion freeze-drying (Ma and Choi 2001), gas foaming/high-pressure processing (Holy et al. 1999), co-continuous blends extrusion techniques, (Guarino et al. 2011) rapid prototyping techniques (Hutmacher et al. 2001), and stereophotolithographic (Grijpma et al. 2005) and inkjet techniques (Zhang et al. 2008). All these techniques are generally conceived to imprint the ordered porous arrays within the monocomponent polymer structure (Figure 13.1). In this context, an important challenge consists of implementing the current techniques of scaffold manufacturing for the use of novel polymer and composites materials, which convey specific molecular signals to cells, able to improve the features of biological microenvironment during *in vitro* experiments, and to promote the natural processes of tissue regeneration *in vivo*.

Besides, the design of advanced multifunctional scaffolds cannot prescind from the peculiar mechanical performances and mass transport properties (permeability and diffusion) of constituent materials, while reproducing the complex 3D anatomical shapes. Indeed, the performance of 3D scaffolds strongly depends not only on the different 3D architectural and geometric features but also on the intrinsic properties of the material, which concur to tailor mechanical and mass transport properties. Although both material chemistry and manufacturing technique strongly affect the functional properties of a scaffold, for example, cell-scaffold interaction, it is the sum of scaffold properties—that is, degradation and mass transport of the resultant porous structure (i.e., materials plus pores)—which provide to cell nutrition and migration. For instance, the addition of bioactive

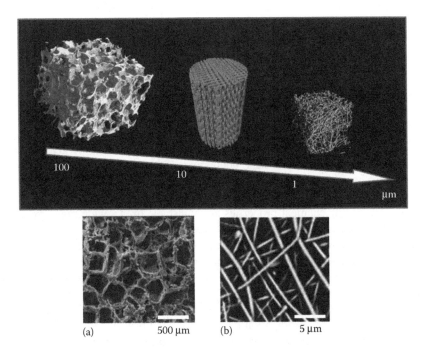

FIGURE 13.1 Characteristic size scale of scaffold processed by different technology: (a) salt leaching and (b) electrospinning.

phases such as calcium phosphates and bioglasses allows stimulating the biological response in osteogenic way, thus promoting a strong bonding to the surrounding tissue, mainly stimulating bone in growth.

13.3   COMPOSITE SCAFFOLDS

13.3.1   POLYMER MATRICES WITH CONTROLLED DEGRADATION

In the last twenty years, resorbable scaffold materials are used as a support matrix or as a substrate for the delivery of cultured cells or for 3D tissue reconstruction (Freed et al. 1994). Likewise, biodegradable polymers may be successfully used in controlled drug delivery strategies from polymer-based carriers (Langer and Tirrell 2004). Hence, it is always mandatory to identify basic criteria to match the mechanical properties and the degradation rate of constituent polymers in scaffold design.

Indeed, the increasing interest toward degradable and resorbable biomaterials mainly rises from their peculiar biological properties (Rezwan et al. 2006) and the opportunity to use them to fabricate multicomponent scaffolds with multiscale degradation kinetics as smart solution to design temporary devices for tissue repair and regeneration (Holy et al. 2003). Indeed, biodegradable polymers could be decomposed into biologically acceptable molecules (without the production of harmful intermediates), which can be metabolized and removed from the body via naturally pathway (metabolism or excretion) (Vert 2005). Despite many advantages offered by materials from natural sources, notably biological recognition, synthetic polymers offer greater advantages than natural ones, in that they can be tailored to give a wider range of properties. In particular, some drawbacks, in terms of mechanical properties, often require the use of synthetic polymers, which combine improved chemical stability to tailored degradation histories, assuring higher durability *in vivo*. The use of synthetic materials has been extensively exploited, for several reasons. First of all, they can be produced under controlled conditions to exhibit in general predictable and reproducible mechanical and physical properties such as tensile strength, elastic modulus, and degradation rate. Moreover, the immunogenic and purification issues relating to natural biomaterials are only partially overcome by recombinant protein technologies. Hence, synthetic materials satisfy this demand, thanks to their highly chemically programmable and reproducible properties. On the contrary, some limitations in terms of cell recognition impose the need to improve their physical and chemical performance, by modification or combination with natural source materials to generate their semisynthetic counterparts (Langer and Tirrell 2004). Possible risks such as toxicity, immunogenicity, and favoring of infections are lower for pure synthetic polymers with constituent monomeric units having a well-known and simple structure (Rezwan et al. 2006).

The most used biodegradable synthetic polymers for 3D porous scaffolds in tissue engineering are saturated polyhydroxyesters, including poly(lactic acid) (PLA) and poly(glycolic acid) (PGA), as well as poly(lactic-*co*-glycolide) (PLGA) copolymers (Seal et al. 2001). Due to their properties, PLA and PGA have been used in products and devices, as degradable; they have been approved by the U.S. Food and Drug Administration. PLA and PGA can be processed easily and their degradation rates, physical, and mechanical properties are adjustable over a wide range by using various molecular weights and copolymers. Indeed, chemical properties of these polymers allow hydrolytic degradation through de-esterification. Once degraded, the monomeric components of each polymer are removed by natural pathways. The body already contains highly regulated mechanisms for completely removing monomeric components of lactic and glycolic acids. However, these polymers undergo a bulk erosion process such that they can cause scaffolds to fail prematurely. In addition, abrupt release of these acidic degradation products can cause a strong inflammatory response (Martin et al. 1996). Polyester degradation occurs by uptake of water followed by the hydrolysis of esters. This mechanism is generally affected by several factors, including chemical composition, processing history, molecular weight, and polydispersity (Mw/Mn), environmental conditions, crystallinity and porosity, especially in the case of 3D foams (Heidemann et al. 2001). In the light of these considerations, aliphatic polyesters can therefore exhibit quite distinct degradation kinetics. PGA, for example, is a stronger acid and is more hydrophilic than PLA, which is partially hydrophobic due to its methyl groups. Hence, PLGA, a copolymer of PLA and PGA, may show intermediate degradation rates that can be modulated as a function of the relative fraction of hydrophobic/hydrophilic phases, crystallinity, and composition of chains (i.e., contents in L-lactic acid and D-lactic acid and/or GA units) (Andrew et al. 2001). Of particular significance for applications in tissue engineering are debris and crystalline by-products, as well as particularly acidic degradation products of aliphatic polyesters such as PLA, PGA, polycaprolactone (PCL) and their copolymers that have been implicated in adverse tissue reactions (Yang et al. 2001). This is the result of the heterogeneous degradation of these polymers, which occurs faster inside than at the exterior by the competition of the next two phenomena: the easier diffusion of soluble oligomers from the surface into the external medium than from inside and the neutralization of carboxylic end groups located at the surface by the external buffer solution (*in vitro* or *in vivo*). The combination of these events contributes to reduce the acidity at the surface, instead enhancing the degradation rate by autocatalysis due to carboxylic end groups in the bulk (Jagur-Grodzinski 1999). It is evident that the advance of hydrolysis reactions is strictly related to the ease of fluids to diffuse into the polymer chains, which is mainly determined by the relative fraction of amorphous/crystalline regions.

Several other factors may affect degradation kinetics, including (1) molecular factors, such as chain orientation, molecular weights (Mw), and polydispersity (Mw/Mn); (2) supramolecular factors, such as crystallinity induced by processing history, pore morphology, and distribution of chemically reactive filler compounds within the matrix (i.e., bioactive fillers or additives); and (3) environmental factors, including mechanical stress and strain, offering a wide range of resorbable properties for custom-made systems.

For these reasons, much effort has been recently addressed to the design a variety of composite materials for tissue engineering scaffolds with tailored degradation properties, by combining different biodegradable polymers. Indeed, it is extremely interesting the idea of controlling the structural properties evolution by the selective degradation of single materials phases, and so, to activate and strictly to control specific biological mechanisms on an appropriate time scale. Besides, degradation behavior of the main poly(α-hydroxyesters and their copolymers (Kohn 1996; Seal et al. 2001), mainly evolving by an erosion mechanism at the surface may release acidic products with adverse response. Although *in vitro* experiments have shown promising outcomes, *in vivo* tests have underlined some biological limitations, including acute inflammatory phenomena ascribable to degradation rates, which are too low (Santavirta et al. 1990).

In this context, recent studies provide to include fast degradable phases in stable polymer matrices—with long degradation times. For instance, Guarino et al. proposed the fabrication of bicomponent scaffolds made of poly(ε-caprolactone) (PCL) and esters of hyaluronic acids with different degradation properties in order to encourage *in vivo* response of scaffold for regenerative medicine (Chiari et al. 2006; Kon et al. 2008; Guarino et al. 2010). Indeed, PCL shows a remarkably slow degradation rate due to its high molecular weight above 50 kDa, requiring three years for its complete removal from the host body (Gabelnick 1983). Long degradation times, coupled with surface erosion mechanisms, offer some key advantages as scaffold material in comparison with other degradable materials by bulk mechanism. First, its capability to preserve the mechanical integrity over the degradation lifetime of the device allows adequate mechanical support during the post-implantation period. Second, a moderate toxicity due to the reduced release of acid products, a decrease in local acidity and a lower solubility and concentration of degradation products, assure an enhanced bone ingrowth into the porous scaffolds due to the increment in pore size as the erosion mechanism advances (Rezwan et al. 2006).

The blending of PCL with semisynthetic materials obtained by chemical modification of purified hyaluronan (HYA)—for example, HYAFF®, formed by the partial or total esterification of the carboxyl groups of glucuronic acid with different types of alcohols—may allow to modulate a wide range of material properties, which directly affect the cell activities. In particular, the extent of chemical esterification may considerably influence the biological properties, promoting the formation of a wide set of polymer-based composites for either favoring or, conversely, inhibiting the adhesion of certain cell types (Campoccia et al. 1998; Grigolo et al. 2001; Turner et al. 2004).

However, it is crucial to properly select biodegradable polymer phases into the scaffolds, which should degrade and resorb at a controlled rate at the same time as the bone tissue regenerates. Along this direction, recently, Ambrosio et al. (2001) have proposed an alternative composite tubular structure by the merging of a polyurethane matrix with continuous fibers of PLA and PGA helically wound by the filament winding technique. By applying the basic theory of continuous fiber-reinforced composites to the scaffold design, a composite material has been developed with an optimal spatial organization of fibers within the polymer matrix, able to mimic the structural organization and performance of the living tissue. More recently, Guarino et al. have proposed a porous system by the integration of resorbable PLA fibers by the filament winding technique, coupled with the conventional salt leaching technique (Guarino et al. 2008b). In these fibrous composites, degradation preferentially occurs at the fiber-polymer interface, resulting in a higher rate of degradation than for either material alone. Usually, the characteristic degradation rate of composites is too high, and therefore, not ideal for clinical applications such as bone fracture fixation, which require strength retention in the long term (i.e., a few weeks up to several months) (Figure 13.2).

In this case, the helicoidal organization of continuous fibers also assures a strong mechanical interlocking between fiber and matrix, thus minimizing the breakdown occurring at the interface. Moreover, the faster degradation of the more hydrophilic component (i.e., PLA fibers) respect to the

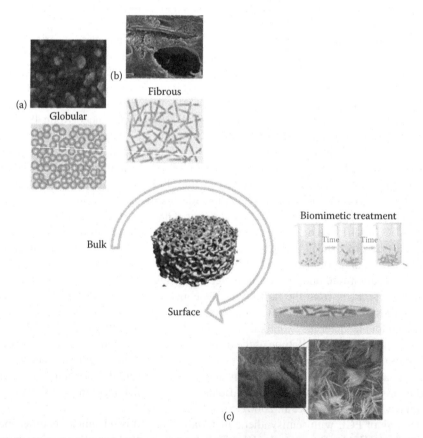

FIGURE 13.2 Bioactivation and reinforcement of scaffold via bulk and surface strategies: (a) Inclusion of HA particles, (b) continuous PLA fibers in PCL porous matrix, and (c) biomimetic treatment of PCL scaffold prepared via solgel technique.

matrix, assures the formation of preferential channels into the porous structure, which may more efficiently support cell colonization and tissue invasion.

13.3.2 CALCIUM PHOSPHATES AS BIOACTIVE FILLERS

The chemical similarity to the mineral component of mammalian bones and teeth has fueled the use of calcium phosphates as bone substitute materials. In fact, they can be employed with different shapes and functionalities within the clinical area. The reason for this popularity is that they are nontoxic, biocompatible, and exhibit a bioactive behavior, which leads to an intimate physicochemical bond between the implant and bone, that is, osteointegration. Additionally, calcium phosphates are also recognized to be osteoconductive, able to provide a scaffold or template for new bone formation, and support osteoblasts adhesion and proliferation (Anselme 2000). Several works dealing with calcium phosphate-based implants have also demonstrated their capability for bone regeneration purposes (Yuan et al. 2001, 2002, 2006). Understanding the solutions applied by nature to different issues, and how said solutions may be a source of inspiration to solve technological problems in the mineral field is of vital importance. Biomimetics can be considered as a technology transfer from nature to the mineral field. A suitable working approach consists of understanding how nature works and trying to copy its mechanisms. For this purpose, getting into the nano-world is required, since many of the biological species we have to handle present nanometric dimensions. Chemistry completely penetrates into the biotechnology and medicine fields. If we look at how nature solves the task of fabricating hard tissue, we will first find that biomineralization processes mainly use calcium and silicon combined with carbonates, phosphates, and oxides (Kuhn et al. 1996). Thus, bone is formed by biomineralization processes, natural sequences of physicochemical reactions that yield the formation of hard tissues in vertebrates or protective tissues in invertebrates, and inferior zoological species. As a result, natural composites are formed. In that way, materials with exceptional mechanical properties that are impossible to obtain with pure materials are reached. Bone apatites can be considered as a basic calcium phosphate. As indicated above, bones of vertebrate animals are organic–inorganic composite materials, whose structure can be described in short as follows: the inorganic component is carbonated and calcium deficient nonstoichiometric hydroxyapatite. These biological apatite crystals exhibit nanometric size, ranging from 25 to 50 nm (LeGeros 1991; Vallet-Regí et al. 2008).

Such crystals grow at the mineralization sites of the collagen molecules, which are grouped together forming collagen fibers. Furthermore, certain hierarchical bone porosity is necessary for several physiological functions performed by bone (Frieb and Werner 2000).

Bone changes its structure and shape continuously in response to its local loading environment within the body. This process is called *bone remodeling*, and is carried out by two types of cells: osteoblasts and osteoclasts. Osteoblasts are responsible for the synthesis of bone matrix and bone formation, while osteoclasts are able to degrade the mineralized matrix, that is, old bone and bone that is not required (bone resorption). Ideally, there should be a balance between the rates of bone gain and bone loss. However, disruption of this equilibrium occurs in metabolic bone diseases such as osteoporosis, which is characterized by increased bone remodeling related to a high bone resorption, leading to an increase in fracture risk. Additionally, when minor damage occurs in bone tissue, it can repair itself by the biochemical activity of osteoblasts. However, when the defect exceeds a critical size, which might be produced from trauma or from the removal of diseased tissue, bone is not able to repair itself. To solve this problem, graft implants and synthetic bone filler materials are currently being used. In a more modern approach, a tissue-engineered scaffold could guide and stimulate bone growth in the so-called approach of regenerative biomedicine. This scaffold should present properties either similar to trabecular bone or the capability of stimulating new bone growth and create a biocomposite with similar structure and properties to trabecular bone. However, one of the main limitations of calcium phosphates is their poor mechanical properties, especially their brittleness and poor fatigue resistance. This is even more evident for highly porous bioceramics and scaffolds, where porosity should be greater than 100 μm. Consequently, for biomedical applications, calcium phosphates are primarily used as fillers and coatings (Boccaccini and Maquet 2003; Webster et al. 2004). In particular, calcium phosphates may be successfully used with biodegradable polymers to combine the advantages of both types of materials into the scaffold (Hutmacher 2000).

Indeed, most synthetic matrices generally show hydrophobic surfaces that makes unfavorable basic cell-interaction mechanisms (i.e., adhesion and proliferation) than on hydrophilic surfaces (Tirrell et al. 2002; Vandiver et al. 2005; Goddard and Hotchkiss 2007; Chen et al. 2008). The inclusion of bioactive solid signals into the polymer matrix may allows supporting the creation of a strong bond with the living host bone at the scaffold/implant interface thanks to an improved wettability ascribable to the presence of the apatite particles (De Aza et al. 2003). As a consequence, synthetic polymer matrices made of biocompatible polyesters (i.e., PCL and polylactide acid) generally have demonstrated a tendency to be inert and to promote the formation of encapsulated fibrous tissues, thereby resulting in significant bone formation. Contrariwise, the addition of calcium phosphate particles to biodegradable porous matrices offers several improvements, which concur to promote the bone osteogenesis, as reported in the case of highly porous composite scaffolds made of PCL and stoichiometric HA particles developed through phase inversion and salt leaching techniques (Guarino et al. 2008b) (Figure 13.2). In this case, results highlighted that the presence of HA enhances the scaffold bioactivity and human osteoblast cell response, evidencing their role as *bioactive solid signals* in the promotion of surface mineralization and, consequently, on the cell-material interaction. In particular, the biological study performed on structures with a double scale of pore size and a fully interconnected porosity, characterized respectively by different PCL/HA volume ratio at the same processing conditions showed that stromal cells from bone marrow (bovine human mesenchimal cells) were able to adhere and grow on PCL-based scaffolds at any HA content, so remarking the their *nature* of precursor with high replicative potential. Indeed, even though cultured *in vitro* in static conditions, without additional stimulants (e.g., growth factors), MSC adhered during the first four weeks of culture, showing a cuboidal appearance on the polymer surface, which is a typical feature of mature osteoblasts. However, in some cases, the presence of HA into PCL scaffolds only slightly affects the biological response and the viability and MSC differentiation seems to be not directly related to the amount of HA in the matrix (Russias et al. 2006). Besides the osteoconductive enhancement, the relative amount of HA is relevant on affecting the intrinsic mechanical response of the composite scaffold and their degradation properties.

Several papers have demonstrated the active role of hydroxyapatite filler on the underlying *in vitro* degradation mechanisms by the simultaneous assessment of the influence of scaffold morphology and the physicochemical properties of the porous scaffolds. The addition of HA particles was found to slightly modify the pore morphology with a slight reduction of the average pore size. More interestingly, other studies on the scaffold mass losses evidenced that the presence of apatite phases embedded in the PCL matrix drastically increases the polymer crystallinity degree promoting the formation of more densely packed crystalline phases within the composite with a lower amount of amorphous regions that are potentially more susceptible to hydrolytic attacks due to a better accessibility of the ester linkage (Guarino et al. 2009). In this case, the increase in crystallinity of polymer matrix in HA-loaded scaffolds hinders the degradation of the composites, preferentially deflecting the fluids at the polymer/ceramic interface, which are more susceptible to hydrolytic attack. Indeed, crystal segments are chemically more stable than amorphous segments and generally reduce water permeation into the matrix in combination with ionic strength, temperature, and pH of the medium. In this context, the addition of other compounds, that is, calcium phosphates or bioactive glasses, may further concur to stabilize the environment conditions surrounding the polymer in order to control its degradation. Meanwhile, the use of rigid bone-like particles embedded into a polymer matrix evidently improve the mechanical properties of the polymer matrix, strengthening the use of composite scaffolds as a substrate for hard tissue replacement (Kikuchi et al. 1997; Khan et al. 2004). However, the contribution of mechanical response due to the ceramic phase may be partially hindered by the presence of macro- and microstructured pores, which even represent a basic requirement to induce the regeneration mechanisms in tissue engineering applications. For this reason, the further integration of biodegradable PLA fibers into the PCL matrix allows improving the mechanical response of the scaffolds, providing spaces required for cellular ingrowth and matrix production. The addition of bioactive apatite-like particles generating needle-like crystals of calcium-deficient hydroxyapatite similar to natural bone apatite also interact with the fiber-reinforced polymer matrix, further enhancing the mechanical response in compression by up to an order of magnitude (Guarino and Ambrosio 2008). Noteworthily, hydroxyapatite-loaded polymer matrices recently show an adverse success due to a lack of homogeneity distribution of ceramic particles in the polymeric matrix which dramatically compromise both the mechanical performance and the bioactive potential of the composite scaffolds (Guarino et al. 2007). The polymer matrix degradation, for example, causes a more frequent escape of HA particles in the time, with the creation of voids within the polymeric structure (Guarino et al. 2010). This evidently often affects the mechanical response of the scaffold, influencing the integrity of the scaffolds at longer times of the *in vitro* culture.

Although recent studies emphasized the crucial role of porosity features (i.e., pore size and shape) on cell-material interaction (Betz et al. 2010). It is recognized that pore architectures with tailored porosity need to be complemented by bone-inducing agents to stimulate the native bone-formation activity (Liu et al. 2010). Recently, it has been demonstrated that chemical modification of hydroxyapatite (HA) crystals, due to Mg^{2+} and CO_3^{2-} ion substitution into porous scaffolds with optimized morphological features (i.e., pore size, surface curvature, and interconnectivity) induce improvements in the biological response of bone-like cells, better triggering the osteogenic response under *in vitro* and *in vivo* conditions (Guarino et al.). Indeed, Mg^{2+}-doped HA is able to retain more water at its surface than stoichiometric HA, with water molecules coordinated to cations and adsorbed as a multilayered (Stayton 2003; Bertinetti et al. 2009). This favors cell anchorage and growth, arising from the ability of calcium phosphate to absorb several important extracellular matrix proteins. More recently, Guarino et al. (2012) demonstrated that ionic substitution also stimulates cells to produce mineral extracellular matter through activation of early osteogenesis mechanisms. Meanwhile, porous architecture with tailored porosity features (i.e., pore size and concavity) supports the *in vivo* growth of mature bone with hierarchical organization by imparting the natural osteon-like structure of bone—with lamellae centripetally assembled from the wall of the macropores toward the central bone marrow cavity—during the first six months of implantation.

Alternatively, chemically inspired approaches based on the solgel transition and colloidal precipitation of calcium phosphates may efficiently improve the particles dispersion by a direct control of precipitated grain sizes though the interaction between calcium and phosphate precursors under controlled temperature and pH conditions (Huang et al. 2000).

13.4 HYBRID SCAFFOLDS

Novel scaffold materials have been developed benefiting from the cell guidance concept that and, hence, from contemporary advances in materials science and biology (Nair and Laurencin 2006). In recent years, research on organic–inorganic hybrid materials has become an important subject of study for materials and biomaterials sciences. The concept of organic–inorganic hybrid materials appeared in the 1980s with the expansion of soft inorganic chemical processes. Organic–inorganic hybrid materials can be generally defined as materials with closely mixed organic and inorganic components (Livage et al. 1988; Sanchez and Gómez-Romero 2004). Furthermore, hybrids are either homogeneous systems derived from monomers and miscible organic and inorganic components or heterogeneous systems (nanocomposites), where at least one of the domains (inorganic or organic) has a dimension ranging from a few angstroms to a few tens of nanometers. It is worth mentioning that the properties of hybrid materials are not only the sum of the individual contributions of both phases but also a large synergy is expected from the intimate coexistence of the two phases through size domain effects and nature of the interfaces (Sanchez et al. 2005). Generally, the behavior of hybrid materials is dependent on the nature and relative content of the constitutive inorganic and organic components, with a close dependence on the experimental conditions. The clinical need for developing bioactive materials for bone regeneration, that is, those materials capable of forming a bioactive bond with natural bone, has motivated much research effort to develop bioactive hybrids. One of the main advantages of these materials aimed at bone implant technologies is that they have the unique feature of combining the properties of traditional materials such as ceramics and organic polymers on the nanometric scale (Novak 1993; Manzano et al. 2006). These strategies involve the fine and homogeneous dispersion of bioactive nano-sized particle in a polymer matrix. Also, because bone can be considered a hierarchical material, with its lowest structural level in the nanoscale range, materials with nanometer structures appear to be natural choices for creating better scaffolds for bone tissue engineering (Du et al. 1998). As reported in the literature, better osteoconductivity can be obtained by using synthetic composite materials, which are similar in the size and morphology to both of the inorganic particles and organic phase of bone (Boccaccini et al. 2002). Bone cell functions can be enhanced by interacting with nanophase ceramics and nanostructured polymers collectively rather than individually (Blaker et al. 2003; Mobasherpour et al. 2007). Furthermore, the induction of new bone formation seems also to be dependent on the presence of apatite crystals on the scaffold surface, which contributes to making the scaffold more bioactive and osteoconductive. Frequently, induction of the apatite crystals deposition on the scaffold surface was stimulated by using the basic Kokubo treatment, which consists of immersion in simulated body fluid (SBF) with a specific ion concentration, not less than that of the human blood plasma. *In vitro* studies have demonstrated that SBF can be used to reproduce *in vivo* apatite layer formation on the surface of various materials after implantation, so confirming its efficacy in improving surface osteoconductivity. Recently, Barrere et al. (2002) have proposed accelerating the classical biomimetic process, reducing its duration from 7 to 14 days to a few days by immersing substrates into supersaturated SBF. This illustrates the interdependence of ionic strength and pH, as well as the influence of the substrate chemistry on the formation of the resultant apatite.

The controlled deposition of HA crystals by accelerating SBF (5xSBF) allows formation of a biologically active surface, which can drastically improve the bonding to living bone (Lisignoli et al. 2001). The proposed integrated approach enables efficient exploitation of the inherent features of composite material bulk and surfaces by modulating the spatial distribution of bioactive signals, moving toward a more efficient replacement of bone tissue. Because the apatite particles are generated from aqueous solution, this technique can be used for various complex porous scaffolds, unlike many other surface modification methods that are limited to flat surfaces or to very thin porous layers.

Furthermore, this technique uses a biocompatible solution and very mild conditions (body temperature reaction), allowing a variety of polymers or composite materials to be used. Also, the hydroxyapatite generated is more similar to natural bone mineral in its low crystallinity and nanocrystal size, which is beneficial to its degradation and remodeling properties. In *in vitro* experiments, mineral crystals were observed to grow on composite materials after seven days in supersaturated biological solution (5xSBF). It was not surprising that higher concentration facilitated apatite formation, since it was reported (Solchaga et al. 1999) that increasing the ionic concentrations of a SBF upper the saturation limit is a practical way to facilitate apatite nucleation and growth.

Recent studies have reported that formation of apatite on artificial materials is induced by functional groups, which could reveal negative charge and further induce apatite via the formation of amorphous calcium phosphate. Apatite coating can be evaluated by scanning electron microscopy/ energy dispersive X-ray spectroscopy image of composite material after treatment in supersaturated solution 5xSBF. After seven days of soaking in 5xSBF, a calcium phosphate crystals layer with a thickness of few microns can be observed. It was also possible to elucidate the presence of HA deposition on the internal pore walls, confirming that the high concentration of SBF solution induced a fast and uniform deposition of hydroxyapatite on the scaffold. In this case, the copious formation of apatite-like globular crystals on the exposed surfaces of the composite scaffolds may be directly correlated to the powerful contribution of nanoscale HA particles (Solchaga et al. 1999).

Different structure of the scaffold, composition, porosity, pore size, and total surface available to cells favored the deposition of mineralized zone along the backbone, rather than among the cells. The cells were well retained inside the scaffold, but h-MSCs establish only few focal contacts with the scaffold, indicating that the mineralization processes were probably mainly induced by the released soluble factors rather than by cell-attachment properties. These data suggest possible *in vitro* trophic effects of h-MSCs (Djouad et al. 2009) in inducing mineralization processes, which can be also influenced by the composition or elasticity of the scaffold. In fact, HA not only enhances hydroxyapatite crystal proliferation and growth, but it can promote mineralization, since it is a prominent extracellular matrix component during the early stage of osteogenesis.

13.4.1 Solgel Approach to Design Bioactive Scaffolds

A solgel method enables the powder less processing of glasses, ceramics and thin films, or fibers, directly from solution. Precursors are mixed at the molecular level and variously shaped materials may be formed at much lower temperatures than it is possible by traditional methods of preparation. One of the major advantages of solgel processing is the possibility to synthesize hybrid organic–inorganic materials. Combination of inorganic and organic networks facilities the design of new engineering materials with exciting properties for a wide range of applications. The organic-inorganic hybrid materials may be prepared in various ways. The simplest one relies on dissolution of organic molecules in a liquid solgel (Catauro et al. 2003; Il-Seok Kim and Kumta 2004). The other way uses the impregnation of a porous gel in the organic solution. In the third type, the inorganic precursor either already has an organic group or reactions occur in a liquid solution to form chemical bonds in the hybrid gel. The solgel process itself leads to formation of gels from mixtures of liquid reagents (sols) at room temperatures. It involves several steps: the evolution of inorganic networks, formation of colloidal suspension (sol), and gelation of the sol to form a network in a continuous liquid phase (gel) (Figure 13.3).

During the *aging* step (after gelation and before drying) the solgel-derived material expulses the liquid phase (solvent which can be water or alcohol) in the process called *syneresis*. Drying of the obtained gels, even at room temperature, produces glass-like materials called *xerogels* (i.e., xeros). The process generates a porous material, where the pore size depends on factors such as time and temperature of the hydrolysis and the kind of catalyst used. The diameter of the pores is directly related to the shrinkage of the *wet* gels. During the drying process, the gel volume decreases even several times (which is the main reason of cracking). In particular, the solgel process for HA preparing usually can produce fine-grain microstructure containing a mixture of nano-to-submicron crystals. These crystals can be better accepted by the host tissue (Catauro et al. 2004). The solgel product is characterized by nano-size dimension of the primary particles. This small domain is a very important parameter for improvement of the contact reaction and the stability at the artificial/natural bone interface. Moreover, the high reactivity of the solgel powder allows a reduction of processing temperature and any degradation phenomena occurring during sintering (Linhart et al. 2001). The low temperature of process allows introduction of bioactive molecules (i.e., growth factors, peptides, dendrimer, and antibiotics) sensible to high temperature (Raucci et al. 2014). The major limitation of the solgel technique application is linked to the possible hydrolysis of phosphates and the high cost of the raw materials (Catauro et al. 2004). On the other hand, most of the solgel processes require a strict pH control, vigorous agitation, and a long time for hydrolysis. These problems were solved by using a nonalkoxide-based solgel approach, where the calcium and phosphate precursors are calcium nitrate tetrahydrate and phosphorus pentoxide, respectively (Linhart et al. 2001). More importantly, gel formation is achieved without the need for any refluxing steps. In this case, the P_2O_5 reacts with alcohol to form $P(O)(OR)_3$ oxyalkoxide with the liberation of water, which is in turn partially hydrolyzes the oxyalkoxide precursors. The presence of phosphorus hydroxyl alkoxide is not sufficient in itself to form a gel, indicating the important role of $Ca(NO_3)_2 \cdot 4H_2O$. It may be speculated that $Ca(NO_3)_2 \cdot 4H_2O$ probably results in the generation of alkoxy-nitrate salts,

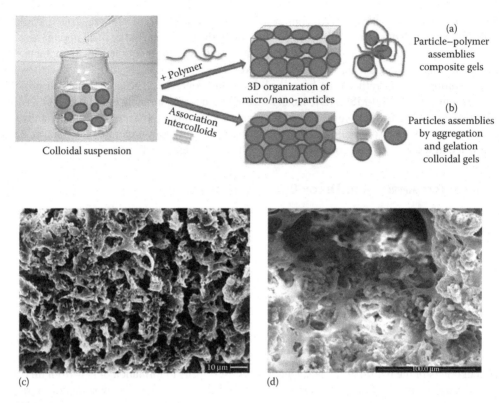

(a)
Particle–polymer
assemblies
composite gels

3D organization of
micro/nano-particles

+ Polymer

Association
intercolloids

Colloidal suspension

(b)
Particles assemblies
by aggregation
and gelation
colloidal gels

(c)

(d)

FIGURE 13.3 Scheme of solgel technology to prepare composite materials (a) and colloidal gels (b). Examples of calcium phosphate/polymer composite material prepared at room temperature by solgel technology (c) and cell-material interaction after 14 days in culture (d).

which participate in a polymerization reaction with the partially hydrolyzed phosphate precursors, the polymerization reaction thereby resulting in the gel. For the synthesis of the organic–inorganic composite material PCL/HA, the polymer may be added during the production of inorganic phase in order to allow the chemical interactions between the components (Raucci et al. 2010; Dessì et al. 2012). Molecular-level mixing of calcium and phosphorous precursors with the polymer chains derived from the solgel process resulted in composites having enhanced dispersion and exhibiting good interaction between the inorganic phase and the polymer matrix. Several studies have demonstrated by Fourier transform infrared spectroscopy and atomic force microscopy analyses the presence of hydrogen bond in the composite materials synthesized by solgel method (Martin et al. 1996; Webster et al. 2000).

The presence of HA particles in the composite material beneficially offsets the acidic release from the polymer through the alkaline calcium phosphate (Shi et al. 2009) and mitigates erosion problems associated with the release of acidic degradation products. *In vivo* and *in vitro* measurements of pH in bone chambers have shown that the pH drop is 0.2 units near the eroding polyesters (Liu and Webster 2007). Furthermore, by solgel process, it is possible to obtain a homogeneous distribution of nHA crystals, with 10–30 nm as diameter and 40–50 nm as length, which were detected into the composite. A homogeneous distribution of nanoscale hydroxyapatite particles in the polymeric matrix allows an increase in bioactive potential of materials. Many investigations of nanophase materials to date have illustrated their potential for bone repair. For example, increased osteoblast adhesion on nano-grained materials in comparison to conventional (micron-grained) materials has been reported (Kokubo 1998; Barrere et al. 2002). Osteoblast proliferation *in vitro* and long-term functions were also enhanced on ceramics with grain or fiber sizes <100 nm (Hench 1991).

In addition to osteoblast responses, modified osteoclast behavior has also been documented on nanophase ceramics and *in vivo* studies have demonstrated increased new bone formation on metals coated with nHA compared to conventional apatite.

13.5 CONCLUSION AND FUTURE TRENDS

This chapter reports recent scaffold-aided approaches to guide the process of tissue regeneration in regenerative medicine. Two main routes have been largely discussed to manufacture multifunctional scaffolds by the revitalization of traditional composite technologies or hybrid materials synthesis in the new biomaterial vision.

A future challenge lies in the reappraisal of the traditional concepts of composite and hybrid materials in the light of recent advances in scaffold technology. Successful implementation of bio-inspired strategies will be absolutely essential to the creation and manipulation of local microenvironments able to promote peculiar biological mechanisms, including tissue morphogenesis, differentiation, and maintenance. Current progresses in composite materials and scaffold design indicate a bright future for their use in tissue engineering and regenerative medicine. However, further significant advances are necessary in order to realize the full potential of new biomaterials, that is, hybrids for clinical use.

REFERENCES

Ambrosio, L., Netti, P.A., Santaniello, B., and L. Nicolais. 2001. *Biomedical Polymers and Polymer Therapeutics. Part 1.* New York: Plenum Publishers. 227–233.

Andrew, S.D., Phil, G.C., and K.G. Marra. 2001. The influence of polymer blend composition on the degradation of polymer/hydroxyapatite biomaterials. *J. Mater. Sci. Mater. Med.* 12: 673–677.

Anselme, K. 2000. Osteoblast adhesion on biomaterials. *Biomaterials.* 21(7): 667–681.

Athanasiou, K.A., Singhal, A.R., Agrawal, C.M., and B.D. Boyan. 1995. In vitro degradation and release characteristics of biodegradable implants containing trypsin inhibitor. *Clin. Orthop. Relat. Res.* 315: 272–281.

Barrere, F., Van, B.C., De, G.K., and P. Layrolle. 2002. Nucleation of biomimetic Ca-P coatings on ti6A14V from a SBF × 5 solution: influence of magnesium. *Biomaterials* 23: 2211–2220.

Bertinetti, L., Drouet, C., and C. Combes. 2009. Surface characteristics of nanocrystalline apatites: Effect of mg surface enrichment on morphology, surface hydration species, and cationic environments. *Langmuir.* 25(10): 5647–5654.

Betz, M.W., Yeatts, A.B., and W.J. Richbourgh. 2010. Macroporous hydrogels upregulate osteogenic signal expression and promote bone regeneration. *Biomacromolecules.* 11: 1160–1168.

Bignon, A., Chouteau, J., Chevalier, J., Fantozzi, G., Carret, J.P., and P. Chavassieux. 2003. Effect of micro- and macroporosity of bone substitutes on their mechanical properties and cellular response. *J. Mater. Sci. Mater. Med.* 14(12): 1089–1097.

Blaker, J.J., Gough, J.E., Maquet, V., Notingher, I., and A.R. Boccaccini. 2003. In vitro evaluation of novel bioactive composites based on Bioglass-filled polylactide foams for bone tissue engineering scaffolds. *J. Biomed. Mater. Res. A.* 67: 1401–1411.

Boccaccini, A.R. and V. Maquet. 2003. Bioresorbable and bioactive polymer/Bioglass® composites with tailored pore structure for tissue engineering applications. *Compos. Sci. Technol.* 63: 2417–2429.

Boccaccini, A.R., Roether, J.A., Hench, L.L., Maquet, V., and R. Jerome. 2002. A composite approach to tissue engineering. *Ceram. Eng. Sci. Proc.* 23: 805–816.

Campoccia, D., Doherty, P., Radice, M., Brun, P., Abatangelo, G., and D.F. Williams. 1998. Semisynthetic resorbable materials fron hyaluronan esterification. *Biomaterials* 19: 2101–2127.

Catauro, M., Raucci, M.G., de Gaetano, F., Buri, A., Marotta, A., and L. Ambrosio. 2004. Sol-gel synthesis, structure and bioactivity of polycaprolactone/CaO•SiO2. *J. Mater. Sci. Mat. Med.* 15: 991–995.

Catauro, M., Raucci, M.G., de Gaetano, F., and A. Marotta. 2003. *J. Mater. Sci.* 38: 3097–3102.

Chen, H., Yuan, L., Song, W., Wu, Z., and D. Li. 2008. Biocompatible polymer materials: Role of protein-surface interactions. *Pro. Polym. Sci.* 33: 1059–1087.

Chiari, C., Koller, U., Dorotka, R., Eder, C., Plasenzotti, R., Lang, S., Ambrosio, L. et al. 2006. A tissue engineering approach to meniscus regeneration in a sheep model. *Osteoarthritis Cartilage.* 14(10): 1056–1065.

Coombes, A.G.A. and J.D. Heckman. 1992. Gel casting of resorbable polymers. 1. Processing and applications. *Biomaterials.* 13(4): 217–224.

De Aza, P.N., Luklinska, Z.B., Santos, C., Guitian, F., and S. De Aza. 2003. Mechanism of bone-like formation on a bioactive implant in vivo. *Biomaterials*. 24: 1437–1445.

Dessì, M., Raucci, M.G., Zeppetelli, S., and L. Ambrosio. 2012. Design of injectable organic–inorganic hybrid for bone tissue repair. *J. Biomed. Mater. Res.* 100(8): 2063–2070.

Djouad, F., Bouffi, C., Ghannam, S., Noel, D., and C. Jorgensen. 2009. Mesenchymal stem cells: innovative therapeutic tools for rheumatic diseases. *Nat. Rev.* 5(7): 392–399.

Du, C., Cui, F.Z., Feng, Q.L., Zhu, X.D., and K. de Groot. 1998. Tissue response to nano-hydroxyapatite/collagen composite implants in marrow cavity. *J. Biomed. Mater. Res.* 42(4): 540–548.

Freed, L.E., Vunjak-Novakovic, G., Biron, R.J., Eagles, D.B., Lesnoy, D.C., Barlow, S.K., and R. Langer. 1994. Biodegradable polymer scaffolds for tissue engineering. *Bio/Technol.* 12: 689–695.

Frieß, W. and J. Werner. 2000. Biomedical applications. In: *Handbook of Porous Solids*. Weinheim, Germany: Wiley-VCH, 2923.

Gabelnick, H.L. 1983. Biodegradable implants: Alternative approaches. In: *Advanced in Human Fertility and Reproductive Endocrinology: Vol. 2: Long Acting Steroid Contraception*. Mishell DR, editor. New York: Raven Press, 149.

Goddard, J.M. and J.M. Hotchkiss. 2007. Polymer surface modification for the attachment of bioactive compounds. *Prog. Polym. Sci.* 32: 698–725.

Grigolo, B., Roseti, L., Fiorini, M., Fini, M., Giavaresi, G., Aldini, N.N., Giardino, R., and A. Facchini. 2001. Transplantation of chondrocytes seeded on a hyaluronan derivative (hyaff-11) into cartilage defects in rabbits. *Biomaterials* 22: 2417.

Grijpma, D.W., Hou, Q., and J. Feijen. 2005. Preparation of biodegradable networks by photo-crosslinking lactide, epsilon-caprolactone and trimethylene carbonate-based oligomers functionalized with fumaric acid monoethyl ester. *Biomaterials*. 26: 2795–2802.

Guarino, V. and L. Ambrosio. 2008. The synergic effect of polylactide fiber and calcium phosphate particles reinforcement in poly ε-caprolactone based composite scaffolds. *Acta Biomater. Cat.* 4: 1778–1787.

Guarino, V., Causa, F., and L. Ambrosio. 2007. Bioactive scaffolds for bone and ligament tissue. *Exp. Rev. Med. Dev.* 4(3): 405–418.

Guarino, V., Causa, F., Netti, P.A., Ciapetti, G., Pagani, S., Martini, D., Baldini, N., and L. Ambrosio. 2008a. The role of hydroxyapatite as solid signal on performance of PCL porous scaffolds for bone tissue regeneration. *J. Biom. Mat. Res. Appl. Biomat.* 86B: 548–557.

Guarino, V., Causa, F., Taddei, P., Di Foggia, M., Ciapetti, G., Martini, D., Fagnano, C., Baldini, N., and L. Ambrosio. 2008b. Polylactic acid fibre reinforced polycaprolactone scaffolds for bone tissue engineering. *Biomaterials* 29: 3662–3670.

Guarino, V., Guaccio, A., and L. Ambrosio. 2011. Manipulating co-continuous polymer blends to create PCL scaffolds with fully interconnected and anisotropic pore architecture. *J. Appl. Biomater. Biomech.* 9(1): 34–39.

Guarino, V., Lewandowska, M., Bil, M., Polak, B., and L. Ambrosio. 2010. *Comp. Sci. Tech.* 70: 1826–1837.

Guarino, V., Scaglione, S., Sandri, M., Alvarez-Perez, M.A., Tampieri, A., Quarto, R., and L. Ambrosio. 2012. MgCHA particles dispersion in porous PCL scaffolds: in vitro mineralization and in vivo bone formation. *J. Tissue Eng. Reg. Med.* doi:10.1002/term.1521.

Guarino, V., Taddei, P., Di Foggia, M., Fagnano, C., Ciapetti, G., and L. Ambrosio. 2009. The influence of hydroxyapatite particles on in vitro degradation behavior of poly epsilon-caprolactone-based composite scaffolds. *Tissue Eng. A.* 15: 3655–3668.

Heidemann, W., Jeschkeit, S., Ruffieux, K., Fischer, J.H., Wagner, M., Kruger, G., Wintermantel, E., and K.L. Gerlach. 2001. Degradation of poly (D,L) lactide implants with or without addition of calciumphosphates in vivo. *Biomaterials* 22(17): 2371–2381.

Hench, L.L. 1991. Bioceramics: From Concept to Clinic. *J. Am. Ceram. Soc.* 74: 1487–1510.

Hing, K.A., Annaz, B., Saeed, S., Revell, P.A., and T. Buckland. 2005. Microporosity enhances bioactivity of synthetic bone graft substitutes. *J. Mater. Sci. Mater. Med.* 16(5): 467–475.

Hing, K.A., Best, S.M., Tanner, K.E., Bonfield, W., and P.A. Revell. 2002. Mediation of bone ingrowth in porous hydroxyapatite bone graft substitutes. *J. Biomed. Mater. Res. A.* 68(1): 187–200.

Hollister, S.J. 2005. Porous scaffold design for tissue engineering. *Nat. Mater.* 4(7): 518–524.

Holy, C.E., Dang, S.M., Davies, J.E., and M.S. Shoichet. 1999. In vitro degradation of a novel poly(lactide-co-glycolide) 75/25 foam. *Biomaterials*. 20: 1177–1185.

Holy, E., Fialkov, J.A., Davies, J.E., and M.S. Shoichet. 2003. Use of a biomimetic strategy to engineer bone. *J. Biomed. Mater. Res. A.* 15; 447–453.

Huang, L.Y., Xu, K.W., and J. Lu. 2000. A study of the process and kinetics of electrochemical deposition and the hydrothermal synthesis of hydroxyapatite coatings. *J. Mater. Sci. Mater. Med. Cat.* 11: 667–673.

Hutmacher, D.W. 2000. Scaffolds in tissue engineering bone and cartilage. *Biomaterials*. 21: 2529–2543.

Hutmacher, D.W., Schantz, J.T., Zein, I., Ng, K.W., Teoh, S.H., and K.C. Tan. 2001. Scaffold design and fabrication technologies for engineering tissues—state of the art and future perspectives. *J. Biomed. Mater. Res. A.* 55: 203–216.

Il-Seok Kim, and P.N. Kumta. 2004. Sol-gel synthesis and characterization of nanostructured hydroxyapatite powder. *Mater. Sci. Eng. B.* 111: 232–236.

Jagur-Grodzinski, J. 1999. Biomedical application of functional polymers. *React. Funct. Polym.* 39: 99–138.

Khan, Y.M., Katti, D.S., and C.T. Laurencin. 2004. Novel polymer-synthesized ceramic composite-based system for bone repair: an in vitro evaluation. *J. Biomed. Mater. Res. A Cat.* 69(4): 728–737.

Kikuchi, M., Cho, S.B., Suetsugu, Y., and J. Tanaka. 1997. In vitro tests and in vivo tests developed TCP/CPLA composites. *Bioceramics.* 10: 407–410.

Kohn, J.R.L. 1996. Bioresorbable and bioerodible materials. In: *Biomaterials Science: An Introduction to Materials in Medicine.* Ratner BD, Hoffman AS, Schoen FJ, JE L, editors. New York: Academic Press, 64–72.

Kokubo, T. 1998. Apatite formation on surfaces of ceramics, metals and polymers in body environment. *Acta Mater.* 46: 2519–2527.

Kon, E., Chiari, C., Marcacci, M., Delcogliano, M., Salter, D.M., Martin, I., Ambrosio, L. et al. 2008. Tissue engineering for total meniscal substitution: animal study in sheep model. *Tissue Eng. Part. A.* 14(6): 1067–1080.

Kuhn, L.T., Fink, D.J., and A.H. Heuer. 1996. Biomimetic strategies and materials processing. In: *Biomimetic Materials Chemistry.* Mann S, editor. Wiley-VCH, 41.

Langer, R. and D.A. Tirrell. 2004. Designing materials for biology and medicine. *Nature.* 428(6982): 487–492.

LeGeros, R.Z. 1991. Calcium phosphates in oral biology and medicine. In: *Monographs in Oral Science.* Myers H, editor. Basel, Switzerland, Vol. 15, Karger.

Le Nihouannen, D., Daculsi, G., Saffarzadeh, A., Gauthier, O., Delplace, S., and P. Pilet. 2005. 36(6): 1086–1093.

Linhart, W., Peters, F., Lehmann, W., Schwarz, K., Schilling, A.F., Amling, M., Rueger, J.M., and M.J. Epple. 2001. Biologically and chemically optimized composites of carbonated apatite and polyglycolide as bone substitution materials. *J. Biomed. Mater. Res.* 54: 166–171.

Lisignoli, G. et al. 2001. Basic fibroblast growth factor enhances in vitro mineralization of rat bone marrow stromal cells grown on non-woven hyaluronic acid based polymer scaffold, *Biomaterials.* 22(15): 2095–2105.

Lisignoli, G., Zini, N., Remiddi, G., Piacentini, A., Puggioli, A., and C. Trimarchi. 2001. *Biomaterials.* 22: 2095–2105.

Liu, H. and T.J. Webster. 2007. Nanomedicine for implants: a review of studies and necessary experimental tools. *Biomaterials* 28(2): 354–369.

Liu, Y., Wu, G., and K. de Groot. 2010. Biomimetic coatings for bone tissue engineering of critical-sized defects. *J. R. Soc. Interf.* 7(Suppl 5): S631–47.

Livage, J., Henry, M., and C. Sanchez. 1988. Sol-gel chemistry of transition metal oxides. *Prog. Solid State Chem.* 18: 259–341.

Ma, P.X. and J.W. Choi. 2001. Biodegradable polymer scaffolds with well-defined interconnected spherical pore network. *Tissue Eng.* 7(1): 23–33.

Manzano, M., Arcos, D., Delgado, M.R., Ruiz-Hernández, E.F., Gil, J., and M. Vallet-Regi. 2006. Bioactive Star Gels. *Chem. Mater.* 18: 5696.

Martin, C., Winet, H., and J.Y. Bao. 1996. Acidity near eroding polylactide-polyglycolide in vitro and in vivo in rabbit tibial bone chamber. *Biomater. Cat.* 17(24): 2373–2380.

Mikos, A.G., Bao, Y., Cima, L.G., Ingber, D.E., Vacati, J.P., and R. Langer. 1993a. Preparation of poly(glycolic acid) bonded fiber structures for cell attachment and transplantation. *J. Biomed. Mater. Res.* 27: 183–189.

Mikos, A.G., Sarakinos, G., Leite, S.M., Vacanti, J.P., and R. Langer. 1993b. Laminated three-dimensional biodegradable foams for use in tissue engineering. *Biomaterials.* 14(5): 323–330.

Mobasherpour, I., Soulati, H.M., Kazemzadeh, A., and M. Zakeri. 2007. Synthesis of nanocrystalline hydroxyapatite by using precipitation method. *J. Alloys Comp.* 430: 330–333.

Moore, M.J., Jabbari, E., Ritman, E.L., Lu, L., Currier, B.L., and A.J. Windebank. 2004. Quantitative analysis of interconnectivity of porous biodegradable scaffolds with micro-computed tomography. *J. Biomed. Mater. Res.* 71A(2): 258–267.

Murphy, W.L., Dennis, R.G., Kileny, J.L., and D.J. Mooney. 2002. Salt fusion: An approach to improve pore interconnectivity within tissue engineering scaffolds. *Tissue Eng.* 8(1): 43–52.

Nair, L.S. and C.T. Laurencin. 2006. Polymers as biomaterials for tissue engineering and controlled drug delivery. *Adv. Biochem. Eng. Biotechnol.* 102: 47–90.

Novak, B.M. 1993. Hybrid Nanocomposite Materials—between inorganic glasses and organic polymers. *Adv. Mater.* 5: 422–433.

Raucci, M.G., Alvarez-Perez, M.A., Meikle, S., Ambrosio, L., and M. Santin. 2014. Poly(ε-lysine) dendrons tethered with phosphoserine increase mesenchymal stem cell differentiation potential of calcium phosphate gels. *Tissue Eng. Part A.* 20(3-4):474–485.

Raucci, M.G., Guarino, V., and L. Ambrosio. 2010. Hybrid composite scaffolds prepared by sol-gel method for bone regeneration. *Compos. Sci. Technol.* 70: 1861–1868.

Rezwan, K., Chen, Q.Z., Blaker, J.J., and A.R. Boccaccini. 2006. Biodegradable and bioactive porous polymer/inorganic composite scaffolds for bone tissue engineering. *Biomaterials.* 27(18): 3413–3431.

Russias, J., Saiz, E., Nalla, R.K., Gryn, K., Ritchie, R.O., and A.P. Tomsia. 2006. Fabrication and mechanical properties of PLA/HA composites. A study of in vitro degradation. *Mater. Sci. Eng. C.* 26: 1289–1295.

Salerno, A., Guarnieri, D., Iannone, M., Zeppetelli, S., Di Maio, E., Iannace, S., and P.A. Netti. 2009. Engineered μ-bimodal poly (ε-caprolactone) porous scaffold for enhanced hMSC colonization and proliferation. *Acta Bio.* 5(4): 1082–1093.

Sanchez, C. and P. Gómez-Romero. 2004. *Functional Hybrid Materials.* Wiley VCH, Weinheim, Germany.

Sanchez, C., Julián, B., Belleville, P., and M. Popall. 2005. Applications of hybrid organic–inorganic nanocomposites. *J. Mater. Chem.* 15: 3559–3592.

Santavirta, S., Konttinen, T.Y., Saito, T., Gronblad, M., Partio, E., Kemppinen, P., and P. Rokkanen. 1990. Immune response to polyglycolic acid implants. *J. Bone Jt. Surg. Br.* 72(4): 597–600.

Seal, B.L., Otero, T.C., and A. Panitch. 2001. Polymeric biomaterials for tissue and organ regeneration. *Mater. Sci. Eng.* 34: 147–230.

Shi, Z., Huang, X., Cai, Y., Tang, R., and D. Yang. 2009. Size effect of hydroxyapatite nanoparticles on proliferation and apoptosis of osteoblast-like cells. *Acta Biomater.* 5(1): 338–345.

Solchaga, L.A., Dennis, J.E., Goldberg, V.M., and A.I. Caplan. 1999. Hyaluronic acid-based polymers as cell carriers for tissue-engineered repair of bone and cartilage. *J. Orthop. Res.* 17(2): 205–213.

Stayton, P.S. 2003. Molecular recognition at the protein-hydroxyapatite interface. *Crit. Rev. Oral. Biol. Med.* 14(5): 370–376.

Tirrell, M., Kokkoli, E., and M. Biesalski. 2002. The role of surface science in bioengineered materials. *Surf. Sci.* 500: 61–83.

Turner, N.J., Kielty, C.M., Walker, M.G., and Canfield, A.E. 2004. A novel hyaluronan-based biomaterial (Hyaff-11) as a scaffold for endothelial cells in tissue engineered vascular grafts. *Biomaterials.* 25(28): 5955–5964.

Vallet-Regí, M. and D. Arcos-Navarrete. 2008. *Biomimetic Nanoceramics in Clinical Use: From Materials to Applications.* Royal Society of Chemistry, Cambridge.

Vandiver, J., Dean, D., Patel, N., Bonfield, W., and C. Ortiz. 2005. Nanoscale variation in surface charge of synthetic hydroxyapatite detected by chemically and spatially specific high-resolution force spectroscopy. *Biomaterials* 26(3): 271–283.

Vert, M. 2005. Aliphatic polyesters: Great degradable polymers that cannot do everything. *Biomacromolecules* 6: 538–546.

Webster, T.J., Ergun, C., Doremus, R.H., Siegel, R.W., and R. Bizios. 2000. Enhanced functions of osteoblasts on nanophase ceramics. *Biomaterials* 21(17): 1803–1810.

Webster, T.J., Waid, M.C., McKenzie, J.L., Price, R.L., and J.U. Ejiofor. 2004. Nano-biotechnology: Carbon nanofibres as improved neural and orthopaedic implants. *Nanotechnology* 15: 48.

Yang, S., Leong, K., Du, Z., and C. Chua. 2001. The design of scaffolds for use in tissue engineering. Part I. Traditional factors. *Tissue Eng.* 7(6): 679–689.

Yuan, H.P., van Blitterswijk, C.A., De Groot, K., and J.D. de Bruijn. 2006. Cross-species comparison of ectopic bone formation in biphasic calcium phosphate (BCP) and hydroxyapatite (HA) scaffolds. *Tissue Eng.* 12(6): 1607–1615.

Yuan, H.P., Van den Doel, M., Li, S.H., van Blitterswijk, C.A., de Groot, K., and J.D. De Bruijn. 2002. A comparison of the osteoinductive potential of two calcium phosphate ceramics implanted intramuscularly in goats. *J. Mater. Sci. Mater. Med.* 13(12): 1271–1275.

Yuan, H.P., Yang, Z.J., de Bruijn, J.D., de Grootm, K., and X.D. Zhang. 2001. Material-dependent bone induction by calcium phosphate ceramics: A 2.5-year study in dog. *Biomaterials* 22(19): 2617–2623.

Zhang, C., Wen, X., Vyavahare, N.R., and T. Boland. 2008. Synthesis and characterization of biodegradable elastomeric polyurethane scaffolds fabricated by the inkjet technique. *Biomaterials* 29(28): 3781–3791.

Zeltinger, J., Sherwood, J.K., Graham, D.A., Mueller, R., and L.G. Griffith. 2001. Effect of pore size and void fraction on cellular adhesion, proliferation, and matrix deposition. *Tissue Eng.* 7(5): 557–572.

14 Bionanocomposite Scaffolds for Tissue Engineering and Gene Therapy

Masami Okamoto, Reika Sakai, Shuichi Arakawa, and Ryoji Ishiki

CONTENTS

14.1 INTRODUCTION

Biopolymers are well-known examples for the renewable source-based environmental benign polymeric materials (Smith 2005). Biopolymers include polysaccharides such as cellulose, starch, alginate, and chitin/chitosan; carbohydrate polymers produced by bacteria and fungi (Chandra and Rustgi 1998); and animal protein-based biopolymers such as wool, silk, gelatin, and collagen. Naturally derived polymers have offered the interesting properties of biocompatibility and biodegradability. One of the advantages of the naturally derived polymers is the biological recognition that may positively support cell adhesion and function, but they have poor mechanical properties. Many of them are also limited in supply and can therefore be costly (Johnson et al. 2003).

On the other hand, polyvinyl alcohol, poly(ε-caprolactone) (PCL), poly(lactic acid) (PLA), poly(glycolic acid) (PGA), poly(hydroxy butyrate) (PHB), and poly(butylene succinate) (PBS) are examples of polymer of synthetic origin but are biodegradable (Platt 2006). In today's commercial environment, synthetic biopolymers have proven to be relatively expensive and available only in small quantities, which include applications in the textile and medical industries, as well as the packaging industry. This has lead to limitations on their application to date. However, synthetic biopolymers can be produced in large scale under controlled conditions and with predictable and reproducible mechanical properties, degradation rate, and microstructure (Sinha Ray and Okamoto 2003a; Platt 2006).

In this direction, one of the most promising synthetic biopolymers is PLA because it is made from agriculture products. PLA is not a new polymer; however, recent developments in the capability to manufacture the monomer economically from agriculture products have placed this material at the forefront of the emerging biodegradable plastics industries.

PLA, PGA, and their copolymers, poly(lactic acid-*co*-glycolic acid) (PLGA), are also extensively used in tissue engineering for treating patients suffering from damaged or lost organ or tissue (Langer and Vacanti 1993; Ma 2004). They have been demonstrated to be biocompatible and degrade into nontoxic components and have a long history of degradable surgical sutures with gained Food and Drug Administration (U.S. Food and Drug Administration) approval for clinical use. PCL and PHB are also used in tissue engineering research.

The task of tissue engineering demands a combination of molecular biology and materials engineering, since in many applications, a scaffold is needed to provide a temporary artificial matrix for cell seeding. In general, scaffolds must meet certain specifications such as high porosity, proper pore size, biocompatibility, biodegradability, and proper degradation rate (Quirk et al. 2004). The scaffold must provide sufficient mechanical support to maintain stresses and loadings generated during *in vitro* or *in vivo* regeneration.

For some of aforementioned applications, the enhancement of the mechanical properties is often needed (Tsivintzelis et al. 2007). This could be achieved by the incorporation of nanoparticles, such as hydroxyapatite (HA) carbon nanotubes (CNTs) and layered silicates. Polymer/layered silicate nanocomposites have recently become the focus of academic and industrial attention (Sinha Ray and Okamoto 2003b). Introduction of small quantities of high aspect ratio nano-sized silicate particles can significantly improve mechanical and physical properties of the polymer matrix (Lee et al. 2003). However, many agents such as the pore structure, porosity, crystallinity, and degradation rate may alter the mechanical properties and, thus, the efficiency of a scaffold. As a consequence, the scaffold fabrication method should allow for the control of its pore size and shape and should enhance the maintenance of its mechanical properties and biocompatibility (Ma 2004; Quirk et al. 2004).

During the last few years, many techniques have been applied for making porous scaffolds. Among the most popular are particulate leaching (Mikos et al. 1993), temperature-induced phase separation (Nam and Park 1999), phase inversion in the presence of a liquid nonsolvent (Van de Witte et al. 1996), emulsion freeze-drying (Whang et al. 1995), electrospinning (Bognitzki 2001), and rapid prototyping (Yavari et al. 2013). On the other hand, foaming of polymers using supercritical fluids is a versatile method in obtaining porous structure (Goel and Beckman 1994; Quirk et al. 2004).

This chapter intends to highlight the *synthetic* biopolymer-based nanocomposites (bionanocomposites) with the aim of producing porous scaffolds in tissue engineering applications during the last decade. The bionanocomposites was introduced several years ago to define an emerging class of biohybrid materials (Darder et al. 2011).

The chapter reviews current research trends on composite and nanocomposite materials for tissue engineering, including strategies for fabrication of the scaffolds with highly porous and interconnected pores. The results of the *in vitro* cell culture analyzed the cell-scaffold interaction using the colonization of stem cells, and degradation of the scaffolds *in vitro* is also discussed.

Another major challenge is a gene-transfer system via the scaffolds, where DNA, a large molecule of over one million Da having a high negative charge, is directed into cells and intracellular compartments for gene expression. This is usually accomplished by adding DNA/material complexes without viral vectors (Salem et al. 2003). Effective nonviral delivery remains a major obstacle to clinical gene therapy. Nonviral vehicles overcome the issues of toxicity, tumorigenicity, and low versatility of viral vectors. Here, we report about the nonviral gene therapy, which still struggle to achieve the high transfection efficiencies (Levine et al. 2013).

14.2 TISSUE ENGINEERING APPLICATIONS

Tissue engineering applies methods from materials engineering and life science to create artificial constructs for regeneration of new tissue (Langer and Vacanti 1993; Ma 2004). With tissue engineering, we can create biological substitutes to repair or replace the failing organs or tissues. One of the most promising approaches toward this direction is to grow cells on highly engineered scaffold structures that act as temporary support for cells, which facilitate the regeneration of the target tissues without losing the three-dimensional (3D) stable structure (Figure 14.1).

Polymeric scaffolds play a pivotal role in tissue engineering through cell seeding, proliferation, and new tissue formation in three dimensions. These scaffolds have shown great promise in the research of engineering a variety of tissues. Pore size, porosity, and surface area are widely recognized as important parameters for a tissue engineering scaffold. Other architectural features such as pore shape, pore wall morphology, and interconnectivity between pores of the scaffolding materials are also suggested to be important for cell seeding, migration, growth, mass transport, and tissue formation (Ma 2004).

The natural scaffolds made from collagen are fast replaced with ultraporous scaffolds from biodegradable polymers. Biodegradable polymers have been attractive candidates for scaffolding materials because they degrade, as the new tissues are formed eventually leaving nothing foreign to the body. The major challenges in the scaffold manufacture lies in the design and fabrication of customizable biodegradable constructs with desirable properties that promote cell adhesion and cell porosity along with sufficient mechanical properties that match the host tissue with predictable degradation rate and biocompatibility (Langer and Vacanti 1993; Mohanty et al. 2000; Ma 2004).

The biocompatibility of the materials is imperative. That is, the substrate materials neither elicit an inflammatory response nor demonstrate immunogenicity of cytotoxicity. Both the surface and the bulk of the scaffolds must be easily sterilizable to prevent infection (Gilding and Reed 1979). For the scaffolds particularly in bone tissue engineering, a typical porosity of 90% with a pore diameter of ca 100 μm is required for cell penetration and a proper vascularization of the ingrown tissue (Karageorgiou and Kaplan 2005).

Another major class of biomaterials for bone repair is bioactive ceramics such as HA and calcium phosphates (CPs) (Hench 1998; Kim et al. 2004). They showed appropriate osteoconductivity and biocompatibility because of their chemical and structural similarities to the mineral phase to native bone. However, the disadvantage is their inherent brittleness and poor shape ability. For this reason, polymer/bioactive ceramic composite scaffolds have been developed in application for the

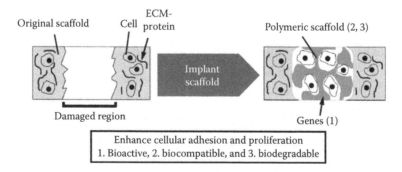

FIGURE 14.1 Schematic representation showing the concept of tissue engineering scaffold. Scaffold material (temporary synthetic extracellular matrix [EMC]) is designed as a three-dimensional mirror image, on which cells grow and regenerate the needed tissue. The scaffold can locally release growth factors (bioactive proteins, enzymes, and nonviral genes [DNAs and RNAs]) or antibiotics and enhance bone ingrowth to treat bone defects and even support wound healing. The composite systems combining advantages of polymers and ceramics seem to be a promising choice for bone tissue engineering.

bone tissue engineering. They exhibit good bioactivity, manipulation, and control microstructure in shaping to fit bone defects (Zhang and Ma 1999).

14.3   BIONANOCOMPOSITES FOR TISSUE ENGINEERING

The extensive literature on bionanocomposites is considered to be a stepping stone toward a greener and sustainable environment. Bionanocomposites are described in detailed studies and reviews elsewhere (Sinha Ray and Bousmina 2005; Okamoto 2006; Bordes et al. 2009).

The most often utilized synthetic biopolymers for 3D scaffolds in tissue engineering are saturated poly(α-hydroxy esters) including PLA, racemic mixture of D,L-PLA (PDLLA) and PGA, as well as PLGA (Kohn 1996; Jagur-Grodzinski 1999; Seal et al. 2001). These polymers degrade through hydrolysis of the ester bonds. Once degraded, the monomeric components of each polymer are removed by natural pathways. The body already contains highly regulated mechanisms for completely removing monomeric components of lactic and glycolic acids. PGA is converted to metabolites or eliminated by other mechanisms. On the other hand, PLA is cleared via the tricarboxylic acid cycle. Due to these properties, PLA and PGA have been used in products and devices, degradable sutures, which have been approved by the U.S. FDA. The cost of PLA for medical use is about 3 \$/g, which is three order of magnitude higher than conventional injection moldable PLA (3 \$/kg) (PLA degradable 2012). PLA and PGA can be processed easily and their degradation rates, physical and mechanical properties are adjustable over a wide range by using various molecular weights and copolymer composition. These polymers undergo a bulk erosion process such that they can cause scaffolds to fail prematurely. The abrupt release of their acidic degradation products can cause a strong inflammatory response (Martin et al. 1996). Table 14.1 shows an overview of the discussed biopolymers and their physical properties. Their degradation rates decreases in the following order:

$$PGA > PDLLA > PLLA > PCL \sim PPF$$

PGA degrades rapidly in aqueous solution or *in vivo*, and loses mechanical integrity between two and four weeks. The extra methyl group in the PLA repeating unit makes it more hydrophobic, reduces the molecular affinity to water, and leads to a slow degradation. Therefore, the scaffolds made from PLA degrade slowly both *in vitro* and *in vivo*, maintaining mechanical properties until several months (Martin et al. 1996).

PCL degrades at a significantly slower rate than PLA, PGA, and PLGA. The slow degradation makes PCL less attractive for biomedical applications, but more attractive for long-term implants

TABLE 14.1
Physical Properties of Synthetic Biopolymers Used as Scaffold Materials

Biopolymers	Thermal Properties		Tensile Modulus (GPa)	Biodegradation Time (Months)
	T_m (°C)[a]	T_g (°C)[b]		
PLLA	173–178	60–65	1.2–3.0	>24
PDLLA	–	55–60	1.9–2.4	12–16
PGA	225–230	35–40	0.5–0.7	3–4
PLGA (50/50)	–	50–55	1.4–2.8	Adjustable: 3–6
PCL	58–63	–60	0.4–0.6	>24
PPF	30–50	–60	2–3	>24

Source: Rezwan, K. et al., *Biomaterials*, 27, 3413–3431, 2006.

[a] Melting temperature.

[b] Glass transition temperature.

and controlled release application (Pitt et al. 1981; Ma 2004). Recently, PCL has been used as a candidate polymer for bone tissue engineering, where scaffolds should maintain physical and mechanical properties for at least six months (Pektok et al. 2008).

Poly(propylene fumarate) (PPF) have been developed for orthopedic and dental applications (Shi and Mikos 2006). The degradation products are biocompatible and readily removed from the body. The double bond of PPF main chains leads to *in situ* cross-linking, which causes a moldable composite. The preservation of the double bond and molecular weight are key issues to control the final mechanical properties and degradation time.

However, PPF and other biodegradable polymers lack the mechanical strength required for tissue engineering of load-bearing bones (Mistry and Mikos 2005). The development of composite and nanocomposite materials combining with inorganic particles, for example, apatite component (i.e., the main constituent of the inorganic phase of bone [Ma 2004]), bioactive glasses, carbon nanostructures (e.g., nanotubes, nanofibers, and graphene), and metal nanoparticles, has been investigated.

14.3.1 CALCIUM PHOSPHATES

CPs are bioactive materials that have wide potential for bone tissue repair and augmentation. They have surface properties that support osteoblast adhesion/prolification (osteoconduction) and stimulate new bone formation (osteoinduction). However, not all types of CPs have the same biological effect *in vivo* (Ghosh et al. 2013). Most are osteoconductive, only certain types are osteoinduction (LeGeros 2008). Four types of CPs are commonly used in bone tissue engineering, namely HA, tricalcium phosphate (TCP), amorphous CPs (ACPs), and biphasic CPs. All CPs show cell differentiation in the presence of osteogenic supplements (such as dexamethasone, ascorbic acid and β-glycerolphosphate), a range of osteoinductive potentials exhibits TCP > BCP ~ HA > ACP. In the absence of osteogenic supplements, their potential decreases in the following order: BCP > TCP > HA (Samavedi et al. 2013) (Table 14.2).

14.3.2 HA-BASE BIONANOCOMPOSITES

HA promotes bone ingrowth and biocompatibility because around 65 wt% of bone is made of HA, $Ca_{10}(PO_4)_6(OH)_2$. Natural or synthetic HA has been intensively investigated as the major component of scaffold materials for bone tissue engineering (Knowles 2003). The Ca/P ratio of 1.50–1.67 is the key issue to promote bone regeneration. Recently, much better osteoconductive properties in HA by changing composition, size, and morphology has been reported (Gay et al. 2009). The nano-sized HA (nHA) may have other special properties due to its small size and huge specific surface area. The significant increase in protein adsorption and osteoblast adhesion on the nano-sized ceramic materials was reported by Webster et al. (2000).

TABLE 14.2

Properties of CPs That Influence Osteoblastic Differentiation

CPs	Solubility	K_{sp}	Ca/P Ratio	Osteoinductivity
HA	Poor	10^{-58}	1.67	+
TCP	Fair	$10^{-25}-10^{-28}$	1.5	++
ACP	High	$10^{-23}-10^{-25}$	1.15–1.67	+++
BCP	Adjustable: TCP/HA ratio		1.4–2.8	Adjustable: ++++

Source: Samavedi, S., et al., *Acta Biomater.*, 9, 8037–8045, 2013.

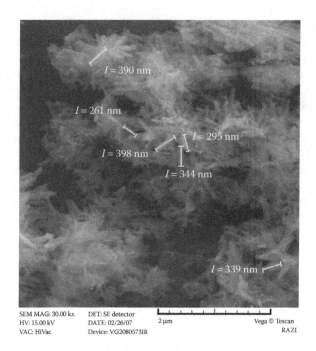

SEM MAG: 30.00 kx DET: SE detector
HV: 15.00 kV   DATE: 02/26/07   2 μm   Vega © Tescan
VAC: HiVac   Device: VG20805731R   RAZI

FIGURE 14.2   Scanning electron micrographs of the synthetic nHA rods. (Reprinted from *Compos. Part A* 39, Nejati, E. et al., Synthesis and characterization of nanohydroxyapatite rods/ poly(L-lactide acid) composite scaffolds for bone tissue engineering, 1589–1596, Copyright 2008, with permission from Elsevier.)

Figure 14.2 shows rod-shaped morphology of the nHA with particle width ranged from 37 to 65 nm and the length from 100 to 400 nm (Nejati et al. 2008). The compressive strength of bioceramics increases when their grain size reduced to nanolevel (El-Ghannam et al. 2004).

Bionanocomposites based on HA particles and biopolymers have attracted much attention for their good osteoconductivity, osteoinductivity, biodegradability, and high mechanical strengths. The Ma group mimicked the size scale of HA in natural bone and showed that the incorporation of nHA improved the mechanical properties and protein adsorption of the composite scaffolds, while maintaining high porosity and suitable microarchitecture (El-Ghannam et al. 2004).

Nejati et al. (2008) reported that the effect of the synthesis nHA on the scaffolds morphology and mechanical properties in poly(L-lactic acid) (PLLA)-based nanocomposites.

The morphology and microstructure of the scaffolds were examined using scanning electron microscope (SEM) (Figure 14.3). The nanocomposite scaffold (Figure 14.3c,d) maintains a regular internal ladder-like pore structure similar to neat PLLA scaffold (Figure 14.3a,b), with a typical morphology processed by thermally induced phase separation (Nam and Park 1999). Rod-like nHA particles are distributed within the pore walls and no aggregation appears in pores (Figure 14.3e,f). However, the bionanocomposite exhibits little effect of nHA on the development of the pore morphology as compared with that of neat PLLA. Pore size of neat PLLA and bionanocomposite scaffolds are in the range of 175–97 μm, respectively. In the case of microcomposite scaffold (Figure 14.3g,h), micro-HA (mHA) particles are randomly distributed in the PLLA matrix. Some are embedded in the pore wall and some are piled together between or within the pores. Among composite and neat PLLA scaffolds, bionanocomposites scaffolds showed highest compressive strength (8.67 MPa) with 85.1% porosity, which is comparable to the high end of compressive strength of cancellous bone (2–10 MPa) (Ramay Hassna and Zhang 2004).

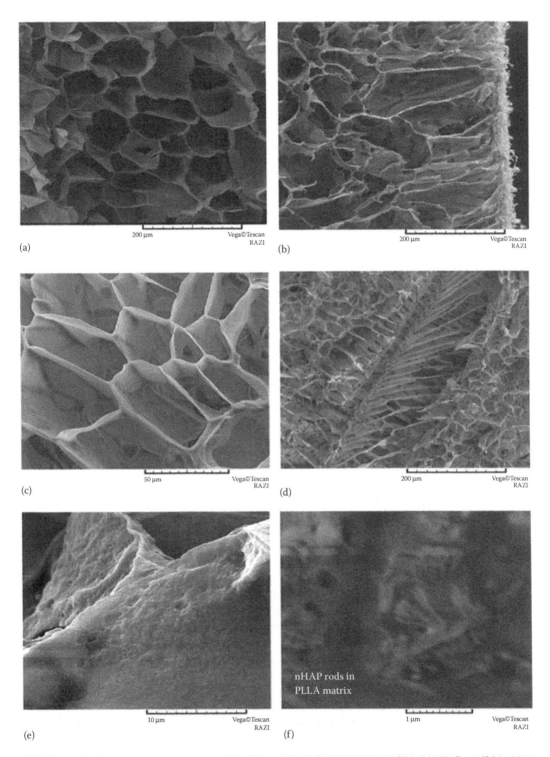

FIGURE 14.3 Scanning electron micrographs of pure PLLA, PLLA/nHA and PLLA/mHAP scaffolds. Neat PLLA and cross section (a, b); PLLA/nHA: 50/50 and cross section (c, d); PLLA/nHAP: 50/50 (e, f). (*Continued*)

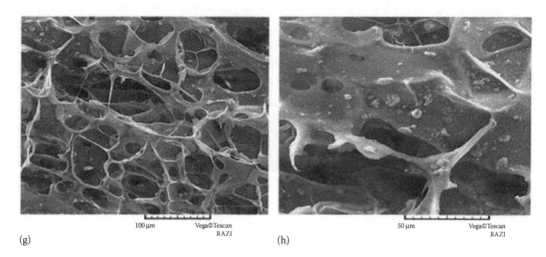

(g)   (h)

FIGURE 14.3 (Continued) Scanning electron micrographs of pure PLLA, PLLA/nHA and PLLA/mHAP scaffolds. PLLA/mHAP: 50/50 scaffold (g, h). (Reprinted from *Compos. Part A* 39, Nejati, E. et al., Synthesis and characterization of nanohydroxyapatite rods/ poly(L-lactide acid) composite scaffolds for bone tissue engineering, 1589–1596, Copyright 2008, with permission from Elsevier.)

PCL/nHA bionanocomposites were prepared and they combine the osteoconductivity and biocompatibility shown by HA ceramic with PCL properties (Wei and Ma 2004; Hong et al. 2005; Rezwan et al. 2006; Bianco et al. 2009). The structural characterization of a novel electrospun nanocomposite and the analysis of cell response by a highly sensitive cell, embryonic stem cell for PCL/Ca-deficient nHA system were investigated (Bianco et al. 2009). For higher Ca-deficient nHA contents (~55 wt%), the mechanical properties significantly decreased; onset decomposition temperature and crystallinity considerably decreased as well.

Due to the brittleness of the HA and the lack of interaction with polymer matrix, the ceramic nanoparticles may present deleterious effects on the mechanical properties, when loaded at high amount. Coupling agents are generally used to overcome the lack of interaction with polymer and nHA aggregation (Hong et al. 2004, 2005; Li et al. 2008). In order to increase the interfacial strength between PLLA and HA, and hence to increase the mechanical properties, the nHA particles were surface-grafted (*g*-HA) with the polymer and further blended with PLLA (Li et al. 2008). The PLLA/g-HA bionanocomposites also demonstrated improved cell compatibility due to the good biocompatibility of the nHA particles and more uniform distribution of the g-HA nanoparticles on the film surface (Hong et al. 2004; Li et al. 2008). These bionanocomposites is of great interest to the biomedical community because the materials have a suitable structure that induces and promotes new bone formation at the required site.

Very recently, Fu et al. (2012) reported the biocompatibility and osteogenesis of electrospun PCL-poly(ethylene glycol)(PEG)-PCL triblock copolymer with nHA of 30 wt%.

In vivo biocompatibility and degradability were investigated by implanting the scaffold into the muscle pouches; the results revealed that the scaffold had good biocompatibility with surrounding tissue, and the fibrous and vascular tissue could grow into the scaffold following with the degradation continued in New Zealand white rabbits (Figure 14.4). At week 20, there was an obvious difference in new bone formation between two groups. In the control (Figure 14.4c), although the newly formed bony tissue increased obviously compared to that of 12 weeks, yet the new bone was mainly composed of cancelous bone, and the bone formation was insufficient to fill the whole defect although a large amount of bone marrow and some vascular tissue could be seen. However, in the treatment group (Figure 14.4f), the defect was completely filled with the newly cortical bone, and the spongy bone hardly could be observed in the defect compared to the Figure 14.4e. It suggested that the spongy bone has been replaced by new cortical bone.

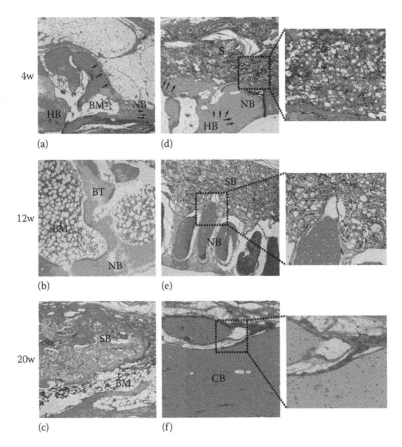

FIGURE 14.4 H & E staining of the cranial bone defect sections in the control group (a–c) and treatment group (d–f) at 4, 12, and 20 weeks postsurgery. The treatment group did not exhibit serious inflammatory response and foreign body reaction. Two groups allowed bone ingrowth, but the treatment group showed faster and more effective osteogenesis at the defect area than that of the control. Abbreviations and signs used: bone marrow (BM), bone trabecula (BT), cortical bone (CB), host bone (HB), new bone (NB), scaffold (S), and spongy bone (SB); the arrows indicate the new bone formed at the margin of host bone. The original magnification of photos (a–d) is 100×. The area shown in higher magnification (400x) is indicated. (Reprinted from *Biomaterials*, 33, Fu, S. et al., In vivo biocompatibility and osteogenesis of electrospun poly(e-caprolactone)-poly(ethylene glycol)-poly(e-caprolactone)/ nano-hydroxyapatite composite scaffold, 8363–8371, Copyright 2012, with permission from Elsevier.)

14.3.3 METAL NANOPARTICLES-BASE BIONANOCOMPOSITES

Currently, metal nanoparticle-based bionanocomposites are used in various biomedical applications such as probes for electron microscopy to visualize cellular structure, drug delivery, diagnosis, and therapy. The unique physical characteristics, gold nanoparticle-based bionanocomposites are used in the optical and photonic fields (Gansel et al. 2009), not medical applications.

Silver (Ag) has been known to have a disinfecting effect and has found applications in traditional medicines. Ag nanoparticles have aptly been investigated for their antibacterial property (Rai et al. 2009). Biopolymer-embedded Ag nanoparticles have been investigated (Lee et al. 2006). The nano-sized Ag permits a controlled antibacterial effect due to the high surface area to contact. For PLLA-based nanocomposite fibers, including Ag nanoparticles, the antibacterial effect longer than 20 days was shown (Xu et al. 2006). PLGA-based nanocomposites were also reported in the scientific papers (Schneider et al. 2008; Xu et al. 2008). The metal nanoparticles induce the thermal conductivity of the bionanocomposites, which leads to enhancement of the degradation rate

(Xu et al. 2006). Furthermore, the Ag nanoparticles change the surface wettability and roughness of the bionanocomposites. For these reasons, it is very difficult to control the bacterial adhesion process.

Of particular interest, it is important to note that very recently Ag nanoparticles are recognized and listed under carcinogenic materials by the World Health Organization. These hazardous signs require immediate and thorough action not only from the environment and human health view points but also from the perspective of socio-economic benefits (WHO 2007).

14.3.4 CARBON-BASE BIONANOCOMPOSITES

Carbon nanostructures in polymer matrix have been extensively investigated for biomedical applications (Harrison and Atala 2007). CNTs have the potential for biomedical scaffold. Honeycomb-like matrices of multiwalled nanotube (MWNT) were fabricated as potential scaffolds for tissue engineering (Mwenifumbo et al. 2007). Mouse fibroblast cells were cultured on the nanotube networks, which was prepared by treating with an acid solution that generates carboxylic acid groups at the defect of the nanotubes. The carbon networks can be used effects as biocompatible mesh for restoring or reinforcing damaged tissues because of no cytotoxicity of the networks. The electrical conductivity of the bionanocomposites including carbon nanostructures is a useful tool to direct cell growth because they can conduct electricity stimulus into the tissue healing process. Osteoblast proliferation on PLLA/MWNT bionanocomposites under alternating current stimulation was investigated (Supronowicz et al. 2002). The results showed an increase in osteoblast proliferation and extracellular calcium deposition on the bionanocomposites as compared with the control samples. Unfortunately, no comparison was made with the currently used orthopedic reference material under electrical stimulation.

Shi et al. (2008) investigate *in vitro* cytotoxicity of single-walled CNT (SWNT)/PPF nanocomposites. The results did not reveal any *in vitro* cytotoxicity for PPF/SWNT functionalized with 4-*tert*-butylphenylene nanocomposites. Moreover, nearly 100% cell viability was observed on the nanocomposites and cell attachment on their surfaces was comparable with that on tissue culture polystyrene. The nature of the functional group at the CNT surface seems to play an important role to improve the dispersion of CNTs in polymer matrix and in the mechanism of interaction with cells. The sidewall carboxylic functionalized SWNTs exhibited a nucleation surface to induce the formation of a biomimetic apatite coating (Armentano et al. 2008).

Nanodiamonds (NDs) synthesized by detonation are one of the most promising materials for multifunctional nanocomposites for various application, including biomedical (Huang et al. 2008; Zhang et al. 2011; Mochalin et al. 2012). To fully benefit from the advantages of NDs as nanofillers for biopolymeric bone scaffolds, they need to be dispersed into single particles. The quality of the filler dispersion in the matrix is of importance, because it determines the surface area of the nanoparticles available for interaction with the matrix. When adequately dispersed, NDs increase strength, toughness, and thermal stability of the bionanocomposites (Karbushev et al. 2008). The purified NDs are composed of particles with 5 nm average diameter. They contain an inert diamond core and are decorated by functional groups such as COOH, OH, and NH_2. (Mochalin et al. 2012).

In a very recent work, Zhang et al. (2012) prepared multifunctional bone scaffold materials composed of PLLA and octadecylamine-functionalized ND (ND-ODA) via solution casting followed by compression molding. Addition of 10 wt% of ND-ODA resulted in a 280% increase in the strain to failure and a 310% increase in fracture energy as compared to neat PLLA. Both of these parameters are crucial for bone tissue engineering and for manufacturing of orthopedic surgical fixation devices. The biomineralization of the bionanocomposite scaffolds in simulated body fluid (SBF) (Kokubo and Takadama 2006) was tested. The apatite nucleation and growth occurred faster on the bionanocomposites than on neat PLLA (Figure 14.5) (Zhang et al. 2012). The increased mechanical properties and enhanced biomineralization make PLLA/ND-ODA nanocomposites a promising material for bone surgical fixation devices and regenerative medicine.

Despite the evidence of research into potential biomedical applications of carbon-based nanocomposites, there have been many published studies on the cytotoxicity of the carbon

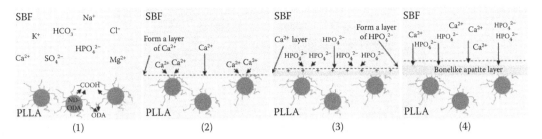

FIGURE 14.5 Schematic representation of a biomineralization process on PLLA/ND-ODA scaffolds in SBF. (1) The initial stage. (2) While in contact with SBF, PLLA is hydrolyzed, resulting in the formation of —COOH groups on the surface of the scaffold. Due to the degradation of PLLA, ND-ODA is exposed to SBF. The exposed —COOH groups of ND-ODA dissociate and form negatively charged —COO— on the surface. In addition, the ND-ODA may speed up the degradation of PLLA to produce more —COOH groups on the PLLA surface. The negatively charged surface attracts Ca^{2+}. (3) The deposited calcium ions, in turn, interact with phosphate ions in the SBF and form bonelike apatite. (4) The bonelike apatite then grows spontaneously, consuming the calcium and phosphate ions to form apatite clusters. (Reprinted from *Biomaterials*, 33, Zhang, Q.W. et al., Mechanical properties and biomineralization of multifunctional nanodiamond-PLLA composites for bone tissue engineering, 5067–5075, Copyright 2012, with permission from Elsevier.)

nanostructures (Smart et al. 2006). Some research groups detected high toxicity in both cells (Shvedova et al. 2003; Cui et al. 2005; Jia et al. 2005; Monterio-Riviere et al. 2005; Chen et al. 2006; Margrez et al. 2006; Nimmagadda et al. 2006; Sayes et al. 2006) and animals (Warheit et al. 2004; Huczko et al. 2005; Muller et al. 2005), and explained mechanisms to cell damage at molecular and gene expression levels (Ding et al. 2005).

Exposure to SWNT resulted in accelerated oxidative stress (increased free radical and peroxide generation and depletion of total antioxidant reserves), loss in cell viability and morphological alterations to cellular structure. It was concluded that these effects were a result of high levels of iron catalyst present in the unrefined SWNT. The possible dermal toxicity in handling unrefined CNT was warned. Similar dermal toxicity warnings were echoed in 2005, in a study, which found that MWNT initiated an irritation response in human epidermal keratinocyte (HEK) cells (Monterio-Riviere et al. 2005). Purified MWNT incubated (at doses of 0.1–0.4 mg/mL) with HEK cells for up to 48 h were observed to localize within cells (Figure 14.6), elicit the production of the proinflammatory cytokine release and decrease cell viability in a time- and dose-dependent manner.

These controversial results reported by different researchers reflect the complex material properties of CNTs such as SWNTs or MWNTs. In addition, different synthesis methods may produce CNTs with different diameters, lengths, and impurities. The results urge caution when handling CNTs and the introduction of safety measures in laboratories should be seriously considered. Most importantly, the success of CNT technology is dependent upon the continuation of research into the toxicology of CNT and CNT-based bionanocomposites. At the same time, the pharmacological development must continue in parallel before providing the guidelines for the safe use in biomedical applications.

14.4 THREE-DIMENSIONAL POROUS SCAFFOLDS

Development of composite scaffolds is attractive as advantageous properties of two or more types of materials can be combined to suit better the mechanical and physiological demands of the host tissue. The tissues in the body are organized into 3D structure as a function of organs. The scaffolds with designed microstructures provide structural support and adequate mass transport to guide the tissue regeneration. To achieve the goal of tissue regeneration, scaffolds must meet some specific needs. A high porosity and an adequate pore size are necessary to facilitate cell seeding and diffusion throughout the whole structure of both cells and nutrients (Rezwan et al. 2006).

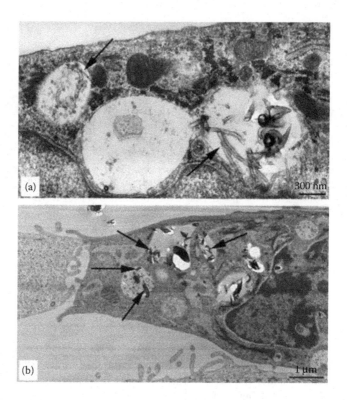

FIGURE 14.6 Transmission electron micrograph of human epidermal keratinocytes (HEKs): (a) intracellular localization of the MWNT—arrows depict the MWNT present within the cytoplasmic vacuoles of a HEK and (b) keratinocyte monolayer grown on a Permanox surface—arrows depict the intracytoplasmic localization of the MWNT. (Reprinted from *Carbon*, 44, Smart, S.K. et al., The biocompatibility of carbon nanotubes, 1034–1047, Copyright 2006, with permission from Elsevier.)

Nanocomposites 3D scaffolds based on biopolymers have been developed by using different processing methods. Most popular techniques include solvent casting and porogen (particulate) leaching, gas foaming, emulsion freeze-drying, electrospinning, rapid prototyping, and thermally induced phase separation (Mikos et al. 1993; Goel and Beckman 1994; Whang et al. 1995; Van de Witte et al. 1996; Nam and Park 1999; Bognitzki et al. 2001; El-Ghannam et al. 2004; Ma 2004; Rezwan et al. 2006; Kretlow and Mikos 2007; Yavari et al. 2013).

14.4.1 Solvent Casting and Particulate Leaching

Organic solvent casting particulate leaching is a very easy way that has been widely used to fabricate biocomposite scaffolds (Mikos et al. 1994). This process involves the dissolution of the polymer in an organic solvent, mixing with nanofillers and porogen particles, and casting the mixture into a predefined 3D mould. The solvent is subsequently allowed to evaporate, and the porogen particles are removed by leaching following the main step (Lu et al. 2000). However, residual solvents in the scaffolds may be harmful to transplanted cells or host tissues. To avoid toxicity effect of the organic solvent, gas foaming can be used to prepare highly porous biopolymer foam (Okamoto 2006). Kim et al. (2007) fabricated PLGA/nHA scaffolds by carbon dioxide (CO_2) foaming and solid porogen (i.e., sodium chloride crystals) leaching (GF/PL) method without the use of organic solvents. Selective staining of nHA particles indicated that nHA exposed to the scaffold surface were observed more abundantly in the GF/PL scaffold than in the conventional solvent casting and

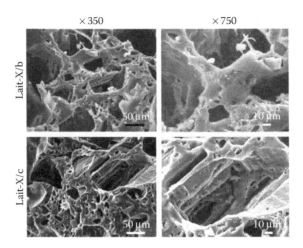

FIGURE 14.7 Scanning electron micrographs of cross-linked PLA porous scaffolds: Lait-X/b and Lait-X/c, for 350X and 750x magnifications. (From Sakai, R. et al., Fabrication of Polylactide-based biodegradable thermoset scaffolds for tissue engineering. *Appl. Macromol. Mater. Eng.*, 2013, 298, 45–52. Copyright Wiley-VCH Verlag GmbH & Co. KGaA. Reproduced with permission.)

particulate leaching scaffold. The GF/PL scaffolds exhibited significant enhanced bone regeneration when compared with conventional one.

Highly porous cross-linked PLA scaffolds were successfully prepared through particulate leaching and foaming followed by leaching methods (Sakai et al. 2013). The scaffolds were porous with good interconnectivity and thermal stability. The SEM images confirmed the pore connectivity and structural stability of the cross-linked PLA scaffold (Figure 14.7).

The scaffolds (Lait-X/b and Lait-X/c) have the same percent of salt particulate with similar particle size; the former was turned into scaffold through simple leaching, while the latter was turned through batch foaming followed by leaching. The qualitative evaluation of the SEM images of Lait-X/b and Lait-X/c showed a well-developed porosity and interconnectivity with pore sizes spanning over a very wide range, from few microns to hundreds of microns.

Figure 14.8 shows the relation between the pore size diameter to the cumulative and differential intrusions of mercury in Lait-X/b and Lait-X/c scaffolds. The maximum intrusion and interconnectivities of Lait-X/b was from 0.1 to 1 μm pore diameter regions and Lait-X/c it was from 0.1 to 10 μm. The porosity, total intrusion volume, total pore area, and median pore diameter (volume) of Lait-X/b calculated by mercury porosimetry were 43%, 0.511 mL g^{-1}, 13.4 m^2 g^{-1}, and 0.520 μm, respectively, whereas for Lait-X/c, it was 49%, 0.688 mL g^{-1}, 27.6 m^2 g^{-1}, and 1.26 μm, respectively. The shift in the values of Lait-X/b and Lait-X/c was due to the effect of batch foaming, which lead to the movement of salt particulate during foaming resulting in the increased porosity and total intrusion volume. The *in vitro* cell culture demonstrated the ability of the scaffold to support human mesenchymal stem cells adhesion confirming the biocompatibility through the cell–scaffold interaction. The *in vitro* degradation of the PLA thermoset scaffolds in phosphate buffered solution was faster for the ones prepared by foaming and subsequent leaching (Sakai et al. 2013). The agglomeration of the smaller crystal (solid porogen) within the 3D polymer matrix enables creation of an interconnected pore network with well-defined pore sizes and shapes.

14.4.2 THERMALLY INDUCED PHASE SEPARATION

3D resorbable polymer scaffolds with very high porosities (~97%) can be produced using the TIPS technique to give controlled microstructures as scaffolds for tissues such as nerve, muscle, tendon, intestine, bone, and teeth (Mikos et al. 1994). The obtained scaffolds are highly porous

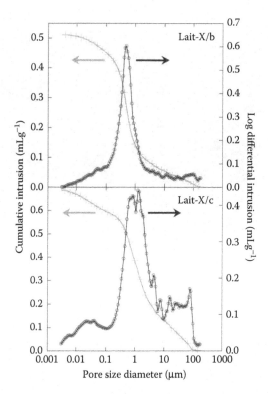

FIGURE 14.8 Pore size distribution in cross-linked PLA porous scaffolds: Lait-X/b and Lait-X/c. (From Sakai, R. et al., Fabrication of Polylactide-based biodegradable thermoset scaffolds for tissue engineering. *Appl. Macromol. Mater. Eng.*, 2013, 298, 45–52. Copyright Wiley-VCH Verlag GmbH & Co. KGaA. Reproduced with permission.)

with anisotropic tubular morphology and extensive pore interconnectivity. Microporosity of TIPS produced foams; their pore morphology, mechanical properties, bioactivity, and degradation rates can be controlled by varying the polymer concentration in solution, volume fraction of secondary phase, quenching temperature, and the polymer and solvent used (Boccaccini and Maquet 2003).

When dioxane was used alone, the porous structure resulted from a solid–liquid phase separation of the polymer solution. During quenching, the solvent was crystallized and the polymer was expelled from the solvent crystallization front. Solvent crystals became pores after subsequent sublimation. To better mimic the mineral component and the microstructure of natural bone, novel nHA nanocomposite scaffolds with high porosity and well-controlled pore architectures were prepared using TIPS techniques (Figure 14.3 [Nejati et al. 2008]). The incorporation of nHA particles into PLLA solution perturbed the solvent crystallization to some extent and thereby made the pore structure more irregular and isotropic. The perturbation by nHA particles, however, was small even in high proportion up to 50% due to their nanometer size scale and uniform distribution. The SEM images showed that the nHA particles were dispersed in the pore walls of the scaffolds and were bound to the polymer very well. PLLA/nHA scaffolds prepared using pure solvent system had a regular anisotropic but open 3D pore structure similar to neat polymer scaffolds, whereas PLLA/mHA scaffolds had an isotropic and a random irregular pore structure.

TiO_2 nanoparticles ($nTiO_2$) have been recently proposed as attractive fillers for biodegradable PDLLA matrix (Boccaccini et al. 2006). 3D PDLLA foams containing both $nTiO_2$ and Bioglass® additions have been synthesized by TIPS. The foams demonstrated the enhancement of the bioactivity and the surface nanotopography.

14.5 *IN VITRO* DEGRADATION

Since the tissue engineering aims at the regeneration of new tissues, hence biomaterials are expected to be degradable and absorbable with a proper rate to match the speed of new tissue formation. The degradation behavior has a crucial impact on the long-term performance of a tissue-engineered cell/polymer construct. The degradation kinetics may affect a range of processes such as cell growth, tissue regeneration, and host response. The mechanism of aliphatic polyester biodegradation is the bio-erosion of the material mainly determined by the surface hydrolysis of the polymer. The scaffolds can lead to heterogeneous degradation, where the neutralization of carboxylic end groups located at the surface by the external buffer solution (*in vitro* or *in vivo*). These phenomena contribute to reduce the acidity at the surface whereas, in the bulk, degradation rate is enhanced by autocatalysis due to carboxylic end groups of aliphatic polyesters. In general, the amount of absorbed water depends on diffusion coefficients of chain fragment within the polymer matrix, temperature, buffering capacity, pH, ionic strength, additions in the matrix, in the medium and processing history. Different polyesters can exhibit quite distinct degradation kinetics in aqueous solutions. For example, PGA is a stronger acid and is more hydrophilic than PLA, which is hydrophobic due to its methyl groups.

Of particular significance for application in tissue engineering are debris and crystalline by-products, as well as particularly acidic degradation products of PLA, PGA, PCL, and their copolymers (Niiranen et al. 2004). Several groups have incorporated basic compounds to stabilize the pH of the environment surrounding the polymer and to control its degradation. Bioglass and CP have been introduced (Rich et al. 2002). Bionanocomposites showed a strongly enhanced polymer degradation rate when compared to the neat polymer (Dunn et al. 2001). As mentioned in PLLA/ND nanocomposites (Figure 14.5), at the same time improvement of osteoconductivity of the nanocomposites, that is, the deposition of the HA crystal on the surface was observed. The fast degradation and the superior bioactivity make these bionanocomposites a promising material for orthopaedic medicine application (Dunn et al. 2001).

14.6 STEM CELL-SCAFFOLDS INTERACTIONS

Synthetic biopolymers are widely used for the preparation of porous scaffolds by different techniques set up for the purpose. The main limitation to the use of PLA-based systems is represented by their low hydrophilicity that causes a low affinity for the cells as compared with the biological polymers. Therefore, the addition of biological components to a synthetic biopolymer represents an interesting way to produce a bioactive scaffold that can be considered as a system showing at the same time adequate mechanical stability and high cell affinity (He et al. 2006; Lazzeri et al. 2007; Allen et al. 2008; Baek et al. 2008; Liu et al. 2009). A variety of extra cellular matrix (ECM) protein components such as gelatin, collagen, laminin, and fibronectin could be immobilized onto the plasma-treated surface of the synthetic biopolymer to enhance cellular adhesion and proliferation. Arg-Gly-Asp (RGD) is the most effective and often employed peptide sequence for stimulating cell adhesion on synthetic polymer surfaces. The RGD cell adhesion sequence was discovered in fibronectin in 1984 (Pierschbacher and Ruoslahti 1984). The RGD amino acid sequence is the minimum unit of a cell adhesive activity domain in adherent proteins, such as fibronectin, fibrin, and vitronectin, which are all ligands of integrins. For example, the aggregation of $\alpha2\beta1$ and $\alpha\nu\beta3$ integrins has been shown to occur specifically during osteogenesis (Mizuno and Kuboki 2001; Schneider et al. 2001). Matsuura et al. (2000) have suggested that the RGD domains of fibronectin and vitronectin have major roles to play in the spreading of osteoblasts on HA surfaces, thus contributing to the osteoconductivity of the surfaces. In support of this theory, Stephansson et al. (2002) have shown that integrin ($\alpha5\beta1$)-fibronectin interactions may be a vital event to the matrix mineralization of MC3T3-E1 cells, while the disruption of these interactions can significantly inhibit osteoblastic differentiation and mineralization.

Zhang et al. (2009) reported that a simple method to immobilize RGD peptide on PCL 3D scaffold surfaces. They demonstrated that rat bone marrow stromal cell (BMSC) adhesion was significantly improved on the RGD-modified PCL scaffolds in a serum-free culture condition.

Surface treatment techniques, such as plasma treatment, ion sputtering, oxidation, and corona discharge, affect the chemical and physical properties of the surface layer without changing the bulk material properties. The effect of the oxygen plasma treatments on the surface of the materials have been shown to charge wettability, roughness, and to enable the selective interaction between PLLA surface and the protein, further improved stem cell attachment (Armentano et al. 2009).

Bone marrow-derived human mesenchymal stem cells are an important cell source for cell therapy and tissue engineering applications. The interactions between stem cell and their environment are very complex and not fully clarified. Previous work showed that cells respond to the mechanical properties of the scaffolds on which they are growing (Discher et al. 2005). Rohman et al. (2007) reported that PLGA and PCL are biocompatible for the growth of normal human urothelial and human bladder smooth muscle cells. Their analysis of the potential mechanism has indicated that differences in degradation behaviors between polymers are not significant, but that the elastic modulus is a critical parameter, where it is relevant to biology at the microscopic (cellular) level and may also have an impact at macroscopic (tissue/organ) scales. They concluded that the elastic modulus is a property that should be considered in the development and optimization of synthetic biopolymers for tissue engineering.

MSCs provided striking evidence that ECM elasticity influences differentiation. Indeed, multipotent cells are able to start a transdifferentiation process toward very soft tissues, such as nervous tissue, when the elastic modulus (E) of the substrate is about 0.5 kPa. Intermediate stiffness (~10 kPa) addresses cells toward a muscle phenotype and harder E ($\geq$30 kPa) to cartilage/bone (Engler et al. 2006). This should address an intelligent design of new biopolymer intended for specific applications (Mitragotri and Lahann 2009). Biopolymers presently used in tissue engineering are extremely stiff. PLA has a bulk elasticity of $E \sim 2$ GPa, that is, 10,000 times stiffer than most soft tissues. Thus, engineering of soft tissue replacements needs to explore biopolymers softer than those presently available.

Poly(butylene/thiodiethylene succinate) block copolymers (PBSPTDGS) were prepared by reactive blending of the parent homopolymers (PBS and PTDGS) in the presence of Ti(OBu)$_4$ (Soccioa et al. 2008). The random copolymer, characterized by the lowest crystallinity degree, exhibits the lowest elastic modulus and the highest deformation at break. When evaluated for indirect citotoxicity, films of block PBSPTDGS30 and random PBSPTDGS240 copolymers appeared entirely biocompatible. In addition, cellular adhesion and proliferation of H9c2 cells (Tantini et al. 2006) (derived from embryonic rat heat) seeded and grown up to 14 days in culture over the same films demonstrate that these new materials might be of interest for tissue engineering applications.

The biocompatibility of neat PLLA, PLLA/nHA, and PLLA/mHA biocomposite scaffolds were evaluated *in vitro* by observing the behavior of the stained MSCs cultured in close contact with the scaffolds (Nejati et al. 2008). Cell growth in material-free organ culture can be distinguished into four stages: cells adhered on the surface of the composite in a round shape during the initial two days. Then the round cell attached, spread, and proliferated into the inner pores of the scaffold, exhibiting morphologies ranging from spindle shaped to polygonal. After one week, the cells reached confluence on the material, while the material-free group did not reach this status (Figure 14.9). The representative cell culture micrographs of cell attachment into the scaffolds after seven days are observed. It is seen that round shape cells attached and proliferated to the scaffolds surface, became spindle-like and migrated through the pores (Figure 14.9a,b,c). The number of round shape cells is noticeable on the surface of pure PLLA scaffold (Figure 14.9a), while proliferated cells on the micro and nanocomposite scaffolds exhibit spindle-shaped morphology (Figure 14.9b,c). The PLLA/HA bionanocomposite scaffolds appeared to be *in vitro* biocompatible and noncytotoxic to cells.

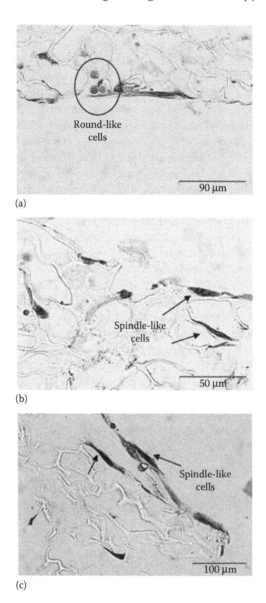

FIGURE 14.9 Optical microscopy photographs of the colored MSCs (H & E staining) attached to the (a) neat PLLA, (b) PLLA/mHAP, and (c) PLLA/nHAP. (Reprinted from *Compos. Part A* 39, Nejati, E. et al., Synthesis and characterization of nanohydroxyapatite rods/ poly(L-lactide acid) composite scaffolds for bone tissue engineering, 1589–1596, Copyright 2008, with permission from Elsevier.)

Clinical trials demonstrate the effectiveness of cell-based therapeutic angiogenesis in patients with severe ischemic diseases; however, their success remains limited. Maintaining transplanted cells in place are expected to augment the cell-based therapeutic angiogenesis.

Use of engineered tissues is becoming commonplace in clinics for thin tissues and for tissues characterized by limited metabolic demand (e.g., cornea [Shah et al. 2008], skin [MacNeil 2007, 2008], and cartilage [Clar et al. 2005; Huang et al. 2010; Safran and Seiber 2010]). For tissue engineering to be a viable treatment option for more complex tissues, it is imperative that they can become vascularized. In fact, the lack of functional vascularization of the engineered constructs can be an indicator of clinical failure (Cassell et al. 2002; Nomi et al. 2002; Koike et al. 2004; Laschke et al. 2006; Hofmann et al. 2008; Yu et al. 2009; Novosel et al. 2011).

In 2012, nHA-coated PLLA microspheres, named nano-scaffold (NS), were, for the first time, generated as a nonbiological, biodegradable, and injectable cell scaffold (Mima et al. 2012). Mima et al. (2012) investigated the effectiveness of NS on cell-based therapeutic angiogenesis. NS are microspheres approximately 100 nm in diameter (Figure 14.10A-a), the surfaces of which are coated with a monolayer of nHA particles with 50 nm in diameter (Figure 14.10A-b, 14.10A-c and 14.10A-d). To assess the cell adhesiveness of NS, SEM was performed after incubation of NS and bare PLLA microspheres (LA) as controls with murine bone marrow mononuclear cells (BMNCs) at 37°C for 8 h *in vitro*. The number of cells adhering to NS was much greater than that to LA (Figure 14.10B-a and 14.10B-b). High-magnification SEM images showed active cell adhesion to NS (Figure 14.10B-c).

BMNCs from enhanced green fluorescent protein (EGFP)-transgenic mice and rhodamine B-containing PLLA microspheres (orange) as a scaffold core or control microspheres were implanted into the ischemic hind limbs of eight-week-old male (C57BL/6NCrSlc) mice to determine the colocalization of implanted cells with injected microspheres (Figure 14.11A). Few implanted BMNCs were observed around LA (Figure 14.11A-a), while markedly larger numbers of cells were seen with NS (Figure 14.11A-b) in ischemic thigh tissue seven days after transplantation. Intramuscular levels of GFP derived from transplanted BMNCs were consistently and significantly higher in the group injected with NS than that injected with LA or BMNCs alone at 3, 7, and 14 days after implantation, while GFP levels were not significantly different between BMNCs alone and LA + BMNCs groups (Figure 14.11B).

Kaplan-Meier analysis demonstrated that NS + BMNC markedly prevented hindlimb necrosis. NS + BMNC revealed much higher induction of angiogenesis in ischemic tissues and collateral blood flow confirmed by 3D-computed tomography angiography than those of BMNC or LA + BMNC groups (Mima et al. 2012). NS-enhanced therapeutic angiogenesis and arteriogenesis showed good correlations with increased intramuscular levels of vascular endothelial growth factor and fibroblast growth factor 2. NS co-implantation also prevented apoptotic cell death of transplanted cells, resulting in prolonged cell retention.

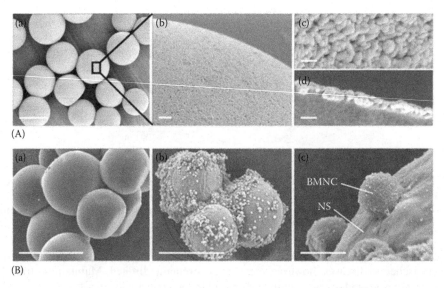

FIGURE 14.10 Scanning electron micrograph of NS (A) and marked cell adhesiveness to NS *in vitro* (B). (A) NS are microspheres approximately 100 μm in diameter (a). The NS surface uniformly coated with nanoscale hydroxyapatite (nHA) crystals was observed at different magnifications (low and high magnification in b and c, respectively). Scanning electron micrograph of an NS cross section indicating a single layer of nHA particles on the NS surface (d). (B) Murine BMNCs were incubated with LA (a) or NS (b, c) at 37°C for 8 h. Large numbers of BMNCs adhered to NS (b, c) but not to LA (a). Scale bars: 100 μm (A-a, B-a, B-b), 5 μm (B-c), 1 μm (A-b), and 100 nm (A-c, A-d). (Reproduced from Mima Y., et al., *PLoS One* 7:e35199, Copyright 2012, with permission from *PLoS One*.)

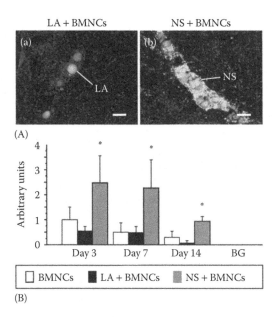

FIGURE 14.11 Prolonged localization of implanted BMNCs in ischemic tissues by NS. (A) Colocalization of BMNCs with NS and LA *in vivo*. Murine BMNCs derived from EGFP-transgenic mice were transplanted together with LA or NS into the thighs in the hind limb ischemic model. Cores of NS and LA containing rhodamine B were used to indicate localization of the injected microspheres in ischemic tissues. Tissue sections seven days after transplantation of LA + BMNCs (a) or NS + BMNCs (b) were counterstained with DAPI, and merged images of DAPI, GFP and rhodamine B are shown. BMNCs were observed as densely clustered around NS (b) but not LA (a). Scale bars: 100 mm. (B) Quantitative evaluation of implanted cells existing in ischemic tissues. Quantitative analysis of intramuscular GFP was performed 3, 7, and 14 days after transplantation. BMNCs were derived from EGFP-transgenic mice. BMNCs were transplanted alone or together with LA or NS into ischemic thigh muscles. Intramuscular GFP values of whole thigh muscles were corrected for total protein and expressed in arbitrary units ($n = 6$ in each group). *$p < .05$ for the NS + BMNCs group compared to the BMNCs alone and LA + BMNCs groups. GFP concentration in normal murine muscle was measured as background (BG). (Reproduced from Mima Y., et al., *PLoS One* 7:e35199, Copyright 2012, with permission from *PLoS One*.)

This nano-scaffold provides promising local environment for implanted cells for the effects on angiogenesis and arteriogenesis through cell clustering, augmented expression of proangiogenic factors, and supporting cell survival without gene manipulation or artificial ECM.

14.7 GENE TRANSFER

The introduction of DNA, RNA, or oligonucleotides into eukaryotic cells is called *transfection*. This process involves the uptake of extracellular molecules through the cell membrane into the cytoplasm and also into the nucleus. When DNA is brought into the nucleus, it can be incorporated into a cell's genetic material and induce the production of specific proteins (McNeil and Perrie 2006). The transient transfection (DNA does not integrate into the host chromosome) and stable transfection (the foreign DNA is integrated into the chromosome and passed over to the next generation) are distinguished. In contrast, the introduction of small-interfering RNA can selectively turn off the production of specific proteins (gene silencing or antisense technology) (Mello and Conte 2004; Mitterauer 2004). Naked DNA itself cannot successfully enter cells; it requires the assistance of a suitable vector. A tail-vein injection of naked DNA into mice did not result in gene expression in major organs (Mahato et al. 1995) because of its rapid degradation by nucleases in the blood (Sakurai et al. 2002).

Scaffold mediated gene delivery is an advantageous strategy for gene transfer. It enables localized delivery of a therapeutic gene. The DNA delivered from the scaffold is principally taken up by the surrounding cells at the implant site, thereby limiting unwanted exposure in other areas (Jang et al. 2005). The scaffold structure allows for sustained gene delivery acting as a reservoir, which gradually releases polyplexes and DNA/CP complex over time. The degradation rate of the scaffold material can be designed, so that the required release rate is achieved (De Rosa et al. 2003). Another advantage of the scaffold is its role in the protection of the polyplexes and DNA/CP complex.

14.7.1 DNA/CP COMPLEX AS DRUG DELIVERY

CP nanoparticles are biocompatible and therefore suitable as drug delivery system, especially for the transfection of cells with DNAs. CP nanoparticles have proven to be effective for nonviral intercellular gene delivery or transfection and to be used for gene silencing. DNA and RNA binding to CP nanoparticles occur through electrostatic interaction between Ca^{2+} in CP carrier and phosphate group in DNA or RNA structure (Chen et al. 2007).

Figure 14.12 shows a schematic representation of gene delivery into cell nucleus through a CP nanoparticle. Cellular uptake of CP nanoparticle loaded with DNAs is caused by endocytosis through lipid bilayer cellular membrane. DNAs escape from the endosome following the dissolution of CPs in the acidic environment of the endocytic vesicle. Degradation of DNAs by lysosomal nucleases could limit the transfection efficiency if endosomal escape of DNAs could not occur before the fusion of endosome with lysosome (in which the pH value is under 5). Production of a certain enzymatic activity, or a stop in synthesizing certain genes, is triggered once the delivered DNAs are transported into the cell nucleus through the nuclear membrane. In general, the transfer of molecules into the nucleus occurs through nuclear pore complexes, that is, large proteins (nucleoporins) that are inserted into the double nuclear membrane that consists of two lipid bilayers. The nuclear pore complexes are highly permeable to small molecules, but they restrict the movement of

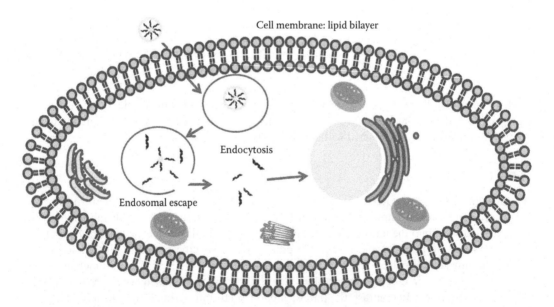

FIGURE 14.12 Schematic of transfection/intracellular delivery of drugs and biomacromolecules by CP nanoparticles. (1) Adsorption on the cell membrane, (2) uptake by endocytosis, (3) escape from endosomes and intracellular release, (4) nuclear targeting, and (5) nuclear entry and gene expression. (Reprinted from *Acta Biomater.*, 8, Bose, S., and S. Tarafder, Calcium phosphate ceramic systems in growth factor and drug delivery for bone tissue engineering: A review, 1401–1421, Copyright 2012, with permission from Elsevier.)

larger molecules across the nuclear envelope. To overcome this barrier, macromolecules that carry a nuclear localization sequence can be recognized by importins and then be actively transported through the pore into the nucleus (Josephson et al. 1999; Bose and Tarafder 2012).

CP chemistry, surface area, surface charge, and crystallinity also play a great role in gene loading efficiency. Mg^{2+} doping into HA increases the surface positive charge of the CP nanoparticle and hence increased its DNA-loading capacity. The presence of β-TCP phase into HA increased gene delivery properties by increasing its solubility inside endosome (Hanifi et al. 2010).

Oyane et al. (2007) reported the successful improvement of the gene transferring by using a laminin/DNA/HA complex. The gene transferring efficiency of the laminin/DNA/HA complex was 1–2 orders of magnitude higher than that of a conventional DNA/CP complex. This is because laminin enhances cell adhesion and spreading, and provides regions of high DNA concentration between the cell and the complex surface.

Biomimetic coating or mineralization is a technique in which an osteoconductive amorphous calcium phosphate (ACP) layer is introduced on the surface of a substratum immersed in a supersaturated SBF solution (Kokubo and Takadama 2006). The ion concentrations of SBF solution are similar to human blood plasma, and the ACP that forms in SBF is calcium-deficient carbonated HA. Table 14.3 shows the ion concentration comparison between SBF and blood plasma adapted from Kokubo and Takadama (2006). Instead of simple surface adsorption, co-precipitation of drug or osteogenic growth factor molecules (DNAs and RNAs) during biomimetic coating is also gaining significant attention from the scientific community (Liu et al. 2010).

Takeshita et al. (2013) examined the computer simulation to provide insight into the structure and stability of DNA while adsorption of charge-balancing ions in the initial stage of the biomineralization. The molecular orbital computer simulation has been used to probe the interaction of DNA with two charge-balancing ions, that is, $CaOH^+$, and $CaH_2PO_4^+$. The adsorption enthalpy of the two ions on double-stranded DNA and/or single-stranded DNA having large negative value (~−60 kcal mol^{-1} per charge-balancing ion) was the evident for the interface in mineralization of HA in SBF. They reported that the ion size affect the conformation of the phosphate backbone and hence the base-base distance and excluded volume correlation.

Shen et al. (2004) examined the elegant study, which demonstrated the effect of mineral properties on plasmid DNA (pDNA) release from CP mineral coatings. Specifically, coatings were formed in SBF solutions in the presence of pDNA, with SBF solutions of varying ion content and concentration. The resulting co-precipitated pDNA-containing mineral coatings had different Ca/P ratios and

TABLE 14.3

Ion Concentration of Human Blood Plasma and SBF Solution

Ions	Blood Plasma (mM)	SBF (mM)
Na^+	142.0	142.0
K^+	5.0	5.0
Mg^{2+}	1.5	1.5
Ca^{2+}	2.5	2.5
Cl	103.0	147.8
HCO_3^-	27.0	4.2
HPO_4^{2-}	1.0	1.0
SO_4^{2-}	0.5	0.5
pH	7.4	7.4

Source: Kokubo, T. and H. Takadama., *Biomaterials*, 27, 2907–2915, 2006.

mineral structures, and these differences resulted in varying pDNA release rates under simulated physiological conditions. The nanocomposite DNA-CP coatings had feature sizes of ~500 nm, and decreasing their size led to an increase in gene delivery, which was believed to be due to the ability of cells to endocytose the smaller particles, which facilitated the release of DNA from CP matrix. In addition, the gene transfer efficiency from the mineralized surface indicated that gene expression was dependent on pDNA amount (Shen et al. 2004; Sun et al. 2012). Choi and Murphy (2010) investigated the release of pDNA from CP coatings formed on PLGA substrates with the intrinsic properties of the CP mineral coating and the surrounding solution conditions.

14.7.2   Polyplexes

Lipoplexes and polyplexes consist of cationic lipids, liposomes, or polymers—poyethyleneimine, poly(L-lysine), poly(D-lysine)—that electrostatically condense nucleic acids into nanoparticles. However, these charged complexes maintain several disadvantages during *in vivo* delivery including rapid aggregation, high clearance from the bloodstream, and inflammatory toxicity (Seow and Wood 2009). Attempts have been made to improve the pharmacokinetics and reduce inflammatory toxicity of these nanoparticles by adding poly(ethylene glycol) (PEG) to the surface of the nanoparticle as a stealth moiety (Balazs and Godbey 2011). While this addition has been shown to increase circulation time of cationic liposomes, PEGylated neutral liposomes containing doxorubicin maintained higher plasma concentration, achieved higher localization within the target tissue, and delivered more effective therapy *in vivo* than the cationic liposomes when directly compared (Zhao et al. 2011). Figure 14.13 illustrates the combinatorial system of scaffold mediated polyplex gene delivery (O'Rorke et al. 2010). The scaffolds used in gene delivery have been developed in different shapes and sizes, from the macro to the nanoscale depending on the targeted clinical application.

14.7.3   Targeting the Nucleus

As discussed previously, the most difficult barrier to successful nonviral gene delivery is entry into the nucleus. *In vivo*, many innate proteins are targeted to the nuclei due to nuclear localization signals (NLS). NLSs are peptide sequences that are recognized by importins. Proteins exhibiting these peptides are bound by the importins and internalized by the nucleus (Josephson et al. 1999; Wong et al. 2007;

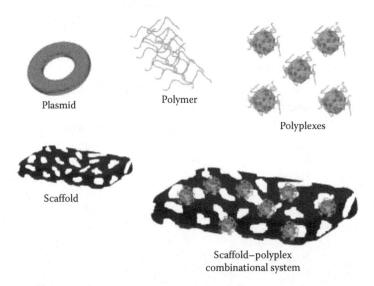

Plasmid   Polymer

Polyplexes

Scaffold

Scaffold–polyplex
combinational system

FIGURE 14.13   The scaffold–polyplex combinatorial system and its components. (Reprinted from *Prog. Polym. Sci.* 35, O'Rorke S., et al., Non-viral polyplexes: Scaffold mediated delivery for gene therapy, 441–458, Copyright 2010, with permission from Elsevier.)

Bose and Tarafder 2012). NLSs are cationic in nature; therefore, it is possible to form complexes between the NLS and the DNA. NLSs can also be conjugated to the polyplex structure although more literature on the use of NLS alone is available. It has been shown that the efficiency of NLSs depends on many factors including the type of NLS, the method of conjugation used, and the size and type of DNA to be delivered (Wong et al. 2007).

It has become clear that polyplexes alone are insufficient to meet the transfection requirements of modern medicine; however, the studies discussed above demonstrate how research is going to meet the demands of nonviral gene delivery.

14.8 CONCLUDING REMARKS

The bionanocomposites reviewed in this chapter are particularly attractive as tissue engineering scaffolds due to their biocompatibility and adjustable biodegradation kinetics. Conventional materials processing methods have been adapted and incorporation of inorganic nanoparticles into porous and interconnected 3D porous scaffolds. The incorporation of nanoparticles and immobilization of biological components on the surface to enhance cellular adhesion and proliferation are promising and currently under extensive research. At the same time, the current research is focused on the interaction between stromal cells and biopolymer interfaces. The bionanocomposite scaffolds with bioactive inorganic phases will be in the center of attention in combination with stem cell seeding.

The fundamentals for biomaterials seem to originate from introducing stem cells. In this direction, the new approach of bionanocomposite enables the scaffold surface to mimic complex local biological functions and may lead in near future to *in vitro* and *in vivo* growth of tissues and organs.

In the present scenario, bioceramics entities have been used for bone tissue engineering scaffolds and drug delivery (Wu et al. 2008; Dvir et al. 2011; Rahaman et al. 2011). Osteomyelitis is a most common medical problem related to bones caused by an inflammatory process, leading to bone destruction caused by infective microorganisms found worldwide in children, where the bone tissue regeneration is required (Gristina et al. 1985).

The important role of biopolymers in scaffold-mediated gene delivery is clear. Biopolymers form the backbone of the structures from the macro sized scaffold to the nano-sized complexes. At present, nonviral vectors are still not as efficient as the established viral vectors. The inclusion of complexes in a bionanocomposite scaffold is a step in the right direction for combating the differences between these methods. The range of specifically designed polymers for DNA transport, the choice of scaffold, and the availability of various targeting ligands have increased alongside the growth in the knowledge base on the mechanism of gene delivery (Wong et al. 2007). The combinational approach of scaffold and complexes is already establishing itself as an important technique in the field of gene therapy.

REFERENCES

Allen B.L., P.D. Kichambare, P. Gou, I.I. Vlasova, A.A. Kapralov, N. Konduru, V.E. Kagan, and A. Star. 2008. Biodegradation of single-walled carbon nanotubes through enzymatic catalysis. *Nano Lett.* 8:3899–3903.

Armentano I., M.A. Alvarez-Pérez, B. Carmona-Rodríguez, I. Gutiérrez-Ospina, J.M. Kenny, and H. Arzate. 2008. Analysis of the biomineralization process on SWNT-COOH and F-SWNT films. *Mater. Sci. Eng. C* 28:1522–1529.

Armentano I., G. Ciapetti, M. Pennacchi, M. Dottori, V. Devescovi, D. Granchi, N. Baldini et al. 2009. Role of PLLA plasma surface modification in the interaction with human marrow stromal cells. *J. Appl. Polym. Sci.* 114:3602–3611.

Baek H.S., Y.H. Park, C.S. Ki, J.C. Park, and D.K. Rah. 2008. Enhanced chondrogenic responses of articular chondrocytes onto porous silk fibroin scaffolds treated with microwave-induced argon plasma. *Surf. Coat. Technol.* 202:5794–5797.

Balazs D., and W. Godbey. 2011. Liposomes for use in gene delivery. *J. Drug Delivery* 2011: 326497.

Bianco A., E. Di Federico, I. Moscatelli, A. Camaioni, I. Armentano, L. Campagnolo, M. Dottori, J.M. Kenny, G. Siracusa, and G. Gusmano. 2009. Electrospun poly(ε-caprolactone)/Ca-deficient hydroxyapatite nanohybrids: Microstructure, mechanical properties and cell response by murine embryonic stem cells. *Mater. Sci. Eng. C* 29:2063–2071.

Boccaccini A.R., J.J. Blaker, M. Maquet, W. Chung, R. Jerome, and S.N. Nazhat. 2006. Poly(D,L-lactide) (PDLLA) foams with TiO$_2$ nanoparticles and PDLLA/TiO$_2$-Bioglass foam composites for tissue engineering scaffolds. *J. Mater. Sci.* 41:3999–4008.

Boccaccini A.R., and V. Maquet. 2003. Bioresorbable and bioactive polymer/Bioglass(R) composites with tailored pore structure for tissue engineering applications. *Compos. Sci. Technol.* 63:2417–2429.

Bognitzki M., W. Czad, T. Frese, A. Schaper, M. Hellwig, M. Steinhart, A. Greiner, and J.H. Wendorff. 2001. Nano-structured fibers via electrospinning. *Adv. Mater.* 13:70–72.

Bordes P., E. Pollet, and L. Averous. 2009. Nano-biocomposites: Biodegradable polyester/nanoclay systems. *Prog. Polym. Sci.* 34:125–155.

Bose S., and S. Tarafder. 2012. Calcium phosphate ceramic systems in growth factor and drug delivery for bone tissue engineering: A review. *Acta Biomater.* 8:1401–1421

Cassell O.C., S.O. Hofer, W.A. Morrison, and K.R. Knight. 2002. Vascularisation of tissue-engineered grafts: The regulation of angiogenesis in reconstructive surgery and in disease states. *Br. J. Plast. Surg.* 55:603–610.

Chandra R., and R. Rustgi. 1998. Biodegradable polymers. *Progr. Polym. Sci.* 23:1273–1335.

Chen W., M. Lin, P. Lin, P. Tasi, Y. Chang, and S. Yamamoto. 2007. Studies of the interaction mechanism between single strand and double-strand DNA with hydroxyapatite by microcalorimetry and isotherm measurements. *Colloids Surf. A: Physicochem. Eng. Aspects* 295:274–283.

Chen X., U.C. Tam, J.L. Czlapinski, G.S. Lee, D. Rabuka, A. Zettle, and C.R. Bertozzi. 2006. Interfacing carbon nanotubes with living cells. *J. Am. Chem. Soc.* 128:6292–6293.

Choi S., and W.L. Murphy. 2010. Sustained plasmid DNA release from dissolving mineral coatings. *Acta Biomater.* 6:3426–3435.

Clar C., E. Cummins, L. Mcintyre, S. Thomas, J. Lamb, L. Bain, P. Jobanputra, N. Waugh 2005. Clinical and cost-effectiveness of autologous chondrocyte implantation for cartilage defects in knee joints: Systematic review and economic evaluation. *Health Technol. Assess.* 9:1–82.

Cui D.X., F.R. Tian, C.S. Ozkan, M. Wang, and H.J. Gao. 2005. Effect of single wall carbon nanotubes on human HEK293 cells. *Toxicol Lett.* 155:73–85.

Darder M., P. Aranda, M.L. Ferrer, M.C. Gutierrez, F.D. Monte, and E. Ruiz-Hitzky. 2011. Progress in bionanocomposite and bioinspired foams. *Adv. Mater.* 23:5262–5267.

De Rosa G., F. Quaglia, A. Bochot, F. Ungaro, and E. Fattal. 2003. Long-term release and improved intracellular penetration of oligonucleotide; polyethylenimine complexes entrapped in biodegradable microspheres. *Biomacromolecules* 4:529–536

Ding L.H., J. Stilwell, T.T. Zhang, O. Elboudwarej, H.J. Jiang, J.P. Selegue, P.A. Cooke, J.W. Gray, and F.Q.F. Chen. 2005. Molecular characterization of the cytotoxic mechanism of multiwall carbon nanotubes and nano-onions on human skin fibroblast. *Nano Lett.* 5:2448–2464.

Discher D.E., P. Janmey, and Y.L. Wang. 2005. Tissue cells feel and respond to the stiffness of their substrate. *Science* 310:1139–1143.

Dunn A.S., P.G. Campbell, and K.G. Marra. 2001. The influence of polymer blend composition on the degradation of polymer/hydroxyapatite biomaterials. *J. Mater. Sci. Mater. Med.* 12:673–677.

Dvir T., B.P. Timko, M.D. Brigham, S.R. Naik, S.S. Karajanagi, O. Levy, H. Jin, K.K. Parker, R. Langer, and D.S. Kohane. 2011. Nanowired three-dimensional cardiac patches. *Nature Nanotech.* 6:720–725.

El-Ghannam A., C.Q. Ning, and J. Mehta. 2004. Cyclosilicate nanocomposite: A novel resorbable bioactive tissue engineering scaffold for BMP and bone-marrow cell delivery. *J. Biomed. Mater. Res.* 71A:377–390.

Engler A.J, S. Sen, H.L. Sweeney, and D.E. Disher. 2006. Matrix elasticity directs stem cell lineage specification. *Cell* 126:677–689.

Fu S., P. Ni, B. Wang, B. Chu, J. Peng, L. Zheng, X. Zhao, F. Luo, Y. Wei, and Z. Qian. 2012. In vivo biocompatibility and osteogenesis of electrospun poly(ε-caprolactone)-poly(ethylene glycol)-poly(ε-caprolactone)/nano-hydroxyapatite composite scaffold. *Biomaterials* 33:8363–8371.

Gansel J.K., M. Thiel, M.S. Rill, M. Decker, K. Bade, V. Saile et al. 2009. Gold helix photonic metamaterial as broadband circular polarizer. *Science* 325:1513–1515.

Gay S., S. Arostegui, and J. Lemaitre. 2009. Preparation and characterization of dense nanohydroxyapatite/PLLA composites. *Mater. Sci. Eng. C* 29:172–177.

Ghosh S.K., S.K. Nandi, B. Kundu, S. Datta, D.K. De, S.K. Roy, and D. Basu. 2013. In vivo response of porous hydroxyapatite and beta-tricalcium phosphate prepared by aqueous solution combustion method and comparison with bioglass scaffolds. *Acta Biomater.* 86:217–227.

Gilding D., and A.M. Reed. 1979. Biodegradable polymers for use in surgery polyglycolic/poly(lactic acid) homo- and copolymers. *Polymer* 20:1459–1464.

Goel S.K., and E.J. Beckman. 1994. Generation of microcellular polymeric foams using supercritical carbon dioxide. I: Effect of pressure and temperature on nucleation. *Polym. Eng. Sci* 34:1137–1147.

Gristina A.G., M. Oga, L.X. Webb, and C.D. Hobgood. 1985. Adherent bacterial colonization in the pathogenesis of osteomyelitis. *Science* 228:990–993.

Hanifi A., M.H. Fathi, H. Mir Mohammad Sadeghi, and J. Varshosaz. 2010. Mg^{2+} substituted calcium phosphate nano particles synthesis for non viral gene delivery application. *J. Mater. Sci. Mater. Med.* 21:2393–2401.

Harrison B.S., Atala A. 2007. Review, carbon nanotube applications for tissue engineering. *Biomaterials* 28:344–353.

He W., T. Yong, Z. Ma, R. Inai, W.E. Teo, and S. Ramakrishna. 2006. Biodegradable polymer nanofiber mesh to maintain functions of endothelial cells. *Tissue Eng.* 12:2457–2466.

Hench LL. 1998. Bioceramics. *J. Am. Ceram. Soc.* 81:1705–1728.

Hofmann A., U. Ritz, S. Verrier, D. Eglin, M. Alini, S. Fuchs, C.J. Kirkpatrick, P.M. Rommens et al. 2008. The effect of human osteoblasts on proliferation and neo-vessel formation of human umbilical vein endothelial cells in a long-term 3D co-culture on polyurethane scaffolds. *Biomaterials* 29:4217–4226.

Hong Z., X.Y. Qiu, J.R. Sun, M.X. Deng, X.S. Chen, and X.B. Jing. 2004. Grafting polymerization of L-Lactide on the surface of hydroxyapatite nano-crystals. *Polymer* 45:6705–6713.

Hong Z., P. Zhang, C. He, X. Qiu, A. Liu, L. Chen, X. Chen, and X. Jing. 2005. Nano-composite of poly (L-lactide) and surface grafted hydroxyapatite: Mechanical properties and biocompatibility. *Biomaterials* 26:6296–6304.

Huang A.H., M.J. Farrell, and R.L. Mauck. 2010. Mechanics and mechanobiology of mesenchymal stem cell-based engineered cartilage. *J. Biomech.* 43:128–136.

Huang H.J., E. Pierstorff, E. Osawa, and D. Ho. 2008. Protein-mediated assembly of nanodiamond hydrogels into a biocompatible and biofunctional multilayer nanofilm. *ACS Nano* 2:203–212.

Huczko A., H. Lange, M. Bystrzejewski, P. Baranowski, H. Grubek-Jaworska, P. Nejman, T. Przybylowski et al. 2005. Pulmonary toxicity of 1-D nanocarbon materials. *Fuller Nanotub. Carbon Nanostruct.* 13:141–145.

Jagur-Grodzinski J. 1999. Biomedical application of functional polymers. *Reactive Funct. Polym.* 39:99–138.

Jang J-H., C.B. Rives, and L.D. Shea. 2005. Plasmid delivery in vivo from porous tissue-engineering scaffolds: Transgene expression and cellular transfection. *Mol. Ther.* 12:475–483.

Jia G., H.F. Wang, L. Yan, X. Wang, R.J. Pei, T. Yan, Y.L. Zhao, and X.B. Guo. 2005. Cytotoxicity of carbon nanomaterials: Single-wall nanotube, multi-wall nanotube, and fullerene. *Environ. Sci. Technol.* 39:1378–1383.

Johnson R.M., L.Y. Mwaikambo, and N. Tucker. 2003. Biopolymer. *Rapra Review Report* 159. London: Rapra Technology, 148.

Josephson L., C. Tung, A. Moore, and R. Weissleder. 1999. High-efficiency intracellular magnetic labeling with novel superparamagnetic-tat peptide conjugates. *Bioconjugate Chem.* 10:186–191.

Karageorgiou V., and D. Kaplan. 2005. Porosity of 3D biomaterial scaffolds and osteogenesis. *Biomaterials* 26:5474–5491.

Karbushev V.V., I.I. Konstantinov, I.L. Parsamyan, V.G. Kulichikhin, V.A. Popov, and T.F. George. 2008. Preparation of polymer-nanodiamond composites with improved properties. *Adv. Mater. Res.* 59:275–278.

Kim H.W., J.C. Knowles, and H.E. Kim. 2004. Hydroxyapatite/poly(epsilon)-caprolactone) composite coating on hydroxyapatite porous bone scaffold for drug delivery. *Biomaterials* 25:1279–1287.

Kim S.S., K.M. Ahn, M.S. Park, J.H. Lee, C.Y. Choi, and B.S. Kim. 2007. A poly (lactide-*co*-glycolide)/hydroxyapatite composite scaffold with enhanced osteoconductivity. *J. Biomed. Mater. Res. Part A* 80A:206–215.

Knowles J.C. 2003. A review article: Phosphate glasses for biomedical applications. *J. Mater. Chem.* 13:2395–2401.

Kohn J.R.L. 1996. Bioresorbable and bioerodible materials. In: B.D. Ratner, A.S. Hoffman, F.J. Schoen, J.E. Lemons, editors. *Biomaterials Science: An Introduction to Materials in Medicine*. New York: Academic Press, 64–72.

Koike N., D. Fukumura, O. Gralla, P. Au, J.S. Schechner, and R.K. Jain. 2004. Tissue engineering: Creation of long-lasting blood vessels. *Nature* 428:138–139.

Kokubo T., and H. Takadama. 2006. How useful is SBF in predicting in vivo bone bioactivity? *Biomaterials* 27:2907–2915.

Kretlow J.D., and A.G. Mikos. 2007. Review: Mineralization of synthetic polymer scaffolds for bone tissue engineering. *Tissue Eng.* 13:927–938.

Langer R., and J.P. Vacanti. 1993. Tissue engineering. *Science* 260:920–926.

Laschke M.W., Y. Harder, M. Amon, I. Martin, J. Farhadi, A. Ring, N. Torio-Padron et al. 2006. Angiogenesis in tissue engineering: Breathing life into constructed tissue substitutes. *Tissue Eng.* 12:2093–2104.

Lazzeri L., M.G. Cascone, S. Danti, L.P. Serino, S. Moscato, and N. Bernardini. 2007. Geratine/PLLA sponge-like scaffolds: Morphological and biological characterization. *J. Mater. Sci. Mater. Med.* 18:1399–1405.

Lee J.H., T.G. Park, H.S. Park, D.S. Lee, Y.K. Lee, S.C. Yoond, and J.D. Nam. 2003. Thermal and mechanical characteristics of poly(l-lactic acid) nanocomposite scaffold. *Biomaterials* 24:2773–2778.

Lee J.Y., J.L.R. Nagahata, and S. Horiuchi. 2006. Effect of metal nanoparticles on thermal stabilization of polymer/metal nanocomposites prepared by a one-step dry process. *Polymer* 47:7970–7979.

LeGeros R.Z. 2008. Calcium phosphate-based osteoinductive materials. *Chem. Rev.* 108:4742–4753.

Levine R.M., C.M. Scott, and E. Kokkoli. 2013. Peptide functionalized nanoparticles for nonviral gene delivery. *Soft Matter* 9:985–1004.

Li J., X.L. Lu, and Y.F. Zheng. 2008. Effect of surface modified hydroxyapatite on the tensile property improvement of HA/PLA composite. *Appl. Surf. Sci.* 255:494–497.

Liu Q., J. Wu, T. Tan, L. Zhang, D. Chen, and W. Tian. 2009. Preparation, properties and cytotoxicity evaluation of a biodegradable polyester elastomer composite. *Polym. Degrad. Stab.* 94:1427–1435.

Liu Y., G. Wu, and K. De Groot. 2010. Biomimetic coatings for bone tissue engineering of critical-sized defects. *J. R. Soc. Interf.* 7:S631–S647.

Lu L., S.J. Peter, M.D. Lyman, H.L. Lai, S.M. Leite, J.A. Tamada, J.P. Vacanti, R. Langer, and A.G. Mikos. 2000. In vitro degradation of porous poly(l-lactic acid) foams. *Biomaterials* 21:1595–1605.

Ma P.X. 2004. Scaffolds for tissue fabrication. *Mater. Today* 7:30–40.

MacNeil S. 2007. Progress and opportunities for tissue-engineered skin. *Nature* 445:874–880.

MacNeil S. 2008. Biomaterials for tissue engineering of skin. *Mater. Today* 11:26–35.

Mahato R.I., K. Kawabata, Y. Takakura, and M. Hashida. 1995. In vivo disposition characteristics of plasmid DNA complexed with cationic liposomes. *J. Drug. Target* 3:149–157.

Margrez A., S. Kasas, V. Salicio, N. Pasquier, J.W. Seo, M. Celio, S. Catsicas, B. Schwaller, and L. Forro. 2006. Cellular toxicity of carbon-based nanomaterials. *Nano Lett.* 6:1121–1125.

Martin C., H. Winet, and J.Y. Bao. 1996. Acidity near eroding polylactidepolyglycolide in vitro and in vivo in rabbit tibial bone chambers. *Biomaterials* 17:2373–2380.

Matsuura T., R. Hosokawa, K. Okamoto, T. Kimoto, and Y. Akagawa. 2000. Diverse mechanisms of osteoblast spreading on hydroxyapatite and titanium. *Biomaterials* 21:1121–1127.

McNeil S.E., and Y. Perrie. 2006. Gene delivery using cationic liposomes. *Expert. Opin. Ther. Pat.* 16:1371–1382.

Mello C.C., and D. Jr. Conte. 2004. Revealing the world of RNA interference. *Nature* 431:338–342.

Mikos A.G., G. Sarakinos, S.M. Leite, J.P. Vacanti, and R. Langer. 1993. Laminated three-dimensional biodegradable foams for use in tissue engineering. *Biomaterials* 14:323–330.

Mikos A.G., A.J. Thorsen, L.A. Czerwonka, Y. Bao, D.N. Winslow, and J.P. Vacanti. 1994. Preparation and characterization of poly(l-lactic acid) foams. *Polymer* 35:1068–1077.

Mima Y., S. Fukumoto, H. Koyamal, M. Okada, S. Tanaka, T. Shoji, M. Emoto, T. Furuzono, Y. Nishizawa, and M. Inaba. 2012. Enhancement of cell-based therapeutic angiogenesis using a novel type of injectable scaffolds of hydroxyapatite-polymer nanocomposite microspheres. *PLoS One* 7:e35199.

Mistry A.S., and A.G. Mikos. 2005. Tissue engineering strategies for bone regeneration. *Adv. Biochem. Eng. Biotechnol.* 94:1–22.

Mitragotri S., and J. Lahann. 2009. Cell and biomolecular mechanics in silico. *Nat. Mater.* 8:15–23.

Mitterauer B. 2004. Gene silencing: A possible molecular mechanism in remission of affective disorder. *Med. Hypotheses* 62(6):907–910.

Mizuno M., and Y. Kuboki. 2001. Osteoblast-related gene expression of bone marrow cells during the osteoblastic differentiation induced by type I collagen. *J. Biochem.* 129:133–138.

Mochalin V.N., O. Shenderova, D. Ho, and Y. Gogotsi. 2012. The properties and applications of nanodiamonds. *Nat. Nano.* 7:11–23.

Mohanty A.K., M. Misra, and G. Hinrichsen. 2000. Biofibers. Biodegradable polymers and biocomposites: An overview. *Macromol. Mater. Eng.* 276:1–24.

Monterio-Riviere N.A., R.J. Nemanich, A.O. Inman, Y.Y.Y. Wang, and J.E. Riviere. 2005. Multi-walled carbon nanotube interactions with human epidermal kerationcytes. *Toxicol Lett.* 155:377–384.

Muller J., F. Huaux, N. Moreau, P. Misson, J.F. Heilier, M. Delos, M. Arras, A. Fonseca, J.B. Nagy, and D. Lison. 2005. Respiratory toxicity of multi-wall carbon nanotubes. *Toxicol Appl. Pharmacol.* 207:221–231.

Mwenifumbo S., M.S. Shaffer, and M.M. Stevens. 2007. Exploring cellular behaviour with multi-walled carbon nanotube constructs. *J. Mater. Chem.* 17:1894–1902.

Nam Y.S., and T.G. Park. 1999. Biodegradable polymeric microcellular foams by modified thermally induced phase separation method. *Biomaterials* 20:1783–1790.

Nejati E., H. Mirzadeh, and M. Zandi. 2008. Synthesis and characterization of nanohydroxyapatite rods/poly(L-lactide acid) composite scaffolds for bone tissue engineering. *Compos. Part A* 39:1589–1596.

Niiranen H., T. Pyhältö, P. Rokkanen, M. Kellomäki, and P. Törmälä. 2004. In vitro and in vivo behavior of self-reinforced bioabsorbable polymer and self-reinforced bioabsorbable polymer/bioactive glass composites. *J. Biomed. Mater. Res. A* 69A:699–708.

Nimmagadda A., K. Thurston, M.U. Nollert, and P.S.F. McFetridge. 2006. Chemical modification of SWNT alters in vitro cell-SWNT interactions. *J. Biomed. Mater. Res. A* 76:614–625.

Nomi M., A. Atala, P.D. Coppi, and S. Soker. 2002. Principals of neovascularization for tissue engineering. *Mol. Aspects. Med.* 23:463–483.

Novosel E.C., C. Kleinhans, and P.J. Kluger. 2011.Vascularization is the key challenge in tissue engineering. *Adv. Drug. Deliv. Rev.* 63:300–311.

Okamoto M. 2006. Biodegradable polymer/layered silicate nanocomposites: A review. In: S. Mallapragada, B. Narasimhan, editors. *Handbook of Biodegradable Polymeric Materials and Their Applications*. Los Angeles, CA: American Scientific Publishers, 153–197.

O'Rorke S., M. Keeney, and A. Pandit. 2010. Non-viral polyplexes: Scaffold mediated delivery for gene therapy. *Prog. Polym. Sci.* 35:441–458.

Oyane A., H. Tsurushima, and A. Ito. 2007. Novel gene-transferring scaffolds having a cell adhesion molecule-DNA-apatite nanocomposite surface. *Gene Therapy* 14:1750–1753.

Pektok E., B. Nottelet, J.C. Tille, R. Gurny, A. Kalangos, M. Moeller, and B.H. Walpoth. 2008. Degradation and healing characteristics of small-diameter poly(ε-caprolactone) vascular grafts in the rat systemic arterial. *Circulation* 118:2563–2570.

Pierschbacher M.D., and E. Ruoslahti. 1984. Cell attachment activity of fibronectin can be duplicated by small synthetic fragments of the molecule. *Nature* 309(5963):30–33.

Pitt C.G., M.M. Gratzel, and G.L. Kimmel. 1981. Aliphatic polyesters. 2. The degradation of poly(DL-lactide), poly(e-caprolactone) and their copolymers in vivo. *Biomaterials* 2:215–220.

PLA degradable. 2012. Kyoto: BMG, 1. http://www.bmg-inc.com/index.html (accessed March 2013).

Platt D.K. 2006. Biodegradable polymers, Market report. London: Rapra Technology, 158.

Quirk R.A., R.M. France, K.M. Shakesheff, and S.M. Howdle. 2004. Supercritical fluid technologies and tissue engineering scaffolds. *Curr. Opin. Solid State Mater. Sci.* 8:313–821.

Rahaman M.N., D.E. Day, B.S. Bal, Q. Fu, S.B. Jung, L.F. Bonewald, and A.P. Tomsia. 2011. Bioactive glass in tissue engineering. *Acta Biomater.* 7:2355–2373.

Rai M., A. Yadav, and A. Gade. 2009. Silver nanoparticles as a new generation of antimicrobials. *Biotechnol. Adv.* 27:76–83.

Ramay Hassna R.R., and M. Zhang. 2004. Biphasic calcium phosphate nanocomposite porous scaffolds for load-bearing bone tissue engineering. *Biomaterials* 25:5171–5180.

Rezwan K., Q.Z. Chen, J.J. Blaker, and A.R. Boccaccini. 2006. Biodegradable and bioactive porous polymer/inorganic composite scaffolds for bone tissue engineering. *Biomaterials* 27:3413–3431.

Rich J., T. Jaakkola, T. Tirri, T. Narhi, A. Yli-Urpo, and J. Seppala. 2002. In vitro evaluation of poly([var epsilon]-caprolactone-*co*-DL-lactide)/bioactive glass composites. *Biomaterials* 23:2143–2150.

Rohman G., J.J. Pettit, F. Isaure, N.R. Cameron, and J. Southgate. 2007. Influence of the physical properties of two-dimensional polyester substrates on the growth of normal human urothelial and urinary smooth muscle cells in vitro. *Biomaterials* 28:2264–2274.

Safran M.R., and K. Seiber. 2010. The evidence for surgical repair of articular cartilage in the knee. *J. Am. Acad. Orthop. Surg.* 18:259–266.

Sakai R., B. John, M. Okamoto, J.V. Seppälä, J. Vaithilingam, H. Hussein, and R. Goodridge. 2013. Fabrication of Polylactide-based biodegradable thermoset scaffolds for tissue engineering. *Appl. Macromol. Mater. Eng.* 298:45–52.

Sakurai F., T. Terada, K. Yasuda, F. Yamashita, Y. Takakura, and M. Hashida. 2002. The role of tissue macrophages in the induction of proinflammatory cytokine production following intravenous injection of lipoplexes. *Gene Ther.* 9:1120–1126.

Salem A.K., P.C. Searson, and K.W. Leong. 2003. Multifunctional nanorods for gene delivery. *Nature Mater.* 2:668–671.

Samavedi S., A.R. Whittington, and A.S. Goldstein. 2013. Calcium phosphate ceramics in bone tissue engineering: A review of properties and their influence on cell behavior. *Acta Biomater.* 9:8037–8045.

Sayes C.M., F. Liang, J.L. Hudson, J. Mendez, W.H. Guo, J.M. Beach, V.C. Moore et al. 2006. Functionalization density dependence of single-walled carbon nanotubes cytotoxicity in vitro. *Toxicol Lett.* 161:135–142.

Schneider G.B., R. Zaharias, and C. Stanford. 2001. Osteoblast integrin adhesion and signaling regulate mineralization. *J. Dent. Res.* 80:1540–1544.

Schneider O.D., S. Loher, T.J. Brunner, P. Schmidlin, and W.J. Stark. 2008. Flexible silver containing nanocomposites for the repair of bone defects: Antimicrobial effect against *E. coli* infection and comparison to tetracycline containing scaffolds. *J. Mater. Chem.* 18:2679–2684.

Seal B.L., T.C. Otero, and A. Panitch. 2001. Polymeric biomaterials for tissue and organ regeneration. *Mater. Sci. Eng. R: Rep* 34:147–230.

Seow Y., and M.J. Wood. 2009. Biological gene delivery vehicles: beyond viral vectors. *Mol. Ther.* 17:767–777.

Shah A., J. Brugnano, S. Sun, A. Vase, and E. Orwin. 2008. The development of a tissue-engineered cornea: Biomaterials and culture methods. *Pediatr. Res.* 63:535–544

Shen H., J. Tan, and W.M. Saltzman. 2004. Surface-mediated gene transfer from nanocomposites of controlled texture. *Nat. Mater.* 3:569–574.

Shi X., and A.G. Mikos. 2006. Poly(propylene fumarate). In: S.A. Guelcher, J.O. Hollinger, editors. *An Introduction to Biomaterials*. Boca Raton, FL: CRC Press, 205–218.

Shi X., B. Sitharaman, Q.P. Pham, P.P. Spicer, J.L. Hudson, L.J. Wilson, J.M. Tour, R.M. Raphael, and A.G. Mikos, 2008. *In vitro* cytotoxicity of single-walled carbon nanotube/biodegradable polymer nanocomposites. *J. Biomed. Mater. Res. Part A* 86A:813–823.

Shvedova A.A., V. Castranova, E.R. Kisin, D. Schwegler-Berry, A.R. Murray, V.Z. Gandelsman, A. Maynard, and P. Baronn. 2003. Exposure to carbon nanotube material: Assessment of nanotube cytotoxicity using human keratinocyte cells. *J. Toxicol Environ. Health A* 66:1909–1926.

Sinha Ray S., and M. Bousmina. 2005. Biodegradable polymer and their layered silicate nanocomposites: In greeing the 21st century materials world. *Prog. Mater. Sci.* 50:962–1079.

Sinha Ray S., and M. Okamoto. 2003a. Biodegradable polylactide and its nanocomposites: Opening a new dimension for plastics and composites. *Macromol. Rapid Commun.* 24:815–840.

Sinha Ray S., and M. Okamoto. 2003b. Polymer/layered silicate nanocomposites: A review from preparation to processing. *Progr Polym Sci* 28:1539–1641.

Smart S.K., A.I. Cassady, G.Q. Lu, and D.J. Martin. 2006. The biocompatibility of carbon nanotubes. *Carbon* 44:1034–1047.

Smith R., editor. 2005. *Biodegradable Polymer for Industrial Applications*. New York: CRC Press.

Soccioa M., N. Lottia, L. Finellia, M. Gazzanob, and A. Munari. 2008. Influence of transesterification reactions on the miscibility and thermal properties of poly(butylene/diethylene succinate) copolymers. *Eur. Polym. J.* 44:1722–1732.

Stephansson S.N., B.A. Byers, and A.J. García. 2002. Enhanced expression of the osteoblastic phenotype on substrates that modulate fibronectin conformation and integrin receptor binding. *Biomaterials* 23:2527–2534.

Sun B., M. Yi, C.C. Yacoob, H.T. Nguyen, and H. Shen. 2012. Effect of surface chemistry on gene transfer efficiency mediated by surface-induced DNA-doped nanocomposites. *Acta Biomater.* 8:1109–1116.

Supronowicz P.R., P.M. Ajayan, K.R. Ullmann, B.P. Arulanandam, D.W. Metzger, and R. Bizios. 2002. Novel current-conducting composite substrates for exposing osteoblasts to alternating current stimulation. *J. Biomed. Mater. Res.* 59:499–506.

Takeshita T., Y. Matsuura, S. Arakawa, and M. Okamoto. 2013. Biomineralization of hydroxyapatite on DNA molecules in SBF: Morphological features and computer simulation. *Langmuir* doi:10.1021/la402589j.

Tantini B., E. Fiumana, S. Cetrullo, C. Pignatti, F. Bonavita, L.M. Shantz, E. Giordano et al. 2006. Involvement of polyamines in apoptosis of cardiac myoblasts in a model of simulated ischemia. *J. Mol. Cell Cardiol.* 40:775–782.

Tsivintzelis I., S.I. Marras, I. Zuburtikudis, and C. Panayiotou. 2007. Porous poly(L-lactic acid) nanocomposite scaffolds prepared by phase inversion using supercritical CO_2 as antisolvent. *Polymer* 48:6311–6318.

Van de Witte P., H. Esselbrugge, P.J. Dijkstra, J.W.A. Van den Berg, and J. Feijen. 1996. Phase transitions during membrane formation of polylactides. I. A morphological study of membranes obtained from the system polylactide-chloroform-methanol. *J. Membr. Sci.* 113:223–236.

Warheit D.B., B.R. Laurence, K.L. Reed, D.H. Roach, G.A.M. Reynolds, and T.R. Webb. 2004. Comparative pulmonary toxicity assessment of single-wall carbon nanotubes in rats. *Toxicol. Sci.* 77:117–125.

Webster T.J., C. Ergun, R.H. Doremus, R.W. Siegel, and R. Bizios. 2000. Enhanced functions of osteoblasts on nanophase ceramics. *Biomaterials* 2:1803–1810.

Wei G., and P.X. Ma. 2004. Structure and properties of nano-hydroxyapatite/polymer composite scaffolds for bone tissue engineering. *Biomaterials* 25:4749–4757.

Whang K., C.H. Thomas, K.E. Healy, and G.A. Nuber. 1995. A novel method to fabricate bioabsorbable scaffolds. *Polymer* 36:837–842.

WHO. 2007. The international decade for action: Water for life-2005-2015. Geneva, Switzerland: WHO: 2 http://www.who.int/water_sanitation_health/wwd7_water_scarcity_final_rev_1.pdf (accessed May 2013).

Wong S.Y., J.M. Pelet, and D. Putnam. 2007. Polymer systems for gene delivery—Past, present, and future. *Prog. Polym. Sci.* 32:799–837.

Wu S., X. Liu, T. Hu, P.K. Chu, J.P.Y. Ho, Y.L. Chan, K.W.K. Yeung et al. 2008. A Biomimetic hierarchical scaffold: natural growth of nanotitanates on three-dimensional microporous ti-based metals. *Nano Lett.* 8:3803–3808.

Xu X., Q. Yang, J. Bai, T. Lu, Y. Li, and X. Jing. 2008. Fabrication of biodegradable electrospun poly(L-lactide-co-glycolide) fibers with antimicrobial nanosilver particles. *J. Nanosci. Nanotechnol.* 8:5066–5070.

Xu X., Q. Yang, Y. Wang, H. Yu, X. Chen, and X. Jing. 2006. Biodegradable electrospun poly (l-lactide) fibers containing antibacterial silver nanoparticles. *Eur. Polym. J.* 42:2081–2087.

Yavari S.A., R. Wauthle, J. van der Stok, A.C. Riemslag, M. Janssen, M. Mulier, J.P. Kruth, J. Schrooten, H. Weinans, and A.A. Zadpoor. 2013. Fatigue behavior of porous biomaterials manufactured using selective laser melting. *Mater. Sci. Eng. C* doi:10.1016/j.msec.2013.08.006.

Yu H., P.J. Vandevord, L. Mao, H.W. Matthew, P.H. Wooley, and S.Y. Yang. 2009. Improved tissue-engineered bone regeneration by endothelial cell mediated vascularization. *Biomaterials* 30:508–517.

Zhang H., C.Y. Lin, and S.J. Hollister. 2009. The interaction between bone marrow stromal cells and RGD-modified three-dimensional porous polycaprolactone scaffolds. *Biomaterials* 30: 4063–4069.

Zhang Q.W., V.N. Mochalin, I. Neitzel, K. Hazeli, J. Niu, A. Kontsos, J.G. Zhou, P.I. Lelkes, and Y. Gogotsi. 2012. Mechanical properties and biomineralization of multifunctional nanodiamond-PLLA composites for bone tissue engineering. *Biomaterials* 33:5067–5075.

Zhang Q.W., V.N. Mochalin, I. Neitzel, I.Y. Knoke, J.J. Han, C.A. Klug, J.G. Zhou, P.I. Lelkes, and Y. Gogotsi. 2011. Fluorescent PLLA-nanodiamond composites for bone tissue engineering. *Biomaterials* 32:87–94.

Zhang R., and P.X. Ma. 1999. Porous poly(l-lactic acid)/apatite composites created by biomimetic process. *J. Biomed. Mater. Res.* 45:285–293.

Zhao W., S. Zhuang, and X.R. Qi. 2011. Comparative study of the in vitro and in vivo characteristics of cationic and neutral liposomes. *Int. J. Nanomed.* 6:3087–3098.

15 Aerated Food Structure and Properties

Martin G. Scanlon

CONTENTS

15.1 INTRODUCTION

Bubbles are an integral part of many food and drink products. Because the bubbles alter the mechanical properties of biomaterials that they are incorporated in, product properties can be significantly affected by aeration (Scanlon 2004). As a result, consumer satisfaction with a variety of food products depends on how well the aeration operations have been conducted (Barham 2001). Given the importance of aeration to product quality, significant amounts of research have been conducted in order to optimize quality in biofoamed food materials (Campbell et al. 1999, 2008). These foods may be manufactured industrially or they may be created as individual items in a restaurant or a kitchen. Regardless, all possess a specific volume fraction of gas, and the structuring of that gas governs food quality, just as the amount and structuring of gas in other biofoams is critical to their performance for their intended application.

The volume fraction of gas bubbles (alternatively described as void fraction) is defined as follows:

$$\phi = \left(1 - \frac{\rho}{\rho_0}\right) \tag{15.1}$$

where:
    ρ is the density of the aerated food or beverage
    ρ_0 is the density of the matrix material of the food

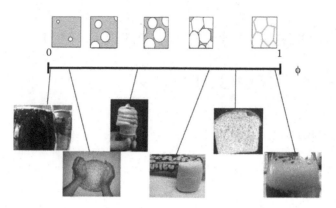

FIGURE 15.1   Range of gas volume fraction for bubbly liquids and foams and illustrated with some typical aerated foods and beverages.

Almost the whole spectrum of the volume fraction of gas is spanned by the wide range of foods and beverages that are consumed (Figure 15.1). In Figure 15.1, void fraction as a criterion for bubble packing is also shown. An important distinction in this regard is the transition from bubbly liquid to foam. This transition, which occurs at a critical gas volume fraction, is interesting not only from a food quality perspective but also as a discovery frontier where the two fluids of a bubbly liquid (liquid food matrix and gas) attain solidity as a foam (Mason 1999).

To show how physical, biological, and chemical effects can be exploited to create a wide variety of aerated food systems, it is worthwhile examining the fundamentals of these two-phase materials.

15.2   FUNDAMENTALS OF TWO-PHASE SYSTEMS

The creation of essentially any food biofoam requires operations to include gas within a liquid matrix. For many beverages the two-phase system that is created is transient; for example, there is only a limited lifetime for tongue tingle in carbonated beverages. For others, such as soufflés, a degree of stability is achieved after aeration by conducting a stabilization operation, usually by supplying or extracting heat. With respect to the gas, in addition to the influence of volume fraction that was discussed above, another structural parameter is the size, or sizes, of the bubbles. Composition is also relevant, with the partial pressures of the gas(es) in the bubbles affecting the properties and the dynamics of the two-phase system. The relevant physical properties of the liquid matrix include its density, viscosity, and composition. In addition to these characteristics of gas and matrix, two further parameters influence the properties of the bubbly liquid system—the interfacial tension between bubbles and matrix, and the solubility of the bubble's gas(es) in the liquid matrix.

One crucial consequence of the difference in the nature of these two phases is that aerated food materials are highly dynamic systems until they are stabilized. Changes in the structure of the aerated foods prior to stabilization can dramatically affect the quality of the food (Barham 2001; Cauvain 2003). Three principal phenomena can occur that are introduced here. Their effects on biofoam properties will be shown later.

First, the difference in density between gas and condensed phase causes the bubbles to cream in bubbly liquids and the liquid to drain in foams. Second, bubble-interface-matrix relations, such as Henry's law and surface rheology considerations, determine how fast gas migrates from the bubbles, or if gas is being generated chemically or biochemically within the liquid matrix, how fast the bubbles grow. A further interface-matrix relation, frequently evident in the complex liquid matrices that are foods, is the preferential adsorption of amphiphilic molecules (either small or polymeric) at the interface after they have diffused through the liquid matrix. Third, the gas in the bubble is at a slightly higher pressure than the pressure in the matrix liquid due to interfacial tension (γ). The pressure difference for a bubble of radius R_1 is defined by Laplace's law:

$$p_b - p_o = \frac{2\gamma}{R_1} \tag{15.2}$$

where:
  p_b is the gas pressure within the bubble
  p_o is the pressure within the matrix

15.3 NUCLEATION AND GROWTH OF BUBBLES IN FOOD BIOFOAMS

15.3.1 BUBBLE NUCLEATION

15.3.1.1 Techniques

The first step in the creation of any food biofoam is the nucleation of bubbles in the liquid matrix. In the food industry, this can be achieved by one or more of three techniques: sparging, supersaturated nucleation, and entrainment of bubbles (Campbell and Mougeot 1999).

In sparging, a gas of specific composition is forced through a frit (membrane) with holes of desired size into a liquid on the other side of the frit. Bubbles nucleate at the frit surface (Figure 15.2). The advantage of sparging is that bubbles of closely matched size are created due to the uniformity of hole sizes in the frit (Kohama et al. 2009). As well as being used for biofoam generation, sparging wines and juices with inert gases such as nitrogen is an effective means of creating nuclei to remove dissolved oxygen, thus preventing the formation of off-flavors during the shelf-life of the beverage.

Supersaturated nucleation is the primary bubble nucleation method for the beverage industry. In this technique, the gas is packaged at an elevated pressure in the headspace of the beverage container. For carbonated beverages such as soda, this gas is carbon dioxide and a typical headspace pressure is about 5 atm. The carbon dioxide dissolves quite readily in the aqueous solution; based on Henry's law, approximately 12 g of carbon dioxide will dissolve in 1 L of diet pop at 10°C, although less in its full-caloried counterpart. When the container is opened, the concentration of dissolved

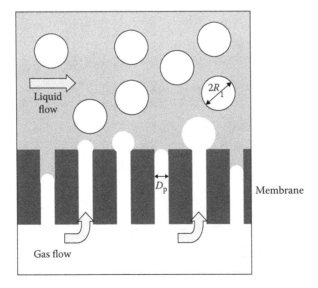

FIGURE 15.2 Creation of a bubbly liquid with a monodisperse bubble size distribution by sparging from a frit with pore size, D_p. (Modified from Kohama, Y. et al., Method for producing monodisperse bubbles. US Patent 7,591,452 [B2], 2009.)

carbon dioxide is well in excess of the equilibrium concentration appropriate for 35 Pa, the partial pressure of carbon dioxide in the atmosphere. As a result, the supersaturated carbon dioxide comes out of solution at nucleation sites within the beverage.

The third method of nucleating bubbles is to entrain the bubbles into the liquid matrix. A typical means of effecting this is by mechanical agitation of the liquid in the presence of a gas. For many foods, this mechanical agitation is induced by mixing. Mixing of foods usually has other objectives (Scanlon and Zghal 2001), but the entrainment of bubbles is often a welcome side-effect, even though it is unwelcome in the manufacture of other materials. During mixing, the liquid captures pockets of gas that are then subdivided into smaller bubbles within the matrix material by further mechanical action. This is especially evident for dough mixed at atmospheric pressure; the longer the mixing process goes on, the greater the amount of air captured by the dough (Figure 15.3). When the dough is mixed under a much reduced pressure, few gas molecules are present in the headspace of the mixer, so that there is little density depression in the dough as it is mixed. Consequently, the amount of gas in the vacuum-mixed dough remains low and invariant with time of mixing (Figure 15.3).

15.3.1.2 Importance of Interfacial Tension

In all three bubble nucleation techniques, interfacial tension plays a prominent role (Ip and Toguri 1994). The formation of bubbles in the matrix that creates the two-phase system incurs an energy penalty (U), which is directly proportional to the interfacial tension.

$$U = 4\pi\gamma R_1^2 \tag{15.3}$$

To overcome this energy penalty, in order that bubbles can be created in the matrix, a driving force must exist. In the case of sparging, the pressure difference across the frit, Δp, is the driving force, with the radius of the bubbles created from a given frit pore size being proportional to the interfacial tension:

$$R_1 \propto \frac{\gamma}{\Delta p} \tag{15.4}$$

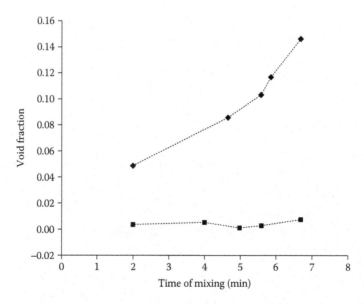

FIGURE 15.3 Change in gas volume fraction with time of mixing for bread dough mixed in air (♦) and in vacuum (■) (Derived from the data of Mehta, K.L. et al., *J. Food Sci.*, 74, E455–E461, 2009, error bars in most cases smaller than symbols.)

In the case of supersaturated nucleation, the driving force is the supersaturation ratio of the dissolved gas (S):

$$S = \frac{c_0}{k_H p_b} - 1 \tag{15.5}$$

where:
c_0 is the equilibrium concentration of the gas for the pressure in the unopened container
p_b is the partial pressure of carbon dioxide in the bubble
k_H is the solubility of the gas in the liquid matrix

The value of S must be such that a free energy term greater than the energy penalty associated with the creation of new bubble surface area is available to drive bubble nucleation. Nucleation is critical: once a stable nucleus is formed, the metastable supersaturated liquid feeds the nucleus, assuring its growth. The supersaturation ratio governs the critical nucleus size, r^*, with a stable nucleus forming at a size that is inversely proportional to S, and with higher interfacial tension demanding larger critical nucleus sizes to ensure bubble growth:

$$r^* = \frac{2\gamma}{p_0 S} \tag{15.6}$$

In a newly opened can of soda, there is a substantial driving force for degassing of the dissolved carbon dioxide since $S \sim 4$. At the supersaturation ratio relevant for such conditions, carbon dioxide can create its own nucleation sites within the liquid matrix. Many nuclei will be formed throughout the liquid, growing as the gas readily comes out of solution and feeds bubble growth. For soda, this nucleus size is about 0.5 μm, so there is a high probability that many bubbles will nucleate in many spots.

The nucleation case for bubbles in bread dough contrasts considerably with that of carbonated beverages. Although there is a ready and continual source of carbon dioxide to feed bubble growth (as the yeast ferment simple sugars), the driving force for nucleation is small since the gas is being generated dynamically. Accordingly, yeast is unable to homonucleate bubbles in the dough. The aerated structure of bread is thus entirely dependent on nucleation occurring in the mixing process, as can be seen in Figure 15.4 (Baker and Mize 1941). Entrained air bubbles are the heteronuclei that are subsequently filled with carbon dioxide, growth being slow due to the very small supersaturation ratio of carbon dioxide within the dough.

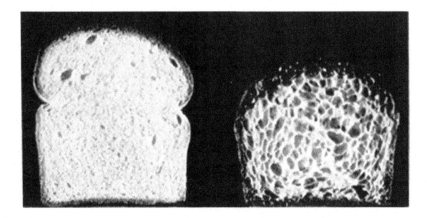

FIGURE 15.4 The effect of nucleation of air bubbles in dough during mixing on the final structure of bread (with permission, AACC International); bread dough mixed at atmospheric pressure (left) and under vacuum (right). (From Baker, J.C., and Mize, M.D., *Cereal Chemistry*, 18, 19–34, 1941.)

This heteronucleation example for dough may also be seen in beverage systems where homonucleation also occurs. A good example evident in a glass of champagne is where a bubble train emerges from a point that floats around the glass. These *fliers* arise from cellulose fibers deposited in the glass by cleaning cloths (Liger-Belair et al. 2008); entrapped air within the cellulose fiber acts as a focal point for outgassing of the supersaturated carbon dioxide in a similar way to how the mixed-in air bubbles allow yeast to leaven a dough. Because foreign particles can act as preferential nucleation sites for bubble formation, beverage manufacturers have engineered their location, typically by point-directed laser ablation of the surface of drinking vessels, so that the bubble train that arises in the beverage spells out a name or mimics a company logo (Kriegel et al. 2010).

Due to the importance of interfacial tension to the nucleation of bubbles (and also their growth), the food industry has developed a number of chemistries to reduce interfacial tension in order that bubbles can be more easily nucleated in liquid matrices (Friberg et al. 2004). These molecules may be relatively small, such as monoglycerides, or large, such as proteins and polysaccharides (Kralova and Sjöblom 2009). In all cases though, the molecule possesses one or more hydrophilic and hydrophobic regions. This amphiphilic character is energetically suited for bubbly liquids and foams because the molecules situate at the gas-aqueous interfaces with hydrophilic regions in the aqueous environment and hydrophobic regions in the bubble. The outcome overall is a lowering of interfacial tension, thus facilitating bubble nucleation. For amphiphilic polymers, there is often reorganization of specific regions of the molecule at the interface to further enhance the surface-active effect. Perhaps the most well-known example is egg white (McGee et al. 1984; This 2010), but other proteins of plant and animal origin undergo conformational changes at the interface, where in addition to the modification of interfacial tension, there can be pronounced surface-rheological effects (Martin et al. 2002).

15.3.2 Bubble Growth

As alluded to in Section 15.3.1, the bubble nuclei that have formed in the liquid matrix grow by one of two techniques, both as a result of being fed from the liquid matrix. The first is where the dissolved molecules of the supersaturated solution desorb at the bubble interface to provide gas for bubble growth. The second is gas provision by leavening agents. In this case, new dissolved molecules of gas are generated *in situ* within the matrix by chemical or biochemical means. These molecules then diffuse to the gas nuclei where their subsequent desorption drives growth of the bubbles.

Bubble growth in champagne is a good example of nuclei growth from a supersaturated solution. Liger-Belair et al. (2008) show a bubble train in champagne where the bubble grows as it rises in the glass. This is not due to a decrease in the pressure head as the bubble rises in the glass, but due to the fact that there is a large supersaturation ratio for carbon dioxide in an opened bottle of champagne. Rapid desorption of the carbon dioxide across the bubble interface grows the bubble.

The most common means of attaining bubble growth in food biofoams is biochemically as a result of the metabolic activities of baker's yeast (*Saccharomyces cerevisiae*). The historical importance of the growth of bubbles in dough by this organism (also known as *leavening*) means that bread is the classic aerated food biofoam (Campbell et al. 1999; Cauvain 2003; Scanlon and Zghal 2001). Chemical leavening agents are also used in the food industry, particularly for targeting specific void fractions in cakes and cookies. A typical reaction scheme is

$$HX + NaHCO_3 = NaX + H_2O + CO_2 \qquad (15.7)$$

where:

HX is the acidulant that reacts with the baking soda

The extent of the reaction is controlled by the amount of sodium bicarbonate, while the type of acidulant and its reactivity as a function of temperature control the rate of the reaction (Figure 15.5).

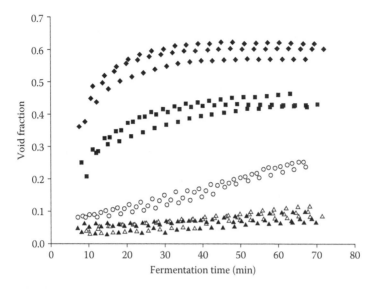

FIGURE 15.5 Approximate change in volume fraction of gas in dough brought about by different leavening systems. Gluconodeltalactone (♦), adipic acid (■), and sodium acid pyrophosphate (▲), each with 2.8 g sodium bicarbonate per 100 g flour leavened at 27°C. Sodium acid pyrophosphate with 4.2 g sodium bicarbonate per 100 g of flour leavened at 27°C (△) and at 37°C (○) is also shown. (Adapted from Bellido, G.G. et al., *J. Cereal Sci.*, 49, 212–218, 2009.)

15.4 STRUCTURE, PROPERTIES, AND DYNAMICS OF FOOD BIOFOAMS

15.4.1 STRUCTURE

15.4.1.1 Transition from Dilute to Concentrated Bubbly Liquids

Regardless of whether a biofoam is being created from the nucleation of many small bubbles or by the growth of a small number of bubbles to larger sizes, the change in volume fraction of the bubbles affects the structure, the properties, and the dynamics of the aerated liquid. For example, in a dilute bubbly system, the viscosity (η) increases with bubble concentration. This increase in viscosity can be appreciated from the Einstein equation, assuming that flow conditions are such that the bubbles maintain sphericity. In this case, the enhanced viscosity is directly proportional to the volume fraction of bubbles in the liquid:

$$\eta = \eta_o \left(1 + 2.5\phi\right) \tag{15.8}$$

where:
η_o is the viscosity of the liquid matrix

Changes in bubble shape brought about by flow conditions will further increase the viscosity enhancement associated with aeration, but differences in bubble size are not a factor.

In most foods where their aerated character adds to our appreciation of them, bubble volume fractions are substantially greater than those used to derive Equation 15.8. As the flow fields around bubbles interact, the bubbly liquid no longer behaves as an ideal Newtonian liquid. This is illustrated in Figure 15.6, where the base material used for making meringues, egg whites and sucrose, has had a specific amount of air entrained into it, so that $\phi = 0.46$ for the bubbly liquid. This volume fraction is not too dissimilar to the volume fraction of air entrained into a soft ice-cream mix (Figure 15.1). Because of the importance of air content to product smoothness and to manufacturer profitability

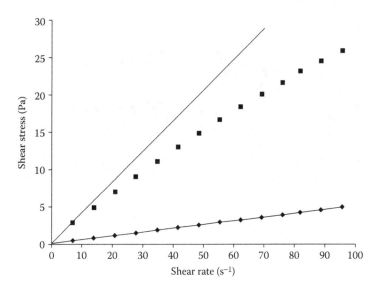

FIGURE 15.6   Changes in the rheology of egg white-sucrose blends caused by increase in void fraction; matrix liquid, $\phi \sim 0$ (♦); bubbly liquid, $\phi = 0.46$ (■).

(Stanley et al. 1996), the ice-cream industry defines it by the term *product overrun*, which is related to gas volume fraction by

$$\text{Overrun} = \frac{\phi}{1-\phi} \times 100 \tag{15.9}$$

From Figure 15.6, it can be seen that the original liquid is Newtonian but with a viscosity some 50 times that of water. The bubbly liquid counterpart is a shear-thinning material: stress does not rise in proportion to shear rate. It is also apparent that the bubbles significantly increase the stress at any given shear rate, thereby requiring greater energy expenditure to pump the bubbly liquid compared to the matrix liquid. For example, if the liquids were being pumped at a rate of 3600 L h⁻¹ through a two-inch external diameter pipe (typical for a holding tube for pasteurization), six times more power is required for this aerated liquid compared to its matrix material (based on work of Steffe (1996)). It is to be emphasized that in addition to the increased energy costs of pumping the bubbles, the processor is also pumping only about half of the liquid per unit time.

In terms of the structure of the bubbly liquid, bubble sizes, as well as volume fraction, govern the rheology of more concentrated bubbly liquids. The potential for bubble flow fields to interact with each other, and thus alter flow properties, depends on bubble size and the relative velocities of bubble centers to each other (Kraynik 1988; Mason 1999).

15.4.1.2 Transition from Concentrated Bubbly Liquids to Foams

As the concentration of bubbles is increased further, the structure changes considerably as the bubbles themselves interact with each other, and they deform from sphericity. At this point, a foam is formed (Weaire and Hutzler 1999). The solid-like characteristics associated with foam formation occur at a critical volume fraction of bubbles. At bubble concentrations somewhat below this critical volume fraction, the *locking* of bubbles in bubbly liquids confers an elastic character to these highly concentrated bubbly liquids (Mason 1999). The specific value for criticality varies with the size distribution of the bubbles, polydispersity increasing the amount of bubbles that can be packed into a bubbly liquid before it transitions to a foam.

Structurally, the bubbly system can now be viewed not from a bubble perspective, but as a network of liquid channels (Plateau borders) at the edges of bubbles, with the channels connected

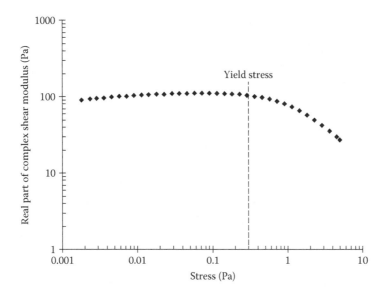

FIGURE 15.7 Yield stress of egg white-sucrose mixture (void fraction = 0.81) determined from a stress sweep experiment (conducted at 10 Hz), where the yield stress is defined as the stress where the stress-independent nature of the real part of the complex shear modulus is transgressed.

to the faces that separate two adjacent bubbles (Kraynik et al. 2003). The greater the volume fraction of bubbles, the thinner the network of Plateau borders. As a result, one can categorize foams as wet, where ϕ is just beyond its critical value, and dry, where ϕ is substantially greater (Cohen-Addad et al. 2013).

Egg whites are good foaming agents, able to structure their constituent liquid into a network that is highly stable (Barham 2001). The surface-active proteins in egg whites are so good at foaming in foods such as meringues that, ironically enough, a drier foam can be attained from egg whites by adding additional water to them, that is, a greater overrun can be attained by diluting their surface-active polymers (This 2010). The solid-like nature of an egg white sucrose foam ($\phi = 0.81$) can be seen in Figure 15.7. The elastic modulus of approximately 100 Pa (at a frequency of 10 Hz) arises from interfacial tension effects; the stress imposed by the rheometer is resisted by restoring elastic forces associated with the stretching of the faces of the bubbles in the foam during the deformation process (Princen and Kiss 1986). When the imposed stress exceeds a value of approximately 0.3 Pa, the foam yields. The values of yield stress and elastic modulus infer that foams have low yield strains, in this case less than half a percent. The yielding happens as the liquid in the Plateau borders causes bubbles to slip relative to one another and the foam exhibits plastic behavior (Cohen-Addad et al. 2013; Durian 1995).

15.4.2 RHEOLOGY

At an industrial food products level, there is the need to process biofoams, and this invariably means deforming these materials and getting them to flow. Therefore, an understanding of the rheology of biofoams permits processing to be performed without total loss of their structural integrity. As will be appreciated from Section 15.4.1, in defining the rheology of biofoams, the distinction between bubbly liquids and foams is a prerequisite. For bubbly liquids, their shear-thinning behavior requires two parameters to define flow properties, the consistency, K, and the flow index, n; for the bubbly liquid in Figure 15.6, $K = 0.53$ Pas$^{0.86}$ and $n = 0.86$. Often, in the food industry, product flow properties are defined by an apparent viscosity. This restricted definition of viscosity is relevant to a shear rate that is imposed on the material in a particular process operation, and for shear-thinning materials, this varies with shear rate. For example, in Figure 15.6, the apparent viscosity of the aerated liquid drops by one-third in tripling the shear rate from 30 to 90 s^{-1}.

For foams, a third parameter, the yield stress (τ_o), must be added to fully capture material properties at all shear rates ($\dot{\gamma}$) because at stresses (τ) below a critical value, the solid-like nature of the foam predominates and the foam does not flow. The Herschel-Buckley equation is appropriate for defining flow properties once the void fraction reaches criticality:

$$\tau = \tau_o + K\dot{\gamma}^n \tag{15.10}$$

The yield stress stabilizes the food biofoam when applied stresses are low and this stability is governed by the structure of the foam. For example, adding a layer of confiture of a given thickness to a sugar frosting may only work if the frosting is made up of small bubbles; a frosting made up of an equivalent volume fraction of larger bubbles may not sustain the load of the confiture layer. This is because the Sauter radius of the bubbles in the foam is a key length scale (Princen and Kiss 1986) dictating the mechanical properties of a foam. The Sauter radius, R_{32}, is the surface-volume mean radius of gas cells in their spherical equal volume state, and is expressed as:

$$R_{32} = \frac{\sum_i n_i R_i^3}{\sum_i n_i R_i^2} \tag{15.11}$$

where:
n_i is the number of bubbles of a given radius R_i

The enhanced mechanical response of a foam due to smaller bubbles can be appreciated from the inverse relationship between yield stress and Sauter radius:

$$\tau_o = \frac{\gamma}{R_{32}} f(\phi) \tag{15.12}$$

where the numerical value of $f(\phi)$ depends on how far the volume fraction of bubbles is from the critical value. The static elastic modulus of the biofoam also scales inversely with bubble size. Thus, a well-whipped foam is not only stiffened because it incorporates a large volume fraction of air into its liquid matrix, but because the substantial subdivision of bubbles that occurs during a longer whipping process stiffens it further.

15.4.3 Dynamics of Aerated Foods and Food Foams

The production of aerated foods is a challenge for food technologists because it is not easy to create high-quality products consistently from materials that are highly prone to changes in structure (Campbell et al. 2008). Structural destabilization occurs in both bubbly liquids and foams because of two types of structural disruption: gravity-induced transformation and biofoam coarsening (Saint-Jalmes 2006).

In bubbly liquids, because the density of gas (in the bubbles) is much less than the density of the liquid matrix, there is a tendency for the bubbles to rise (the creaming phenomenon). This may be particularly noticeable in muffin mixes when the viscosity of the continuous phase decreases as the batter temperature rises when the muffins are placed in the oven. The ensuing tunneling that arises from the creaming of bubbles is a noticeable defect in the finished product.

In foams, the bubbles are locked in place by topological constraints (Weaire and Hutzler 1999). In this case, gravity acts on the denser component of the foam, and the liquid moves from the faces and down through the Plateau borders that define the boundaries of neighboring bubbles. As a result of the drainage of liquid, the face separating a pair of bubbles can thin, especially toward the top

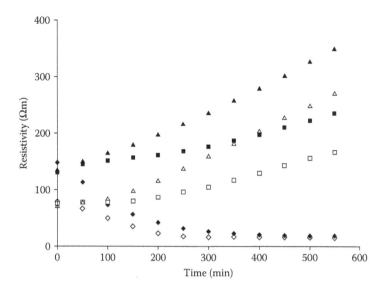

FIGURE 15.8 Change in electrical resistivity of wet ($\phi = 0.66$, open symbols) and dry ($\phi = 0.79$, closed symbols) egg white-sucrose foams measured toward the top ($\triangle,\blacktriangle$), middle ($\blacksquare,\square$), and bottom ($\blacklozenge,\lozenge$) of a column of the foam.

of the foam, and this can accelerate structural changes due to coarsening. The wetter the foam, the easier it is for the draining liquid to overcome capillarity effects in the Plateau borders. This can be seen in Figure 15.8, where two foams made from sucrose egg whites mixes have their electrical resistance measured as a function of height. As the liquid drains, the loss of liquid higher up the foam does not allow effective transport of electrical current and so resistivity rises with time. The rate of change is faster in the wet foam ($\phi = 0.66$) since the reduced capillary forces make it easier for the liquid to move down through the network of Plateau borders. Toward the bottom of the foam, the draining liquid from the top layers causes growth in the thickness of the walls and Plateau borders in the bottom portion of the foam and the electrical resistance falls. As with the drying of the top layers, enhanced conductivity occurs at a faster rate at the bottom of the wetter foam because downward transport of liquid is easier.

The second destabilization phenomenon, coarsening, is a process where the median bubble size grows with time. In both foams and bubbly liquids, coarsening is driven by disproportionation, while in foams, bubble coalescence further hastens coarsening.

Coalescence occurs when loss of liquid from the faces is such that a critical thickness is reached where there is an abrupt destabilization. The rupture of the face between adjacent bubbles allows two smaller bubbles to become one larger bubble. Coalescence and drainage are synergistic since the loss of liquid from the upper portions of the foam thins the walls increasing the probability of coalescence. Coalescence may be diminished if stabilizing molecules are left at the thinned face as the liquid drains (Murray and Ettelaie 2004).

In disproportionation, diffusive-driven transport of gas occurs due to the slight pressure differences that exist between bubbles of different sizes (Equation 15.2). The rate (kg s^{-1}) at which gas diffuses ($\dot{m}$) is given by

$$\dot{m} = kA\Delta p \tag{15.13}$$

where:
k is a permeability constant
A is the surface area through which the gas diffuses (Weaire and Hutzler 1999)

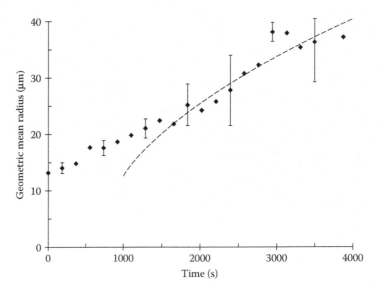

FIGURE 15.9 Evolution of the geometric mean bubble radius for a well-whipped egg white sucrose foam ($\phi = 0.81$). R^2 dependence on time ($R \sim [t\text{-}t_0]^{0.5}$, with $t_0 = 675$ s) drawn for longer times.

In bubbly liquids, k is lower than in foams due to the larger diffusive path length, which the molecules must cross to go from the smaller to the larger bubbles. Conversely, in foams, where bubbles are separated by thin faces, the rate of growth in the median bubble size due to disproportionation is fast. The solubility of the gas in the aqueous matrix affects the value of k, with higher solubilities speeding the rate of bubble growth (Cohen-Addad et al. 2013). Differences in gas solubility that retard disproportionation are exploited in the brewing industry: nitrogen, which at room temperature is about 85 times less soluble than carbon dioxide in water, prolongs the life of the head on a glass of beer due to its lower permeability slowing the rate of disproportionation. Nitrogen in the Guinness widget is a key feature in this technology in order that a vital quality attribute for the beverage is retained in the canned product (a stable head for the lifetime of the beer). A fascinating tale of government policy changes, market forces and the technology that led to the invention of the widget is described in the book *Bubbles in Food* (Campbell et al. 1999, Chapter 31). Also in the book is a neat study (Chapter 36), showing why it is essential that not all of the carbon dioxide is substituted by nitrogen in order that the beer maintains a desirable mouthfeel.

Because disproportionation is a diffusively driven process, bubble size grows according to a square root relationship with time:

$$R^2 \sim \left(t - t_0\right) \tag{15.14}$$

Changes in bubble size over time can be pronounced, as seen in Figure 15.9, where median bubble size in a meringue mix increases threefold over the course of one hour at room temperature. The expected quadratic dependence is not manifest till later times, likely due to both drainage and disproportionation occurring in the foam.

15.5 STABILIZATION OF FOAMS AND AERATED FOODS

Because of the dynamic changes that can disrupt the structure of biofoams, the food technologist or the culinary artist usually adds a stabilization step so that additional time-dependent changes in biofoam structure are minimized. Viscosity enhancement by a drop in temperature may suffice to stabilize a biofoam for long enough that its structure is maintained and the quality of the product is assured.

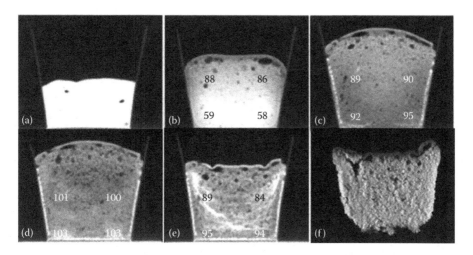

FIGURE 15.10 Cake collapse during baking and cooling due to use of untreated flour. Subparts show X-ray tomography images, while F is a C-cell image of the resulting cake crumb structure. (Data from Whitworth, M., X-ray tomography of structure formation in bread and cakes during baking, in: Campbell, G.M., Scanlon, M.G., Pyle, D.L., Eds., *Bubbles in Food 2: Novelty, Health and Luxury,* Eagan Press, St. Paul, MN, 273–286, 2008.)

This stabilization may be enhanced by phase transitions, for example, water in ice cream or lipids in aerated chocolate, or by physical gelation, as in the stabilization of marshmallows. Another common means of stabilization is through thermosetting. A denaturation transition in the proteins is a typical mechanism that permits such thermal stabilizations, for example, in a meringue, but stabilization may be further enhanced by a very large increase in viscosity due to starch gelatinization, as occurring in the baking stabilization of cakes and bread.

In thermosetting, the process itself can actually cause destabilization of the aerated structure. This is evident in bread and cake manufacture, where the rising temperature in the oven increases bubble size as a result of more gas being generated in the matrix that then desorbs into the bubbles; thermal expansion of the gas in the bubbles also contributes to bubble size growth. A poor-quality loaf with large holes can then arise if the viscosity of the dough matrix is insufficiently high to resist the face thinning that can cause significant coalescence of adjacent bubbles. In cakes, gluten *strength* is weaker (so as not to impart a rubbery texture to the cake) and so the consequences of low viscosity during bubble expansion may be readily apparent. Crust collapse at the top of the aerated structure can be one such consequence in cake production. This is seen in Figure 15.10, where the gas in the bubbles contracts as temperature drops when the cake is removed from the oven. If the walls separating adjacent bubbles have been substantially thinned they cannot withstand the load of the cake's superstructure, and so the cake densifies as the walls buckle (Gibson and Ashby 1997).

15.6 CONCLUSIONS

Bubbles are an integral part of many foods and beverages, where they modify food properties to enhance product quality. The inherent instability of bubbles represents a significant research challenge for food scientists and technologists striving to control bubbles to create foamy food systems. Control of bubble size and interfacial tension are critical parameters in this regard.

ACKNOWLEDGMENTS

I am the grateful to the Natural Sciences and Engineering Research Council, Canada, for research support for a number of years to study aerated food systems. I also acknowledge my collaborator

Dr John Page in the Department of Physics & Astronomy and our former students Guillermo Bellido, Daiva Daugelaite, and Jeremy Spencer. I am also grateful to AACC International and Dr Martin Whitworth for permission to include Figures 15.4 and 15.10 in this chapter.

REFERENCES

Baker, J.C., Mize, M.D., 1941. The origin of the gas cell in bread dough. *Cereal Chemistry* 18, 19–34.

Barham, P., 2001. *The Science of Cooking*. Springer-Verlag, Berlin, Germany.

Bellido, G.G., Scanlon, M.G., Page, J.H., 2009. Measurement of dough specific volume in chemically leavened dough systems. *Journal of Cereal Science* 49, 212–218.

Campbell, G.M., Mougeot, E., 1999. Creation and characterization of aerated food products. *Trends in Food Science & Technology* 10, 283–296.

Campbell, G.M., Scanlon, M.G., Pyle, D.L. (Eds.), 2008. *Bubbles in Food 2: Novelty, Health and Luxury*. Eagan Press, St. Paul, MN.

Campbell, G.M., Webb, C., Pandiella, S.S., Niranjan, K. (Eds.), 1999. *Bubbles in Food*. Eagan Press, St. Paul, MN.

Cauvain, S.P. (Ed.), 2003. *Breadmaking: Improving Quality*. Woodhead Publishing, Cambridge.

Cohen-Addad, S., Höhler, R., Pitois, O., 2013. Flow in foams and flowing foams. *Annual Reviews in Fluid Mechanics* 45, 241–267.

Durian, D.J., 1995. Foam mechanics at the bubble scale. *Physical Review Letters* 75, 4780–4783.

Friberg, S.E., Larsson, K., Sjöblom, J., 2004. *Food Emulsions*, 4th ed. Marcel Dekker, New York.

Gibson, L.J., Ashby, M.F., 1997. *Cellular Solids: Structure and Properties*, 2nd ed. Cambridge University Press, Cambridge.

Ip, S.W., Toguri, J.M., 1994. The equivalency of surface tension, energy and surface free energy surface. *Journal of Materials Science* 29, 688–692.

Kohama, Y., Kukizaki, M., Nakashima, T., 2009. Method for producing monodisperse bubbles. US Patent 7,591,452 (B2).

Kralova, I., Sjöblom, J., 2009. Surfactants used in food industry: A review. *Journal of Dispersion Science and Technology* 30, 1363–1383.

Kraynik, A.M., 1988. Foam flows. *Annual Reviews in Fluid Mechanics* 20, 325–357.

Kraynik, A.M., Reinelt, D.A., van Swol, F., 2003. Structure of random monodisperse foam. *Physical Review E* 67, 031403.

Kriegel, R., Huang, X., Grant, R.P., Radhakrishna, H., 2010. Bottles with controlled bubble release. US Patent Application Publication US 2010/0104697 (A1).

Liger-Belair, G., Polidori, G., Jeandet, P., 2008. Recent advances in the science of champagne bubbles. *Chemical Society Reviews* 37, 2361–2580.

Martin, A., Bos, M., Cohen Stuart, M., van Vliet, T., 2002. Stress-strain curves of adsorbed protein layers at the air/water interface measured with surface shear rheology. *Langmuir* 18, 1238–1243.

Mason, T.G., 1999. New fundamental concepts in emulsion rheology. *Current Opinion in Colloid and Interface Science* 4, 231–238.

McGee, H.J., Long, S.R., Briggs, W.R., 1984. Why whip egg whites in copper bowls? *Nature* 308, 667–668.

Mehta, K.L., Scanlon, M.G., Sapirstein, H.D., Page, J.H., 2009. Ultrasonic investigation of the effect of vegetable shortening and mixing time on the mechanical properties of bread dough. *Journal of Food Science* 74, E455–E461.

Murray, B.S., Ettelaie, R., 2004. Foam stability: Proteins and nanoparticles. *Current Opinion in Colloid and Interface Science* 9, 314–320.

Princen, H.M., Kiss, A.D., 1986. Rheology of foams and highly concentrated emulsions III. Static shear modulus. *Journal of Colloid and Interface Science* 112, 427–437.

Saint-Jalmes, A., 2006. Physical chemistry in foam drainage and coarsening. *Soft Matter* 2, 836–849.

Scanlon, M.G., 2004. Biogenic cellular solids, in: Dutcher, J.R., Marangoni, A.G., Eds., *Soft Materials: Structure and Dynamics*. Marcel Dekker, New York, pp. 321–349.

Scanlon, M.G., Zghal, M.C., 2001. Bread properties and crumb structure. *Food Research International* 34, 841–864.

Stanley, D.W., Goff, H.D., Smith, A.K., 1996. Texture-structure relationships in foamed dairy emulsions. *Food Research International* 29, 1–13.

Steffe, J.F., 1996. *Rheological Methods in Food Process Engineering*, 2nd ed. Freeman Press, East Lansing, MI.

This, H., 2010. Solution to the whipped egg white challenge. *Analytical and Bioanalytical Chemistry* 398, 1845–1845.

Weaire, D., Hutzler, S., 1999. *The Physics of Foams*. Clarendon Press, Oxford.

Whitworth, M., 2008. X-ray tomography of structure formation in bread and cakes during baking, in: Campbell, G.M., Scanlon, M.G., Pyle, D.L., Eds., *Bubbles in Food 2: Novelty, Health and Luxury*. Eagan Press, St. Paul, MN, pp. 273–286.

16 Formation and Stability of Food Foams and Aerated Emulsions

Sarah Santos-Murphy, Alistair Green, and Philip Cox

CONTENTS

16.1 INTRODUCTION

Foods in the form of foams are immensely common and recognizable. Foam or froth on drinks or in whipped cream and the cell structure of bread are probably the most familiar examples. This chapter hopes to identify why food foams are attractive to the consumer and the manufacturer. It will examine the prominent mechanisms behind food foam stability, their function, and finally highlight possible new trends. It should be noted that our familiarity with foamed foods belies the complexity of structure and function that even the most common foods must have and deliver.

This innate complexity is further exacerbated by the golden rules that control mass food manufacture, distribution, storage, and consumption, which are that food must be safe, it must be palatable, it must be affordable, and it must be safe (Norton et al. 2007). Among the priorities the overriding concern for the mass manufacturer is safety, although this is well covered elsewhere, second to this, possibly, is palatability. This somewhat nebulous term is then concerned with various factors. The most pressing factor of which is the mismatch of timescale of preservation (weeks to months) to the rate at which a food must be functional in the mouth (seconds). It is this required immediacy of response and the realization that multiple factors, for example, shear, melting, dilution, and compression are required to deliver all of its hedonic properties (Norton et al. 2006; Le Reverend et al. 2010). It is these competing factors that make food foams so fascinating. Such interesting concepts are, of course, not limited to food foams; thus, this chapter may stay away from strictly using food foams as examples, but by the end of the process, the integration of foam structures into new food should be apparent.

16.1.1 Food Challenges and the Hard Life Experienced by Foods

The primary challenge faced by either conventional or in this case a product with a significant gas phase volume is probably postmanufacture stability; this is particularly true during distribution and storage. But what should not be underestimated are the mechanisms of use. Microbial considerations aside it is all too common for products to experience intermittent consumption (Norton et al. 2008). This irregular process is not to be taken too lightly as with the common examples of whipped cream, butter, or mayonnaise, each experiences a prolonged history of use over days or even months. This extended period of use will be studded with episodes of temperature and physical abuse as the product is removed from chilled storage, scraped, mixed, spooned, and then eventually returned to a thermally shocking environment. These repeated and variable actions should have no or an appreciably small effect upon the product or its physical or oral properties. In a nutshell, the last spoon of mayonnaise, which has sat at the back of the refrigerator for the past three months, should be as physically or orally satisfying as when the jar was opened. This is no mean feat but is commonly expected from the food industry using only a *natural palette* of ingredients, however, as an industry it commonly succeeds. What has not been considered here are the physical mechanisms that affect aerated food and its longevity. Most notably these include coalescence, ripening, and film drainage. However, most of these effects are system specific; for example, beer foam will coarsen through drainage around the air cells and coalescence. Bread due to the durability of the cell walls in the structure does not suffer from these effects. Thus, it is difficult to generalize over destabilization mechanism; suffice it to say that these processes are always present due to phase density differences and the continuous presence of gravity.

As introduced above, the greater challenge to novel or existing foods along with the maintenance of stability over time is the persistence of meta-stability, so that once eaten, the foods can perform over a very short space of time to deliver all of their hedonic properties (Norton et al. 2006; Le Reverend et al. 2010). Indeed, not all of the properties may be sensorial responses in mouth. More commonly, novel foods are expected to provide either controlled delivery/functionality/structuring further down the gastrointestinal tract and again controlled meta-stability is the key to this and once more this has to be achieved within the boundaries of food production unit operations and the *palette* of ingredients available to the food engineer.

An exceedingly apt example of the effects of meta-stability is the phase inversion of butter or its analogues in the mouth. This change of phase produces the pleasant sensorial response and associated flavor release. However, such functionality is only possible if the product has been engineered to have the correct stability, and in the case of oral response, a meta-stability from kinetically trapped structures that may be appropriately disrupted in the mouth. In addition, the functional properties of a product are also defined by its microstructure and stability, that is, flow behavior and stability. Thus, the competing demands of functionality, meta-stability, performance, microbial safety, and so on must all be engineered at the point of manufacture. In the case of the product classes considered here, the manufacturing process is either predominantly emulsification and or phase inversion.

If a very common foamed food, such as ice cream, is then used as an example with the above processes then the physics of these effects (i.e., eating!) stay the same but the complexity of the system increases dramatically, so that when the equally familiar product of ice cream replaces butter, instead of having an essentially two phases of fat crystals and water to consider (incidentally, there is a surprising amount of air incorporated into butter), there are now five distinct phases. The complexity of the argument is therefore best expressed through the analogous foam—that is, butter!

16.1.2 MOUTH MACHINE

Not surprisingly, any discussion on food interaction, microstructure, and perception must consider the mouth and the action of it upon the food placed within it. However, the mouth is a complex *machine* (see Figure 16.1), and its interaction with food structure is more complicated again. Possibly, the perception of these interactions is the most complex of all. In a recent study into ice-cream formulation and consumer perception, Santos-Murphy (2014) found that consumers' perception of ice cream was not determined by formulation; in fact, participants ate similar amounts of low- and high-quality ice creams. Indeed, all of the above concerns are inherently variable and by definition ever changing with time (Lillford 1991). The mouth volume itself differs between individuals and on average between the sexes. Reported values, based on the held mass of food, for adults are 30 ± 10 g for men and 25 ± 8 g for women (Medicis and Hiiemae 1998), with a typical *mouthful* being given as 18 ± 5 g for men and 13 ± 4 g for women (Medicis and Hiiemae 1988), respectively. But what happens to this 15–20 g of food when it enters the mouth? Upon ingestion, the epiglottis and the action of the lips (Kohyama et al. 2004) bind the oral cavity for the duration of the mastication process until swallowing is induced (Prinz and Lucas 1995). Initially, the action of the teeth comminutes foods into smaller pieces that are then more amenable to swallowing. However, this simple statement belies the complexity of the processes involved in eating, swallowing, and very importantly the perception of food (Hutchings and Lillford 1988; Lillford 2000).

The amount of mastication, and therefore, the duration of food in the mouth depends upon the amount of food present, and most importantly upon the type of food. Bourne (2002), based on the work of Lee and Camps (1991), described the mastication time and mastication pattern for a variety of foods. Water showed the shortest residence time and oral movement and was quickly swallowed after just one second. Honey, a more viscous Newtonian fluid, had an increased residence time (4 s with an increased level of oral processing by the tongue). Not surprisingly, neither of these fluids had any mastication involved in their consumption; when solid foods were tested, the patterns of mastication increased and the residence times rose considerably. Hard sweets showed profound movement around the oral cavity and a concomitantly longer holding time in the mouth (48 s), whereas a slightly more friable food, potato crisps, showed an intermediate level of mastication and residence of 5 s. But what is the purpose of this complex movement within the oral cavity and what are the processes that are acting upon food (especially solid or soft solid ones) during this time?

The teeth break the structure of foods into smaller pieces (comminution). At the same time, the process should provide gratification (Bourne 2002); clearly, this has to be an important consideration as we no longer eat merely to survive, and this has been the case for some considerable time (Garnsey 1999). This enjoyment of food is mediated by the texture or mouthfeel of the food, flavor released from it, and probably the appreciation of the social context within which the food is consumed, that is, the family dinner or a swish little restaurant on the Champs Elysees. The comminution process also allows for saliva to wet and lubricate the food and the moistened increased surface area provides improved access for amylase to begin its attack on starch (Bongaerts et al. 2007). In addition, heat transfer from or to the food also takes place as the newly ingested food either heats or cools toward body temperature. Mass transfer may also take place as soluble ingredients dissolve in the increasingly available saliva, for example, sugars

(Kilkast 2004) and flavors to be released (Taylor and Linforth 1996). It is the particle size reduction and lubrication of the food that provides the feedstock for the final process (Lillford 2000; Peyron et al. 2004; Bongarts et al. 2007). This is where the broken, lubricated, but presumably adhesive, food particles reform into a bolus of suitable size and lubricity to be easily swallowed (Hutchings and Lillford 1988; Rosenthal 1999; Lillford 2000). It is therefore between the points of ingestion and swallowing that perception of food quality may take place. But this simplified approach may well be confounded as there is an old phrase, which remains very true: "we eat with our eyes."

McClements and Decker (2009) makes a good point: "Food undergoes many events on its way from the plate to the stomach, e.g., spooning, stirring, ingestion, mastication, and swallowing." Here, the treatment the food receives during these processes will change it dramatically. But what is sensed during these dramatic changes? If the complications of flavor, acoustics expectation, and context are conveniently ignored, due to their immense complexity, what remains is texture. However, even this singular concept remains difficult to accurately disentangle. The reader is now required to re-consider the previous few paragraphs and their implications but to consider the fate of contrasting foods through these processes. For example, how would, indeed does, beer foam differ from the response of ice cream or bread to the processes described above? Beyond this, how is the structure of the foodstuff maintained and how is it affected? More specifically, what is perceived by the consumer? Ideally, and as we shall see in more detail later, the beer foam is a delicate protein-stabilized structure that will not withstand mastication and will collapse rapidly. Conversely, bread with its protein and carbohydrate cell walls will moisten and collapse to a bolus for swallowing. Here, air cell size and wall strength play an important part in the process. Finally, in our short, but important, list of aerated model foods, what happens to the ice cream? Simply, the ice within the structure will melt, the fat from the air cell walls will become mobile the texture will become *creamier* and the sweet viscous fluid will pass over the taste buds on its way south.

16.1.3 Perception of Texture

The next step in considering foamed foods is to try and define texture and then to place this concept in a context of differing mechanisms to incorporate air into food. Lillford (2001) successfully used the concept of food material science and linked it to perception of foodstuffs via the physics encountered in the mouth. Such a powerful combination points to the need to correlate or integrate data from taste panels (Section 16.2) with instrumental measurements derived in the laboratory. As such, this is a sound route to take when trying to examine perception and, although both concepts are inherently complex, they are inextricably linked. In addition, most foods and indeed mouths are inhomogeneous; as such, median values for most pertinent parameters can only be considered (Lillford 2001). However, the number of possible parameters associated with perception is huge, as is the list for the potential factors defining the material properties of foods themselves. This is in addition to the confounding factors mentioned above and the action of the mouth (see Figure 16.1). As Kilkast (2004) points out, texture perception is composed of macrostructural and microstructural composition. As mastication and therefore comminution conveniently destroys the macrostructure (Peyron et al. 2004) a consideration of the effects of microstructure is important and approachable here and as pointed out by Lillford (2001), "texture is a consequence of the microstructure."

Le Reverand et al. (2010) concisely describe how the development of texture, rheology, and a variety of texture-related attributes are reliant upon a colloidal product's microstructure, that is, the ice cream. They have also summarized many of the milestones in the development of the theories to describe the interaction of emulsions and colloids within the mouth and their perception. Principally, Shama and Sherman (1973) and Shama et al. (1973) and then Stanley and Taylor (1993) and Akhtar et al. (2000) have defined and then refuted the concept of a typical shear rate within the mouth under which the rheology of food should be considered. In less engineering terms, this is a measure of the typical stress the tongue and pallet exert upon food during movement through the mouth (see Figure 16.1) and is indicative of the condition under which food is sensed

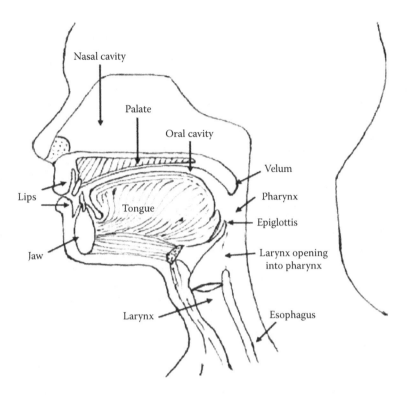

FIGURE 16.1 The anatomy of the mouth and oral cavity. (Reproduced from Le Reverend, B.J.D. et al., *Curr. Opin. Coll. Interf. Sci.*, 15, 84–89, 2010.)

(Van Aken et al. 2007). In the case of our foams, how much gas has been lost to create a condensed structure. This further leads to possible mathematical descriptors of the perception of colloidal foods and hydrocolloid thickened foods, for example, Kokini and Cussler (1983), Morris (1987), and Cutler et al. (1983). Importantly, Kokini (1987a, 1994) developed a correlation between a hedonic concept and instrumental measurements of a food's material properties. Equation 16.1 shows the general form for the equation:

$$\text{Creaminess} = A \times \log(\text{thickness}) + B \times \log(\text{smoothness}) - C \times \log(\text{slipperiness}) \qquad (16.1)$$

where:
A, B, and C are experimentally found constants

This approach sets a framework within which the data generated from sensory panels can be used to interpret instrumental measurements or vice versa. Here, creaminess is a hedonic characteristic that intuitively involves oil content and viscosity and has a mean numerical value readily gained from a sensory panel. However, as discussed below, panels or the use of them do have limitations, and it is the other parameters that are perhaps of interest here as they can be derived in the laboratory. Thickness is the first and most obvious parameter and has been measured for liquids or soft solids for some considerable while (Shama and Sherman 1973; Terpstra et al. 2009). Similarly, smoothness is a parameter that has a logical and instrumentally measurable basis (Strassburg et al. 2007). For example, Can a gased foam be considered smooth? The limits of sensitivity to texture are in some ways quite large, given the range of sensations the mouth can deliver with. Strassburg et al. (2007) suggested a lower limit for detecting an individual particle as a reasonably large 25 μm. When it is considered that most emulsified foods have a particle size range of approximately 1 μm, it is clear that for the majority of fluids, this term can rapidly be ignored. What is left is the somewhat vague term of *slipperiness*. Perhaps a better term to use is *lubrication*. Here, either term is used to describe the textural interaction of foods as they pass between the tongue and the palate during mastication and swallowing (see Figure 16.1). However, before moving on a passing mention should be made here to the action of saliva in providing lubrication beyond its other roles of wetting, dilution, adhesion, and digestion (Bongaerts et al. 2007). The most pertinent example is bread, think about toast. Saliva is a complex fluid composed of proteins and ions in an aqueous medium (Wilkinson et al. 2000). In the context of texture perception considered here, it is generally overlooked in many studies, but as stressed by Wilkinson et al. (2000), its presence in the mouth should not be ignored when developing instrumental measures of food texture. Most notably, saliva has three main confounding properties: (1) its production rate is highly variable between individuals (Mackie and Pangborn 1990) and therefore its dilution effects upon foods is variable; (2) it has a strong lubricating property itself and this affects and in turn is affected by the food present in the mouth (Davies et al. 2009; Vingerhoeds et al. 2009); and (3) its structure, and here its lubricating properties, may not remain constant or may be affected by the presence of astringent compounds (Vardhanabhuti and Foegeding 2010). Thus, all papers in this field should be read with the presence of saliva in mind, owing to the fundamental nature of saliva's purpose and presence (Bongaerts et al. 2007).

Phew, the concepts above form quite a heavy introduction. However, all of these processes happen in a coordinated fashion whenever a bite of food is taken. The job of the reader is now to consider the relative fates of the air in food and how it impinges on personal perception and enjoyment. A simple experiment is to take some ice cream and allow it to melt. Upon re-freezing, after the loss of air from the product try eating it, that's if you can get it out of the container! Small but amusing experiments aside what comes next is a discussion of how common food foams are constructed and stabilized. This is then continued into the use of air as a functional ingredient and to aid low fat formulations.

16.2   ICE CREAM—BUILDING A DESTROYING COMPLEX STRUCTURE

Ice cream is a commonly consumed frozen dessert, the term of which is used to cover a broad range of different types. These range from dairy ice creams, which are a frozen aerated mixture of dairy ingredients sugars and flavors, to sorbets, which are fruit-based aerated sugar syrup solutions that contain neither fat nor milk.

The legal definition of an ice cream varies from country to country. In America, for a product to be called an ice cream, the product must contain no less than 10% milk fat, and Canada, no less than 8%. However, in the United Kingdom, the industry is a little more complicated, with several categories of ice cream being available to consumers.

The structure of an ice cream is highly complex and is of great importance to the perception of in-mouth texture when it is consumed. This complex structure is formed during the aeration and freezing processes. The structural phases of the ice cream include ice crystals, fat globules, air bubbles, and a serum phase known as the *matrix*, consisting of sugars and stabilizers. This discrete phase is formed by secondary nucleation in a scrapped surface freezer (Figure 16.2).

Ice crystals can vary in size from 1 to 150 μm (Cook and Hartel 2010). It is vital to the mouthfeel and texture of the ice cream that the sizes of the ice crystals are small, with a diameter of no more than 50 μm, preferably around 35 μm, for limited detection. Above this, the ice crystals will be detected in the mouth and the ice cream will have a gritty mouthfeel (Marshall et al. 2003). The desired quality of an ice cream is via small ice crystals to give a smooth and palatable texture. It is therefore vital to control the rate of crystallization in order to develop ice crystals with correct size, shape, and distribution qualities. This can be achieved during the nucleation freezing process, which takes place during the freezing stage of manufacture (Clarke 2008). In order to produce small ice crystals, the nucleation must take place at a low temperature. The shorter time the ice-cream mixture is present in the freezer, coupled with slower dasher speeds, also aid in producing ice crystals with

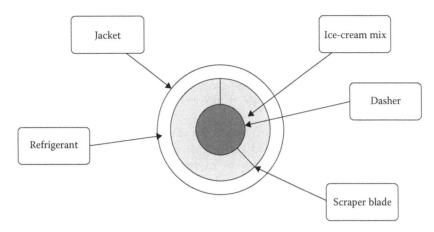

FIGURE 16.2 Diagram of a dasher. (Reproduced from Clarke, C., *The Science of Ice Cream*, 3rd edn, RSC Publishing, Cambridge, 2008.)

smaller mean size. This is the reason that Freon® and ammonia are used as a jacketed refrigerant in the freezing process, as it promotes rapid nucleation (Goff 1995). The whipping process also allows the ice crystals to remain discrete.

When an ice cream melts in the mouth, the structural elements change, as the ice melts and the fat stabilized foam structure collapses. The outside temperature and also the rate of heat transfer affect the rate at which the ice melts. However, this cannot occur until the fat-stabilized foam collapses, which is function of the physical properties and structure of partial coalescence of the fat globules, at the periphery of the air cells.

16.2.1 Air Incorporation and Structure

The fat structure is formed during homogenization in the manufacturing process and is vital to stabilizing the air phase of the formulation (Clarke 2008).

During homogenization, a pre-emulsion is formed by dispersing molten liquid fat in order to obtain droplet sizes below 2 µm (Eisner et al. 2007). However, research has indicated that the vital role of pasteurization also helps to melt the fat in order to create a well-homogenized fat emulsion (Marshall and Arbuckle 1996). Following high-pressure homogenization, discrete and partially coalesced fat droplets are present in both the matrix phase and at the surface of the air phase. High-pressure homogenization (200 atm) and double homogenization can produce smaller fat droplets allowing for an increased fat surface area (Clarke 2008). This is beneficial in low-fat ice-cream mixes as the air bubbles become more stable.

The emulsifiers control the structure of these fat globules and the degree of the partial coalescence in the system. Emulsifiers are added to the ice cream mix to increase the stability of the fat emulsion during freezing, and improve desirable qualities in the ice cream such as whipping ability and allow for a slower meltdown (Baer et al. 1997; Goff 1997). Also, the type of emulsifier, such as lecithin found in egg yolks, or Polysorbate 80, and the level of incorporation used can impact on the partial coalescence (Goff 1997; Davies et al. 2000). Generally, according to Tharp and Young (2012), in higher fat ice creams, lower levels of emulsifier are needed. Above a fat content of 15%, no emulsifier is usually required. Here, the food engineers' restricted palette of compounds reasserts itself and the excellent emulsifier known as Polysorbate 80 has a restriction of incorporation at around 0.06% due to its potential to taint flavor. Mono dygliceride/polysorbate blends are commonly used to address this problem (Tharp and Young 2012), as are stabilizers and emulsifier blends. The concentration of the total solids of the ice cream influences the amount of the blend that can be incorporated into the ice cream mix, and also depends on the type of the ice cream desired, that is, soft serve or low fat. Typical levels of emulsifier/stabilizer blend are between 1% (nonfat ice creams) and 0.4% (soft-scoop ice creams) (Naresh and Merchant 2006).

Fat, as well as being vital for structure, has a profound impact upon flavor and mouth coating of the ice creams. Some volatiles are soluble in oil and not in water and the solid fat particles allow an increase in viscosity of the matrix phase, which contributes to a decreased meltdown rate (Koxholt et al. 2001).

As an ice cream is both an emulsion and a foam, the size and distribution of air cells play a vital role in forming the structure of the ice cream (Ronteltap and Prins 1990). It is also intrinsic to the sensorial aspect of creaminess, a heavily desirable quality of the ice cream (Wildmoser et al. 2004), with smaller bubbles sizes producing a more pronounced sensation of creaminess. Defining *creaminess* is complex, and defining human sensitivity to creaminess is even more complex. It is generally accepted that creaminess has a hedonic level and is a key component of sensory appeal, especially in foods such as ice cream and yoghurt (Folkenberg and Martens 2003). Kokini (1987b) suggested that there was a relationship between thickness and creaminess attributes and shear stress on the palate. The perception of creaminess is a function of smoothness and thickness, which is related to rheological properties (Frøst and Janhøj 2007). Air is incorporated into the mixture during the freezing stage in the scraped surface heat exchanger, and small air bubbles (around 20–80 μm in diameter) are produced. The fat droplets are vital to the air interface, and during the freezing and aeration of the mix, the homogenized milk fat emulsion undergoes partial coalescence, causing the fat droplets to cluster and aggregate, which then form around the air bubbles and stabilize them.

16.2.2 AIR DESTABILIZATION

The purpose of air in an ice cream is to soften it. Without air, it would be hard and inedible. The air phase also allows for light to scatter, affecting color and appearance, and also hinders the separation of ice crystals reducing the risk of accretion, such as Ostwald ripening (Clarke 2008).

There are two coarsening mechanisms that involve air cells in ice cream: disproportionation and coalescence. These two mechanisms are somewhat interconnected, due to the rate of one on another (Walstra 1996). Disproportionation is analogous to Ostwald ripening of ice crystals; the Laplace pressure inside an air bubbles is larger than that of the outside. The smaller the bubbles, the larger the pressure; hence, there is a net transfer of air from the smaller bubbles to the larger, causing the smaller bubbles to disappear. The serum phase also allows for gas mobility. Coalescence occurs when two bubbles come into contact and the film separating them ruptures.

A way of inhibiting disproportionation is increasing the viscosity of the matrix phase (Sofjan and Hartle 2003). This increased viscosity reduces the rate of diffusion between bubbles, eliminating disproportionation. The absorption of emulsifiers at the bubble surface interface reduces coalescence by lowering the surface tension (Clarke 2008).

Overrun is the measurement of air whipped into the ice-cream mix during freezing and is calculated as a percentage increase of the finished product. Hartel (1996) stated that low overrun causes coarser ice crystal formation compared to the same formulation made with a higher overrun. This is because the air cells may aid the impediment of ice crystals during freezing. Flores and Goff (1999) suggested that a low amount overrun does not influence ice crystal size and but that around 70% is necessary to have a noticeable impact on microstructure. However, when the volume of air reached critical volume, increasing overrun had less impact on the overall ice-cream structure.

16.2.3 INTRODUCING MORE AIR

Dressaire et al. (2008) described that incorporating more air into an ice cream mix may be a way to increase the volume of the ice cream without adding caloric value. The incorporation of air is vital to the overall eventual microstructure of the ice cream, and the smaller the air bubbles, the more palatable it is to the consumer. It therefore would be beneficial to consider the use of micro-bubbles in the ice cream. However, the Laplace pressure always tries to destabilize such structures. The use of micro-bubbles in foods has lead to research that indicates improved and longer shelf life of products.

16.2.4 MATRIX PHASE

This is a highly viscous freeze-concentrated continuous serum phase in which the air bubbles, fat globules, and ice crystals are embedded. Containing a solution of dissolved colloidal sugars, proteins, and stabilizers, the concentration of these solutes is considerably higher than that of the mix due to 75% of the water in the mix being frozen. (Clarke 2008). The matrix phase also lends itself in aiding the mouthfeel of the ice cream. Due to the water in the phase becoming frozen in the form of ice crystals, the concentration of the dissolved sugars and stabilizers in the matrix phase will aid in determining the viscosity of the ice cream.

So far, we are trying to build the picture that the ice cream is a complex polyphasic system that comprises ice crystals, fat globules, air bubbles, and an unfrozen serum phase known as the *matrix*. These individual phases when combined provide the properties of the underlying microstructure to an ice cream. Ice crystals and air bubbles usually range between 20 and 50 μm, while the air bubbles are partially coated with fat droplets, which themselves are coated with an emulsified layer. The matrix phase consists of sugars and polysaccharides in a freeze-concentrated solution. Structural development then continues during manufacturing processes such as blending, pasteurization, homogenization, aging, freezing, and hardening (Clarke 2008).

Above this incredibly complex structure is the manufacturing processes, which are vital to the development and stability of the microstructure. Generally, the manufacturing can be broken down into two stages: mix preparation and freezing operations (Marshal et al. 2003; Clarke 2008). Mix preparation consists of blending ingredients, batch or continuous pasteurization, homogenization, cooling, and aging. The freezing operations then begin once the mixture has aged. This creates two discrete phases; millions of tiny air bubbles and ice crystals are dispersed into the concentrated mix via batch or continuous freezing, and then the ice cream is packaged, left to harden, and then stored, ready to be distributed.

16.2.5 CAN AIR HELP IN REDUCING FAT?

Fat is vital to the structural and sensorial properties of the ice cream. The fat in dairy products increase richness of flavor, carries flavor compounds, allows for lubrication of the palette, aids the desirable melting properties, and provides structure for foam stabilization (Marshall et al. 2003). While premium ice creams contain between 10% and 18% fat, low-fat alternatives are now being sought out by consumers. However, consumers still wish to have all the perceived and sensory qualities of a premium full-fat ice cream, as was seen in the study by Yilsay et al. (2005), into fat replacers. Results emphasized the importance of fat as a flavor enhancer.

Fat in dairy ice creams comes from milk fat, such as cream, buttermilk, and anhydrous milk fats (Marshall et al. 2003). It is the volatile fatty acid chains of the triglycerides in the milk fat that lead to the flavor of milk fat, and allow for a wide melting range. (Goff 2008). This melting range causes the ice crystals to melt and the fat-stabilized foam structure to collapse. When this occurs in the mouth, it aids the mouth coating properties of the ice cream, leading to an increased sensation of creaminess.

This sensation of creaminess is of the upmost importance when it comes to ice cream acceptance by consumers. Santos-Murphy (2014) found that there was very little difference in consumer palatibity ratings between two differing-quality ice creams. There was also no difference in the overall intake of the two different ice creams (Figure 16.3).

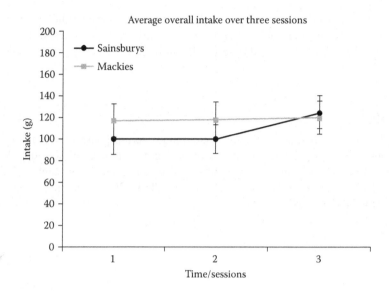

FIGURE 16.3 Overall intake of ice creams over sessions. (From Santos-Murphy, S.M., *Ice Cream: An Approach Considering Formulation Microstructure and Eating Behaviour* A thesis submitted at the University of Birmingham, Birmingham, 2014.)

16.2.6 REPLACING SUGARS

Sugar has several vital functions in ice cream: not only does it give the product a sweet flavor and desirable taste and it affects the viscosity of the matrix phase, but it also depresses the freezing point of the matrix phase, which in turn affects the foam properties and stability (Clarke 2008). Commonly used sugars include sucrose. By manipulating the quantity and type of sugar, a harder or softer ice cream can be produced. Literature has indicated that polyols may be used as a sugar replacement in ice cream formulation (Bordi et al. 2005). Polyols are sugar alcohols that still structurally resemble sugars, but have the advantage of a lower calorific value when compared to sucrose and a reduced insulin response (Zumbe et al. 2001; Livesey 2003). However, Koutsou et al. (1996) and Clarke (2008) have both documented that the use of polyols in food can lead to gastroenteric problems when consumed in large amounts.

Hopefully, this section has convinced the reader of the absolute complexities components of structure, formulation, and processing incorporated in ice cream. The next section continues this story of fascinating complexity.

16.3 BREAD

Despite an immensely long history and familiarity, bread research continues to fascinate scientists and engineers alike. As well as providing a complex but utterly familiar set of scientific/material properties challenges, it also provides a secondary, or primary, depending upon your perspective, set of trials for the process engineer. Simplistically, bread is a biphasic solid, which is soft, and flavorsome. To others, bread is a polyphasic material, which is changeable in its raw material state, difficult to process; however, it is wholly dependent upon its structure to deliver its functionality but produced in staggeringly large amounts for very low costs (Scanlon et al. 2002).

The structural side of cereal and bread science is nicely described by the review of Goesaert et al. (2005). But to provide a more *digestible* context: sliced white bread, suitable for sandwich making, has a daily production within the United Kingdom of 14 million loaves. This is the formulation, proofing, moving, dividing, baking, packaging, distribution, retailing, and consumption of a fresh product for under $2 and all in perfect condition. Thus, the two paradigms of food research, mentioned above, are satisfied as bread offers an incredibly interesting set of materials to work with but with the need to manufacture staggeringly large quantities of complex and fragile materials with very low unit costs. Layered on top of this complex foam is the fact that its primary ingredient varies between type batch and age (Gupta et al. 1992). Plus, the scale of machinery used in the formation present potential differences in process history. Finally, and for some most interestingly, bread dough can spontaneously rebel against the company and unexplainably become sticky and refuse to be processed (Chen and Hoseney 1995).

16.3.1 Overview of the Inflation of Bread

The cell walls of bread are primarily constructed from wheat flour and 70%–75% of this material is starch with 10%–12% protein. The balance of the mass is made from a variety of polysaccharides and lipids (around 5%) (Goesaert et al. 2005).

However, prior to processing, the dough contains in the region of 70% water and a low initial gas content; post process, the water content of fresh bread is 14% and the gas content is in the region of 70% (by volume) (Mills et al. 2003; Goesaert et al. 2005).

Thus, there is a considerable transformation in the structure when processing from mixed wet dough to expanded *dry* bread. The four main aids to this gross morphological change are the starch, gluten, CO_2, and heat. Gluten is a special case and almost unique in its properties. As such it will be discussed in slightly more detail later. However, the role of starch in the bread process is not to be underestimated. During the mixing stage of the dough, the starch from the flour is present as swollen mobile granules. These are free to move during mixing and in the proofing where proto air cells are incorporated by the mixing process and the viscous paste hydrates Goesaert et al. (2005). Obviously, starch is not the only macromeolecule present within the dough; however, it does provide a significant portion of the mechanical strength of the bread we eat, but also during the proofing and cooking stages. As just mentioned, air is incorporated into the wet viscous mass of the dough by mixing. It is these *seed air cells* that now inflate during proofing; therefore, no mixing no rise (Trinh 2013). The air bubbles that are incorporated into the dough during the mixing impact substantially on the overall final texture of the bread (Leroy et al. 2008) During mixing, gas is concentrated in the liquid phase of the dough in the form for small nuclei and these expand due to the release of fermentation gases and during baking, these gases continue to increase due to temperature increases. The foam structure of fermenting bread dough occurs due to a series of mechanisms relating to the dispersion of gas cells and consequently as certain amount of loss of the gas during baking. It is believed that the role of gluten proteins and non-starch poplar lipids can be used to control this (Carlson 1981). The expansion in the dough volume as a result of yeast fermentation and expansion continues until approximately 55°C, during the baking process (Pérez-Nieto 2010). This expansion of dough affects the finished loaf height, and is most rapid during the first stage of baking.

As the temperature approaches 55°C, there is a significant decrease in the expansion rate until the bread dough enters the final cooking stage, where the height of the bread remains for a short period until the centre of the dough continues to rise as it is at a lower temperature and the rest of the dough reduces its height slightly and then the loaf height reduces at a slow rate, with less than a 5% loss in weight during this stage (Pérez-Nieto 2010).

Upon cooling, the gelatinized starch retrogrades back into a more crystalline state (the involvement of the starch's intimate relationship with the gluten is discussed in a moment). This stiffens of wall structure then forms the mechanically strong but pliable structure of the bread crumb (Gan et al. 1995; Mills et al. 2003).

The microstructure of bread is obviously complex and goes through several different transitional structural phases from dough to baked good. Bread dough, made from yeast, flour and water, exhibits viscoelastic behavior and each of these ingredients plays a vital role in the structure formation of the dough. However, it is not just the role of ingredients, but the formation of the gas bubbles in the dough that aid in the creation of what will become the breadcrumb structure in the final baked product.

16.3.2 BREADCRUMB STRUCTURE

The final crumb structure of a baked loaf of bread is determined through the formation and retention of gas bubbles during the dough phase. Breadcrumb is porous consisting of two main components: crumb cell wall structure and gas bubbles. The crumb wall structure—often referred to as the *matrix* (Keetles et al. 1996)—contributes to the bread's structure and mechanical properties. The size and distribution of gas bubbles varies from loaf to loaf, but are of generally a small size, around 70–80 μm, which then expand to reach up to a few millimeters in diameter in the final product (Shimiya and Nakamura 1997). For a consumer acceptable product, the optimal structure of the matrix is required to be thin, but resilient to deformation, as being squeezed is often a consumer test for freshness.

To assess breadcrumb structure, digital imaging is often used (Zghal et al. 2001). Characterizations can often include density and crumb grain features, such as gas bubble size, crumb wall thickness, and void fraction. Measures of cell size uniformity and fraction of missing cell walls can also often be determined. Cauvain et al. (1999) assessed how different stages of the bread making process affected gas bubbles formation and structure and how this then impacted on the finished baked product.

16.3.3 MECHANICAL PROPERTIES OF BREAD

The texture of bread is intrinsically linked with the its mechanical properties, of which thick firmness and resilience are two of the most critical properties (Cauvain and Young 2004); the relationship between structural and mechanical properties of bread are also of important note, creating difference between bread dough, in which structural and mechanical differences could be tested and obtained using different flour types, as different flours can have different absorption rates, which lead to differences in the cell walls (Zghal et al. 2001). Specific mechanical properties of measuring properties of bread structure can be gathered from tensile testing, including stress strain, failure stress, and energy to failure.

As described, 70% of the final loaf volume is the gas phase, and has vital impact on its sensorial attributes. One major challenge of bread producers is controlling this gas phase during the early stages of baking and through the proving process, as the gas must be captured within the dough with the aim of it being released at the end of baking. The gas cell structure is delicate and many factors determine the gas cell formation. These include the initial formation of the foam structure during mixing and the stabilization of the foam structure through factors such as disproportionation and coalescence.

16.3.4 GLUTEN AND ITS PARTICULAR PROPERTIES

Gluten plays a rather important role in the structure and cell wall formation of bread. The cell walls of the dough have swollen starch granules, which are interlinked by a mixture of aggregated starch and gluten proteins. Gluten provides important elasticity to the air cells to allow for expansion. However, it retains its role even after baking. Bread quality decreases during storage, because breadcrumb firms and loses resilience (Goesaert et al. 2009). However, Gray and Bemiller (2003) have debated the exact nature of flour components in bread firming. Lagrain et al. (2007) investigated the role of gluten and starch in crumb structure and texture of fresh and stored straight bread dough. This was done by specifically targeting gluten functionality with redox agents and changing starch properties with maltogenic exo-amylase.

Results indicated that while the elastic modulus of the breadcrumb strongly increased on storage, its cellular structure did not change. Changing the gluten properties in the dough has a profound effect on the overall density and its foam structure, while not impacting on its rheological properties of the crumb cell wall. Inclusion of antifirming amylase did not alter the density or the cell structure but strongly impacted on storage of the crumb texture.

Obviously, this is a gross oversimplification of how the cellular foam structure is built up by the material properties of flour and the accumulation of CO_2. Indeed, the surface chemistry of air cells is worthy of a chapter in its own right (see Figure 16.2 in Mills et al. 2003!), but it is hoped that it encapsulates the complexity of the structure and the chemistry. It is then the job of the process and the process engineer to control these competing activities (e.g., starch gelatinization, retrogradation, CO_2 evolution, and gluten elasticity) and regulate them for the production of large numbers of loaves. In addition, the engineer has to take the variability of materials and process into account and still deliver. For example, not all wheat strains are the same and most show different gluten contents.

16.3.5 Engineering Foam Structure

Figure 16.4 at first glance appears relatively obvious for most foam systems. However, this excellent figure from Zghal et al. (2002) when viewed as the response of different breads can be seen to encompass a vast array of the concepts presented in this chapter. Each of the symbols (○, ●, and ▲) represents a different variety of flour (more exactly the strength of the gluten in them). Thus, the mechanical responses for different varieties can be used to tailor a particular response, that is, the variety represented by the squares has a low fracture stress (softer) than the variety with circle, which cluster higher up the curve and demonstrate a stronger (firmer) texture. The open symbols represent the same flour varieties but in these instances (not the squares here), the flours have been hydrated to slightly different levels (3%–5% difference). Now the differences between varieties are less marked. Thus, the process it receives can affect the final product. However, the model used to develop the predictions of the food response relied upon knowledge of the size of the air cells and

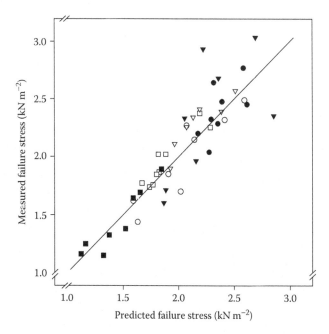

FIGURE 16.4 The correlation between a predicted mechanical response and a measured fracture force and that predicted from the observed structure. (Modified from Zghal, M.C. et al., *J. Cereal Sci.* 36, 167–176, 2002.)

the lamella walls and the air cell sizes and their relative degree of coalescence. Thus, bread can be described mechanically and *a priori* knowledge of the structure can inform all parts of the production and supply chain. Finally, all the varieties and hydration levels cluster about the 45° line, that is, despite the care with which the experiments were performed and the accuracy of the prediction and measurements bread resolutely refuses to give up its natural variance and this happens 14 million times a day on one small island in western Europe alone.

Bread and cake production is essentially hydrating the starch and protein to form an elastic matrix, which subsequently expands and is heat fixed (Zanoni et al. 1993; Mondal and Datta 2008). Indeed, the following quote (from Highfield 2015) encapsulates the uniqueness of bread and bread gluten encapsulates gas quite succinctly:

> Bread is special because of its bubbles. It's got these bubbles because wheat, when mixed with water, salt and yeast is the only cereal that can trap the carbon dioxide and give us raised bread.

16.4 WET PROTEIN FOAMS—BLURRING THE EDGES

Emulsions and foams are, for the purposes of this part of the chapter, synonymous; both are defined as fine dispersions of one fluid (either an immiscible liquid or a gas) in another fluid, most commonly water. Although foams generally have a larger size distribution, the mechanisms and geometries that govern their stability are, especially in the case of wet foams, identical, all be it often on somewhat different magnitude scales.

Food foams and emulsions, although gaining popularity in recent years, are widely taken for granted. Milk is an indispensable part of our early years (Dettwyler 1995) and while long-term stability is not an issue, during breast feeding, many of us continue to drink milk (cow's, goat's, or sheep's) for most of our lives. The stability of milk then becomes greatly important, to retailers and producers even if the consumer still takes the lack of separation for granted.

Remaining with the example of infants, the stability of milk becomes yet more complicated when dealing with children who are not being breast fed, but are still too young to be recommended cow's milk. When this is the case, formula milk is often used and the ability to mix a dried powder into a reasonably stable emulsion is surprisingly complex, even if only on the micron scale.

Other examples of common food foams and emulsions include mayonnaise, whipped cream, salad dressings, meringue, milkshakes, butter/margarine, and beer head. However, this list also shows a progression from fat-stabilized systems to protein structures. All these require careful balancing of ingredients, and correct processing to ensure they are as stable as possible, so as to ensure maximum shelf life, while maintaining the correct breakdown characteristics to ensure consumer satisfaction throughout the full length of the shelf life.

As is especially obvious with beer head, froths and foams are inherently unstable, and therefore are relatively rare; especially in nature, one of the reasons for this is the high energy requirement to break up the dispersed phase to produce the foam or emulsion (Cooper and Kennedy 2010). Another issue to natural foams and emulsions is creating a mono-disperse internal phase. This is difficult, and when not achieved, it results in a relatively short-lived foam or emulsion.

16.4.1 Stabilizing Materials

Many man-made emulsions, such as paint, skin cream, and shampoo, use chemical surfactants to stabilize the dispersed phase. Other materials that stabilize emulsions and foams include gels, as with champagne jelly using gelatin, or cakes and breads using a carbohydrate gel set during baking. Sugars are also used, primarily to increase the viscosity of the continuous phase; however, on the case of honeycomb, the sugar sets around the gas bubbles, forming a solid foam structure. In nature, however, the use of proteins dominates (Cooper and Kennedy 2010), being reproducible by plants and animals much more easily than any other material.

Proteins, unlike chemical surfactants and emulsifiers, tend to form an elastic film on the surface. This film forms through electrostatic, hydrophobic, and/or covalent interactions between the proteins, which have already been seen with bread. (MacRitchie 1978; Dickinson and Euston 1990; Krägel et al. 1995). The resultant film can be of variable thickness and elasticity depending upon the protein present and the prevailing conditions.

Many emulsion and foam stabilizers can be used in combination; for example, a meringue uses a combination of egg white proteins and sugar to produce foam stable enough to hold its shape until baked. The sugar adds viscosity to the continuous phase, while the proteins stabilize the air bubbles. Proteins and chemical emulsifiers, however, prevent each other from stabilizing the dispersion efficiently. This is due to the similarities between the way proteins and chemical emulsifiers form emulsions and foams and the differences between how they stabilize the dispersed phase. Both adsorb to the surface via their amphipathic nature, while proteins form a strong film across the surface of the droplet or bubble, while the surfactants form a mobile raft, moving across the surface to quickly fill vulnerable areas. When used together, the surfactants' mobility is reduced while the protein film's elasticity and stability are compromised (Coke et al. 1990; Wilde 1996).

Protein foams are also used in a variety of applications, from fluorinated whey and soy being used in fire-fighting foams (Perkowitz 2000) to egg whites being used in a meringue and barley and yeast proteins helping stabilize a beer's head (Cooper and Kennedy 2010).

16.4.2 FOAMING MECHANISM

It has been shown that random coil proteins are better at foaming than globular proteins (Graham and Phillips 1976, 1979; Townsend and Nakai 1983; Velissariou 1992); whether denatured by pH (Llopis and Albert 1959) or heat (O'Neill et al. 1989), the unfolded proteins show more surface activity. The more flexible protein structure is being more mobile and covers a greater area than the more compact globular proteins, adsorbing to the surface more quickly and forming dilute films more readily (Evans et al. 1970). This tertiary structure and the hydrophobicity of the amino acids that make up the protein chain determine the proteins' tendency to foam (Velissariou 1992); the more hydrophobic a section of the protein is, the more likely it is to adsorb to an interface, and the more unwound the protein, the more likely it is to encounter an interface. The same logic implies that a larger polypeptide will produce a more stable foam structure (Slack and Bamforth 1983).

Globular proteins, however, promote the best stability (Graham and Phillips 1976). This is due to the energy required to detach a globular structure from a surface, which is considerably higher than that required to remove an amino acid strand (Green et al. 2013). The surface tension of a droplet or air bubble, especially if the diameter is very small, often is sufficient to disrupt macromolecular conformation of a protein (Clarkson et al. 1999); however, some proteins, such as the hydrophobins, which are soon to be discussed, have strong enough tertiary structures to be able to resist damage to some or all of the protein, maintaining a globular conformation even at the interface.

It is this tendency for proteins to unfold, or remain globular, that goes some way toward explaining why some proteins foam well but make poor stabilizers, while others stabilize foams but make poor foaming agents (Wilde and Clark 1996; Ferreira et al. 2005). Some proteins also show conformational changes or aging at the interface (Douillard and Teissie 1991; Fillery-Travis et al. 2000), which can cause a slower surfactant response than with chemical surfactants and emulsifiers (Wilde and Clarke 1996), while others show reversible adsorption (Gonzalez and MacRitchie 1970; Herrington and Sahi 1987).

Due to the generally larger size and slower surfactant property response, protein emulsions and foams tend to have a larger size distribution, per unit of energy input, than those found with chemical emulsifiers and surfactants (Fillery-Travis et al. 2000).

Generally, to produce an acceptable structure, much higher quantities of protein are required than for chemical emulsifiers because at low concentrations, the adsorption is diffusion controlled (De Feijter and Benjamins 1987) and proteins diffuse much slower than their smaller molecular weight counterparts. It is therefore arguable that the solution should be moved away from the protein's isoelectric point (pI) as proteins are expected to be least soluble at the protein's pI. However, the protein is also likely to have maximum molecular association and greatest rate of adsorption to an interface (Poole et al. 1984; Kim and Kinsella 1985; Waniska and Kinsella 1985) and the surface viscoelasticity is at a maximum at the protein's pI (Kim and Kinsella 1985; Maeda et al. 1991). It is therefore a playoff between solubility and the protein's ability to adsorb into the interface and, once there, form an elastic film. However, high charge density, as one would expect from a protein away from its pI, is a disadvantage to foam stability (Townsend and Nakai 1983), as charge interactions between proteins increase the stability of foam (Velissariou 1992), while a large charge density proves to stabilize, electrostatically repulsing other particles (Husband et al. 1997; Green et al. 2013). Large hydrophilic loops or tails can also provide steric hindrance to emulsion particle interactions and film drainage (Fillery-Travis et al. 2000).

Protein adsorption can also be competitive, with the potential for displacement from the surface by a more concentrated protein in solution (Velissariou 1992); however, assuming that the two proteins are equally surface active, and both have the ability to contribute to the elastic film, the complex protein mixture will appear to behave like a single protein solution (Velissariou 1992).

Another method used to aid surface activity is salt addition, which promotes hydrophobic interactions, particularly at the gas-liquid interface (Llopis and Albert 1959), sacrificing solubility. The addition of short-chain alcohols does not affect surface tension, but increases surface viscosity. It does however reduce the film elasticity (Bucholtz et al. 1979). Increased addition of alcohol results in higher protein solubility, and foam stability, reaching a maximum before declining (Bumbullis et al. 1979; Velissariou 1988; Ahmed and Dickinson 1990; Crompton and Hagarty 1991).

16.4.3 Beer Foam—Combination of Theory and Practice

Beer foam is a dynamic foam that people are familiar with, and opinions on what the ideal beer head varies both generationally and geographically, as well as from person to person. Nevertheless, the importance of the head on a pint of beer, both aesthetically and organoleptically, on consumer satisfaction cannot be denied (Bamforth 1985; Langstaff and Lewis 1993a; Blasco et al. 2012) as the beer's head is perceived as an aesthetic representation of the beer's quality (Bamforth 1985; Langstaff and Lewis 1993b; Stewart 2004) assessed on stability, quantity, lacing, whiteness, creaminess, and strength (Roberts et al. 1978; Evans and Bamforth 2009). It has been shown that a beer with a fine foam is perceived as having a *well balanced* taste, while a course foam gives the perception of the beer flavor being *too soft and dull* (Ono et al. 1983).

To achieve appreciably *soft* foam, the air bubbles need to be as mono-dispersed as possible (Wilde and Clark 1996). Thus, the balance of bubble nucleation, growth, separation, rise, and foaming must be regulated to minimize variations in the bubble size. This, given the dynamic nature of the foam, the change in the height the bubbles must rise through, the intermittent turbulence within the glass, and the decreasing supersaturation of the gas dissolved within the beer, becomes far more complex than it initially appeared. This coupled with the variability of the ingredients, even from the same source, adds yet another factor to the equation.

16.4.4 Stability

Beer foam stability is dependent on proteins, metal ions, polysaccharides, and melanoides. (Blasco et al. 2012). With proteins being the key foaming factor (Roberts et al. 1978; Stewart 2004) plus glycoproteins (Velissariou 1992) and mannoproteins (Cameron et al. 1988) also playing a major role in the foam stability. Isohumulones, together with metal ions, also interact with proteins to improve the foamability of the beer (Roberts et al. 1978).

A fine balance is required during beer production, as the head on the beer is important; however, too much foaming during production can result in unwanted phenomenon (Blasco et al. 2011), such as the unnecessary loss of product, a large increase in the size of the vessel required, reducing the yield, and can affect the efficiency of the yeast fermentation. The foam also has the potential to foal the top of the production vessel and block the carbon dioxide exit tube, causing a build-up of pressure in the vessel, which poses a hazard to the manufacturer. Other notable examples of nature trying to prevent foam production are now mentioned below.

Van Nierop et al. (2004) reported that at differing heights above sea level, the boiling temperature of the wort changes, even if the same components are present, resulting in the amount of lipid transfer protein 1 (LTP1) changing. This can result in insufficient foaming if the LTP1 concentration is low while free fatty acid (FFA) levels are high. Conversely if the LTP1 levels are higher, the beer will foam correctly even in the presence of high FFA levels.

Similarly, non-specific lipid transfer protein (nsLTP) found in cereals, such as barley, has been shown to be important for beer foam stability (Sorensen et al. 1993). This is possible due to the binding of lipids such as FFAs (Fillery-Travis et al. 2000). Lipids are detrimental to foam stability, occupying the hydrophobic regions of the proteins and competitively adsorbing to the air–water interface (Roberts et al. 1978; Clark et al. 1994). This is why the head on a beer might remain present for a long time if left, but once drinking commences, the foam disperses quickly, as the oils from the skin around the mouth destabilize the foam.

16.4.5 GUSHING

Throughout this section, maximizing foam and emulsion stability has been the primary focus, with the possible exception of during the fermentation of beer. However, beer foam can become too stable, and build up too quickly, known as *gushing*. Gushing is when, in the absence of other causes, such as shaking a beer, and then opening it, rapidly forms a large volume of foam, causing the foam to overflow from the bottle (Stewart 2004), which results in a waste of the beer entrained within the foam and a significant degassing of the beer left in the bottle, thus affecting the mouthfeel.

There are two main categories of gushing, although the names are not particularly consistent. The first is primary gushing, or epidemic gushing (Stewart 2004), which is the result of something going wrong with the ingredient source, such as mould growth on the barley (Gjertsen et al. 1965; Amaha et al. 1973; Gyllang and Martinson 1976; Haikara 1980; Weideneder 1992; Niessen 1993; Zepf 1998), on an imbalance between the foaming promoters and suppressors (Carrington et al. 1972), which can also lead to unnaturally flat beer. The second type, secondary or sporadic gushing (Stewart 2004), is a result in a processing or storage error.

It has been noted that adding *Fusarium* to barley causes a gushing malt (Gjertsen et al. 1965), while addition to a mash of mycelial extracts or to the culture filtrate failed to cause gushing. The authors therefore concluded that *Fusarium* did cause gushing, however, not directly.

This supports the most commonly accepted cause of gushing, currently. This suggests that the presence of hydrophobins, secreted by filamentous fungi, is the cause of the phenomenon. Hippeli and Elstner (2002) suggested that hydrophobins were the Nigrospora-*gushing factor* isolated by Kitabatake and Amaha (1974). It is possible that hydrophobins are the cause, and could act as nucleation sites for the gas bubbles, while the highly surface-active nature of the proteins are more than capable of creating highly stable foam.

16.5 FOAMS AS FAT

We have just mentioned hydrophobins in the context of beer and Tchuenbou-Magaia et al. (2009b) introduced the concept of air-filled emulsions for fat reduction strategies. An air-filled emulsion is an encapsulation of air (typically 1–3 μm in diameter), which has the physical properties of air-filled emulsions (flowability, near colloidal stability, etc.). Indeed, Tchuenbou-Magaia et al. (2009b) concisely describe the routes to achieving these structures using hydrophobins. Figure 16.5 shows a confocal image of an air-filled emulsion, which has been stained to reveal the presence of protein.

FIGURE 16.5 A confocal laser scanning microscopy (CLSM) image of an air filled emulsion. (Courtesy of F. Tchuenbou-Magaia.)

The exterior of the air cells, the protein, shines brightly, and the water/protein matrix between the air cells shows a dull glow due to residual, unbound protein. With the air cells themselves, there is no signal from the air and so they appear black.

The original air cells and air-filled emulsions were all constructed using a sonication technique and hydrophobin proteins. Hydrophobins are a class of protein found exclusively within filamentous fungi; however, due to their self-assembly properties at interfaces, they are attracting considerable academic and industrial interest (Cox et al. 2007). These molecules are attracting a lot of interest from food and nonfood applications and although as molecules they have a long history of consumption through their natural occurrence, it currently remains unclear in the literature what their characterization is in terms of food use, or indeed if they need to be treated as a novel ingredient. The main reason for this interest in the inherent strength of the molecules when assembled into layers at interfaces. Figure 16.6 shows a layer of hydrophobin protein supporting close to gram quantities of water with ease.

Due to this strong protein coat when conceptually viewed from the continuous phase, an air-filled emulsion should be undistinguishable from oil-filled emulsions as both should have equally sized droplets with a protein-stabilized interface. Both emulsions should (and did) have a similar rheology and tribology and were resistant to Ostwald ripening (Tchuenbou-Magaia and Cox 2011). Unfortunately, hydrophobins are still quite expensive and the design criteria to replicate air-filled emulsions are therefore quite exacting as the physical properties of the proteins themselves are extraordinary. However, it should be possible to achieve the same microstructure with cheaper *palette* acceptable alternative proteins. In practice, the search for alternatives centered about food grade proteins with a high occurrence of cysteine in their structures.

16.5.1 Waste Proteins and Protein Modification

While searching for alternatives, Tchuenbou-Magaia et al. (2011) identified the physical properties that typified the action of the hydrophobin and used a set of criteria based upon the action of hydrophobins to screen or other proteins. Ideally, the replacement proteins should still be strong,

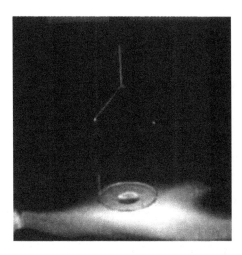

FIGURE 16.6 A hydrophobin film self-assembled across an interfacial measuring geometry.

self-assembling, and provide an accurate mimic for oil and if possible, contain air for fat replacement and buoyancy reasons.

Tchuenbou-Magaia et al. (2011) did indeed identify a number of food grade proteins that could act as hydrophobin replacers. All were sourced from either waste industry streams or abundant sources. In most cases, the replacement protein was very much cheaper than hydrophobins. However, all showed interfacial activity (Damodaran et al. 1998) and free cysteine residues, which can mediate the cross-bonding of the protein to form the air cells (Cox and Hooley 2009; Green et al. 2013; Littlejohn et al. 2012).

However, to tailor and produce microbubbles with desirable size and stability in order is still a challenge due to natural instabilities driven by their high Laplace pressures and the solubility of air in water. However, Figure 16.7 shows the result (unpublished data) of varying the solvent quality of the native protein solution to cause a change in the number and size of microbubbles formed from a consistent delivered process. For instance, easily manipulated parameters such as pH or ionic strength could be used to generate materials with differing material properties.

16.5.2 Fat Replacement Strategy and Beyond

Tchuenbou-Magaia et al. (2009, 2011) used the strength of HFBII-stabilized and BSA-stabilized air cells by making a stable air-filled triphasic (air/oil/water). Here, up to 50% of the oil from an oil-in-water emulsion has been replaced by air cells. Figure 16.8 shows a triphasic emulsion, which even after several months of ambient storage retains its structural integrity. In this case, only 50% of the oil has been removed from the formulation. This is to allow the remaining oil to reinforce the oral properties, but also to continue to carry flavor.

Once part of food foam engineering that hasn't been mentioned (or has it?) is Pickering stabilization of foams. Here, solid particles are placed at the air/water interface, where they have a strong energy of attachment and produce very stable foams (see Binks 2002; Spyropoulos et al. 2008).

The odd statement above is because this is dependent upon your point of view as no (or not to be so quickly dated) or very few Pickering-based foods available on the market. This is despite very favorable physical characteristics. However, if readers consult reviews on the area (e.g., Lam et al. 2014), there are a wealth of probable routes for potential manufacture. But also at the same time, there are numerous other examples where the reader will go "well that is familiar! How can that be a new technology?" For example, fat crystals in ice cream interfaces. In addition, there are

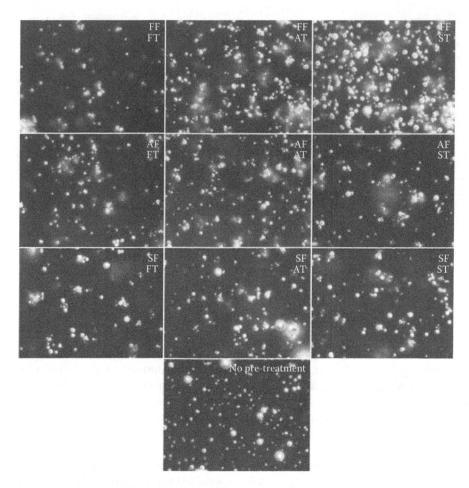

FIGURE 16.7 Design of a bubble. (With permission.)

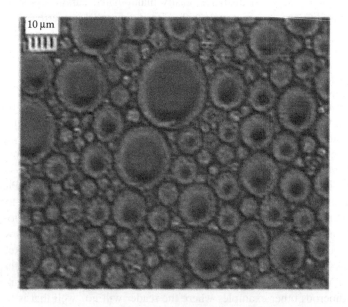

FIGURE 16.8 A triphasic emulsion. (Courtesy of F. Tchuenbou-Magaia.)

things where the margin between molecule and particles blurs. Here, hydrophobins form a good example (Green et al. 2013), where the sub 10 nm hydrophobins act like larger Pickering particles at interfaces due to the rigidity of the molecule. However, is a rigid protein a particle or a molecule?

16.6 CONCLUSIONS

In this short chapter, attempts have been made to provide a context of where food foams exist (in mouth processing). Examples have been given of several different mechanisms and molecules that are commonly used to form food foams. The intention was to highlight the variety of structures and structural molecules that are so familiar to us but are still capable of producing the vast array of structures that we enjoy eating; for instance, a beer foam is inherently different from a bread but both use protein as their stabilizer. All the contributors to this chapter have attempted to express their passion for the subject area because even though food foams are ubiquitous, they still give endless fascination to the scientist and engineer. As a final thought: Garnsey (1999) pointed out that 2000 years ago, bread was mass manufactured and commonly sold at the road side (Walmarticus?) but in the first nine months of 2014, there were still 3500 hits for bread research on Web of Science.

REFERENCES

Ahmed, M. and Dickinson, E. (1990) Effect of ethanol on the foaming of aqueous protein solutions. *Colloids and Surfaces* 47: 353–365.

Akhtar, M., Brent, S., Murray, B.S., and Dickinson, E. (2006) Perception of creaminess of model oil-in-water dairy emulsions: Influence of the shear-thinning nature of a viscosity-controlling hydrocolloid. *Food Hydrocolloids* 20: 839–847.

Amaha, M. et al. (1973) Gushing inducers produced by some mould strains. *Proceedings of the European Brewery Convention Congress, Salzburg*. Elsevier: Amsterdam, the Netherlands.

Baer, R. J., Wolkow, M. D., and Kasperson, K. M. (1997) Effect of emulsifiers on the body and texture of low fat ice cream. *Journal of Dairy Science* 80: 3123–3132.

Bamforth, C. W. (1985) The foaming properties of beer. *Journal of the Institute of Brewing* 91(6): 370–383.

Binks, B. P. (2002) Particles as surfactants-similarities and differences. *Current Opinion in Colloid and Interface Science* 7: 21–41.

Blasco, L., Veiga-Crespo, P., Sánchez-Pérez, A., and Villa, T.G. (2012) Cloning and characterization of the beer foaming gene CFG1 from *Saccharomyces pastorianus*. *Journal of Agricultural and Food Chemistry* 60(43): 10796–10807.

Blasco, L., Viñas, M., and Villa, T.G. (2011) Proteins influencing foam formation in wine and beer: The role of yeast. *International Microbiology* 14(2): 61–71.

Bongaerts, J., Rossetti, D., and Stokes, J. (2007) The lubricating properties of human whole saliva. *Tribology Letters* 27: 277–287.

Bordi, P., Cranage, D., Stokols, J., Palchak, T., and Powell, L. (2005) Effect of polyols versus sugar on acceptability of ice cream among a student and adult population. *Foodservice Research International* 15(1): 41–50.

Bourne, M.C. (2002) *Food Texture and Viscosity: Concept and Measurement*, M.C. Bourne (ed.), Academic Press, San Diego, CA.

Bucholz, H., Kalischewski, K., and Schügerl, K. (1979) Foam behavior of biological media. *European Journal of Applied Microbiology and Biotechnology* 7(4): 321–331.

Bumbullis, W., Kalischewski, K., and Schügerl, K. (1979) Foam behavior of biological media. *European Journal of Applied Microbiology and Biotechnology* 7(2): 147–154.

Cameron, D.R., Cooper, D.G., and Neufeld, R.J. (1988) The mannoprotein of *Saccharomyces cerevisiae* is an effective bioemulsifier. *Applied and Environmental Microbiology* 54(6): 1420–1425.

Carlson, T. (1981) *Law and Order in Wheat Flour Dough. Colloidal Aspects of the Wheat Flour Dough and its Lipid and Protein Constituents in Aqueous Media*. Thesis, University of Lund, Lund, Sweden.

Carrington, R., Collett, R.C., Dunkin, I.R., and Halek, G. (1972) Gushing promoters and suppressants in beer and hops. *Journal of the Institute of Brewing* 78(3): 243–254.

Cauvain, S.P., Whitworth, M.B., and Alava, J.B. (1999) The evolution of cell structure in bread doughs and its effect on bread structure, in G.M. Campbell, C. Webb, S.S. Pandiella, and K. Niranjan (eds.), *Cells in Food*, Eagan Press, St. Paul, MN, pp. 85–88.

Cauvain, S.P. and Young, L.S. (2004) *Baking Problems Solved*. Woodhead Publishing, Cambridge, p. 25.

Chen, W.Z. and Hoseney, R.C. (1995) Development of an objective method of dough stickiness. *Food Science and Technology* 28(5): 467–473.

Clark, D.C., Wilde, P.J., and Marion, D. (1994) The protection of beer foam against lipid-induced destabilization. *Journal of the Institute of Brewing* 100(1): 23–25.

Clarke, C. (2008) *The Science of Ice Cream*, 3rd edn. RSC Publishing, Cambridge.

Clarkson, J. R., Cui, Z.F., and Darton, R.C. (1999) Protein denaturation in foam: II. Surface activity and conformational change. *Journal of Colloid and Interface Science* 215(2): 333–338.

Coke, M., Wilde, P.J., Russell, E.J., and Clark, D.C. (1990) The influence of surface composition and molecular diffusion on the stability of foams formed from protein/surfactant mixtures. *Journal of Colloid and Interface Science* 138(2): 489–504.

Cook, K.L.K. and Hartel, R.W. (2010) Mechanisms of ice crystallization in ice cream production. *Comprehensive Reviews in Food Science and Food Safety* 9: 213–222.

Cooper, A. and Kennedy, M.W. (2010) Biofoams and natural protein surfactants. *Biophysical Chemistry* 151(3): 96–104.

Cox, A.R., Cagnol, F., Russell, A.B., and Izzard, M.J. (2007) Surface properties of class II hydrophobins from *Trichoderma reesei* and influence on bubble stability. *Langmuir* 23: 7995–8002.

Cox, P.W. and Hooley, P. (2009) Hydrophobins: New prospects for biotechnology. *Fungal Biology Reviews* 23: 40–47.

Crompton, I.E. and Hegarty, P.K. (1991) *Proceedings of the 3rd Conference on Halting*. Brewing and Distilling, Oxford, p. 277.

Cutler, A.N., Morris, E.R., and Taylor, L.J. (1983) Oral perception of viscosity in fluid foods and model systems. *Journal of Texture Studies* 14: 377–395.

Damodaran, S., Anand, K., and Razumovsky, L. (1998) Competitive adsorption of egg white proteins at the air-water interface: Direct evidence for electrostatic complex formation between lysozyme and other egg proteins at the interface. *Journal of Agricultural and Food Chemistry* 46: 872–876.

Davies, E., Dickinson, E., and Bee, R.D. (2000) Shear stability of sodium caseinate emulsions containing monoglyceride and triglyceride crystals. *Food Hydrocolloids* 14: 145–153.

Davies, G.A., Wantling, E., and Stokes, J.R. (2009) The influence of beverages on the stimulation and viscoelasticity of saliva: Relationship to mouthfeel? *Food Hydrocolloids* 23: 2261–2269.

De Feijter, J.A. and Benjamins, J. (1987) Adsorption kinetics of proteins at the air-water interface. *Food Emulsions and Foams* 72–85.

Dettwyler, K. (1995) A natural age of weaning. *Breastfeeding: Biocultural Perspectives*. Hawthorne, New York.

Dickinson, E. and Euston, S.R. (1990) Simulation of adsorption of deformable particles modelled as cyclic lattice chains. A simple statistical model of protein adsorption. *Journal of Chemical Society, Faraday Transactions* 86(5): 805–809.

Douillard, R. and Teissei, J. (1991) Surface pressure and fluorescence study of ribulose-1, 5-bisphosphate carboxylase/oxygenase adsorption at an air/buffer interface. *Journal of Colloid and Interface Science* 143(1): 111–119.

Dressaire, E., Bee, R., Bell, D.C., Lips, A., and Stone, H.A. (2008) Interfacial polygonal nanopatterning of stable microbubbles. *Science* 320(5880): 1198. doi: 10.1126/science.1154601.

Eisner, M.D., Jeelani, S.A.K., Bernhard, L., and Windhab, E.J. (2007) Stability of foams containing proteins, fat particles and non-ionic surfactants. *Chemical Engineering Science* 62: 1974–1987.

Evans, D.E., and Bamforth, C.W. (2009) *Beer Foam: Achieving a Suitable Head*. Academic Press, Burlington, MA.

Evans, M.T.A., Mitchell, J., Mussellwhite, P.R., and Irons, L. (1970) The effect of the modification of protein structure on the properties of proteins spread and adsorbed at the air-water interface. *Surface Chemistry of Biological Systems*. Springer, New York, pp. 1–22.

Ferreira, I.M.P.L.V.O., Jorge, K., Nogueira, L.C., Silva, F., and Trugo, L.C. (2005) Effects of the combination of hydrophobic polypeptides, iso-α acids, and malto-oligosaccharides on beer foam stability. *Journal of Agricultural and Food Chemistry* 53(12): 4976–4981.

Fillery-Travis, A., Mills, E.N.C., and Wilde, P. (2000) Protein-lipid interactions at interfaces. *Grasas y aceites* 51(1–2): 50–55.

Flores, A.A. and Goff, H.D. (1999) Ice crystal size distributions in dynamically frozen model solutions and ice cream as affected by stabilizers. *Journal of Dairy Science* 82: 1399–1407.

Folkenberg, D.M. and Martens, M. (2003) Sensory properties of low fat yoghurts. Part B: Hedonic evaluations of plain yoghurts by consumers correlated to fat content, sensory profile and consumer attitudes. *Milchwissenschaft-Milk Science International* 58: 154–157.

Frøst, M.B. and Janhøj, T. (2007) Understanding creaminess. *International Dairy Journal* 17: 1298–1311.

Gan, Z., Ellis, P.R., and Schofield, J.D. (1995) Mini review gas cell stabilisation and gas retention in wheat bread dough. *Journal of Cereal Science* 21: 215–230.

Garnsey, P. (1999) *Food and Society in Classical Antiquity*. Cambridge University Press, Cambridge.

Gjertsen, P., Trolle, B., and Andersen, K. (1965) Gushing caused by microorganisms, specially *Fusarium* species. *Proceedings of the 10th European Brewery Convention Congress*, Stockholm, Sweden.

Goesaert, H., Brijs, K., Veraverbeke, W.S., Courtin, C.M., Gebruers, K., and Delcour, J.A. (2005) Wheat flour constituents: How they impact bread quality, and how to impact their functionality. *Trends in Food Science and Technology* 16: 12–30.

Goesaert, H., Slade, L., Levine, H., and Delclour, J.A. (2009) Amylases and bread firming–an integrated view. *Journal of Cereal Science* 50: 345–352.

Goff, D. (2008) *Dairy Science and Technology Education Series*. University of Guelph, Guelph, Canada.

Goff, H.D. (1995) *Dairy Science and Technology Education*. University of Guelph, Guelph, Canada.

Goff, H.D. (1997) Partial coalescence and structure formation in dairy emulsions. *Food Proteins and Lipids*, American Chemical Society. J.E. Kinsella Memorial, S. Damodaran (ed.), Plenum Press, New York, pp. 137–148.

Gonzalez, G. and MacRitchie, F. (1970) Equilibrium adsorption of proteins. *Journal of Colloid and Interface Science* 32(1): 55–61.

Graham, D.E. and Phillips, M.C. (1976) The conformation of proteins at the air-water interface and their role in stabilizing foams, in *Foams*, R.J. Akers (eds.), Academic Press, New York, pp. 237–255.

Graham, D.E. and Phillips, M.C. (1979) Proteins at liquid interfaces: III. Molecular structures of adsorbed films. *Journal of Colloid and Interface Science* 70(3): 427–439.

Gray, J.A. and Bemiller, J.N. (2003) Bread staling: Molecular basis and control. *Comprehensive Reviews in Food Science and Food Safety* 2: 1–21.

Green, A.J., Littlejohn, K.A., Hooley, P., and Cox, P.W. (2013) Formation and stability of food foams and aerated emulsions: Hydrophobins as novel functional ingredients. *Current Opinion in Colloid and Interface Science* 18(4): 292–301.

Gupta, R.B., Batey, I.L., and MacRitchie, F. (1992) Relationships between protein composition and functional properties of wheat flours. *Cereal Chemistry* 69: 125–131.

Gyllang, H. and Martinson, E. (1976) *Aspergillus fumigatus* and *Aspergillus amstelodami* as causes of gushing. *Journal of the Institute of Brewing* 82(3): 182–183.

Haikara, A. (1980) Gushing induced by fungi. *Symposium on the Relationship between Malt and Beer*, Monograph VI European Brewing Convention Symposium, Brauwelt-Verlag Nurnberg, Germany, pp. 251–259.

Hartel, R.W. (1996) Ice crystallization during the manufacture of ice cream. *Trends in Food Science and Technology* 7: 315–321.

Herrington, T.M. and Sahi, S.S. (1987) Desorption of bovine serum albumin from the air-water interface. *Food Emulsions and Foams: Based on the Proceedings of an International Symposium Organised by the Food Chemistry Group of the Royal Society of Chemistry at Leeds from 24th–26th March 1986*. No. 58. Royal Society of Chemistry, London.

Highfield, R. (2015) Bubbles the key to a healthier diet. *Telegraph* article, http://www.telegraph.co.uk/news/science/science-news/3341689/Bubbles-the-key-to-a-healthier-diet.html.

Hippeli, S. and Elstner, E.R. (2002) Are hydrophobins and/or non-specific lipid transfer proteins responsible for gushing in beer? New hypotheses on the chemical nature of gushing inducing factors. *Zeitschrift Fur Naturforschung C* 57(1/2): 1–8.

Husband, F.A., Wilde, P.J., Mackie, A.R., and Garrood, M.J. (1997) A comparison of the functional and interfacial properties of β-casein and dephosphorylated β-casein. *Journal of Colloid and Interface Science* 195(1): 77–85.

Hutchings, J.B. and Lillford, P.J. (1988) The perception of food texture— The philosophy of the food break-down path. *Journal of Texture Studies* 19: 103–115.

Keetles, S.M., Vliet, T.V., and Walstra, P. (1996) Gelation and retrogradation of concentrated starch systems: Effect of concentration and heating temperature. *Food Hydrocolloids* 10: 363–368.

Kilkast, D. (2004) Measuring the consumer perceptions of texture, in D. Kilcast (ed.), *Texture in Food: Solid Foods*. Woodhead Publishing: Cambridge.

Kim, S.H. and Kinsella, J.E. (1985) Surface activity of food proteins: Relationships between surface pressure development, viscoelasticity of interfacial films and foam stability of bovine serum albumin. *Journal of Food Science* 50(6): 1526–1530. *Kitabatake & Amaha*.

Kitabatake, K. and Amaha, M. (1974) Production of gushing factor by a Nigrospora sp. in liquid culture media. *Bulletin of Brewing Science* 20: 1–8.

Kohyama, K., Hayakawa, F., Sasaki, T., and Azuma, T. (2004) Direct measurement of lip pressure when ingesting semi liquid food. *Journal of Texture Studies* 35: 554–569.

Kokini, J. (1987a) Viscoelastic properties of semisolid foods and their biopolymeric components. *Food Technology* 41: 89–95. In I. Norton, P.W. Cox, and Spyropoulos (2010) *Practical Food Rheology: An Interpretive Approach.* John Wiley & Sons.

Kokini, J. (1994) Predicting the rheology of food biopolymers using constitutive models. *Carbohydrate Polymers* 25: 319–329.

Kokini, J.L. (1987b) The physical basis of liquid food texture and texture—Taste interactions *Journal of Food Engineering* 6: 51–81.

Kokini, J.L. and Cussler, E.L. (1983) Predicting the texture of liquid and melting semi-solid foods *Journal of Food Science* 48: 1221–1225.

Koutsou, G.A., Storey, D.M., Lee, A., Zumbe, A., Flourie, B., Lebot, Y., and Olivier P. (1996) Dose-related gastrointestinal response to the ingestion of either isomalt, lactitol, or maltitol in milk chocolate. *European Journal of Clinical Nutrition* 50(1): 17–21.

Koxholt, M.M.R., Eisenmann, B., and Hinrichs, J. (2001) Effect of the fat globule sizes on the meltdown of ice cream. *Journal of Dairy Science* 84: 31–37.

Krägel, J., Wüstneck, R., Clark, D., Wilde, P., and Miller, R. (1995) Dynamic surface tension and surface shear rheology studies of mixed β-lactoglobulin/Tween 20 systems. *Colloids and Surfaces A: Physicochemical and Engineering Aspects* 98(1): 127–135.

Lagrain, B., Thewissen, B.G., Brijs, K., and Delcour, J.A. (2007) Impact of redox agents on the extractability of gluten proteins during bread making. *Journal of Agricultural and Food Chemistry* 55(13): 5320–5325.

Lam, S., Velikov, K.P., and Velev, O.D. (2014) Pickering stabilization of foams and emulsions with particles of biological origin. *Current Opinion in Colloid & Interface Science* 19(5): 490–500.

Langstaff, S.A. and Lewis, M.J. (1993a) The mouthfeel of beer—A review. *Journal of the Institute of Brewing* 99(1): 31–37.

Langstaff, S.A. and Lewis, M.J. (1993b) Foam and the perception of beer flavor and mouthfeel. *Technical Quarterly-Master Brewers Association of the Americas* 30: 16–17.

Lee, W.E. and Camps, M.A. (1991) Tracking foodstuff location within the mouth in real time: A sensory method. *Journal of Texture Studies* 22: 277–287.

Le Reverend, B.J.D., Norton, I.T., Cox, P.W., and Spyropoulos, F. (2010) Colloidal aspects of eating. *Current Opinion in Colloid Interface Science* 15: 84–89.

Leroy, V., Fan, Y., Strybulevych, A.L., Bellido, G.G., Page, J.H., and Scanlon, M.G. (2008) Investigating the cell size distribution in dough using ultrasound, in *Cells in Food 2 Novelty, Health and Luxury*, G.M. Campbell, M.G. Scanlon, and D.L. Pyle (eds.), Eagan Press, St. Paul, MN, pp. 51–61.

Lillford, P.J. (1991) Texture and acceptability of human foods, in *Feeding and the Texture of Food*, J.F.V. Vincent and P.J. Lillford (eds). Cambridge University Press, Cambridge, pp. 231–243.

Lillford, P.J. (2000) The materials science of eating and food breakdown. *MRS Bulletin* 25: 38–43.

Lillford, P.J. (2001) Mechanisms of fracture in foods. *Journal of Texture Studies* 32: 397–417.

Littlejohn, K., Hooley, P., and Cox, P.W. (2012) Bioinformatics predicts diverse *Aspergillus* hydrophobins with novel properties. *Food Hydrocolloids* 27: 503–516.

Livesey, G. (2003) Health potential of polyols as sugar replacers, with emphasis on low glycaemic properties. *Nutrition Research Reviews* 16: 163–191.

Llopis, J. and Albert, A. (1959) The influence of temperature on monolayers of bovine γ-globulin. *Archives of Biochemistry and Biophysics* 81(1): 159–168.

Mackie, D.A., and Pangborn, R.M. (1990) Mastication and its influence on human salivary flow and alpha-amylase secretion. *Physiology and Behaviour* 47(3): 593–595.

MacRitchie, F. (1978) Proteins at interfaces. *Advances in Protein Chemistry* 32: 283.

Maeda, K., Yokoi, S., Kamada, K., and Kamimura, M. (1991) Foam stability and physicochemical properties of beer. *Journal of the American Society of Brewing Chemists* 49.

Marshall, R.T. and Arbuckle, W.S. (1996) *Ice Cream*, 5th edn. International Thomson Publishers, New York.

Marshall, R.T., Goff, H.D., and Hartel, R.W. (2003) *Ice Cream*, 6th edn., Kluwer Academic, New York.

McClements, D.J. and Decker, E.A. (2009) *Designing Functional Foods: Measuring and Controlling Food Structure Breakdown and Nutrient Absorption.* Woodhead Publishing: Cambridge, pp. 15–16.

Medicis, S.W. and Hiiemae, K.H. (1998) Natural bite sizes for common foods. *Journal of Dental Research* 77: 295–295.

Mills, E.N.C., Wilde, P.J., Salt, L.J., and Skeggs, P. (2003) Bubble formation and stabilization in bread dough. *Food and Bioproducts Processing* 81: 189–193.

Mondal, A. and Datta, A.K. (2008) Bread baking—A review. *Journal of Food Engineering* 86: 465–474.

Morris, E.R. (1987) Organoleptic properties of food polysaccharides in thickened systems, in *Industrial Polysaccharides: Genetic Engineering, Structure/Property Relations and Applications*, M. Yalpani (ed). Elsevier Science, Amsterdam, the Netherlands, pp. 225–238.

Naresh, L. and Merchant, S.U. (2006) Stabilizer blends and their importance in the ice cream industry— A review. *New Zealand Food Magazine*, 44–51.

Niessen, M.L. (1993) Entwicklung und Anwendung immunchemischer Verfahren zum Nachweis wichtiger Fusarium-Toxine bei der Bierbereitung sowie mykologische Untersuchungen im Zusammenhang mit dem Wildwerden (Gushing) von Bieren.

Norton, I., Fryer, P., and Moore, S. (2006) Product/process integration in food manufacture: Engineering sustained health. *AIChE Journal* 52(5): 1632–1640.

Norton, I.T., Moore, S., and Fryer, P.J. (2007) Understanding food structuring and breakdown: Engineering approaches to obesity. *Obesity Reviews* 8: 83–88.

Norton, I.T., Spyropoulos, F., and Cox, P.W. (2008) Emulsions, emulsification and phase inversion. *Food Hydrocolloids* 23: 1521–1526.

O'Neill, E., Morrissey, P.A., and Mulvihill, D.M. (1989) The effects of pH and heating on the surface activity of muscle proteins. *Food Chemistry* 34(4): 295–307.

Ono, M., Hashimoto, S., Kakudo, Y., Nagami, K., and Kumada, J. (1983) Foaming and beer flavor. *Journal of American Society of Brewing Chemists* 41: 19–23.

Perkowitz, S. (2000) *Universal Foam: From Cappuccino to the Cosmos*. Walker & Company, New York.

Peyron, M., Mishellany, A., and Woda, A. (2004) Particle size distribution of food boluses after mastication of six natural foods. *Journal of Dental Research* 83: 578–582.

Poole, S., West, S.I., and Walters, C.L. (1984) Protein–protein interactions: Their importance in the foaming of heterogeneous protein systems. *Journal of the Science of Food and Agriculture* 35(6): 701–711.

Prinz, J.F. and Lucas, P.W. (1995) Swallow thresholds in human mastication. *Archives of Oral Biology* 40(5): 401–403.

Roberts, R.T., Keeney, P.J., and Wainwright, T. (1978) The effects of lipids and related materials on beer foam. *Journal of the Institute of Brewing* 84(1): 9–12.

Ronteltap, A.D. and Prins, A. (1990) The role of surface viscosity in gas diffusion in aqueous foams. II. Experimental. *Colloids and Surfaces* 47: 285–298.

Rosenthal, A.J. (1999) Food Texture: Measurement and Perception. Springer: Beverley hills, CA.

Santos-Murphy, S.M. (2014) *Ice Cream: An Approach Considering Formulation Microstructure and Eating Behaviour* A thesis submitted at the University of Birmingham, Birmingham.

Scanlon, M.G., Elmehdi, H.M., and Page, J.H. (2002) Probing gluten interactions with low-intensity ultrasound, in *Wheat Quality Elucidation*: The Bushuk Legacy, P.K.W. Ng and C.W. Wrigley (eds). AACC Press, St. Paul, MN, pp. 170–182.

Shama, F., Parkinson, C., and Sherman, P. (1973) Identification of stimuli controlling the sensory evaluation of viscosity I. Non-oral methods. *Journal of Texture Studies* 4: 102–110.

Shama, F. and Sherman, P. (1973) Identification of stimuli controlling the sensory sensory cation of stimuli controlling method. *Journal of Texture Studies* 4: 111–118.

Shimiya, Y. and Nakamura, K. (1997) Changes in size of gas cells in dough and bread during breadmaking and calculation of critical size of gas cells that expand. *Journal of Texture Studies* 28: 273–288.

Slack, P.T. and Bamforth, C.W. (1983) The fractionation of polypeptides from barley and beer by hydrophobic interaction chromatography: The influence of their hydrophobicity on foam stability. *Journal of the Institute of Brewing* 89(6): 397–401.

Sofjan, R.P. and Hartle, R.W. (2003) Effects of overrun on structural and physical characteristics of ice cream. *International Dairy Journal* 14(3): 255–262.

Sorensen, S.B., Bech, L.M., Muldbjerg, M., Beenfeldt, T., and Breddam, K. (1993) Barley lipid transfer protein 1 is involved in beer foam formation. *Technical Quarterly (USA)* 30(4): 136–145.

Spyropoulos, F., Cox, P.W., Fryer, P.J., and Norton, I.T. (2008) Immiscible liquid-liquid mixing, in *Food Mixing—Principles and Applications*, P.J. Cullen (ed). Blackwell, Oxford.

Stanley, N.L. and Taylor, L.J. (1993) Rheological basis of oral characteristics of fluid and semi-solid foods: A review. *Acta Psychologica* 84(1): 79–92.

Stewart, G.G. (2004) The chemistry of beer instability. *Journal of Chemical Education* 81(7): 963.

Strassburg, J.A., Burbidge, A.S., Hartmann, C., and Delgado, A. (2007) Geometrical resolution limits and detection mechanisms in the oral cavity. *Journal of Biomechanics* 40: 3533–3540.

Taylor, A.J. and Linforth, R.S.T. (1996). Flavour release in the mouth. *Trends in Food Science and Technology* 7(12): 444–448.

Tchuenbou-Magaia, F.L., Al-Rifai, N., Mohamed Ishak, N.E., Norton, I.T., and Cox, P.W. (2011) Suspensions of air cells with cysteine-rich protein coats: Air-filled emulsions. *Journal of Cellular Plastics* 47(3): 217–232.

Tchuenbou-Magaia, F.L. and Cox, P.W. (2011) Tribological study of suspensions of cysteine-rich protein stabilized microbubbles and subsequent triphasic A/O/W emulsions. *Journal of Texture Studies* 42(3): 185–196.

Tchuenbou-Magaia, F.L., Norton, I.T., and Cox, P.W. (2009) Hydrophobins stabilised air-filled emulsions for the food industry. *Food Hydrocolloids* 23(7): 1877–1885.

Terpstra, M.E.J., Jellema, R.H., Jansen, A.M., de Wijk, R.A., Prinz, J.F., and Van Der Linden, E. (2009) Prediction of texture perception of mayonnaises from rheological and novel instrumental measurements. *Journal of Texture Studies* 40: 82–108.

Tharp, B.W. and Young, S.L. (2012) *Tharp & Young on Ice Cream: An Encyclopedic Guide to Ice Cream Science and Technology*. DEStech Publications: Lancaster, PA.

Townsend, A.A. and Nakai, S. (1983) Relationships between hydrophobicity and foaming characteristics of food proteins. *Journal of Food Science* 48(2): 588–594.

Trinh, L. (2013) *Gas Cells in Bread Dough*. PhD thesis, School of Chemical Engineering and Analytical Science, University of Manchester.

Van Aken, G.A., Vingerhoeds, M.H. and de Hoog, E.H. (2007). Food colloids under oral conditions. *Current Opinion in Colloid and Interface Science* 12: 251–262.

Van Nierop, S.N.E., Evans, D.E., Axcell, B.C., Cantrell, I.C., and Rautenbach, M. (2004) Impact of different wort boiling temperatures on the beer foam stabilizing properties of lipid transfer protein 1. *Journal of Agricultural and Food Chemistry* 52(10): 3120–3129.

Vardhanabhuti, B. and Foegeding, E.A. (2010) Evidence of interactions between whey proteins and mucin: Their implication on the astringency mechanism of whey proteins at low pH, in *Gums and Stabilisers for the Food Industry 15*, G.O. Phillips, P.A. Williams, and D.J. Wedlock (eds). RSC Publishing, London, pp. 137–146.

Velissariou, M. (1988). *WSc Examination Thesis*, School of Chemical Engineering, University of Birmingham, Birmingham.

Velissariou, M. (1992). *PhD Examination Thesis*, Faculty of Chemical Engineering, University of Birmingham, Birmingham.

Vingerhoeds, M.H., Silletti, E., de Groot, J., Schipper, R.G., and van Aken, G.A. 2009. Relating the effect of saliva-induced emulsion flocculation on rheological properties and retention on the tongue surface with sensory perception. *Food Hydrocolloids* 23: 773–785.

Walstra, P. (1996) Dispersed systems: Basic consideration, in *Food Chemistry*, O.R. Fennema (ed), 3rd edn. Marcel Dekker, New York, pp. 95–155.

Waniska, R.D. and Kinsella, J.E. (1985) Surface properties of beta-lactoglobulin: Adsorption and rearrangement during film formation. *Journal of Agricultural and Food Chemistry* 33(6): 1143–1148.

Weideneder, A. (1992) *Untersuchungen zum malzverursachten Wildwerden (Gushing) des bieres*.

Wilde, P.J. (1996) Foam measurement by the microconductivity technique: An assessment of its sensitivity to interfacial and environmental factors. *Journal of Colloid and Interface Science* 178(2): 733–739.

Wilde, P.J. and Clark, D.C. (1996) Foam formation and stability. *Methods of Testing Protein Functionality* 1: 110–152.

Wildmoser, H., Sheiwiller, J., and Windhab, E.J. (2004) Impact of disperse microstructure on rheology and quality aspects of ice cream. *Lebensmittel Wissenschaft & Technologie* 37: 881–891.

Wilkinson, C., Dijksterhuis, G.B., and Minekus, M. (2000) From food structure to texture. *Trends in Food Science & Technology* 11: 442–450.

Yilsay, T.O., Yilzman, L., and Bayizit, A.A. (2005) The effect of using whey protein fat replacer on textural and sensory characterisation of low fat vanilla ice cream. *European Food Research and Technology* 222(1): 171–175.

Zanoni, B., Peri, C., and Pierucci, S. (1993) A study of the bread-baking process I: A phenomenological model. *Journal of Food Engineering* 19: 389.

Zepf, M. (1998) Gushing-Ursachenfindung anhand von Modellversuchen. Dissertation an der Technischen. Universität München, München, Germany.

Zghal, M.C., Scanlon, M.G., and Sapirstein, H.D. (2001) Effects of flour strength, baking absorption, and processing conditions on the structure and mechanical properties of bread crumb. *Cereal Chemistry* 78: 1–7.

Zghal, M.C., Scanlon, M.G., and Sapirstein, H.D. (2002) Cellular structure of bread crumb and its influence on mechanical properties. *Journal of Cereal Science* 36: 167–176.

Zumbe, A., Lee, A., and Storey, D. (2001) Polyols in confectionery: The route to sugar-free, reduced sugar, and reduced calorie confectionery. *British Journal of Nutrition* 85(Suppl 1): S31–S45.

Index

Note: Locator followed by '*f*' and '*t*' denotes figure and table in the text